MICROSCOPY APPLIED TO MATERIALS SCIENCES AND LIFE SCIENCES

MICROSCOPY APPLIED TO MATERIALS SCIENCES AND LIFE SCIENCES

Edited by
Sabu Thomas, PhD
Ajay Vasudeo Rane, MTech
Nandakumar Kalarikkal, PhD
Krishnan Kanny, PhD

Apple Academic Press Inc.
3333 Mistwell Crescent
Oakville, ON L6L 0A2 Canada

Apple Academic Press Inc.
9 Spinnaker Way
Waretown, NJ 08758 USA

© 2019 by Apple Academic Press, Inc.
Exclusive worldwide distribution by CRC Press, a member of Taylor & Francis Group
No claim to original U.S. Government works
International Standard Book Number-13: 978-1-77188-672-7 (Hardcover)
International Standard Book Number-13: 978-1-351-25158-7 (eBook)

All rights reserved. No part of this work may be reprinted or reproduced or utilized in any form or by any electric, mechanical or other means, now known or hereafter invented, including photocopying and recording, or in any information storage or retrieval system, without permission in writing from the publisher or its distributor, except in the case of brief excerpts or quotations for use in reviews or critical articles.

This book contains information obtained from authentic and highly regarded sources. Reprinted material is quoted with permission and sources are indicated. Copyright for individual articles remains with the authors as indicated. A wide variety of references are listed. Reasonable efforts have been made to publish reliable data and information, but the authors, editors, and the publisher cannot assume responsibility for the validity of all materials or the consequences of their use. The authors, editors, and the publisher have attempted to trace the copyright holders of all material reproduced in this publication and apologize to copyright holders if permission to publish in this form has not been obtained. If any copyright material has not been acknowledged, please write and let us know so we may rectify in any future reprint.

Trademark Notice: Registered trademark of products or corporate names are used only for explanation and identification without intent to infringe.

Library and Archives Canada Cataloguing in Publication

Microscopy applied to materials sciences and life sciences / edited by Ajay Vasudeo Rane, Sabu Thomas, PhD, Nandakumar Kalarikkal, PhD, Krishnan Kanny, PhD.

This book contains selected research and review papers presented at the World Congress on Microscopy (WCM 2015) held at the International and Inter-university Center for Nanoscience and Nanotechnology, Mahatma Gandhi University, Kottayam, Kerala, India.
Includes bibliographical references and index.
Issued in print and electronic formats.
ISBN 978-1-77188-672-7 (hardcover).--ISBN 978-1-351-25158-7 (PDF)

1. Microscopy--Congresses. 2. Polymers--Microscopy--Congresses. 3. Nanocomposites (Materials)--Microscopy--Congresses. I. Rane, Ajay Vasudeo, editor II. Thomas, Sabu, editor III. Kalarikkal, Nandakumar, editor IV. Kanny, Krishnan, 1964-, editor V. World Congress on Microscopy (2015 : Kottayam, India)

QH201.M53 2018	620.1'92	C2018-903742-3	C2018-903743-1

Library of Congress Cataloging-in-Publication Data

Names: Rane, Ajay V., 1987- editor. | Thomas, Sabu, editor. | Kalarikkal, Nandakumar, editor. | Kanny, Krishnan, editor. | World Congress on Microscopy (2015 : Kerala, India)
Title: Microscopy applied to materials sciences and life sciences / editors Ajay Vasudeo Rane, Sabu Thomas, Nandakumar Kalarikkal, Krishnan Kanny.
Description: Toronto; New Jersey: Apple Academic Press, 2019. | Includes bibliographical references and index.
Identifiers: LCCN 2018030071 (print) | LCCN 2018031570 (ebook) | ISBN 9781351251587 (ebook) | ISBN 9781771886727 (hardcover : alk. paper)
Subjects: | MESH: Microscopy--methods | Microscopy--instrumentation | Biopolymers | Nanostructures | Congresses
Classification: LCC QH201 (ebook) | LCC QH201 (print) | NLM QH 207 | DDC 502.8/2--dc23
LC record available at https://lccn.loc.gov/2018030071

Apple Academic Press also publishes its books in a variety of electronic formats. Some content that appears in print may not be available in electronic format. For information about Apple Academic Press products, visit our website at **www.appleacademicpress.com** and the CRC Press website at **www.crcpress.com**

ABOUT THE EDITORS

Sabu Thomas

Professor Thomas is currently Pro-Vice Chancellor of Mahatma Gandhi University and the Founder Director and Professor of the International and Inter University Centre for Nanoscience and Nanotechnology. He is also a full professor of Polymer Science and Engineering at the School of Chemical Sciences of Mahatma Gandhi University, Kottayam, Kerala, India. Prof. Thomas is an outstanding leader with sustained international acclaims for his work in Nanoscience, Polymer Science and Engineering, Polymer Nanocomposites, Elastomers, Polymer Blends, Interpenetrating Polymer Networks, Polymer Membranes, Green Composites and Nanocomposites, Nanomedicine and Green Nanotechnology. Dr. Thomas's groundbreaking inventions in polymer nanocomposites, polymer blends, green bionano-technological and nano-biomedical sciences, have made transformative differences in the development of new materials for automotive, space, housing, and biomedical fields. In collaboration with India's premier tyre company, Apollo Tyres, Professor Thomas's group invented new high performance barrier rubber nanocomposite membranes for inner tubes and inner liners for tyres. Professor Thomas has received a number of national and international awards which include: Fellowship of the Royal Society of Chemistry, London FRSC, Distinguished Professorship from Josef Stefan Institute, Slovenia, MRSI Medal, Nano Tech Medal, CRSI medal, Distinguished Faculty Award, *Dr. APJ Abdul Kalam Award* for Scientific Excellence—2016, Mahatma Gandhi University—*Award for Outstanding Contribution*—November 2016, *Lifetime Achievement Award of the Malaysian Polymer Group, Indian Nano Biologists* award 2017 and Sukumar Maithy Award for the best polymer researcher in the country. He is in the list of most productive researchers in India and holds a position of No. 5. Recently, because of the outstanding contributions to the field of Nanoscience and Polymer Science and Engineering, Prof. Thomas has been conferred Honoris Causa (DSc) Doctorate by the University of South Brittany, Lorient, France and University of Lorraine, Nancy, France. Very recently, Prof. Thomas has been awarded Senior Fulbright Fellowship to visit 20 Universities in the United States. Professor Thomas has published over 750 peer reviewed research papers, reviews and book chapters. He has co-edited 72 books published by Royal Society, Wiley, Wood head, Elsevier, CRC Press, Springer, Nova, etc. He is the inventor of 6 patents. The H index of Prof. Thomas is 84 and has more than 33,000 citations. Prof. Thomas has delivered over 300 Plenary/Inaugural and Invited lectures in national/international meetings over 30 countries. He has established a state of the art laboratory at Mahatma Gandhi University in the area of Polymer Science and Engineering and Nanoscience and Nanotechnology through external funding form DST, CSIR, TWAS, UGC, DBT, DRDO, AICTE, ISRO, DIT, KSCSTE, BRNS, UGC-DAE, Du Pont, USA, General Cables, USA, Surface Treat Czech Republic, MRF Tyres and Apollo Tyres. Professor Thomas has several international collaborative projects with a large number of countries abroad. He has already supervised 85 PhD theses. The former co-workers of Prof. Thomas occupy leading positions in India and abroad.

Ajay Vasudeo Rane

Ajay Vasudeo Rane is currently a doctoral research fellow at Durban University of Technology, South Africa in Composites Research Group. His current work is based on biopolymers-based nanocomposites. His area of study and readings include morphology of filled polymers—interface and interphase study, structural activity of reinforcement in polymer composites. Previous research work includes polymer recycling, polymer composites for structural and functional applications.

About the Editors

Nandakumar Kalarikkal

Nandakumar Kalarikkal, PhD, is an Associate Professor of Physics at the School of Pure and Applied Physics as well as the Joint Director of the Centre for Nanoscience and Nanotechnology, Mahatma Gandhi University, India. He received his PhD in Semiconductor Physics from Cochin University of Science and Technology, Kerala, India. Dr. Kalarikkal's research group is specialized in areas of nanomultiferroics, nanosemicondustors and nanophosphors, nanocomposites, nanoferroelectrics, nanoferrites, nanomedicine, nanosenors, ion beam radiation effects, phase transitions, etc. Dr. Kalarikkal's research group has extensive exchange programs with different industries and research and academic institutions all over the world and is performing world-class collaborative research in various fields. Dr. Kalarikkal's center is equipped with various sophisticated instruments and has established state-of-the-art experimental facilities that cater to the needs of researchers within the country and abroad.

Krishnan Kanny

Krishnan Kanny, PhD, is a Professor of Material Science and Engineering in the Department of Mechanical Engineering, Durban University of Technology, South Africa, as well as the Director of the Composites Research Group at that university. Prof. Kanny is a seasoned engineer and scientist with over 20 years of experience in management, leadership, and human resources development. He received his PhD in Material Science and Engineering from Tuskegee University, Alabama, United States. He is a member of the Engineering Council of South Africa (ECSA), South African Institute of Mechanical Engineers (SAIMechE), and American Society of Mechanical Engineers (ASME), 2002–2003, American Institute of Aeronautics and Astronomics (AIAA), 2002. He is a National Research Foundation-rated scientist.

Prof. Kanny's professional interests include designing, processing, and testing of composite material systems, reinforced thermoset and thermoplastic structures, and nano-infused structures for aerospace, naval, and automotive applications. Other areas of interest are failure analysis via computational and analytical modeling, and characterization and morphological analysis using scanning electron microscopy (SEM), transmission electron microscopy (TEM), and X-ray photoelectron spectroscopy (XPS). Prof. Kanny has more than 150 international peer-reviewed publications.

CONTENTS

Contributors .... *xiii*

Abbreviations .... *xix*

Preface .... *xxiii*

PART I: Materials Sciences 1

1. **Scanning Electron Microscopy as a Powerful Morphological Characterization Technique for Polymer Blends and Polymer Nanocomposites** 3

   Saju Daniel and Sabu Thomas

2. **Atomic Force Microscopy: Principles and Applications in Polymer Composites** 35

   Abitha V. K., Ajay Vasudeo Rane, and Sabu Thomas

3. **Barium Strontium Titanate: Preparation, Dielectric, Ferroelectric, and Microscopic Studies** 57

   Anuradha Kumari, Saumya Shalu, and Barnali Dasgupta Ghosh

4. **Preparation and Characterization of Copper Thin Films for Antimicrobial Applications** 91

   Udaya Bhat K., Arun Augustin, Suma Bhat, and Udupa K. R.

5. **Fabrication, Scanning Electron Microscopic, and Electrochemical Studies of Nickel Hydroxide Battery Electrodes Modified with Zinc Oxide** 133

   B. J. Madhu

6. **Comparative Spectroscopic Study of Polyindole/Poly(Vinyl Acetate) Composites** 147

   Deepak J. Bhagat and Gopal R. Dhokane

7. **Welding of Alloy C-276** 167

   Manikandan M., Arivazhagan N., and Nageswara Rao M.

8. **Surface Plasmon Resonance-Based Sensors** 191

   Raj Kumar Gupta, Devanarayanan V. P., and V. Manjuladevi

9. **Design Issues in High Strain Rate Dynamic Compressive Failure of Structural Ceramics, Polymers, and Composites** 207

   Saikat Acharya, K. S. Ghosh, D. K. Mondal, and A. K. Mukhopadhyay

10. **Applications of Electron Backscatter Diffraction in Materials Science** 227

    Ravi Chandra Gundakaram

11. **Light Microscopy in Studying of Formation and Decomposition of Wood** 257

    Galina F. Antonova and Victoria V. Stasova

12. **Microstructural Analysis of Polymer Blends, Composites, and Nanocomposites** 275

    M. Fathima Rigana, Simi Annie Tharakan, P. Thirukumaran, A. Shakila Parveen, R. Balasubramanian, S. Balaji, C. P. Sakthi Dharan, and M. Sarojadevi

13. **Graphene-Modified Carbon Microsurfaces in Voltammetric Sensing Applications** 303

    Gururaj Kudur Jayaprakash, Bananakere Nanjegowda Chandrashekar, and Bahaddurghatta Eshwaraswamy Kumara Swamy

PART II: Life Sciences 317

14. **Functionalized Nanomaterials for Biological and Catalytic Applications** 319

    K. Anand, R. M. Gengan, and A. A. Chuturgoon

15. **Removal of Organic Cancer Carcinogens from Wastewater Using Green Synthesis Nanoparticles** 341

    K. Anand, K. G. Moodley, and A. A. Chuturgoon

16. **Assessing Adipose Tissue Engineering In Vitro and In Vivo: A Microscopic Approach** 355

    Balu Venugopal, Francis B. F., Susan Mani, Harikrishnan V. S., Varma H. K., and Annie John

17. **The Contribution of Light Microscopy to Study Male Reprotoxicity of Cadmium** 379

    Maria de Lourdes Pereira, José Coelho, Renata Tavares, Henrique M. A. C. Fonseca, Virgília Silva, Paula P. Gonçalves, and Fernando Garcia e Costa

18. **Microscopy Assessment of Emerging Contaminants' Effects on Aquatic Species** 397

    Ângela Barreto, Ana Violeta Girão, Maria de Lourdes Pereira, Tito Trindade, Amadeu Mortágua Velho da Maia Soares, and Miguel Oliveira

19. **Chromium: The Intriguing Element. The Biological Role Cr(III)-Tris-Picolinate: Is It Safe or Not?** 427

    Teresa Margarida dos Santos, Manuel Ferreira, and Maria de Lourdes Pereira

20. **Treatment of Pharmaceutical Wastewater: A Case Study on Degradation in Electrochemical Oxidation** 461

    Saptarshi Gupta, Leichombam Menan, and Srimanta Ray

Index *493*

CONTRIBUTORS

Abitha V. K.
School of Chemical Sciences, Mahatma Gandhi University, Kottayam 686560, Kerala, India.
E-mail: abithavk@gmail.com

Saikat Acharya
Advanced Mechanical and Materials Characterization Division, CSIR-Central Glass and Ceramic Research Institute, Kolkata 700032, India

K. Anand
Discipline of Medical Biochemistry and Chemical Pathology, School of Laboratory Medicine and Medical Sciences, College of Health Sciences, University of KwaZulu-Natal, Durban, South Africa.
E-mail: organicanand@gmail.com

Galina F. Antonova
V. N. Sukachev Institute of Forest, SB RAS Krasnoyarsk, 660036, Russian Federation.
E-mail: antonova_cell@mail.ru

Arivazhagan N.
School of Mechanical Engineering, VIT University, Vellore, India

Arun Augustin
Department of Metallurgical and Materials Engineering NITK Surathkal, Srinivasanagar, 575025, India

S. Balaji
Department of Chemistry, Anna University, Chennai 600025, Tamil Nadu, India

R. Balasubramanian
Department of Chemistry, Anna University, Chennai 600025, Tamil Nadu, India

Ângela Barreto
Department of Biology and CESAM, University of Aveiro, Campus Santiago, 3810-193 Aveiro, Portugal

Deepak J. Bhagat
Department of Physics, Nehru Mahavidyalaya, Nerparsopant 445102, Maharashtra, India.
E-mail: bhagatd@rediffmail.com

Suma Bhat
Department of Mechanical Engineering, Srinivasa School of Engineering, Mukka, Srinivasanagar, 575025, India

Udaya Bhat K.
Department of Metallurgical and Materials Engineering NITK Surathkal, Srinivasanagar, 575025, India. E-mail: udayabhatk@gmail.com

Bananakere Nanjegowda Chandrashekar
Department of Materials Science and Engineering, Southern University of Science and Technology, Shenzhen, Guangdong 518055, P. R. China

A. A. Chuturgoon
Discipline of Medical Biochemistry and Chemical Pathology, School of Laboratory Medicine and Medical Sciences, College of Health Sciences, University of KwaZulu-Natal, Durban, South Africa.
E-mail: chutur@ukzn.ac.za

José Coelho
Department of Biology and CICECO—Aveiro Institute of Materials, University of Aveiro, Aveiro 3810-193, Portugal

Fernando Garcia e Costa
Department of Morphology and Function, CIISA—Interdisciplinary Centre of Research in Animal Health, Faculty of Veterinary Medicine, University of Lisbon, Lisbon, Portugal

Saju Daniel
International and Inter University Centre for Nanoscience and Nanotechnology, Mahatma Gandhi University, Kottayam, Kerala, India
Department of Chemistry, St. Xavier's College Vaikom, Kottayam, Kerala, India.
E-mail: sajudanielalpy@gmail.com

Devanarayanan V. P.
Department of Physics, Birla Institute of Technology and Science, Pilani 333031, Rajasthan, India

C. P. Sakthi Dharan
Department of Chemistry, Anna University, Chennai 600025, Tamil Nadu, India

Gopal R. Dhokane
Department of Physics, Arts, Science and Commerce College, Chikhaldara 444807, Maharashtra, India

Manuel Ferreira
CICECO—Aveiro Institute of Materials, University of Aveiro, 3810-193 Aveiro, Portugal
Baixo Vouga Hospital Centre, 3810-193 Aveiro, Portugal

Francis B. F.
Transmission Electron Microscopy Lab, Biomedical Technology Wing, Sree Chitra Tirunal Institute for Medical Sciences and Technology, Trivandrum 695012, Kerala, India

Henrique M. A. C. Fonseca
Department of Biology and GeoBioTec, University of Aveiro, 3810-193 Aveiro, Portugal

R. M. Gengan
Faculty of Applied Science, Department of Chemistry, Durban University of Technology, Durban, South Africa

Barnali Dasgupta Ghosh
Department of Chemistry, Birla Institute of Technology, Mesra, Ranchi 835215, Jharkhand, India.
E-mail: barnali.iitkgp@gmail.com

K. S. Ghosh
Department of Metallurgical and Materials Engineering, National Institute of Technology Durgapur, Durgapur 713209, India

Ana Violeta Girão
Department of Chemistry and CICECO—Aveiro Institute of Materials, University of Aveiro, Campus Santiago, 3810-193 Aveiro, Portugal

Paula P. Gonçalves
Department of Biology and CESAM, University of Aveiro, 3810-193 Aveiro, Portugal

Contributors

Ravi Chandra Gundakaram
Centre for Materials Characterization and Testing, International Advanced Research Centre for Powder Metallurgy and New Materials (ARCI), Balapur P.O., Hyderabad 500005, Telangana, India

Raj Kumar Gupta
Department of Physics, Birla Institute of Technology and Science, Pilani 333031, Rajasthan, India.
E-mail: raj@pilani.bits-pilani.ac.in

Saptarshi Gupta
Department of Chemical Engineering, National Institute of Technology, Agartala, Barjala, Jirania, Tripura (West), India

Harikrishnan V. S.
Division of Laboratory Animal Science, Biomedical Technology Wing, Sree Chitra Tirunal Institute for Medical Sciences and Technology, Trivandrum 695012, Kerala, India

Gururaj Kudur Jayaprakash
Departamento de Ingeniería de Proyectos, Centro Universitario de Ciencias Exactas e Ingenierías, Blvd. Marcelino García Barragán 1421, C. P. 44430, Guadalajara Jal., México.
E-mail: rajguru97@gmail.com

Annie John
Transmission Electron Microscopy Lab, Biomedical Technology Wing, Sree Chitra Tirunal Institute for Medical Sciences and Technology, Trivandrum 695012, Kerala, India.
E-mail: karippacheril@gmail.com

Anuradha Kumari
Department of Chemistry, Birla Institute of Technology, Mesra, Ranchi 835215, Jharkhand, India

B. J. Madhu
Department of Physics, Government Science College, Chitradurga 577501, Karnataka, India.
E-mail: bjmadhu@gmail.com

Susan Mani
Transmission Electron Microscopy Lab, Biomedical Technology Wing, Sree Chitra Tirunal Institute for Medical Sciences and Technology, Trivandrum 695012, Kerala, India

Manikandan M.
School of Mechanical Engineering, VIT University, Vellore, India. E-mail: mano.manikandan@gmail.com

Leichombam Menan
Department of Civil Engineering, National Institute of Technology, Agartala, Barjala, Jirania, Tripura (West), India

D. K. Mondal
Department of Metallurgical and Materials Engineering, National Institute of Technology Durgapur, Durgapur 713209, India

K. G. Moodley
Faculty of Applied Science, Department of Chemistry, Durban University of Technology, Durban, South Africa

A. K. Mukhopadhyay
Advanced Mechanical and Materials Characterization Division, CSIR-Central Glass and Ceramic Research Institute, Kolkata 700032, India

Nageswara Rao M.
School of Mechanical Engineering, VIT University, Vellore, India

Miguel Oliveira
Department of Biology and CESAM, University of Aveiro, Campus Santiago, 3810-193 Aveiro, Portugal

A. Shakila Parveen
Department of Chemistry, Anna University, Chennai 600025, Tamil Nadu, India

Maria de Lourdes Pereira
CICECO—Aveiro Institute of Materials, University of Aveiro, 3810-193 Aveiro, Portugal
Department of Biology, University of Aveiro, Campus de Santiago, 3810-193 Aveiro, Portugal.
E-mail: mlourdespereira@ua.pt

Ajay Vasudeo Rane
Composite Research Group, Department of Mechanical Engineering, Durban University of Technology, Durban, South Africa

Srimanta Ray
Department of Chemical Engineering, National Institute of Technology, Agartala, Barjala, Jirania, Tripura (West), India. E-mail: rays.nita@gmail.com

M. Fathima Rigana
Department of Chemistry, Anna University, Chennai 600025, Tamil Nadu, India

Teresa Margarida dos Santos
Department of Chemistry, University of Aveiro, Campus de Santiago, 3810-193 Aveiro, Portugal.
E-mail: teresa@ua.pt
CICECO—Aveiro Institute of Materials, University of Aveiro, 3810-193 Aveiro, Portugal

M. Sarojadevi
Department of Chemistry, Anna University, Chennai 600025, Tamil Nadu, India.
E-mail: msrde2000@yahoo.com

Saumya Shalu
Department of Chemistry, Birla Institute of Technology, Mesra, Ranchi 835215, Jharkhand, India.

Virgília Silva
Department of Biology and CESAM, University of Aveiro, 3810-193 Aveiro, Portugal

Amadeu Mortágua Velho da Maia Soares
Department of Biology and CESAM, University of Aveiro, Campus Santiago, 3810-193 Aveiro, Portugal

Victoria V. Stasova
V. N. Sukachev Institute of Forest, SB RAS Krasnoyarsk, 660036, Russian Federation

Bahaddurghatta Eshwaraswamy Kumara Swamy
Department of P.G. Studies and Research in Industrial Chemistry, Kuvempu University, Shankaraghatta 577451, Shimoga, Karnataka, India. E-mail: kumaraswamy21@gmail.com

Renata Tavares
Biology of Reproduction and Stem Cell Research Group, Center for Neuroscience and Cell Biology, Department of Life Sciences, University of Coimbra, 3004-517 Coimbra, Portugal

Simi Annie Tharakan
Department of Chemistry, Anna University, Chennai 600025, Tamil Nadu, India

P. Thirukumaran
Department of Chemistry, Anna University, Chennai 600025, Tamil Nadu, India

Sabu Thomas
School of Chemical Sciences and International and Inter University Centre for Nanoscience and Nanotechnology, Mahatma Gandhi University, Kottayam, Kerala, India 686560

Tito Trindade
Department of Chemistry and CICECO—Aveiro Institute of Materials, University of Aveiro, Campus Santiago, 3810-193 Aveiro, Portugal

Udupa K. R.
Department of Metallurgical and Materials Engineering NITK Surathkal, Srinivasanagar, 575025, India

Varma H. K.
Bio-ceramic Laboratory, Biomedical Technology Wing, Sree Chitra Tirunal Institute for Medical Sciences and Technology, Trivandrum 695012, Kerala, India

Balu Venugopal
Tissue Culture Lab, Biomedical Technology Wing, Sree Chitra Tirunal Institute for Medical Sciences and Technology, Trivandrum 695012, Kerala, India

ABBREVIATIONS

AA	ascorbic acid
AFM	atomic force microscopy
AN	acrylonitrile
AOP	advanced oxidation process
API	active pharmaceutical ingredient
ARCI	International Advanced Research Centre for Powder Metallurgy and New Materials
ASCs	adipose-derived mesenchymal stem cells
ATE	adipose tissue engineering
AuNPs	gold nanoparticles
BBB	blood–brain barrier
BDD	boron-doped diamond
BEI	backscattered electron imaging
BN	boron nitride
BSA	bovine serum albumin
BSEs	backscattered electrons
BST	barium strontium titanate
BTO	barium titanate
CCD	charged coupled detector
CD	circular dichroism
CGDS	cold gas dynamic spray
CL	cathodoluminescence
CLSM	confocal laser scanning microscope
CNFs	cellulose nanofibers
CNT	carbon nanotubes
COM	crystal orientation map
CPCSEA	Committee for the Purpose of Control and Supervision of Experiments on Animals
CSL	coincident site lattice
CTBN	carboxyl-terminated (butadiene-*co*-acrylonitrile)
CV	cyclic voltammetry
CVD	chemical vapor deposition
DA	dopamine
DRAM	dynamic random access memory

DRDO	Defence Research Development Organization
EBIC	electron beam induced current
EBSD	electron backscatter diffraction
EBW	electron beam welding
EC	emerging contaminants
ECO	electrochemical oxidation
EDS	energy dispersive spectroscopy
EDX	energy dispersive X-ray
EPA	Environmental Protection Agency
eSBS	epoxidized styrene-block-butadiene-block-styrene
EU	European Union
FBS	fetal bovine serum
FE	field emission
FE-SEM	field emission scanning electron microscopy
FGS	functionalized graphene oxide sheets
FISH	fluorescent in situ hybridization
FSH	follicle stimulating hormone
FTIR	fourier-transform infrared spectroscopy
GBCD	grain boundary character distribution
GCE	glassy carbon electrode
GFRP	glass fiber-reinforced polymer
GFT	glucose tolerance factor
GMAW	gas metal arc welding
GOS	grain orientation spread
GPCR	G protein-coupled receptor
GPR	general-purpose unsaturated polyester resin
GTA	gas tungsten arc
GTAW	gas tungsten arc welding
h-BN	hexagonal boron nitride
HAZ	heat affected zone
HE	hematoxylin–eosin
HOPG	highly ordered pyrolytic graphite
IACE	Institutional Animal Ethics committee
IGF	insulin-like growth factor
ILs	ionic liquids
IPF	inverse pole figure
IPN	interpenetrating polymer network
IQ	image quality
ITC	isothermal titration calorimetry
LB	Langmuir–Blodgett

LH	luteinizing hormone
LMB	leucomethylene blue
LMWCr	low-molecular weight chromium
MB	methylene blue
MCR	multicomponent reaction
MF	microfiltration
MLCC	multilayer ceramic capacitor
MMP	matrix metalloproteinases
MPMR	modified polymerization of monomeric reactant
MRSA	methicillin-resistant *Staphylococcus aureus*
MWCNTs	epoxy-multiwalled carbon nanotubes
NF	nanofiltration
Ni/MH	nickel–metal hydride
NP	nanoparticles
NR	natural rubber
OAPS	octa(aminophenyl)silsesquioxane
OER	oxygen evolution reaction
OIM	orientation imaging microscopy
OT	organothiol
PA	polyamide
PBz	polybenzoxazines
PCGTA	pulsed current gas tungsten arc
PCGTAW	pulsed current gas tungsten arc welding
PEC	Predicted Environmental Concentrations
PL	photoluminescence
PLA	poly(lactic acid)
PMDA	pyromelltic dianhydride
PMMA	poly(methyl methacrylate)
POSS	polyhedral oligomeric silsesquioxane
POSS-Cy	cyanate ester functionalized POSS
PP	polypropylene
PPCPs	personal care products
PSA	prostate-specific antigen
PTCR	positive temperature coefficients of resistivity
PTFE	polytetrafluoroethylene
PVAc	poly(vinyl acetate)
PVD	physical vapor deposition
PVP	polyvinylpyrrolidone
PZT	piezoelectric scanner tube
RBCs	red blood cells

rBST	recombinant bovine somatotropin
RF	reflectometry
RI	refractive index
RIU	refractive index unit
RO	reverse osmosis
ROS	reactive oxygen species
SAM	self-assembled monolayer
SBS	styrene-block-butadiene-block-styrene
SE	Schottky emission
SE	spectroscopic ellipsometry
SEI	secondary electron imaging
SEM	scanning electron microscopy
SHPB	split Hopkinson pressure bar
SP	surface plasmon
SPM	scanning probe microscopy
SPR	surface plasmon resonance
SQ	squarylium dye
STAR	steroidogenic acute-regulatory protein
STM	scanning tunneling microscopy
STO	strontium titanate
STPs	sewage treatment plants
SWE	single wavelength ellipsometry
TCP	topologically closed phases
TE	thermionic emission
TE	tissue engineering
TOC	total organic carbon
UA	uric acid
UF	ultrafiltration
UPR	unsaturated polyester
UV	ultraviolet
UV–Vis	ultraviolet–visible
VUV	vacuum ultraviolet radiation
WWTP	wastewater treatment plants
XNBR	carboxylated nitrile rubber
XRD	X-ray diffraction
ZOI	zone of inhibition

PREFACE

This book contains selected research and review papers presented at the World Congress on Microscopy (WCM 2015) held at the International and Inter University Center for Nanoscience and Nanotechnology, Mahatma Gandhi University, Kottayam, Kerala, India. This conference brought together a panel of highly accomplished experts in the field of microscopy. Technical talks consisted of basic studies and applications and addressed topics of novel issues in the field of microscopy. The 3-day conference featured a series of keynote addresses, a number of plenary sessions, invited talks, and contributed lectures focusing on recent advances, difficulties, and breakthrough in field of microscopy. Additionally, there was a poster session, and four best poster presentations were selected for awards.

The main attention in this collection of scientific papers is focused on recent theoretical and practical advances in polymers—their blends, composites, and nanocomposites related to their microscopic characterization. This volume provides important original and theoretical experimental results, which use nonroutine methodologies often unfamiliar to the usual readers. Review and research papers in this book present novel application of more familiar experimental techniques and analyses of composite processing problems, which indicate the need for new experimental approaches. Better understanding of the properties of polymer composites may be possible by utilizing new techniques to measure microstructure property relationships and by utilizing techniques and expertise developed in the conventional filled polymer composites. Characterization techniques, particularly microstructural characterization, have proven to be extremely difficult because of the range of length-scales associated with these materials.

This book highlights recent accomplishments and trends in the field of polymer nanocomposites and filled polymers related to microstructural characterization. We anticipate that this book will be of significant interest to scientists working on the basic issues surrounding polymers, nanocomposites, and nanoparticulate-filled polymers, as well as those working in industry on applied problems, such as processing. Because of the multidisciplinary nature of this research, this book will attract a broad audience, including chemists, materials scientists, physicists, chemical engineers, and processing specialists who are involved and interested in the future frontiers of blends.

The book has a wide variety of topics related to microscopy techniques in materials sciences. Topics include:

- Instrumentation and Techniques: advances in scanning probe microscopy, scanning electron microscopy, transmission electron microscopy, and optical microscopy, 3D imaging and tomography, electron diffraction techniques, analytical microscopy, advances in sample preparation techniques in situ microscopy, correlative microscopy in life and material sciences, low-voltage electron microscopy.
- Life Sciences: structure and imaging of biomolecules, live cell imaging, neurobiology, organelles and cellular dynamics, multidisciplinary approaches for medical and biological sciences, microscopic application in plants, microorganism, and environmental science, and super-resolution microscopy in biological sciences.
- Material Sciences: materials for nanotechnology, metals alloys and intermetallic, ceramics, composites, minerals and microscopy in cultural heritage, thin films, coatings, surfaces and interfaces, carbon-based materials, polymers, and soft materials, self-assembled materials, semiconductors, magnetic materials, and polymers and inorganic nanoparticles.

This book gives insight and better understanding into the development in microscopy as a tool for characterization. The book reports on recent research going on in the field of microscopy in life sciences and material sciences mainly related to its synthesis, characterizations, and applications. The book discusses the application of microscopic techniques in life sciences and material sciences and their applications and state of current research carried out.

The editors are especially grateful for the dedication and effort of our colleagues who participated and contributed to the symposium. Additionally, without the tireless effort of the publication staff of Apple Academic Press, this book would have not been possible. Finally, it is our pleasure to acknowledge financial contributions by the Department of Science and Technology, Anton Paar India Pvt. Ltd., to support this international conference.

PART I
Materials Sciences

CHAPTER 1

SCANNING ELECTRON MICROSCOPY AS A POWERFUL MORPHOLOGICAL CHARACTERIZATION TECHNIQUE FOR POLYMER BLENDS AND POLYMER NANOCOMPOSITES

SAJU DANIEL[1,2*] and SABU THOMAS[1,3]

[1]*International and Inter University Centre for Nanoscience and Nanotechnology, Mahatma Gandhi University, Kottayam, Kerala, India*

[2]*Department of Chemistry, St. Xavier's College Vaikom, Kottayam, Kerala, India*

[3]*School of Chemical Science, Mahatma Gandhi University, Kottayam, Kerala, India*

*Corresponding author. E-mail: sajudanielalpy@gmail.com

ABSTRACT

The scanning electron microscope (SEM) is fundamentally a microscope with high magnification power which gives valuable information such as external morphology, chemical composition, crystalline structure, and orientation of materials making up the sample. This information is provided by focusing high-energy electron beam to the surface of the sample so as to generate a variety of signals due to the electron–sample interaction. The advantages of SEM over traditional microscopes such as large depth of field, high resolution, control over the degree of magnification due to the usage of electromagnets rather than lenses, distinctly clear images make the scanning electron microscopy one of the most useful characterization techniques in research today. Because of this reason, a detailed study of SEM including fundamental principles of scanning electron microscopy, basic design of SEM—electron gun, comparison of three types of electron guns, sample stage, electromagnetic lenses, detectors for all signals, image display and recording and vacuum system, interaction volume, factors affecting the image quality, sample preparation, advantages and disadvantages of SEM, and applications of SEM are reviewed in this chapter. In addition to this, illustration of SEM as a powerful tool for the morphological characterization of polymer nanocomposites and polymer–blend nanocomposites with suitable examples taken from recently reported research works have also been included in this chapter.

1.1 INTRODUCTION TO SCANNING ELECTRON MICROSCOPY

A scanning electron microscope (SEM) is a powerful magnification tool that utilizes focused beams of electrons to obtain high-resolution, three-dimensional images that provide topographical, morphological, and compositional information, which makes them invaluable in a variety of science and industry applications. The SEM was developed by Professor Dr. Charles Oatley and his coworkers in late 1950s at Cambridge University. The SEM was developed as a result of the limitations of optical microscopy. Two inherent factors, rather large wavelength of visible light and rather poor depth of field, are responsible for the limitations. According to the theory, the limit of resolution of a microscope must be equal to the half the wavelength of the imaging energy. Resolving power of a microscope is inversely proportional to the limit of resolution, that is, if the limit of resolution is very small, the microscope is able to differentiate between two very closely spaced points of the specimen. In optical microscope, visible light is used for imaging, the wavelength of visible light is high so that the limit of resolution of optical microscope is high and hence the resolving power or resolution of the optical microscope is low. In SEM, electrons are used instead of visible light and hence the resolution is high as the wavelength of electrons is much smaller than wavelength of light. The main parameter effecting depth of field is the aperture angle. The problem with optical microscopy is that a high-power objective lens has a short focal length, which increases the aperture angle and decreases the depth of field as shown in Figure 1.1.[1] SEM has a long working distance and a small aperture opening making a very shallow aperture angle and hence a good depth of field as shown in Figure 1.2.[1]

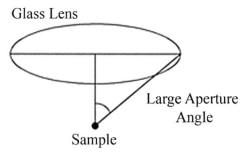

FIGURE 1.1 Large aperture angle of optical microscope.[1]

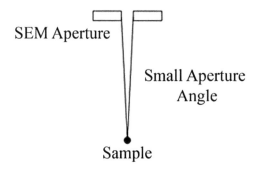

FIGURE 1.2 Small aperture angle of SEM.[1]

1.2 FUNDAMENTAL PRINCIPLES OF SCANNING ELECTRON MICROSCOPY

The fundamental principle is that a beam of electrons is created by thermionic emission gun, field emission (FE) gun, or Schottky emission (SE) gun. The electron beam is accelerated by applying high accelerating voltage and allowed to pass through a system of apertures and electromagnetic lenses to generate a thin beam of electrons. This thin beam of electrons is focused to a fine point at the specimen surface and this point is scanned across the specimen under the control of currents in the scan coils. Electrons are radiated from the specimen by the action of the scanning beam and assembled by a suitably located detector. Figure 1.3[4] is a schematic diagram that demonstrates various signals emitted from the specimen when the incident electron beam interacts with the specimen. These signals comprise secondary electrons, backscattered electrons (BSEs), diffracted backscattered electrons (EBSDs), X-rays, visible light (cathodoluminescence, CL), and heat. The SEM makes use of BSEs and secondary electrons for imaging samples: secondary electrons are most valuable for showing morphology and topography on samples and BSEs are most valuable for finding contrasts in composition in multiphase samples. The secondary electron imaging (SEI) is used mainly to image fracture surfaces and gives a high-resolution image. The backscattered electron imaging (BEI) is used typically to image a polished section; the brightness of the BEI depends on the atomic number of the specimen. X-ray generation is produced by inelastic collisions of the incident electrons with electrons in discrete orbitals of atoms in the sample. When the excited electrons return to lower energy states, X-rays of a definite wavelength are emitted. Thus, characteristic X-rays are produced for each element in a mineral that is excited by the electron beam. Detectors collect these X-rays, BSEs, and secondary electrons and convert them into a signal that is sent to a screen similar to a television screen to produce the image. Dot by dot, row by row, an image of the original object is scanned onto a monitor for viewing. The detector registers different levels of brightness on a monitor depending on the number of electrons that reach the detector. The magnification of the image is the ratio of the size of the screen to the size of the area scanned on the specimen.[1–6]

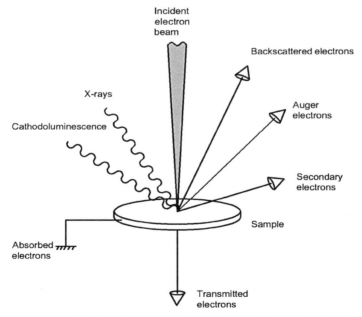

FIGURE 1.3 Emission of various electrons and electromagnetic waves from the specimen.[4]

1.3 BASIC DESIGN OF SEM

Figure 1.4 illustrates the basic design of SEM.[4]
Essential components of all SEMs include the following:

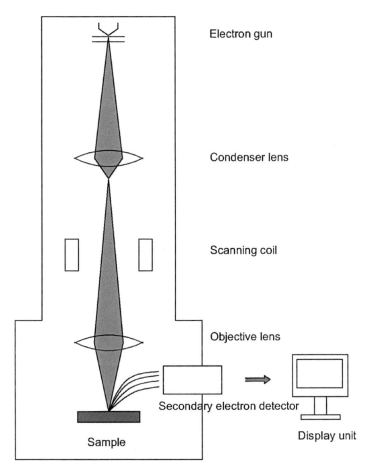

FIGURE 1.4 Basic design of scanning electron microscope.[4]

1.3.1 ELECTRON GUN

A beam of electrons is produced at the top of the microscope by an electron gun. There are three types of electron gun:

1. thermionic emission electron gun,
2. FE electron gun, and
3. SE electron gun.

1.3.1.1 THERMIONIC EMISSION ELECTRON GUN

Thermo-electrons are emitted from a filament (cathode) made of a thin tungsten wire by heating a filament at high temperature about 2800 K, and this electron beam flows through the hole of the anode. The current of the electron beam can be adjusted by placing an electrode called Wehnelt electrode between the cathode and anode and applying a negative voltage to it. LaB_6 can also be used as the cathode. W filaments are very simple and inexpensive, whereas LaB_6 filaments are brighter than W and expensive.[4-6] Figure 1.5 explains the working principle of the thermionic emission electron gun, which has been used in scanning electron microscope.[4]

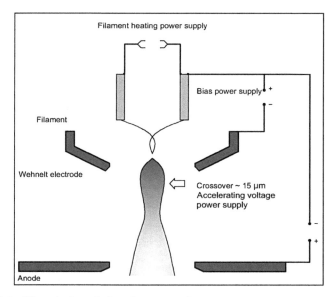

FIGURE 1.5 Thermionic emission electron gun.[4]

1.3.1.2 FE ELECTRON GUN

The FE gun employs the FE effect that takes place when a high electric field is applied to a metal surface. The cathode is made of a thin tungsten wire.

A tungsten single crystal is soldered to this tungsten wire, and the tip of the tungsten single crystal is bent to be a curvature radius of about 100 nm. This is called the emitter. In contrast to the thermionic emission (TE) gun and SE gun, FE gun requires no heating of the emitter for the emission of the electron. When a positive voltage is applied to a metal plate (the extracting electrode), the tunneling effect occurs and electrons are emitted from the emitter. The emitted electron beam flows through the holes of the extracting electrode. An electron beam having certain energy is obtained when an accelerating voltage is applied to the electrode located below the extracting electrode.[4-6] Figure 1.6 represents the FE electron gun, which has been used in scanning electron microscope.[4]

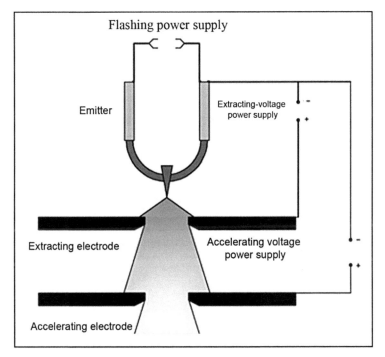

FIGURE 1.6 Field emission electron gun.[4]

1.3.1.3 SE ELECTRON GUN

The SE gun utilizes the Schottky-emission effect that takes place when a high electric field is applied to a heated metal surface. The cathode is a ZrO/W emitter. Coating of tungsten with ZrO greatly decreases the work function; thus, a large emission current can be obtained at a relatively low cathode temperature of about 1800 K. To shield the thermo-electrons from the emitter, a negative voltage is applied to an electrode, called the suppressor.[4] Figure 1.7 illustrates the basic principle of Schottky-emission electron gun, which has been used in scanning electron microscope.[4]

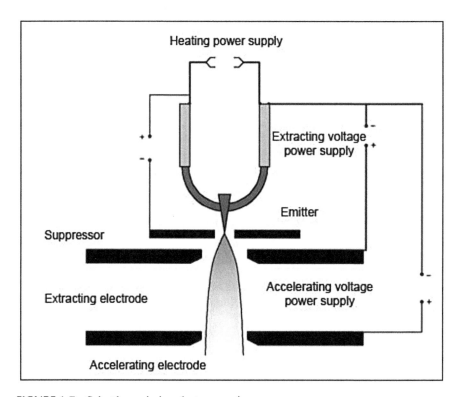

FIGURE 1.7 Schottky-emission electron gun.[4]

1.3.1.4 COMPARISON OF THE FEATURES OF THE TE GUN, FE GUN, AND SE GUN

The FE gun is the outstanding one when we consider the electron source size, brightness, life time, and energy width of the electron beam. The TE gun is superior in terms of the probe current and current stability. By considering these features, we can conclude that the FE gun is appropriate for morphological observation gun at high magnifications and the TE gun is suitable for versatile applications such as analysis where high magnification is not required. In terms of application, the position of SE gun lies between these two electron guns.[4] Table 1.1 compares the three types of electron guns, which have been used in scanning electron microscope.[4]

TABLE 1.1 Comparison of Three Types of Electron Guns.

	TE gun		FE gun	SE gun
	Tungsten	LaB$_6$		
Electron source size	15–20 μm	10 μm	5–10 nm	15–20 nm
Brightness (A/cm^2/rad^2)	10^5	10^6	10^8	10^8
Energy speed (eV)	3–4	2–3	0.3	0.7–1
Lifetime	50 h	500 h	Several years	1–2 years
Cathode temperature (K)	2800	1900	300	1800
Current fluctuation (per hour)	<1%	<2%	>10%	<1%

Note that the brightness is obtained at 20 kV. (Reprinted from Ref. 23.)

1.3.1.5 ELECTROMAGNETIC LENSES

A fine electron beam is required for the SEM. For this purpose, the diameter of the electron beam is adjusted by placing a lens below the electron gun. An electron microscope applies electromagnetic lenses to control the flow of electrons. Electromagnetic lenses consist of a huge bundle of windings of insulated copper wire, a soft iron cast and pole piece. A magnetic field is induced by the current in the winding and reaches its main strength at the pole piece of the lens. The accelerated electrons entering the magnetic field are deviated following the law of a charge passing a magnetic field according to Lorentz magnetic force equation and the direction of deflection is given by Fleming's left hand rule. The resultant force is always perpendicular to the plane defined by the direction of the magnetic field and the direction of the electrons. In conclusion, the electrons take a circular path through the lens system. The advantage of electromagnetic lens in an electron microscope is that the focal length of the lens can be varied by changing the current through the wires.[1,4,7] Figure 1.8 illustrates the working principle of an electromagnetic lens, which has been used in SEM.[1]

A typical SEM has two lenses, a condenser lens and an objective lens. If the diameter of the filament is d_0 and it is placed a distance u_1 from the condenser lens and the demagnified filament image is formed at a distance v_1 on the other side of the condenser lens then the diameter of the demagnified filament image, $d_1 = d_0 v_1/u_1$. From this equation, it is clear that the stronger the condenser lens, the shorter will be the value of v_1 and the beam diameter d_1. The function of the objective lens in an SEM is to demagnify the beam again. The diameter of the final probe, d, on the specimen is

$$d = d_1\, v_2/u_2 \qquad d = d_1\, WD/u_2$$

where u_2 is the distance between the demagnified source image (cross-over below the condenser lens with a diameter d_1) and the objective lens and v_2 is the working distance. Working distance is defined as the distance between the objective lens and the specimen. The distance between condenser and objective lenses ($v_1 + u_2$) is a constant for a given SEM. Hence, a strong condenser lens produces small value for v_1 which in turn results in a larger u_2 and smaller d value. The probe size in SEM is decreased either by increasing the strength of the condenser lens or by decreasing the working distance.[9]

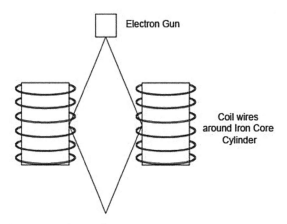

FIGURE 1.8 Electromagnetic lens.[1]

1.3.1.6 SAMPLE STAGE

In an electron microscope, high magnification is used for observing the specimen. Therefore, a specimen stage that solidly supports the specimen and moves easily is required. The specimen stage for SEM is capable of performing the following movements: horizontal movement (X, Y), vertical movement (Z), specimen tilting (T), and rotation (R). The selection of a field of view can be carried out by the X and Y movements, whereas the change of image resolution and the depth of focus can be provided by the Z movement. Most SEMs use the eucentric specimen stage. The advantage of this specimen area is that the observation area is not shifted by the tilting of the specimen and also the focus on the specimen is not altered by shifting the field of view due to the tilting of the specimen.[4]

1.3.1.7 DETECTORS FOR ALL SIGNALS

1.3.1.7.1 Secondary Electron Imaging

Secondary electrons are those that emitted from the specimen with energies below 50 eV and produced by knocking out them from their respective orbits around an atom by an incident electron and their generation is shown in Figure 1.9.[1] Because of their low energies and low penetration depth, it is possible to attain secondary electron imaging with high magnification and high resolution. The secondary electron detector captures these electrons and forms an image of the surface of the sample. The direction of the emission of the secondary electrons depends on the orientation of the features of the surface; as a result, the image formed will reflect the characteristic feature of the region of the surface that was exposed to the electron beam.[1,4–10]

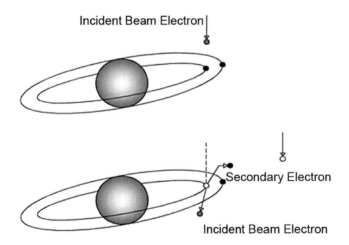

FIGURE 1.9 Generation of secondary electrons.[1]

The secondary electron detector is used for detecting the secondary electrons emitted from the specimen. A high voltage of about 10 kV is applied to a scintillator which is coated on the tip of the detector. When the secondary electrons from the specimen strike the scintillator, light is emitted. The light guide leads the emitted light to a photo-multiplier tube which converts light to electrical signal in the amplified form. A supplementary electrode, called the collector, is placed before the scintillator. The number of secondary electrons can be controlled by changing the voltage of the collector.[4] Figure 1.10 represents a secondary electron detector, which has been used in scanning electron microscope.[4]

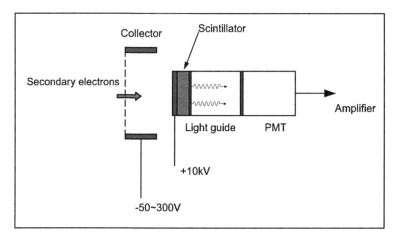

FIGURE 1.10 Secondary electron detector.[4]

1.3.1.7.2 Backscattered Electron Imaging

BSEs are those incident electrons that approach the nucleus of an atom sufficiently closely to be scattered through a large angle and reemerge from the surface. Number of BSEs emitted is less than number of secondary electrons emitted and their energies are much higher than that of secondary electrons. BSE images have slightly less resolution than secondary electron images. This is due to the reason that the area giving rise to the signal is larger than the probe size as they come from slightly deeper in the specimen. The number of BSEs emitted increase with the increase in atomic number of the constituent atoms in the specimen. As a result, an area that contains heavy atom appears bright in the BSE image. Therefore, BEI is mainly used for getting compositional information. BSEs can also provide crystallographic information, as electron channeling occurs [1,4–10] Figures 1.11[1] and 1.12[14] represent the generation of back-scattered electrons and back-scattered electron detector, which have been used in scanning electron microscope.

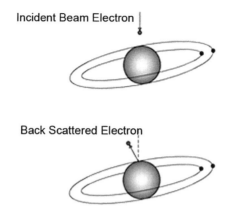

FIGURE 1.11 Generation of back-scattered electrons.[1]

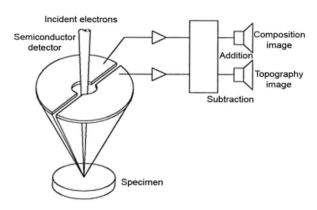

FIGURE 1.12 Back-scattered electron detector.[14]

1.3.1.7.3 *Electron Beam Induced Current Image*

For every incident electron, hundreds of electron–hole pairs are generated in the specimen. Normally, most of them recombine within about 10–12 s. In spite of this, if an applied electric field separates the electrons and holes before they recombine, an induced current flow between the electrodes, forming an electron beam induced current (EBIC) image. The current flowing from each point or voltage across each point will depend on the conductivity of the specimen at that point, the lifetime of the electrons and holes, and their mobilities. This technique can be used for recording varying concentrations of electrically active defects and determining failure points in devices.[6,9]

1.3.1.7.4 Cathodoluminescence Detection and Imaging

For every incident electron, hundreds of electron–hole pairs are generated in the specimen. Normally, most of them recombine within about 10–12 s. When the electrons recombine with the holes, light is emitted. The wavelength of light emitted depends on the band-gap energy of the specimen which in turn depends on the composition. The signal may be passed through a spectrometer before being measured by a suitable detector. This technique is excellent for revealing defects that degrade radiative properties. CL signals come from the entire specimen beam interaction volume.[5-7,9]

1.3.1.7.5 Voltage-Contrast Imaging

When a voltage is applied across a semiconductor in the SEM, the secondary electron image will be different from that with no voltage, the potential developed across the active regions changes the number of secondary electrons emitted from those areas. More electrons can escape from active regions where a negative voltage develops, so these appear brighter regions than the region where a positive voltage develops. This is another technique that is useful in failure analysis.[4,7]

1.3.1.7.6 Auger Electrons

The production of Auger electron is an alternative to characteristic X-ray emission after ionization of an inner shell. Energy is released when an electron from an upper shell de-excites and fills the vacancy in the ionized shell. This energy can either be converted to an X-ray quantum of energy $hv = E_2 - E_1$ or transferred to another atomic electron, which leaves the specimen as an Auger electron with a characteristic kinetic energy determined from the difference between $E_2 - E_1$ and the energy necessary to overcome the ionization energy and the work function.

Auger electrons are emitted from atomic layers very close to the surface and give valuable information about the surface chemistry. Because of the low numbers of Auger electrons and the need to measure their energies with precision, Auger electron imaging is usually performed in dedicated instruments and requires advanced detectors and instrumentation.[4-7]

1.3.1.7.7 Energy Dispersive X-Ray Spectroscopy

Characteristic X-rays are emitted from the atom when an incident high-energy electron knocks out an inner-shell electron and an outer-shell electron moves into the empty orbit and the principle of the generation of characteristic X-rays is shown in Figure 1.13.[4] There is now a progression of electron jumps from higher to lower energy states until all the electron states are refilled. At each stage, X-rays are emitted to conserve energy. Measurement of the energies or wavelengths of these X-rays gives information about the chemical composition of the specimen. The X-rays are detected either by energy dispersive or by a wavelength dispersive spectrometer. Energy dispersive X-ray spectroscopy (EDS or EDX) is the more common attachment to SEMs as it provides rapid qualitative analysis of the specimen.

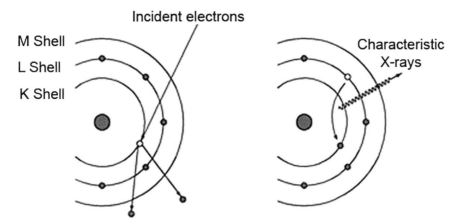

FIGURE 1.13 Principle of the generation of characteristic X-rays.[4]

The EDS is used to analyze characteristic X-ray spectra by measuring the energies of the X-rays. X-rays emitted from the specimen enter the semiconductor detector and generate electron–hole pairs whose quantities correspond to the X-ray energy. The values of X-ray energies are obtained by measuring these quantities. The electric noise can be reduced by cooling the detector by liquid nitrogen. The advantage of the EDS is that the X-rays from a wide range of elements from B to U are analyzed simultaneously.[4-7,10] Figure 1.14 demonstrates the working principle of energy dispersive X-ray spectroscopy.[4]

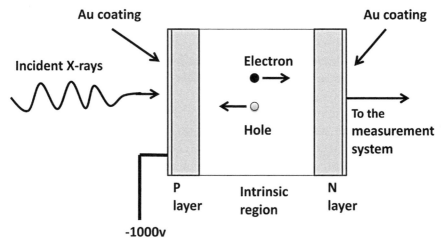

FIGURE 1.14 EDS semiconductor detector.[4]

1.3.2 IMAGE DISPLAY AND RECORDING

The output signals from the detector are amplified and then conveyed to the display unit. Since the scanning on the display unit is synchronized with the electron probe scan, brightness variation, which depends on the no of electrons, appears on the monitor screen on the display unit, thus forming SEM image. A cathode ray tube was used for many years as display unit. In recent years, liquid crystal display has been widely used.[4,6,7]

1.3.3 VACUUM SYSTEM

For the successful operation of a scanning electron microscopy, the column must be kept under high vacuum. There are several reasons for providing high vacuum in SEM. First, a hot tungsten filament will oxidize and burn out in the presence of air. Second, the column must be kept clean if the beam is to be well focused: moisture in the air will cause corrosion and dust particles in the beam path can block the beam or may become charged and deflect the beam. Third, air molecules will scatter electrons. The use of more than one type of vacuum pump is required for attaining the vacuum necessary for the operation of SEM. The pumping system generally utilizes a mechanical pump in conjunction with either a diffusion pump or a diffusion pump-assisted ion pump.[4,6,8]

1.4 INTERACTION VOLUME

The "interaction volume" is the area of the sample excited by the electron beam to produce a signal. The penetration of the electron beam into the sample is affected by the accelerating voltage used, the higher the kV, the greater the penetration. The effective interaction volume can be calculated using the electron range, $R = 0.0276\, A\, E_0^{1.67} / \rho Z^{0.89}$ (µm), where A is the atomic weight (g/mol), Z is the atomic number, ρ is the density (in g/cm³), and E_0 is the energy of the primary electron beam (in kV) and is shown in Figure 1.15.[13] SE, BSE, and X-ray are produced throughout the interaction volume, provided the beam electrons still have enough energy to generate it. The SE signal is readily absorbed and therefore can only be detected relatively close to the surface (i.e., less than 10 nm).

The BSE signal is of higher energy and able to escape from a more moderate depth within the sample. The X-ray signal can escape from a greater depth, although the X-ray signal absorption is actually variable depending upon its energy.[7,8,10]

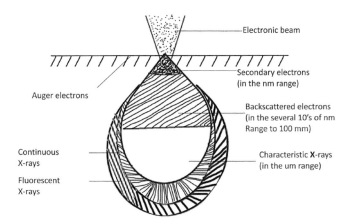

FIGURE 1.15 Interaction volume of the electron beam with a bulk sample.[13]

1.5 FACTORS AFFECTING THE IMAGE QUALITY

1.5.1 INFLUENCE OF ACCELERATING VOLTAGE ON IMAGE QUALITY

In SEM, lower accelerating voltages must be used to obtain fine surface structure. This is because at higher accelerating voltages, the beam penetration and diffusion area is large enough to produce unnecessary signals from the bulk of the specimen. And these signals reduce the image contrast and veils fine surface structures.[4,8]

1.5.2 PROBE CURRENT, PROBE DIAMETER, AND IMAGE QUALITY

In SEM, higher magnification and resolution of the image is attained by decreasing the electron probe diameter on the specimen. The probe current is directly proportional to probe diameter. It is therefore necessary to select a probe current suited for the magnification and observation conditions (accelerating voltage, specimen tilt, etc.) and the specimen.[4,7,9]

1.5.3 INFLUENCE OF EDGE EFFECT ON IMAGE QUALITY

If there exist uneven steps or thin protrusions on the specimen surface, the edges of the steps or protrusions appear bright and the bright parts have certain width instead of getting sharp lines. This phenomenon is called the edge effect. The degree of the edge effect depends on the accelerating voltage. The lower the accelerating voltage, the smaller the penetration depth of incident electrons into the specimen. This reduces bright edge portions, thus resulting in the microstructures present in them being seen more clearly.[4]

1.5.4 DEPENDENCE OF IMAGE QUALITY ON TILT ANGLE

The objective of the specimen tilt is to (1) improve the quality of secondary electron images, (2) obtain information different form that obtained when the specimen is not tilted, that is, to observe topographic features and observing specimen sides, and (3) obtain stereomicrographs. When the specimen is tilted, lengths observed are different from their actual values. Therefore, when measuring pattern widths, it is necessary to measure without specimen tilting or to correct values obtained from a tilted state. It is sometimes difficult to correctly judge their topographical features from the SEM images. In such a case, observation of stereo SEM images makes it easy to understand the structure of the specimen. In stereo observation, after a field of interest is photographed, the same field is photographed again with the specimen tilted from 5° to 15° and viewing these two photos using stereo-glasses with the tilting axis held vertically provides a stereo-image.[4]

1.5.5 DETECTOR POSITION AND SPECIMEN DIRECTION

When the specimen is illuminated with an electron beam, the amount of secondary electrons produced depends on the angle of incidence theoretically. However, there arises a difference in the image brightness depending on whether the tilted side of the specimen is directed to the secondary electron detector or the opposite side. In the case of a long specimen, to avoid the above problem, the longitudinal axis of the specimen must be directed to the detector to make the brightness uniform.[4]

1.5.6 INFLUENCE OF CHARGE-UP ON IMAGE QUALITY

When a nonconductive specimen is directly illuminated with an electron beam, the specimen gets charged and thus preventing normal emission of secondary electrons and the charge developed on the surface repels the electron beam that scans the specimen surface which results in the positional shift of the electron probe. As a result, this charge-up will lead to some unusual phenomena such as abnormal contrast and image deformation. Generally, the following methods are used to reduce specimen charge-up. (1) Reducing the probe current, (2) lowering the accelerating voltage, and (3) tilting the specimen to find a balanced point between the amount of incident electrons and the amount of electrons that go out of the specimen. Usually, the surface of a nonconductive specimen is coated with some conductive metal prior to observation to avoid specimen charge up.[4,7,9,10]

1.5.7 EFFECT OF WORKING DISTANCE

Working distance greatly influence the image quality. In most cases, the working distance has to be reduced to achieve better resolution especially at lower accelerating voltages.[4,8]

1.5.8 EFFECT OF SPOT SIZE

Spot size basically confines the beam current and will thereby cause for brightness and contrast compensations. Typically, smaller spot sizes allow for higher resolution and a greater depth of field.[4,8]

1.6 SAMPLE PREPARATION

Since the SEM operates at high vacuum conditions and utilizes electrons to form an image, special preparation of samples must be required. All water should be removed from the samples before doing SEM as the water vaporizes in the vacuum. The electrons that entered the specimen lose their energy and they are absorbed in the specimen. In the case of conductive specimens, the electrons flow through the specimen stage. All metals are conductive and hence no preparation is required before being used. If the specimen is nonconductive, the electron stays in the specimen where it is and charging up of the specimen occurs. If the charging occurs, the electron probe that scans over the specimen is deflected by the repulsive force from a charged potential, resulting in a positional shift of the electron probe. This leads to the distortion of the image. All nonmetals need to be made conductive by covering the sample with a thin layer of conductive material commonly carbon, gold, or some other metal or alloy. The choice of material for conductive coatings depends on the data to be acquired: carbon is most suitable for elemental analysis while metal coatings are most effective for high-resolution electron imaging applications. Sputter coater is a device that helps to coat metal on the surface of a nonconductive specimen. The sputter coater uses an electric field and argon gas. The sample is placed in a small chamber that is at a vacuum. An electric field cause an electron to be removed from the argon, making the atoms positively charged. These positively charged argon ions are attracted by a negatively charged gold foil and the argon ions knock gold atoms from the surface of the gold foil. These gold atoms fall and settle onto the surface of the sample producing a thin gold coating.

The other method of coating is evaporation of a metal in a vacuum evaporation. A specimen is placed in a vacuum chamber and an appropriate amount of coating metal is placed in a tungsten wire basket. When the chamber is evacuated and the tungsten wire filament is heated to the vaporization point of the coating metal, the metal condenses on the specimen. Since the metal stream from the filament is in a straight line, the specimen must be tilted and rotated during the evaporation. Alternatively, an electrically insulating sample can be examined without a conductive coating in an instrument capable of "low vacuum" operation.[4,6,7,9,10]

Special treatment is not required to observe the specimen surface. After cutting the specimen with a suitable size for observation, expose a surface to observe. It is necessary to prepare a cross-section to observe the internal structure. The methods used are as follows.

Fracturing

If a specimen is hard, it is fractured to prepare a cross-section. If a specimen is soft at normal temperature but hard at low temperature, freeze fracturing is applied to this material in liquid nitrogen.

Cutting

If a specimen is soft like a polymer, it can be cut by using an ultramicrotome as in the case of TEM. The cross-section obtained by this method is very flat. A specimen with a few scars can be accepted if low magnification is used. In such a case, razor blade can be used for preparing cross-section.[4]

1.7 SEM APPLICATIONS

SEMs have a wide range of applications both in industry as well as in scientific fields. Main applications of SEM include topographical, morphological and compositional investigations, analysis of surface fractures, provide information in microstructures, detection of crystalline structures, provide qualitative chemical analysis, examine surface contaminations, and reveal spatial variations in chemical compositions, semiconductor inspection, and assembly of microchips for computers. SEMs are significant research tool in fields such as life science, biology, gemology, medical and forensic science, and metallurgy.[2,12]

1.8 ADVANTAGES OF SEM

Advantages of SEM comprise its wide range of applications such as different types of imaging and versatile information collected from different detectors. SEMs can easily be operated with the proper training and advances in computer technology and associated software make operation user-friendly. The working of the instrument is fast enough to complete SEI, BSE, and EDS analyses within 5 min. In addition, the technological advances in modern SEMs provide the data in digital form. Even though all samples must be prepared before placing in the vacuum chamber, most SEM samples require minimal preparation action.[2,12]

1.9 DISADVANTAGES OF SEM

The main disadvantage of SEM is related to size and cost. SEMs are expensive, large, and must be housed in an area free of any possible electric, magnetic, or vibration interference. A steady voltage and current to electromagnetic coils and circulation of cool water must be maintained throughout the experiment. Special training is required to operate an SEM as well as prepare samples. The preparation of samples can result in artifacts. SEMs are limited to solid materials. SEMs carry a small risk of radiation exposure associated with the electrons that scatter from beneath the sample surface.[2,12]

1.10 CHARACTERIZATION OF POLYMER BLENDS AND POLYMER NANOCOMPOSITES BY SEM

Some examples for the morphological characterization of polymer blends and polymer nanocomposites by SEM are illustrated below. These illustrations prove that SEM is a powerful technique for the morphological characterization of polymer blends and polymer nanocomposites.

Poornima Vijayan et al. correlated the rheological parameters obtained and the outcomes from phase separation studies using optical microscopy with the domain size of the phase separated carboxyl-terminated (butadiene-*co*-acrylonitrile) (CTBN) in epoxy/CTBN blend and epoxy/clay/CTBN ternary nanocomposites. For this purpose, they measured the domain parameters of the phase separated CTBN from the SEM images of etched smooth cut surface of the samples shown in Figure 1.16[15] and found that average diameter of CTBN in epoxy/clay/CTBN ternary nanocomposite is lower than in epoxy/CTBN blend. They observed a lower relaxation time for epoxy/clay/CTBN system due to the dominating catalytic activity of nanoclay on epoxy-anhydride reaction. They concluded that most of the phase separation process occurs well before the gelation which resulted in phase separated CTBN with lower domain size in the ternary system due to the shorter time available for the growth of domains.[15]

Maria et al. revealed that the improved mechanical property and the slow-stress relaxation of O2Mt filled natural rubber (NR)/nitrile rubber (NBR) nanocomposites are due to the strong interaction between O2Mt and NR and confirmed this interaction by scanning electron microscopic studies and the SEM images are shown in Figure 1.17.[14] In their studies, it is also proved that O1Mt shows affinity toward NBR predominantly and explained the observation as follows. The polar interaction between NBR and O1Mt localizes the O1Mt predominantly in the NBR phase. This results in the increase of NBR phase viscosity, which ultimately decreases the polymer chain mobility and suppresses the coalescence of the NBR domains, leading to a fine morphology in the NR matrix for the 70/30 NR/NBR system which can be evidenced from Figure 1.18.[14]

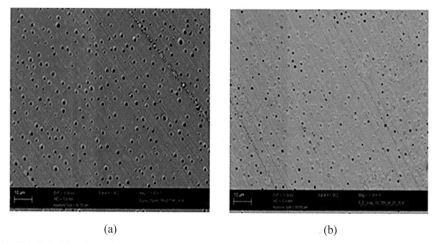

FIGURE 1.16 Scanning electron microscopic (SEM) images of smooth cut surface of (a) epoxy/15 phr CTBN and (b) epoxy/3 phr clay/15 phr CTBN.[15]

FIGURE 1.17 SEM images showing the O2Mt clay localized in the NR phase. (Reprinted from Ref. 14. © 2014 Elsevier B.V.)

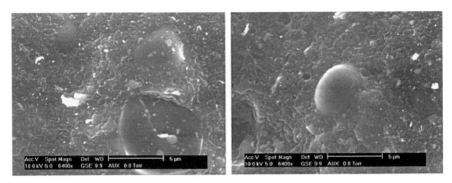

FIGURE 1.18 SEM images showing the decrease in domain size of NBR-dispersed phase. (Reprinted from Ref. 14. © 2014 Elsevier B.V.)

Geroge et al. studied the morphology of the fractured surface of cross-linked epoxy and epoxy block copolymer blends by using FESEM and their observation and interpretation are as follows. Figure 1.19a[17] shows the SEM micrographs of the neat epoxy system, which reveal a single phase. The fracture surface was typically flat and smooth and the cracks spread freely and regularly and orient in the direction of the loading, typical for a brittle material. For the blends containing unmodified styrene-block-butadiene-block-styrene (SBS) that is with 0 mol% epoxidation degree, a heterogeneous morphology with phase separated structure is observed which is shown in Figure 1.19b.[17] Figure 1.19b specifies that the phase separation occurs in the macroscopic length scale. The styrene and butadiene domains are coagulated in the epoxy matrix. Figure 1.19c[17] shows the sample with the lowest degree of epoxidation epoxidized styrene-block-butadiene-block-styrene (eSBS26), which has a better compatibility of block copolymer in the epoxy matrix due to the epoxidation of butadiene and in which the size of the immiscible domain is decreased. As the epoxidation degree increases, the heterogeneous morphology changes to homogeneous morphology. It can be seen that when the epoxidation degree reaches 39 mol% (Fig. 1.19d, e, and f),[17] the domain size of the block copolymers in epoxy matrix again decreases and form domains in the size of nanometer scale. On the other hand, for 47 mol% epoxidation (Fig. 1.19g and h),[17] the blend surface looks homogeneous, but the magnified image (Fig. 1.19i)[17] reveals PS nanophase; however, these are too small. It is difficult to find out the exact morphologies of the blends with 39 and 47 mol% epoxidized SBS by SEM images. However, it can be concluded that there is a change in morphology from macroscopic level to nanolevel (Fig. 1.19),[17] or in other words, epoxidized copolymers at highest degree of epoxidation could be able to create nanostructured templates in epoxy resin.[17]

Scanning Electron Microscopy as a Powerful Morphological Technique 27

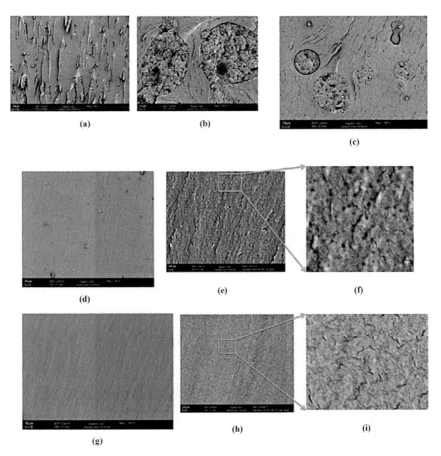

FIGURE 1.19 SEM micrographs of fracture surface of (a) neat epoxy, (b) epoxy blends with 10 wt% SBS, (c) epoxy blends with 10 wt% eSBS26, (d–f) epoxy blends with 10 wt% eSBS39 ((f) high magnification (1000 × 800 nm^2)), (g–i) epoxy blends with 10 wt% eSBS47 ((i) high magnification (800 × 600 nm^2)).[17]

Chirayil et al. prepared nanofibril reinforced unsaturated polyester nanocomposites and studied morphology, mechanical and barrier properties, viscoelastic behavior, and polymer-chain confinement. The SEM images of the fracture surfaces of the tensile-tested composites and neat are shown in Figure 1.20.[16] The microscopic aspects of the fracture surface of the composite samples revealed good explanation for the improvement in mechanical properties. Maximum improvement in tensile propertics were observed at 0.5 wt% composites, where effective stress transfer takes place at this optimum loading as the matrix particles are seen sticking to the filler surfaces.[18]

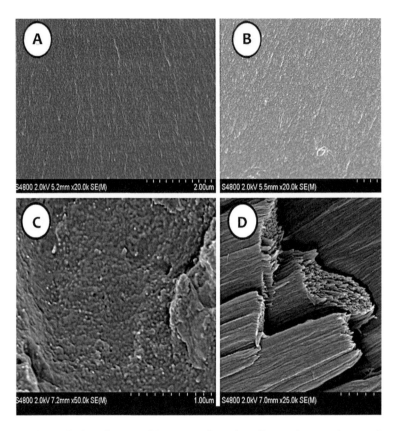

FIGURE 1.20 The SEM images of fracture surface of tensile tested composite samples: neat polyester, (b) 0.5 wt% INF, (c) 1 wt% INF, and (d) 5 wt% INF. (Reprinted with permission from Ref. 16. © 2014 Elsevier.)

Chandran et al. studied the morphological features of the PP/NR (70/30) blend and their nanocomposites by using scanning electron microscopy. The SEM images of PP/NR blend (70/30) and their nanocomposites are displayed in Figure 1.21.[19] By strongly analyzing the images, they could conclude the following points. The images reveal droplet morphology for the uniformly distributed NR particles. The holes indicate the position of the toluene extracted NR phase in the polypropylene (PP) matrix. It is obvious from the images that the size of the dispersed NR domains reduced systematically upon the addition of Cloisite 20A loading (Fig. 1.21a–e).[19]

Rajisha et al. prepared potato starch nanocrystal reinforced natural rubber nanocomposites and studied the morphology of the NR/starch nanocrystal material by using FE scanning electron microscopy. The distribution level of the filler within the matrix was evaluated by observing the FESEM image of the surface fractured films. Figure 1.22a[18] and b shows the nanocomposite films of NR nanocomposite with 5 and 10 wt% of nanocrystals, respectively. In the FESEM micrograph, a homogenous dispersion of starch nanocrystals which appears like white dots can be observed. A uniform distribution of the nanocrystal in the matrix is clearly seen for the compositions. Such an even and uniform distribution of the filler in the matrix is essential for obtaining optimum properties. As the filler loading is increased, the nanocrystals show a tendency to agglomerate which can be observed in Figure 1.22b[18] image. The particle size of the nanocrystal was observed to be in the range 12 nm.[20]

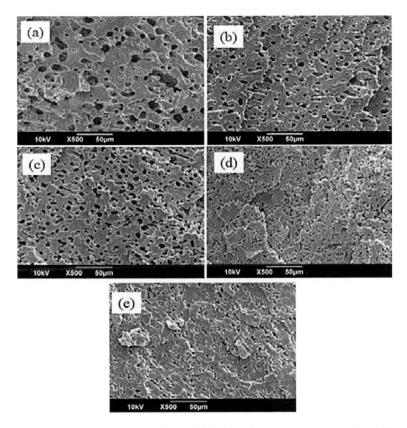

FIGURE 1.21 SEM micrographs of PP/NR/Cloisite 20A nanocomposites where NR phase has been preferentially extracted by toluene: (a) 70/30/0, (b) 70/30/1, (c) 70/30/3, (d) 70/30/5, and (e) 70/30/7.[19]

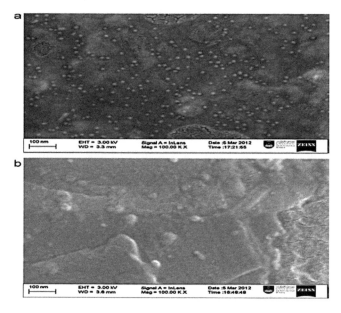

FIGURE 1.22 FESEM micrograph of (a) NR95 starch nanocomposite and (b) NR90 starch nanocomposites. (Reprinted with permission from Ref. 18. © 2014 Elsevier B.V.)

Abraham et al. prepared green nanocomposites of natural rubber/nanocellulose and studied the morphology of the composite by SEM. The SEM images of the nanocomposite with different filler content are shown in Figure 1.23.[19] Figure 1.23 shows the SEM image of (a) unfilled NR matrix, (b) the tensile fracture surface of 2.5 wt% nanocomposite, (c) 10% nanocomposite after equilibrium toluene swelling, and (d) the tensile fracture surface of 10 wt% nanocomposite. Figure 1.23A[21] gives the clear information about the gum matrix. The SEM of the tensile fracture surface in Figure 1.23b[21] and d[21] clearly explains the homogenous dispersion of nanocellulose in NR matrix and the voids present in the 10% nanocomposite is probably due to the agglomerated nanocellulose present at the higher weight percentage. The void present in the nanocomposite too has some contribution to the uptake of solvent at higher filler concentration. The agglomeration and the presence of voids are more clearly explained by the SEM of 10% nanocomposite taken after equilibrium toluene swelling which is shown in Figure 1.23c.[21] The unfilled NR and the uniform dispersion of nanocellulose in NR latex matrix with different concentration is clear from the SEM images as seen in Figure 1.23.[21]

Scanning Electron Microscopy as a Powerful Morphological Technique   31

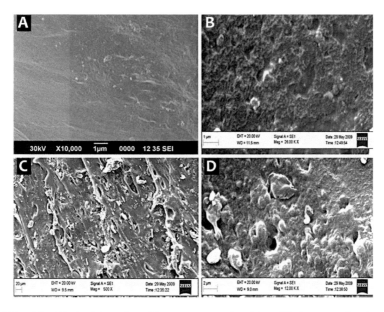

FIGURE 1.23   SEM of the NR and its nanocomposites (a) unfilled NR matrix, (b) tensile fracture surface of 2.5 wt% nanocomposite, (c) 10% nanocomposite after equilibrium toluene swelling, and (d) tensile fracture surface of 10 wt% nanocomposites. (Reprinted with permission from Ref. 19. © 2013 Elseiver B.V.)

Jose et al. studied the dynamic mechanical properties of immiscible polymer systems with and without compatibilizer. The SEM micrographs shown in Fig. 1.24[22] reveal that all the blends exhibit two-phase, non-uniform and unstable morphology, typical of immiscible and incompatible multiphase systems. The SEM micrographs demonstrated that as the amount of dispersed phase increases, morphology becomes more coarse and, therefore, it can be argued that the uncompatibilized blends with co-continuous phase structure possess least stable morphology. Thus, N50 is believed to be maximum incompatible, as it possesses a co-continuous phase structure. The emulsifying action of the compatibilizer can be evaluated from the SEM micrographs showing the morphology of polyamide12 (PA12)/PP blends in the presence of compatibilizer in Figure 1.25.[22] Note that the particle size registered a dramatic decrease with the addition of compatibilizer. This clearly establishes the occurrence of interfacial chemical reactions between the amine group of polyamide and maleic anhydride group of the compatibilizer, which decrease the segmental mobility and increase the internal friction, nearer to the interface region, and enhanced the storage modulus of the blends.[22]

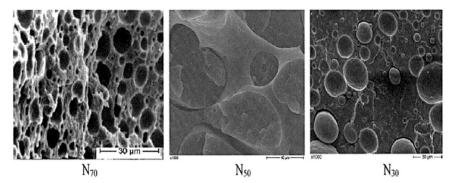

FIGURE 1.24 SEM micrographs showing the morphology of uncompatibilized PA12/PP blends. (Reprinted with permission from Ref. 20. © 2015 Elsevier Ltd.)

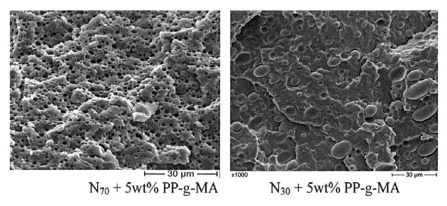

FIGURE 1.25 SEM micrographs showing the morphology of compatibilized N_{70} and N_{30} blends. (Reprinted with permission from Ref. 20. © 2015 Elsevier Ltd.)

1.11 SUMMARY

In this chapter, we have covered almost all the points regarding the fundamentals of scanning electron microscopy and the information that can be gained from this technique. So this chapter provides a guide to SEM and the analytical techniques based on them. Scanning electron microscopy can offer a variety of imaging techniques with resolutions in the range 1 μm–1 nm, depending on the microscope and the signal used to form the image. Since this chapter comprises some examples for characterization of polymer blends and nanocomposites by SEM and the way of analyzing and concluding some valuable points by observing the images, it will surely lend a hand to the researchers to solve their problem in their research field related to morphology.

KEYWORDS

- **scanning electron microscope**
- **secondary electrons**
- **back scattered electrons**
- **energy dispersive X-ray spectroscopy**
- **polymer nanocomposites**
- **polymer blend nanocomposites**

REFERENCES

1. http:/www.eag.com/mc/.
2. *Geochemical Instrumentation and Analysis, Browse Collection.* serc.carleton.edu.
3. www.understanding-cement.com/sem-introduction.html.
4. www.unamur.be/universite/services/microscopie/sme.../sme-documents-semlight.
5. https://optiki.files.wordpress.com/2013/03/electron-microscopythe-basics.pdf.
6. https://imf.ucmerced.edu/downloads/semmanual.pdf.
7. cfamm.ucr.edu/documents/sem-intro.pdf.
8. www.sjsu.edu/people/anastasia.micheals/courses/MatE143/s1/SEM_GUIDE.pdf.
9. coen.boisestate.edu/faculty-staff/files/2012/01/SEM.pdf.
10. www.zmb.uzh.ch/static/bio407/assets/Script_AK_2014.pdf.
11. www.charfac.umn.edu/sem primer.pdf.
12. www.microscopemaster.com/scanning-electron-microscope.html.
13. Poornima Vijayan, P.; Puglia, D.; Kenny, J. M.; Thomas, S. *Soft Matter* **2013,** *9*, 2899–2911.
14. Maria, H. J.; Lyczko, N.; Nzihou, A.; Joseph, K.; Mathew, C.; Thomas, S. Stress Relaxation Behavior of Organically Modified Montmorillonite Filled Natural Rubber/Nitrile Rubber Nanocomposites. *Appl. Clay Sci.* **2014,** *87*, 120–128.
15. George, S. M.; Puglia, D.; Kenny, J. M.; Causin, V.; Parameswaranpillai, J.; Thomas, S. Morphological and Mechanical Characterization of Nanostructured Thermosets from Epoxy and Styrene-block-butadiene-block-styrene Triblock Copolymer. *Ind. Eng. Chem. Res.* **2013,** *52*, 9121–9129.
16. Chirayil, C. J.; Joy, J.; Mathew, L.; Koetz, J.; Thomas, S. Nanofibril Reinforced Unsaturated Polyester Nanocomposites: Morphology, Mechanical and Barrier Properties, Viscoelastic Behavior and Polymer Chain Confinement. *Ind. Crops Prod.* **2014,** *56*, 246–254.

17. Chandran, N.; Chandran, S.; Maria, H. J.; Thomas, S. Preparation and Characterization of Potato Starch Nanocrystal Reinforced Natural Rubber Nanocomposites *RSC Adv.* **2015,** *5,* 86265–86273.
18. Rajisha, K. R.; Maria, H. J.; Pothan, L. A.; Ahmad, Z.; Thomas, S. *Int. J. Biol. Macromol.* **2014,** *67,* 147–153.
19. Abraham, E.; Thomas, M. S.; John, C.; Pothen, L. A.; Shoseyov, O.; Thomas, S. Green Nanocomposites of Natural Rubber/Nanocellulose: Membrane transport, Rheological and Thermal Degradation Characterisations. *Ind. Crops Prod.* **2013,** *51,* 415–421.
20. Jose, S.; Thomas, S.; Parameswaranpillai, J.; Santhosh Aprem, A.; Karger-Kocsis, J. Dynamic Mechanical Properties of Immiscible Polymer Systems with and Without Compatibilizer. *Polym. Test.* **2015,** *44,* 168–176.
21. *Electron Microscopy and Analysis*, 3rd ed.; In Goodhew, J., Humphreys, J., Beanland, R. Eds.; Taylor and Francis, 2001; p 254.
22. *Scanning Electron Microscopy and X-ray Microanalysis*, third ed. In Goldstein, J., Newbury, D. E., Joy, D. C., Lyman, C. E., Echlin, P., Lifshin, E., Sawyer, L., Michael, J. R. Eds.; Springer, 2012; p 689.
23. SEM: Scanning Electron Microscope A To Z: Basic Knowledge for Using The SEM. https://www.jeol.co.jp/en/applications/pdf/sm/sem_atoz_all.pdf

CHAPTER 2

ATOMIC FORCE MICROSCOPY: PRINCIPLES AND APPLICATIONS IN POLYMER COMPOSITES

ABITHA V. K.[1*], AJAY VASUDEO RANE[2], and SABU THOMAS[1,3]

[1]School of Chemical Sciences, Mahatma Gandhi University, Kottayam 686560, Kerala, India

[2]Composite Research Group, Department of Mechanical Engineering, Durban University of Technology, Durban, South Africa

[3]International and Inter University Centre for Nanoscience and Nanotechnology, Mahatma Gandhi University, Kottayam 686560, Kerala, India

*Corresponding author. E-mail: abithavk@gmail.com

ABSTRACT

Atomic force microscopy (AFM) belongs to the family of scanning probe microscopy. It is not a technique based on electron microscopy. It has been normal practice in the last couple of years to supplement electron microscopy with AFM as it provides clear surface characterization of polymers, blends, nanoparticles, and nanocomposites. In this chapter, we are mainly focusing on different modes of operations of AFM and their applications in the area of polymers and their blends and nanocomposites.

2.1 INTRODUCTION

Advances in modern science cannot develop without reliable instruments for characterization of microstructure, physical, and chemical properties of micro and nanocomposites. Atomic force microscopy (AFM) belongs to the class of scanning probe microscopy (SPM) in which the surface of the samples was analyzed using sharp probes. SPM technique differs in the type of probe used and also the nature of the interaction with the probe and the surface of the sample. Depending upon the interactions, different types of SPMs are available in the market; they are as follows: scanning tunneling microscopy (STM), AFM, and scanning near-field optical microscope. Structural information of materials can be obtained by scanning and transition electron microscopes. For imaging the surfaces with very high magnification in 1981, Gerd Binning and Heinrich Rohrer developed the STM (Fig. 2.1). For this invention, they got the Nobel Prize for Physics in 1986.[3] In 1985, Binning invented the AFM. STM measures the tunneling current while AFM measures the interaction forces between the surface and the tip. The invention and development of commercial microscopes increased the speed of research. AFM and STM are the most successful instruments in the field of SPM for the past years. The likelihood to apply AFM to conductive and nonconductive materials has pulled in researchers and scientists from different areas of the world. Commercial AFM can be routinely applied with efficiency comparable to electron microscopes.[12]

Gerd Binning　　　　　　　　　　　　Heinrich Rohrer

FIGURE 2.1 Inventors of AFM and STM.

The main advantage of AFM and STM techniques are that the samples can be measured without any special surface preparation. The instrument portrays the morphology of the sample with quantitative information such as roughness or height distributions.[2]

2.2 PRINCIPLES OF AFM

The working principle of an AFM is based on the deflection of a very sensitive cantilever due to repulsive forces between atoms on the sample surface and atoms at the cantilever tip. The surface forces include van der Waals forces, dipole–dipole forces, electrostatic forces, magnetic forces, capillary forces, etc. The force exerting is of pico-Newton range. So when a sharp tip is brought closer to the surface, it interacts with the surface forces. Attractive interaction occurs between 1 nm and 0.3 nm; however, below 0.3 nm, there occurs a repulsive interaction. So the critical distance is the distance between attractive interaction and repulsive interaction.[1] AFM probe consists of a force sensing cantilever with a very sharp tip integrated at the lower side of its free end. When the tip is brought into the proximity of the sample surface, attractive forces between the tip and the sample lead to a decrease in the height between the tip and the surface. This leads to a downward deflection of the cantilever governed by Hooke's law $F = -kz$, where F is the force, k is the stiffness of the cantilever, and z is the deflection experienced by the cantilever. A cantilever with a low spring constant is needed to sense small forces (<0.1 nN). Deflection is smaller than 0.1 nm. This deflection is a measure of the force experienced by the tip and hence a measure of the subsequent variation in tip height. The deflection is measured by using a laser beam when the sample is scanned.

2.3 INSTRUMENTATION

In this section, we are going to discuss about the different components of an AFM instrument. AFM instrument of APE Research Nanotechnology [A-100 AFM] is shown in Figure 2.2.

A typical block diagram of AFM is given in Figure 2.3. It consists of motorized X–Y stage, sample, cantilever, piezoelectric scanner tube (PZT), feedback system, and detector devices.[17]

- **Motorized X–Y stage**: In most of the AFM measurements, sample is moved in the X–Y directions under a fixed position of cantilever. This is done with the help of a motorized X–Y stage driven by a piezoelectric component.
- **Sample**: The sample generally has the dimension of 10 mm × 10 mm. The surface of the sample should be free from dirt and other impurities. The main aspect to consider is that the surface should be flat with minimum thickness.

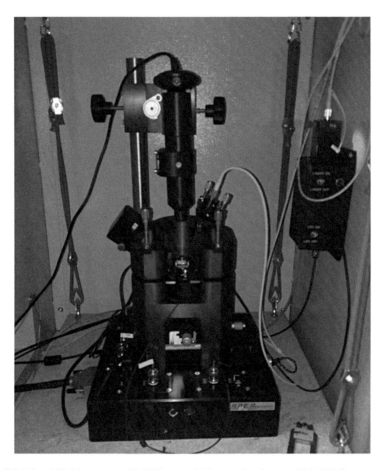

FIGURE 2.2 AFM instrument [APE Research Nanotechnology A-100 AFM].

- **Cantilever**: Cantilevers are prepared from silicon or silicon nitride made by photolithographic and etching techniques.[10] A cantilever with a low spring constant is needed for high resonant frequency (about 10–100 kHz) to sense small forces (0.1 nm or lower). Rectangular single crystal silicon cantilevers with pyramidal tips are most commonly used. Tips with radius of curvature of 5–50 nm are commonly available.
- **PZT scanner**: Piezoelectric scanner is the core of STM. A piezo-element consists of three mutually perpendicular piezoelectric transducers. Upon applying voltage, piezo-electric transducer will contract or expand. Nowadays, AFM uses the tube scanner. A tube made of PZT, metalized on the outer and inner surfaces, is poled in the radial

direction. The remaining quadrants are connected to a certain direct current.[8]
- **Feedback and detector devices**: The force on the tip due to its interaction with the sample is sensed by detecting the deflection of the lever. Deflection is smaller than 0.1 nm. Laser beam deflection is used for measuring lever deflection. Split-diode photo detector is used for detection.

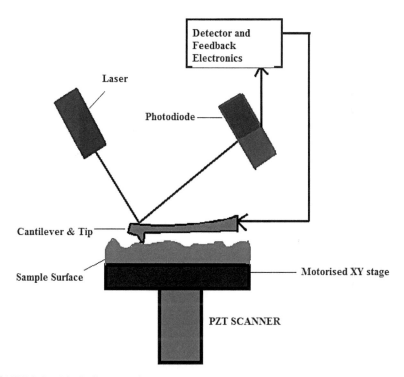

FIGURE 2.3 Block diagram of AFM.

2.4 INTERACTION FORCES IN AFM

When the tip-to-sample spacing is large, the force between the tip and the sample is attractive. A noncontact atomic force microscope is operated within this range of forces, where the total net force on the cantilever is negative. As the tip approaches the sample, the force between atoms on the tip and atoms on the sample eventually becomes repulsive. The phenomenon is shown in Figure 2.2. When the cantilever tip is "in contact" with the sample (in the absence of liquid layers on the sample surface), the repulsive force dominates, exerting a positive net force on the cantilever. An atomic force microscope in contact mode is operated within this range of forces, where the total net force on the cantilever is positive (Fig. 2.4).[15,19]

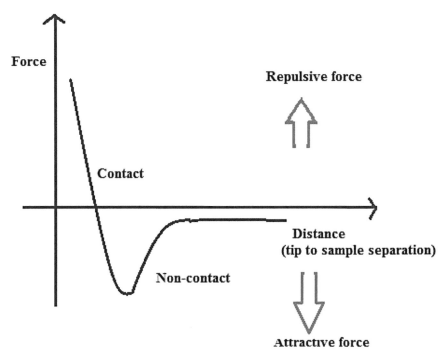

FIGURE 2.4 Interatomic force versus distance curve.

2.5 DIFFERENT MODES OF MEASUREMENT IN AFM

There are two different modes of measurement in AFM, they are

1. Constant height mode
2. Constant force mode

2.5.1 CONSTANT HEIGHT MODE

Scan is initiated with a preselected deflection signal, that is, a preselected interaction force known as the set-point value. The topographical features will cause corresponding local changes in the deflection signal. This is measured and recorded against the tip location (deflection image). Samples with low roughness can be scanned in this way. This mode will damage the cantilever with samples of high roughness.

2.5.2 CONSTANT FORCE MODE

A preselected deflection signal (set-point value) is used as the reference value for the feedback loop. Topographical structures lead to variation in deflection signal which is compensated by adjusting the height of the tip, thereby keeping the force constant. Height is adjusted by the output voltage of a feedback loop. This output voltage is therefore a linear response of the height variation which is recorded (height image). The constant force mode gives better results compared to constant height mode.

2.6 DIFFERENT IMAGING MODES OF AFM

AFM can be operated in a number of imaging modes. The basic imaging modes are as follows.

2.6.1 STATIC MODE (CONTACT MODE)

As the tip approaches the sample to within a few nanometers, the tip experiences an attractive force which causes the cantilever to bend downward. If the attractive force is high enough (exceeds the withdrawing force of the cantilever), the cantilever is pulled toward the sample surface and contact occurs. The tip experiences different types of interactions during its way to the surface which include the attractive force, repulsive force, capillary force (due to surface layer of condensed water), etc. The force on the tip is repulsive with a mean value of 10^{-9} N. Each type of interaction gives the respective characteristic property of the sample such as topography, local stiffness, adhesive property, etc. For topographic studies in contact mode, AFM usually works with a set-point force that corresponds to a small repulsive interaction. In contact mode, the deflection of the cantilever is sensed and compared in a DC feedback amplifier to some desired value of deflection. An example of a contact mode AFM mounted on the cantilever is shown in Figure 2.5. If the measured deflection is different from the desired value, the feedback amplifier applies a voltage to the piezo to raise or lower the sample relative to the cantilever to restore the desired value of deflection. Problems with the contact mode are excessive tracking forces applied by the probe to the sample. Under ambient conditions, sample surfaces are covered by a layer of adsorbed gases consisting of water and nitrogen. When probe touches this contaminant layer, a meniscus forms and the cantilever is pulled by surface tension toward the sample surface. This meniscus forces and other attractive forces may be neutralized by operating with the probe and sample partly immersed in a liquid.

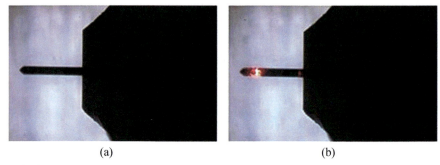

FIGURE 2.5 (a) Contact mode tip used for AFM and (b) laser focused on the contact mode tip.

2.6.2 DYNAMIC MODE (NONCONTACT MODE AND TAPPING MODE)

Dynamic mode is usually carried out at certain height above the sample. Hence, it is also called the noncontact mode. The cantilever is externally oscillated at its resonance frequency. As it is brought closer to the surface the oscillation parameters (amplitude, frequency, and the phase) are modified by interaction forces. These changes in oscillation with respect to the external reference oscillation serve as measurement parameters. This provides information about the sample's characteristics. The noncontact mode with tip is as shown in Figure 2.6. In this mode, the tip remains at about 50–150 Å above the sample surface. Attractive van der Waals force acting between the tip and the sample are detected and topographic images are constructed by scanning the tip above the surface. The attractive forces from the sample are substantially weaker than the forces used in contact mode. For highest resolution, it is necessary to measure force gradients from van der Waals forces which may extend only a nanometer from the sample surface. The attractive forces from the sample are weaker than the forces used by contact mode. The fluid contaminant layer in this case is thicker than the range of van der Waals force gradient and attempts to image the true surface with the noncontact AFM fails as the tip become trapped inside the contaminant layer.[17]

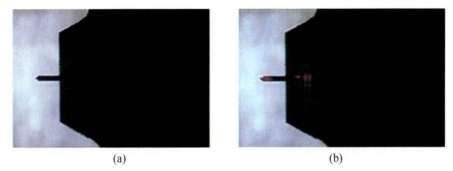

FIGURE 2.6 (a) Noncontact mode tip used for AFM and (b) laser focused on the noncontact mode tip.

Tapping Mode AFM: It is a modified dynamic mode which is also called intermittent contact mode. Most sample surfaces develop a liquid meniscus layer at ambient temperature. The tip at close ranges may stick to the liquid layer. To bypass this, intermittent contact mode (tapping mode) is developed. The cantilever is driven to oscillate up and down with higher amplitude of 100–200 nm so that intermittent contact with the sample occurs. During each vibration cycle, the tip taps the sample only for a very short time. Thus, surface damage is avoided. So this gentle AFM mode is suitable for soft materials and polymers. Tapping mode overcomes the problems associated with friction, adhesion, electrostatic forces, and other difficulties for contact mode, by alternatively placing the tip in contact with the surface to provide high resolution and then lifting the tip off the surface to avoid dragging the tip across the surface. When the tip contacts the surface, the high frequency makes the surfaces stiff. Tapping mode prevents the tip from sticking to the surface and causes damage during scanning. Compared to contact and noncontact modes, when the tip contacts the surface it has sufficient oscillation amplitude to overcome the tip sample adhesion forces. The surface material is not pulled by sideways by shear forces because the applied force is always vertical. Another advantage of this mode of technique is its large and linear operating range.[16]

2.7 APPLICATIONS OF AFM IN POLYMER COMPOSITES

AFM can be used as an effective tool for the surface analysis and filler distribution in polymer composites. There are mainly two types of images, height image and phase image. Typically, height and interaction images are collected simultaneously. In general, the surface topography is better presented by height image, whereas the nanostructures and fine morphological features are better distinguished in amplitude or phase images.

Jeon, Kim, and Kim[11] has studied on the natural rubber/styrene butadiene and rubber/butadiene rubber blends; they had found that the height image captures well the surface roughness of the sample, whereas the amplitude and phase images shows nearly flat and uniform structures.

Satyanarayana, Bhowmick, and Dinesh Kumar[18] investigated on the natural rubber (NR)/carboxylated nitrile rubber (XNBR) clay nanocomposites by solution mixing. AFM phase images for 50/50 NR/XNBR shows that the dark continuous phase is the XNBR phase and the bright dispersed phase is due to the NR phase. NR shows spherical coalesced droplets of 4–5 μm in size as shown in Figure 2.7. The morphological studies of neat NR/XNBR show the incompatibility between NR and XNBR.

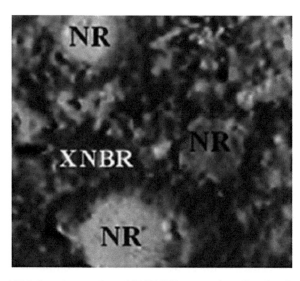

FIGURE 2.7 AFM phase image of neat NR/XNBR composites. (Reprinted with permission from Ref. 18. © 2016 Elsevier Ltd.)

Satyanarayana et al.[18] also studied on the Cloisite 15A (8 phr)-filled NR/XNBR composites with and without cross-linking. From the AFM images shown in Figure 2.8a and b, the major portions of Cloisite 15A particles are preferentially seen in the dispersed brighter NR phase (circles in Fig. 2.8a) which mutilates the morphology of the dispersed NR phase. This image further reveals that the average thickness of the nanoclay particle is in the range of 50–60 nm.

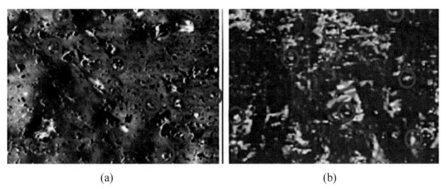

(a)   (b)

FIGURE 2.8 AFM phase images of (a) 8 phr Cloisite 15A-filled uncross linked NR/XNBR blend and (b) 8 phr Cloisite 15A-filled cross-linked NR/XNBR blend. (Reprinted with permission from Ref. 18. © 2016 Elsevier Ltd.)

Figure 2.8b shows the AFM phase image of cross-linked 50XNBR–50NR blend with 8 phr of Cloisite 15A. It is interesting to note that the NR phase is completely mutilated and the nanoclay particles are preferentially enriched in the NR phase (as seen in Fig. 2.8b). The thickness of the Cloisite 15A nanoclay particles present in the NR phase of cross-linked 50XNBR–50NR blend is around 50–60 nm which is similar to the thickness of the Cloisite 15A nanoclay particles present in the NR phase of uncross linked 50XNBR–50NR blend. This clearly confirms that cross-linking has no effect on the morphology of the 50XNBR–50NR blend nanocomposites.

Eda and Chhowalla[6] studied on the graphene-filled polystyrene nanocomposites and their electrical properties. They have used phenyl isocyanate-treated graphene oxide and also they functionalized graphene oxide sheets (FGS). The size of FGS was found to remain unchanged during the reduction and deposition processes as shown in Figure 2.9.

The size of FGS is dependent on the size of initial graphite crystals but can be adjusted by ultrasonication of the suspension to break up the individual sheets. In this study, composite thin films consisting of FGS with three different average sizes were prepared from two graphite sources. Because of viscous flow of polymer solution during spin-coating, most FGS were found to be lying nearly parallel to the substrate surface.

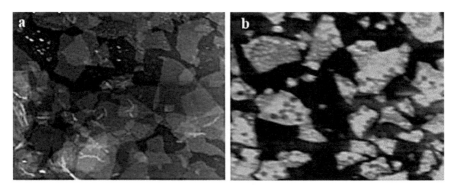

FIGURE 2.9 (a) Phenyl isocyanate-treated GO and (b) FGS composite thin films. (Reprinted with permission from Ref. 6. © 2009 American Chemical Society.)

AFM analysis of epoxy-multiwalled carbon nanotubes (MWCNT) nanocomposites consisting of different filler loading was observed by Guadagno et al.[9] Epoxy 0.5-phr MWCNT at different magnifications reveals the presence of well-dispersed MWCNT nanotubes (Fig. 2.10).

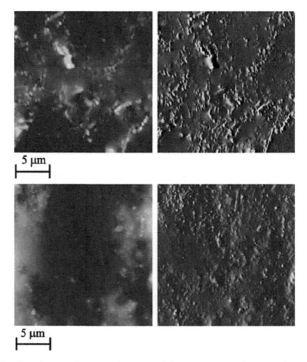

FIGURE 2.10 Tapping mode AFM images of the epoxy-MWCNT (0.5) composite. Left, height images; right, amplitude images. (Reprinted with permission from Ref. 9. © 2009 Elsevier.)

From Figure 2.11, we can observe the magnified image of two epoxy with 0.5-phr MWCNT which form both individual well-dispersed nanotubes and also the bamboo-like structure of the nanotubes with larger diameter inside the epoxy matrix.

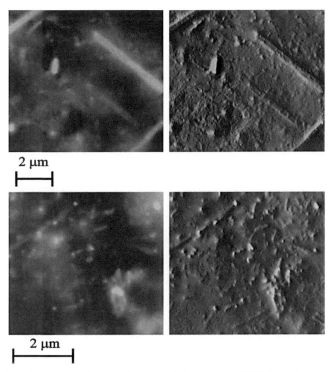

FIGURE 2.11 Tapping mode AFM images of the epoxy-MWCNT (0.5) composite. Left, height images; right, amplitude images. (Reprinted with permission from Ref. 9. © 2009 Elsevier.)

AFM image of epoxy MWCNT (3 phr) at different magnifications are shown in Figure 2.12. The images show large smooth zones corresponding to the MWCNT, and small entities with a grain aspect identifiable as ordered domains of epoxy resin, forming a different phase. It is worth noting that in samples with a larger percentage of nanotubes (3%), they are localized in bundles. Moreover, the inside nanotubes of a bundle maintain their orientation. Therefore, AFM proves that the carbon nanotubes are well embedded into the epoxy matrix and singularly dispersed or in bundles, depending on their concentration. As a matter of fact, the composites do not show significant nanotube aggregates at low percentage of filler (0.5%), whereas at higher percentage of filler (3.0%), the nanotubes appear organized in bundles. However, also in this last case, no large bundles of nanotubes are present, indicating that the interaction between MWCNTs and the epoxy chains helps to reduce the possibility of large agglomerations.

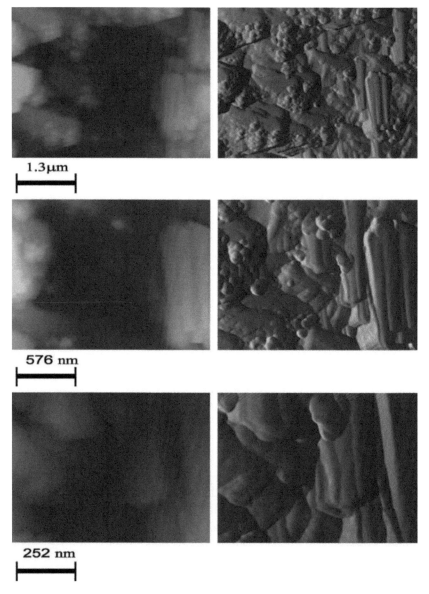

FIGURE 2.12 Tapping mode AFM images of the epoxy-MWCNT (3.0) composite. Left, height images; right, amplitude images. (Reprinted with permission from Ref. 9. © 2009 Elsevier.)

In another study, the dispersion of cellulose nanofibers (CNFs) in the poly(lactic acid) (PLA) matrix was investigated by ScanAsyst AFM technique.[7] The images (256 × 256) were recorded using scanning rates of 1.4 Hz and a scan angle of 90°. The nominal spring constant and nominal resonant frequency was 0.4 N/m and 70 kHz, respectively. Composite films with a uniform surface and a thickness of approximately 30 μm were used for AFM investigations. AFM height and amplitude images showed (Fig. 2.13) bundles of CNFs with ribbon-like shape structure. From these pictures, it is obvious that a considerable reduction in fiber size has been accomplished by the acid treatment considering an average size of 20 μm for the raw microcrystalline cellulose. The diameter of individual fibers ranges from 11 to 44 nm with an average of 20 ± 6 nm. To ensure the repeatability and the accuracy of the AFM scanning, different zones of the same sample were analyzed and the presented images are representative of the totality of the sample.

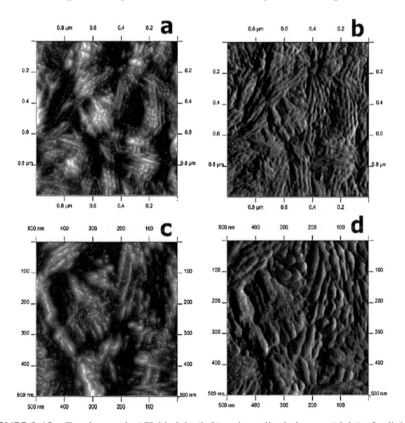

FIGURE 2.13 Tapping mode AFM height (left) and amplitude images (right) of cellulose nanofiber type: (a) and (b) 1 μm × 1 μm (c) and (d) 500 nm × 500 nm. (Reprinted with permission from Ref. 7. © 2013 Elsevier.)

Figure 2.14a and b shows typical topographic images of PLA composites containing silane-treated and -untreated CNFs. The comparative observation of these figures shows a better dispersion in the case of PLA composite containing CNFs. Several areas with CNF agglomerations marked with red circles (light gray in printed version) and individual nanofibers covered by the polymer (marked with black arrows) can be observed.

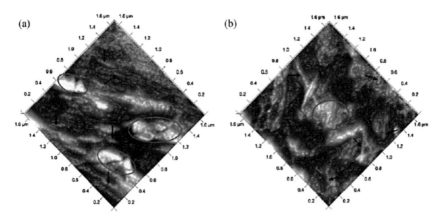

FIGURE 2.14 AFM images of untreated (a) and treated (b) CNF-reinforced PLA composites at a scan size of 1.7 μm × 1.7 μm. (Reprinted with permission from Ref. 7. © 2013 Elsevier.)

A distinct improvement of the dispersion quality is observed when the nanofibers are silane treated (Fig. 2.14b). CNFS are more uniformly distributed on the surface of PLA/CNFS composite and less agglomerated fibers can be seen in Figure 2.14b as compared to Figure 2.14a. Most of CNF agglomerates are located at the surface of the material as opposite to CNFS which are located deeper in the polymer, suggesting a better cellulose fiber–matrix interface in the second case.

Cellulose whisker and their nanocomposites have been evaluated by Kvien, Tanem, and Oksman.[13] AFM analysis of the cellulose whiskers showed to be a good alternative to electron microscopy. The whiskers appeared significantly broader having a rounded shape (Fig. 2.15a and b). This broadening effect can be explained by the tip used for imaging. In general, the AFM tips have a finite size and shape. As the tip passes over a sample with surface features of comparable size as that of the tip, the shape of the tip will contribute to the image that is formed.

AFM analysis of the bulk film prepared at cryogenic temperatures enabled detailed information of the whiskers in the matrix (Fig. 2.16). The whiskers partly protruded from the PLA matrix, as shown in the topography image, most probably due to differences in thermal expansion during heating from cryogenic temperatures. These images were recorded directly on the cryomicrotomed surface without any further treatment of the sample. However, the images would probably benefit from the use of high-resolution probes. AFM topography and phase-contrast imaging have the necessary resolution capabilities without the need for staining. AFM could therefore be a powerful alternative to conventional BF TEM in such composite materials where contrast between the whiskers and the matrix is limited and the beam sensitivity is a major challenge.

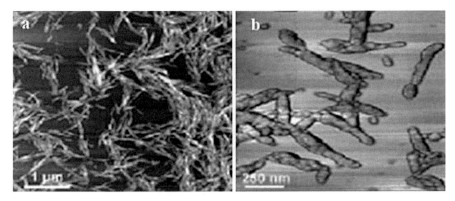

FIGURE 2.15 (a) AFM topography image of cellulose whiskers after drying and (b) AFM phase image of cellulose whiskers showing broadening effect. (Reprinted with permission from Ref. 13. © 2005 American Chemical Society.)

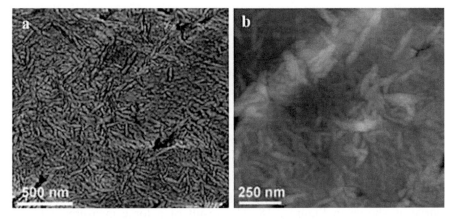

FIGURE 2.16 (a) AFM phase contrast image and topography image (b) of a cryomicrotomed surface of the solution-cast cellulose whiskers–PLA nanocomposite. (Reprinted with permission from Ref. 13. © 2005 American Chemical Society.)

AFM can be also used to evaluate the elasticity of different materials. The force displacement curves are mainly used for this. Young's moduli of soft materials can be evaluated by using the linear elastic contacts theory. Young's moduli agreed well with macroscopic compression tests when indentation strains did not exceed the linear elastic limit. These results are consistent with the generally accepted view that small-strain deformation of many rubber-like materials is virtually a linear elastic process and can be modeled accordingly. Lin, Dimitriadis, and Horkay[14] find out the local elasticity of rubber-like materials using different models such as Hertz model, Mooney–Rilvin model. The extracted values of Young's modulus show good agreement with those obtained by both macroscopic compression testing and by fitting truncated portions of the force curves with the Hertz equation. Lin et al. also explained the Young's modulus maps (Fig. 2.17) which is obtained directly during the AFM imaging. Over regions of comparable size, local Young's moduli of the PVA gels varied over a much narrower range approximately 16–24 kPa for the 6% gel and 90–120 kPa for the 12% gel than in the cartilage (<1–120 kPa). The variability in stiffness in the cartilage corresponds to a high degree of local inhomogeneity, which is a characteristic feature of many biological tissues.

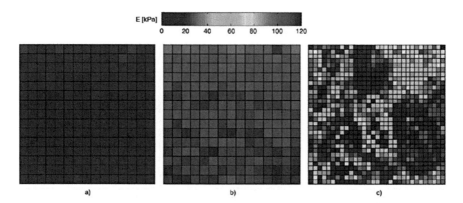

FIGURE 2.17 Young's modulus (E) maps from the "force-volume" indentation of (a) a 6% PVA-gel-probed using, (b) a 12% PVA-gel-probed, (c) a 1-day-old mouse articular cartilage sample. (Reprinted with permission from Ref. 14. © 2007 eXPRESS Polymer Letters.)

Butt, Cappella, and Kappl[4] had given a schematic of force deflection curve as shown in Figure 2.18. The result of a force measurement is a measure of the cantilever deflection, Z_c, versus position of the piezo, Z_p, normal to the surface. To obtain a force-versus-distance curve, Z_c and Z_p

have to be converted into force and distance. The force F is obtained by multiplying the deflection of the cantilever with its spring constant k_c: $F = k_c Z_c$. The tip–sample separation D is calculated by adding the deflection to the position: $D = Z_p + Z_c$. We call this tip–sample separation "distance."

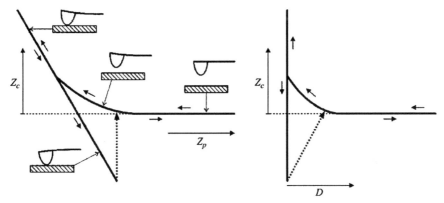

FIGURE 2.18 Schematic of a typical cantilever deflection-versus-piezo height (Z_c-vs.-Z_p) curve (left) and corresponding Z_c-vs.-D plot, with $D = Z_c + Z_p$. (Reprinted with permission from Ref. 4. © 2005 Elsevier.)

Dimitriadis, Horkay, Maresca, Kachar, and Chadwick[5] also found out the modulus measurement using PVA. For AFM analysis of each polymer concentration, a relatively thick sample was made (1 mm) and data were collected to ascertain the validity of the Hertz model and to explore the errors involved when using sharp tips.

2.8 CONCLUSIONS

AFM is an efficient tool for the polymer composites and nanocomposite applications.

The surface characteristics of the polymers can be identified using height and phase images gives the nanostructure and fine morphological images. Tapping mode imaging gives the good identification of polymer surfaces and filler phase dispersion. AFM can be used in different areas such as material science, life sciences, electrical, and electronics for the identification of surfaces. In general, AFM can be used in the surface identification and modulus analysis of polymer composites. In future, the AFM connected with spectroscopic analysis will be a major breakthrough in the field of polymer composites.

KEYWORDS

- atomic force microscopy
- scanning probe microscopy
- electron microscopy
- polymers
- nanocomposites

REFERENCES

1. Binnig, G. Atomic Force Microscope and Method for Imaging Surfaces with Atomic Resolution. *US Patent 4,724,318* (3), 1988; pp 1–8.
2. Binnig, G.; Quate, C. F. Atomic Force Microscope. *Phys. Rev. Lett.* **1986,** *56* (9), 930–933.
3. Binnig, G.; Rohrer, H.; Gerber, C.; Weibel, E. 7 × 7 Reconstruction on Si(1 1 1) Resolved in Real Space. *Phys. Rev. Lett.* **1983,** *50* (2), 120–123.
4. Butt, H. J.; Cappella, B.; Kappl, M. Force Measurements with the Atomic Force Microscope: Technique, Interpretation and Applications. *Surf. Sci. Rep.* **2005,** *59* (1–6), 1–152.
5. Dimitriadis, E. K.; Horkay, F.; Maresca, J.; Kachar, B.; Chadwick, R. S. Determination of Elastic Moduli of Thin Layers of Soft Material Using the Atomic Force Microscope. *Biophys. J.* **2002,** *82* (5), 2798–2810.
6. Eda, G.; Chhowalla, M. Graphene-Based Composite Thin Films for Electronics. *Nano Lett.* **2009,** *9* (2), 814–818.
7. Frone, A. N.; Berlioz, S.; Chailan, J. F.; Panaitescu, D. M. Morphology and Thermal Properties of PLA–Cellulose Nanofibers Composites. *Carbohydr. Polym.* **2013,** *91* (1), 377–384.
8. Binning, G.; Smith, D. P. E. Single Tube Three Dimensional Scanner for Scanning Tunneling Microscopy. *Rev. Sci. Instrum.* **1986,** *57* (8), 1688–1689.
9. Guadagno, L.; Vertuccio, L.; Sorrentino, A.; Raimondo, M.; Naddeo, C.; Vittoria, V.; et al. Mechanical and Barrier Properties of Epoxy Resin Filled with Multi-Walled Carbon Nanotubes. *Carbon* **2009,** *47* (10), 2419–2430.
10. Ibe, J. On the Electrochemical Etching of Tips for Scanning Tunneling Microscopy. *J. Vaccum Sci. Technol.* **1990,** *8* (4), 3570–3575.
11. Jeon, I. H.; Kim, H.; Kim, S. G. Characterization of Rubber Micro-Morphology by Atomic Force Microscopy (AFM). *Rubber Chem. Technol.* **2003,** *76* (1), 1–11.
12. Kushmerick, J. G.; Weiss, P. S. Scanning Probe Microscopes. In *Encyclopedia of Spectroscopy and Spectrometry*, 2010; pp 2464–2472.
13. Kvien, I.; Tanem, B. S.; Oksman, K. Characterization of Cellulose Whiskers and their Nanocomposites by Atomic Force and Electron Microscopy. *Biomacromolecules* **2005,** *6* (6), 3160–3165.

14. Lin, D. C.; Dimitriadis, E. K.; Horkay, F. Elasticity of Rubber-Like Materials Measured by AFM Nanoindentation. *Express Polym. Lett.* **2007,** *1* (9), 576–584.
15. Lin, D. C.; Dimitriadis, E. K.; Horkay, F. Robust Strategies for Automated AFM Force Curve Analysis—I. Nonadhesive Indentation of Soft, Inhomogeneous Materials. *J. Biomech. Eng.* **2007,** *129*, 430–440.
16. Morita, S.; Giessibl, F. J.; Sugawara, Y.; Hosoi, H.; Mukasa, K.; Sasahara, A.; Onishia, H. Noncontact Atomic Force Microscopy and Its Related Topics. In *Springer Handbook of* Nanotechnology; Bhushan, B., Eds.; Springer: Berlin, Heidelberg, 2010.
17. Zhang, S.; Li, L.; Kumar, A. Scanning Tunneling Microscopy and Atomic Force Microscopy. In *Materials Characterisation Techniques*; CRC Press, 2008; pp 95–123.
18. Satyanarayana, M. S.; Bhowmick, A. K.; Dinesh Kumar, K. Preferentially Fixing Nanoclays in the Phases of Incompatible Carboxylated Nitrile Rubber (XNBR)–Natural Rubber (NR) Blend Using Thermodynamic Approach and Its Effect on Physicomechanical Properties. *Polymer (United Kingdom)* **2016,** *99*, 21–43.
19. Ducker, W. A.; Cook, R. F.; Clarke, D. R. Force Measurement Using an Atomic Force Microscopy. *J. Appl. Phys.* **1990,** *67* (9), 4045–4052.

CHAPTER 3

BARIUM STRONTIUM TITANATE: PREPARATION, DIELECTRIC, FERROELECTRIC, AND MICROSCOPIC STUDIES

ANURADHA KUMARI, SAUMYA SHALU, and
BARNALI DASGUPTA GHOSH[*]

Department of Chemistry, Birla Institute of Technology, Mesra, Ranchi 835215, Jharkhand, India

[*]Corresponding author. E-mail: barnali.iitkgp@gmail.com

ABSTRACT

Since the discovery of ferroelectricity in perovskite ceramic materials, there has been continuous development of materials and techniques to convert their advanced properties into novel technological applications. Barium titanate and strontium-doped barium titanate are excellent materials for microwave applications because of their high dielectric constant and very low dissipation factor. In this work, we review the dielectric and ferroelectric properties of $BaTiO_3$ and barium strontium titanate highlighting the effect of Sr^{2+} doping. Various synthesis routes have been described. A brief discussion of the fundamentals of ferroelectricity and dielectric behavior of solids has been presented. Various microscopic techniques employed to provide a better insight into the properties of ferroelectric ceramic materials have been outlined. The potential of ferroelectric ceramic materials in integrated electronics as passive components and nonvolatile memory applications has been briefly discussed.

3.1 INTRODUCTION

In recent years, the interest in complex oxides and their electrical properties has revived. The most interesting properties associated with these oxides are their ferroelectric and high dielectric property. These properties were first identified decades ago but the recent interest is due to the advanced techniques that can now be employed to produce high-quality materials and novel device applications. Barium titanate ($BaTiO_3$) is the most widely studied perovskite ferroelectric material.[1,2] The crystal structure of barium titanate can hold various dopant ions, depending on their ionic size. Doped $BaTiO_3$ has wide electrical usage as in sensors with positive temperature coefficients of resistivity (PTCR), capacitors, ferroelectric thin-film memories, and transducers (Fig. 3.1). In the perovskite structure (ABO_3) of ceramics, A-site cation is larger from alkaline metal groups, whereas B-site cation is smaller from transition metal groups. Due to the intrinsic capacity of the perovskite structure to host dissimilar ions, many different metal dopants can be easily fit in the structure of barium titanate.[3] Depending on application, various dopants (strontium, iron, nickel, calcium, manganese, etc.) can be used to change the chemical and physical characteristics of barium titanate. Doping on A-site increases the dielectric constant but decreases the piezoelectric coefficient and dissipation factor, it also affects the heat generation; because of this reason, they are mainly used in high-frequency applications and ultrasonic applications.

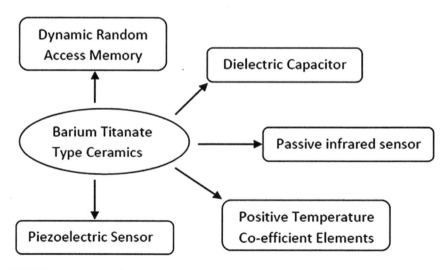

FIGURE 3.1 Recent advances in ceramic applications.

B-site doping on barium titanate increases the dielectric constant value, decreases the dielectric loss (dissipation factor), and also increases the piezoelectric coefficient; therefore, this type of ceramic can be used as actuators in vibration and noise control, optical applications, benders, etc.[4] Traditionally, ceramic method (also called the mixed oxide route) is used for the preparation of barium titanate, in which barium and titanium carbonates, the metal precursors, are heated at temperatures as high as 1200°C.[5,6] This route is not very fruitful for the synthesis of good-quality ceramics because the product obtained by this method has high impurity contents, is nonhomogeneous, and has very large particle sizes.[7-9] There are various methods employed for the preparation of doped barium titanate powder, for example, sol–gel, hydrothermal, spray pyrolysis, and coprecipitation, etc.[10-13]

Strontium-doped barium titanate has been extensively studied as an advanced ceramic for microwave applications because of its excellent electrical properties and good electric field tunabilities depending on the concentration of doping elements in barium titanate.[14] Doping with three-dimensional (3D) transition elements in barium titanate always acts as acceptors in B-sites while strontium acts as acceptor in A-site. It has been found that a proper concentration of dopant ions is effective to improve the dielectric properties of $BaTiO_3$. Barium strontium titanate (BST) is an attractive ceramic material because of its high permittivity, polarizations, and large induced strains that are attainable.[15]

3.2 PEROVSKITE STRUCTURE AND PHASE TRANSITION

$BaTiO_3$ adopts the perovskite structure consisting of the ABO_3-type unit cell. It can be visualized as a 3D framework of BO_6 octahedra as shown in Figure 3.2. In the unit cell structure, Ba^{2+} ions are situated in A sites, Ti^{4+} ions occupy B sites, and oxygen is present in the center of the face edge. $BaTiO_3$ can exist in five different crystal structures, namely, cubic, tetragonal, hexagonal, orthorhombic, and rhombohedral. The cubic and hexagonal structures are paraelectric in nature while the tetragonal, orthorhombic, and rhombohedral forms are ferroelectric. It is reported that the hexagonal structure of barium titanate is stable above 1460°C. On cooling below 1460°C, a transformation occurs from hexagonal phase to cubic phase of barium titanate. At Curie temperature, paraelectric–ferroelectric phase transition occurs, which means that the cubic phase of barium titanate transforms into the tetragonal phase and this phase is stable up to 0°C. At 0°C, the tetragonal phase changes into the orthorhombic phase by elongation along a phase diagonal. On further lowering the temperature, the orthorhombic phase of BaTiO$_3$ transforms to the rhombohedral phase at −90°C.[16]

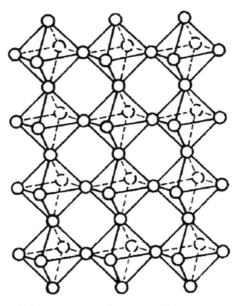

FIGURE 3.2 BO$_6$ octahedra arrangement in a perovskite-type structure (from Moulson, A. J.; Herbert, J. M. *Electroceramics*, 2nd ed.; John Wiley & Sons: England, 2003. Reproduced with permission).

In BST (Ba$_{1-x}$Sr$_x$TiO$_3$), partial substitution of Ba ions by Sr ions takes place, which modifies the nature and paraelectric–ferroelectric transition temperature of barium titanate. The nature of Ba$_{1-x}$Sr$_x$TiO$_3$ is dramatically dependent on the value of x.[9,17] Sr ion doping on barium titanate usually shifts the Curie point T_C to lower temperature. For Ba$_{1-x}$Sr$_x$TiO$_3$ bulk ceramic, the Curie temperature varies from 120 to −240°C. The unit cell of Ba$_{1-x}$Sr$_x$TiO$_3$ adopts a cubic structure above its Curie point but below this point the structure becomes distorted to the tetragonal phase. Other phase transformations can also occur near 0°C and −80°C: unit cell is orthorhombic below 0°C and it is rhombohedral in structure below −80°C.[18]

3.3 PROCESSING METHODOLOGY

Generally, the first step in the production of bulk dielectric ceramics is powder synthesis. The prepared powder is then pressed with the help of die into the required shapes and sizes. The next steps are sintering, electroding, and to study the dielectric properties of the ceramic. Several methods, such as solid-state reaction, sol–gel method, hydrothermal method, citrate process, etc., are used for the preparation of barium titanate-type ferroelectric ceramics.

3.3.1 SOLID-STATE METHOD

The most common practice of ceramic powder preparation is through the solid-state reaction, traditionally the mixed oxide route (Fig. 3.3). Stoichiometric amounts of the metal oxides or carbonates are mixed. The powders can be mixed by either of these two routes: dry ball-milling or wet ball-milling. Wet ball-milling is advantageous as it is faster than dry-milling, though an additional step of liquid removal is involved.

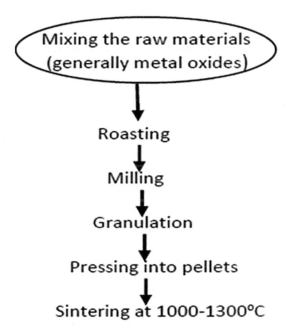

FIGURE 3.3 A schematic representation of the mixed oxide route.

In solid-state synthesis of BST, barium and strontium carbonates along with titanium oxide are mixed together followed by calcination at high temperature. It is a simple and low-cost processing method for the preparation of ceramics, but the disadvantages of this method are that it requires high temperature for the calcination process and the obtained ceramic is of large grain size and therefore high dielectric constant ceramic can't be obtained by this method.[19] This method can produce a mixture of phases as a side product. To overcome these problems, chemical reaction methods have been developed to obtain uniform fine particles and to control grain growth of the ceramics.

3.3.2 HYDROTHERMAL SYNTHESIS

Hydrothermal synthesis method is also used for the preparation of perovskite compound.[20] Very low temperature is required to synthesize the desired product. Boulos and coworkers employed the hydrothermal method to prepare $BaTiO_3$ ceramic from two different titanium precursors $TiCl_3$ and TiO_2 and $BaCl_2 \cdot 2H_2O$. The reaction was performed at two different temperatures 150°C and 250°C. The grain size of the sample prepared using $TiCl_3$ at 150°C was in the range 40–70 nm, and at 250°C, it was in the range 80–120 nm. The powder obtained from TiO_2 showed average grain size in the range 40–70 nm.[21] Due to incomplete reactions and loss of metal ions in precipitate, an excess of precursor ions is to be supplied which maintains the chemical stoichiometry in the obtained product in this method.

Microwave hydrothermal method showed certain advantages over the conventional method, namely, cost-effectiveness due to faster kinetics of the reaction and rapid internal heating. Tan C reported the synthesis of $BaTiO_3$ thin films by the microwave hydrothermal route.[22] Guo and coworkers compared the conventional and microwave hydrothermal methods and concluded that $BaTiO_3$ powder can be prepared in shorter time and at lower temperature in the microwave route.[23]

3.3.3 COPRECIPITATION METHOD

In the coprecipitation method, the metal cations are taken in the form of their soluble salts (e.g., nitrates) and coprecipitated from a common medium. The medium may be hydroxides, carbonates, oxalates, or citrates. For all practical applications, oxides or carbonates of the required metals are first digested with an acid, and then, the precipitating agent is added to the solution. The precipitate thus obtained is dried and then heated to the required temperature in appropriate atmosphere to yield the final product.

The temperature required to decompose the precipitates is generally much less than the temperatures used in the solid-state reaction. However, if all metal ions do not form insoluble precipitates, it becomes difficult to control the stoichiometry. Coprecipitation is a simplified approach for the synthesis of powders.

For the preparation of BST by chemical coprecipitation method, barium carbonate is dissolved in acetic acid and warmed for complete dissolution. This is followed by mixing of stoichiometric amounts of strontium acetate and titanium tetra-isopropoxide. Finally, distilled water is used as the precipitating agent. Mahata and coworkers have reported the synthesis of $BaTiO_3$ and its doped analogues by this method.[24]

3.3.4 SOL–GEL METHOD

The sol–gel method or chemical solution deposition has been used to prepare a number of perovskite ceramics. In the sol–gel method, many alkoxides can be used for the preparation of ceramics. In the first step, the metal oxide precursors are dissolved in an appropriate solvent to form a stable solution. A number of chemical reactions, namely, hydrolysis, chelation, etc., are involved during this step. Next, a stable sol is formed by slow hydrolysis of this solution and the stable sol is then slowly polymerized to form a gel. The gel is then dried and calcined to make ceramic powders (Fig. 3.4).

Yang and coworkers have reported the synthesis of BST by the sol–gel method.[25] Wodecka-Dus and coworkers prepared $Ba_{0.6}Sr_{0.4}TiO_3$ by sol–gel method. They used barium acetate, strontium acetate, and tetra-butyl titanate for the preparation of ceramic. Cubic phase of the obtained ceramic was confirmed by XRD measurement. Dielectric properties of the BST were studied with the variation of temperature and frequency. Electrical permittivity was measured in the temperature range of −100–100°C. From the dielectric measurements, they measured Curie–Weiss temperature and Curie–Weiss constant for $Ba_{0.6}Sr_{0.4}TiO_3$ ceramics.[26]

The advantages of sol–gel technique include low-cost, thickness, and composition control, uniformity, greater purity, short fabrication cycle, and low temperature required for calcination.[27] A major limitation of the sol–gel process is that it does not yield the perovskite phase directly. The final annealing yields the ceramic but this step is preceded by a nonferroelectric phase.[28]

Spray drying and roasting techniques are also chemical methods employed for the preparation of ceramics. In this method, solution of Ba^{2+}, Sr^{2+}, and Ti^{4+}

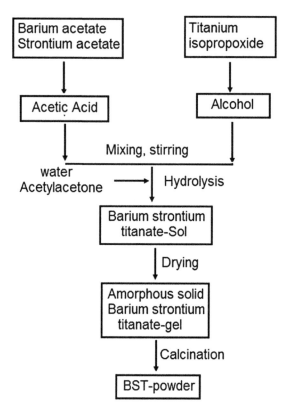

FIGURE 3.4 A schematic representation of barium strontium titanate prepared by sol–gel method (from Yang, W.; Chang, A.; Yang, B. J. *Mater. Synth. Process* **2002**, *10 (6)*, 303–309. Reproduced with permission).

are mixed together at high temperature, and after thermal decomposition, it forms BST.[19] This process provides less deviation from stoichiometry. The disadvantages of this method are the formation of agglomerated product which requires grinding and Cl⁻ ion occupies the face center position in barium titanate.[29]

The polymeric precursor method is another extensively used route, employed to synthesize ceramics. In this process, a solution of ethylene glycol, citric acid, and metal ions is polymerized to form polyester-type resin. The advantage of this method is due to its simplicity for obtaining powders of high purity and possibility to maintain the initial stoichiometry. However, the processing time is greater in this method.

3.4 DIELECTRIC PROPERTIES

The electrical properties in solids can be explained by considering that the valence electrons supplied by the atoms in solids spread through their entire structure. This concept can be expressed more formally by making a simple extension of the MO theory in which solids are treated like indefinitely large molecules known as the *tight-binding approximation*. This description in terms of delocalized electrons can also be used to describe nonmetallic solids. Large number of atomic orbitals in solids overlaps to form large number of closely spaced molecular orbitals forming an almost continuous *band* of energy levels. These bands are separated by *band gaps*, which are the disallowed values of MO energies.

A dielectric (or insulator) can be viewed as a semiconductor with a large energy band gap. In an insulator, the band of highest energy that contains electrons (at $T = 0$) is normally termed the *valence band*. The next higher band (which is empty at $T = 0$) is called the *conduction band*. Semiconductors have small energy band gap and thermal excitation provides sufficient energy to excite electrons from the valence band to an upper empty conduction band. In a *dielectric*, charge carriers are mainly injected from the electrical or other external sources simply because a dielectric's energy band gap is relatively large, so a higher amount of energy is required for such band to band transitions.[30] The band structure of solids is represented in Figure 3.5.

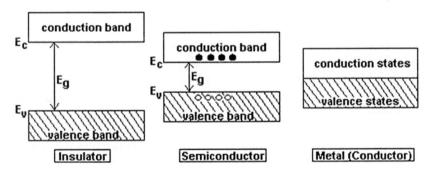

FIGURE 3.5 The energy band structure of an insulator, semiconductor, and conductor: there is a significant gap between the filled and empty bands of the insulator (dielectric).

3.4.1 FUNDAMENTAL KNOWLEDGE ABOUT DIELECTRIC

The valence shell electrons of the atoms in a solid are the most exposed to external forces such as electric fields, magnetic fields, electromagnetic waves, mechanical stress, etc. The interactions between these electrons and the external electrical force result in the occurrence of various dielectric properties. For instance, electromagnetic waves induce polarization in dielectric materials hence generating new intrinsic fields within these materials which interact with external fields, resulting in a rich show of dielectric phenomena.[31]

Electric polarization, as the name suggests, refers to the relative separation of the positive and negative charges in atoms or molecules, that is, generation of electric dipoles in the material, in the presence of an external electric field (Fig. 3.6). Alternatively, it can also refer to the reorientation of existing dipoles along the direction of the external field or the displacement of mobile charge carriers at the interfaces of impurities/dopants or other defect boundaries when exposed to an external electric field. The work done for this charge separation is performed using the potential energy released from this polarization process because the total potential energy of the system in an electric field decreases after electric polarization.[31] The five basic types of electric polarization mechanisms occurring within a dielectric material are as follows:

1. Electronic polarization
2. Ionic or atomic polarization
3. Dipolar polarization
4. Spontaneous polarization
5. Interface or space-charge polarization

However, electronic, ionic, and orientational polarizations are the three major mechanisms of electric polarization. These mechanisms predominate at moderate electric field strength (i.e., when the external field is much lower than the intrinsic atomic or molecular fields) and for materials with low conductivities so that the effect of charge carriers be neglected.

The polarization processes do not show much dependence on temperature as the restoring force against the displacement is temperature independent. The polarization due to the migration of charge carriers to form space charges at interfaces or grain boundaries is called space-charge polarization. Thus, there would be four components that make up the total polarizability of any material, represented by a.

$$a = a_e + a_i + a_o + a_d$$

where a_e denotes electronic polarization, a_i, a_o, and a_d are the ionic, orientational, and space-charge polarizations, respectively.

Barium Strontium Titanate

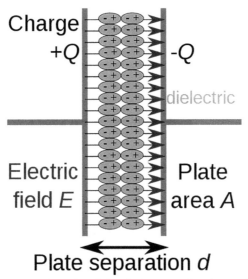

FIGURE 3.6 The polarization of a dielectric between the plates of a capacitor on the application of electric field. (Source: Papa November, https://commons.wikimedia.org/wiki/File:Capacitor_schematic_with_dielectric.svg)

3.4.2 DIELECTRIC CONSTANT

Permittivity or dielectric constant is an important electrical property of ceramic material. Below a certain critical field, for most of the ceramic materials, the electric field strength does not affect its dielectric properties. The dielectric constant value mainly depends on the frequency of the alternating electric field. It also depends on the arrangement of atoms and the defects of the material. In addition to these factors, dielectric constant also gets affected by temperature, pressure (physical parameters), etc. To have a full understanding of the concept of dielectric constant, it is imperative that we know what a capacitor is. A capacitor is a device to store charge. There are two conducting plates in a capacitor with a dielectric material between them. The capacitance of a capacitor is denoted by C, whose standard unit is Faraday (F).

It is defined as

$$Q = CV$$

where Q is the charge on each capacitor plate and V is the electric potential applied between the plates. Hence, $1\ F = 1\ CV^{-1}$.

Capacitance of a material is affected by a number of factors such as the geometry of the capacitor (directly related to area of capacitor), the dielectric material used, etc. Here, we are concerned with effect of the dielectric material.

The dielectric constant (K) of a substance can be explained as the fraction of capacitance using that material as the dielectric in a capacitor to the capacitance of an empty capacitor. It can also be defined in terms of permittivity. Permittivity of a material is its ability to polarize in the presence of an electric field.[32]

Dielectric constant (K) can be calculated by using the formula:

$$K = \frac{C}{C_0}$$

C_0 is the capacitance of an empty capacitor, $C_0 = \varepsilon_0 A/d$ and C is the capacitance with the dielectric between the plates of the capacitor, $C = \varepsilon \varepsilon_0 A/d$.

Alternatively, dielectric constant K can also be given by K is a $\varepsilon/\varepsilon_0$; ε_0 is a permittivity of free space (8.85×10^{-12} F m^{-1}); E is a permittivity of the dielectric material; A is an area of the plate of the capacitor; and d is a distance between the plates of the capacitor.

3.4.3 DIELECTRIC LOSS

An efficient dielectric supports a varying charge with minimal dissipation of energy in the form of heat. There are two main forms of loss that may dissipate energy within a dielectric.

In conduction loss, a flow of charge through the material causes energy dissipation. Dielectric loss is the dissipation of energy through the movement of charges in an alternating electromagnetic field as polarization switches direction. If we reverse the direction of the applied field, as in the case of an alternating field, the direction of polarization would also reverse so as to align with the direction of the field. The switching or movement of charges takes place over a period of time called relaxation time, typically in the order of ~10^{-11} s. If the frequency of the applied external field is higher than ~10^{11} Hz, the dipoles of the dielectric material cannot keep up with field, that is, they lag behind. Consequently, the material's net polarization decreases and so does its dielectric constant (Fig. 3.7). Dielectric loss is especially high around the relaxation or resonance frequencies of the polarization mechanisms as the polarization lags behind the applied field, causing an interaction between the field and the dielectric's polarization that results in heating.

Barium Strontium Titanate 69

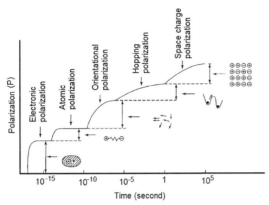

FIGURE 3.7 Trends in the variation of different polarization mechanisms against time in the presence of step-function electric field (from Kao, K. C. Dielectric Phenomena in Solids, 1st ed.; *Elsevier*: North-Holland, **2004**. Reproduced with permission).

3.5 FERROELECTRIC PROPERTIES

The term ferroelectricity is derived from its analogy to ferromagnetics. Valasek was the first to observe the ferroelectric effect in the Rochelle salt (KNaC$_4$H$_4$O$_6$·4H$_2$O) in 1921. However, it was only in the 1950s, when the perovskite compound BaTiO$_3$ was found to show this property that the field of ferroelectrics received a boost. Dielectric materials may be classified into two major categories: nonferroelectric (normal dielectric) materials and ferroelectric materials. The polarization induced by an externally applied field in normal dielectric materials is very small, with the dielectric constant usually less than 100, and its effects on other physical properties are also very small. However, ferroelectrics exhibit a large polarization, with the dielectric constant up to 105, under certain conditions. This large magnitude of polarization has attracted attention from many researchers to study it theoretically and to develop various practical applications. A diverse range of materials shows ferroelectric properties, namely, triglycerine sulfate and isomorphous materials, potassium dihydrogen phosphates and isomorphous materials, Rochelle salt, perovskite-structured materials like barium tiatanate BaTiO$_3$, BST, lead zirconatetitanate Pb(Ti,Zr)O$_3$, lead titanate PbTiO$_3$, KNbO$_3$, etc. Among these, the perovskite materials have been widely studied.

Ferroelectrics are crystalline materials possessing a net separation of charges in the absence of an electric field. An important requirement for this is that the material should lack a center of symmetry.[4] There is a critical temperature, called the Curie temperature, T_C, which marks the transition from the ordered to the disordered phase. At this temperature, the dielectric constant may reach values three to four orders of magnitude higher than in the disordered phase. A ferroelectric crystal shows a reversible spontaneous electric polarization and a hysteresis loop that can be observed in certain temperature regions, delimited by this transition point T_C. At temperatures above T_C, the crystal is no longer ferroelectric and exhibits normal dielectric behavior.

It is important to note that even in the absence of an external electric field, the polarization in ferroelectric materials has a nonzero value, and when an external field is applied in a direction opposite to this polarization, the direction of polarization in the ferroelectric material is reversed, hence the term-reversible spontaneous polarization.[33]

3.5.1 BASICS OF FERROELECTRICITY: DOMAIN THEORY

A single crystal or a system that is ferroelectric will undergo spontaneous polarization below the Curie temperature. The electrostatic energy of the system becomes high and the system becomes very unstable if all the dipoles of the polarization are pointing in one direction. Therefore, a system always tends to minimize its potential energy by forming regions in the crystal, each containing a large number of dipoles all aligned in the same direction. These regions are known as the domains. Domains are arranged in such a way that the polarization of different domains will compensate for each other and the net polarization of the whole crystal along any direction will cancel out. Domains are separated by domain walls, which are the loci of the dipole orientation from one direction of the domain to another of the neighboring one.[31] A single crystal may contain many domains with zero net polarization, but it can be changed to a single crystal with a single domain by a strong poling electric field, in which the dipoles of all domains point in the same direction, that is, all domains will join to form a large domain. When a phase transformation begins, the domains will be nucleated at several places within the crystal and the nuclei of domains will grow along the ferroelectric axes until the transformation of the new phase in the whole volume is completed.[31]

On increase in field E, all domains gradually orient themselves in the direction of the field reaching a saturation point at the tip of the loop where almost all the domains are aligned along the field lines. On extrapolation of the tip of loop to the polarization axis, that is, at zero field, we get the region of spontaneous polarization P_s. The region of the polarization axis from

Barium Strontium Titanate 71

the origin to the point where the hysteresis loop meets the polarization axis represents the region of remnant polarization P_r. P_r is less than P_s because after the field is removed, some domains may revert back to their original positions; hence, the net polarization decreases (Fig. 3.8).

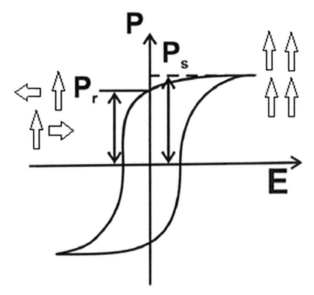

FIGURE 3.8 A typical hysteresis loop of a ferroelectric solid.

3.6 MICROSCOPIC STUDIES

Microscopic studies include scanning electron microscope (SEM), transmission electron microscope (TEM), and atomic force microscope. SEM and TEM along with energy dispersive X-ray analysis are the excellent tools for surface topography and composition study of the sample. SEM generates images of the sample to examine with a focused beam of electrons, and in TEM, images are formed by interaction of the electrons transmitted through the ultrathin sample. Surface morphology and composition of the samples can be detected by microscopic studies. SEM and TEM provide images at higher resolution than the light microscopes. Metal composition of the particles can be examined by energy dispersive X-ray data.[34] TEM can generate a diffraction pattern of a sample. A pattern of dots can be formed in case of single crystal but in case of an amorphous or polycrystalline solid material, a series of rings forms. The diffraction pattern for the single crystal is dependent upon the orientation of the specimen. The TEM images of the sample give the information about the space group symmetries in the crystal and also the orientation of the crystal.[35] SEM can provide a high-resolution image of the sample but the resolution is not high enough as is possible in the TEM. In SEM, the surface of the sample is scanned by a focused beam of electrons having energy in the range between 0.5 and 30 keV. These primary electrons when interact with the sample produce the electrons and photons, most importantly secondary electrons, backscattered electrons, and photons of characteristic X-rays. These electrons and photons carry various information of the surface morphology and the element distribution of the specimen.[36] Microscopic studies help in the measurement of grain size of the ceramics. The particle size of ceramic plays a significant role to study about the electrical properties (dielectric, ferroelectric, and piezoelectric) of the ceramics.

3.7 STUDY OF BARIUM TITANATE

Ferroelectric behavior was first observed in barium titanate, $BaTiO_3$ ceramic material. It also shows high dielectric constant value and low dissipation factor (good electrical properties). As a consequence, it was the pioneer of ceramic use for dielectric ceramic capacitors.

3.7.1 DIELECTRIC PROPERTY

The dielectric constant value of $BaTiO_3$ depends on temperature, frequency, dopants, grain size, etc.[40] The tetragonal structure of barium titanate shows excellent dielectric properties, which make it a good candidate for electronic application such as in dynamic random-access memory (DRAM), dielectric field tunable elements for high frequency devices, decoupling capacitors, etc.[37]

At room temperature, the dielectric constant value 1500–2000 was reported for the pure barium titanate with the grain size in the range of (20–50 μm).[38] The dielectric constants of barium titanate rise with reduce in grain size (down to 0.8 μm) because 90° domain wall cannot exist below 0.8 μm grain size.[39] It is very low for pure barium titanate having grain size lower than 700 nm due to phase change of ceramic from tetragonal to pseudocubic.[40–42]

Kim et al. prepared barium titanate by Pechini method and studied about its dielectric properties with the variation of grain size. The grain size of the obtained ceramic was in the range of 0.86–10 μm. The dielectric constant was 4500 and 1800 at room temperature, and at Curie temperature, it was 6200 and 7000, for 0.86 and 10 μm grain size of the barium titanate, respectively. The experiment showed that with the increase in grain size of ceramic, dielectric constant decreases.[43]

Lokare reported the variation of dielectric constant value with temperature at different frequency (1 kHz, 10 kHz, 100 kHz, and 1 MHz) for barium titanate ceramic. It was notice that the dielectric constant value of ceramic increases gradually with the rise in temperature at all frequencies and the maximum value at its Curie temperature (T_C) observed, after that it decreases. This indicates that at Curie temperature, phase changes from ferroelectric to paraelectric state. Dipolar, electronic, ionic, and interfacial polarizations affect the dielectric constant value. Due to the formation of crystal defects, interfacial polarization increases and the dipolar polarization decreases because of increase in randomness in the dipoles of the ceramics. The increase in dielectric constant with temperature up to Curie point indicates the increase in interfacial polarization. Electrons of the ceramic are thermally activated at Curie temperature, which results in an increase in dielectric constant. It was reported that with increase in frequency dielectric constant, value of barium titanate ceramic decreases. The very high value of dielectric constant is observed at low frequency region which is attributed to interfacial polarization. This is a general characteristic of ceramic materials.[44] In weak external field, domains of the crystal do not participate in polarization.[45]

Benlahrache and coworkers obtained the dielectric constant value of pure barium titanate prepared by milling and calcinations process. Figure 3.9 represents the variation of dielectric constant with frequency at different temperature. Dielectric constants were measured with the applied bias of 100 V for the pure barium titanate sintered at 1500°C for 2 h. At 30°C, dielectric constant decreases below 1 kHz but for higher frequencies slightly changes in dielectric constant were observed. At higher temperature, similar results were observed.[46]

Lokare discussed that how the dielectric loss (tan δ) varies with the frequency of barium titanate.[44] In the dielectric system, energy dissipation is termed as dielectric loss and it is directly related to the imaginary part of the dielectric constant of ceramic. At low frequency region, higher value of dielectric loss (dissipation factor) was reported and with increase in frequency, tan δ decreases. This type of variation is also reported by various researchers.[47,48]

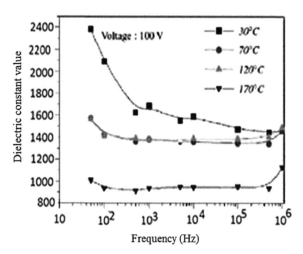

FIGURE 3.9 Frequency dependence of dielectric constant of pure BaTiO$_3$. (Adapted from Benlahrache, M. T.; Barama, S. E.; Benhamla, N.; Achour, S. *Mater. Sci. Semicond. Proc.* **2006**, *9*, 1115–1118. Reproduced with permission)

3.7.2 FERROELECTRIC PROPERTY

BaTiO$_3$ has a perovskite (CaTiO$_3$)-type structure. Perovskites in general have the structural formula ABO$_3$, with a cubic close-packed arrangement of composition AO$_3$ with the A-ion coordinated to 12 oxygen ions and the B-ion in the octahedral interstices (Fig. 3.10). In an ideal packing, the ionic radii holds the following relation[49]:

$$R_A + R_O = 2(R_B + R_O)$$

In BaTiO$_3$, Ti belongs to the 3D transition series and has vacant *d* orbitals for electrons to form covalent bonds with its neighboring elements. Ti^{4+} has an ionic radius of about 0.68 Å, and the radius of Ba^{2+} is about 1.35 Å. These ions form nice octahedral cages, with the O^{2-} in the cages.

However, this structure becomes unstable at a certain transition temperature and must transform to a more stable one. At the Curie's temperature T_C, the crystal transforms into a tetragonal form, resulting from the stretching of the cubic unit cells along one particular edge where the Ba^{2+} ions shift 0.05 Å upward from their original position in the cubic structure, Ti^{4+} ions shift upward by 0.1 Å, and the O^{2-} ions downward by 0.04 Å to form the tetragonal structure. As a result of the ion shifts (shown in Fig. 3.11), the centroid of the positive charges no longer coincides with the centroid of the negative charges; therefore, the unit cells become permanently polarized and behave as permanent dipoles, leading to spontaneous polarization.[31] The phase transformations in BaTiO$_3$ with respect to temperature are shown in Figure 3.12a and the corresponding changes in spontaneous polarization are shown in Figure 3.12b. The direction of spontaneous polarization is always along the direction of the unit cell's elongation, that is, the stretching direction. This is commonly referred to as the ferroelectric polar axis.

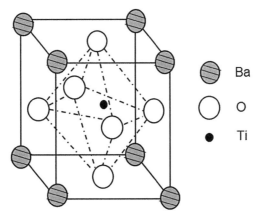

FIGURE 3.10 The unit cell of BaTiO$_3$.

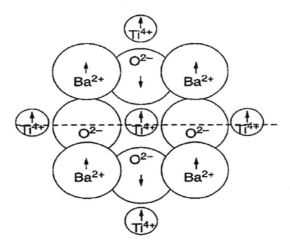

FIGURE 3.11 The ion displacements due to the cubic–tetragonal distortion in BaTiO$_3$. (Adapted from Kao, K. C. Dielectric Phenomena. In *Solids*, 1st ed. Elsevier: North-Holland, 2004. Reproduced with permission.)

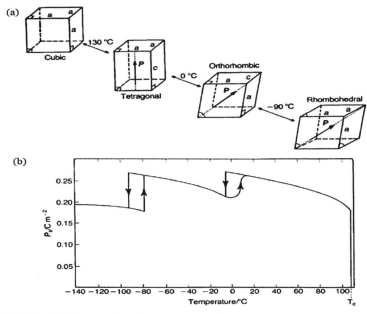

FIGURE 3.12 (a) The transformations of the BaTiO$_3$ structure with temperature and (b) variation of spontaneous polarization with respect to temperature in single-crystal BaTiO$_3$. (Adapted from Moulson, A. J.; Herbert, J. M. *Electroceramics*, 2nd ed.; John Wiley & Sons: England, 2003. Reproduced with permission.)

3.7.3 MICROSCOPIC STUDIES OF BARIUM TITANATE, BaTiO$_3$

TEM study can be used to measure the average grain size and to obtain the microstructure of ceramics. Figure 3.13 shows the TEM image of barium titanate. Sakabe and coworkers observed fine subrounded particles of barium titanate calcined at 700–1200°C with narrow particle-size distribution.[50] TEM study shows that the grain size of the pure barium titanate increases with increasing calcinations temperature. Barium titanate powders are in cubic perovskite structure with paraelectric phase, when the grain size is less than 85 nm.[51]

Figure 3.14 represents the SEM image of barium titanate after sintering at 1150°C. It was reported that after sintering at 1150°C, the average grain size of barium titanate was 1800 nm but sintering at 1050°C decreases the grain size to 280 nm. The average grain size of barium titanate ceramic was determined by SEM study. It is reported that with increasing sintering temperature, the density and grain size of ceramic increase, and on decreasing sintering temperature, the grain size of barium titanate became finer with higher porosity.[52]

Barium Strontium Titanate 77

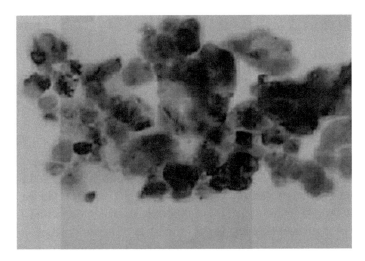

FIGURE 3.13 TEM image of barium titanate. (Adapted from Hsiang, H. I.; Lin, K. Y.; Yen, F. S.; Hwang, C. Y. *J. Mater. Sci.* **2001**, *36*, 3809–3815. Reproduced with permission. © 2001 Springer.)

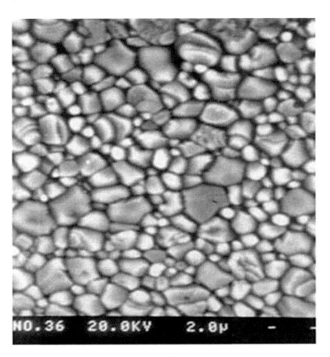

FIGURE 3.14 SEM of $BaTiO_3$ after sintering at 1150°C for 30 min in air. (Adapted from Luan, W.; Gao, L.; Guo, J. *Ceram. Int.* **1999**, *25*, 727–729. Reproduced with permission. © 1999 Elsevier.)

3.7.4 EFFECT OF DOPANTS ON BARIUM TITANATE

Barium titanate ($BaTiO_3$) is a common electronic ceramic material, widely used for multilayer ceramic capacitors (MLCCs) because of its very high dielectric constant (the ratio between the permittivity of the medium to the permittivity of free space), ferroelectric properties, PTCR, high permittivity, and piezoelectric properties. High dielectric constant value, low dissipation factor, and good temperature stability can be achieved by doping. With the help of doping ferroelectric properties of $BaTiO_3$, ceramic can be effectively managed. $BaTiO_3$ can be doped by different elements like Ni, Sr, La, Mg, Fe, etc. Doping elements can either occupy Ba^{2+} site which is dodecahedrally coordinated or Ti^{4+} site, which is octahedrally coordinated in $BaTiO_3$ crystal structure. A significant change in the properties of barium titanate is reported due to different types of dopants (A-site or B-site dopants).

3.7.4.1 EFFECTS OF Sr DOPING

Sr doping substitutes the Ba^{2+} ion in the perovskite structure of barium titanate.[53] Original characteristics of barium titanate are maintained on doping with Sr but under various conditions, it decreases its Curie temperature (T_c).[29] It was reported that each mol% of $SrTiO_3$ doping on barium titanate decreases the T_C point of ceramic from 125°C by −3%.[54] $Ba_{1-x}Sr_xTiO_3$ exhibits cubic perovskite structure for $0.3 \leq x \leq 1$ at a room temperature and tetragonal structure for $0 \leq x \leq 0.3$.[55] S. Bobby Singh and coworkers prepared $Ba_{0.7}Sr_{0.3}TiO_3$ ceramic from barium 2-ethylhexanoate, strontium 2-ethylhexanoate, and titanium(IV) isopropoxide precursors through a modified sol–gel method. They observed tetragonal phase after heating the product at 550°C for 1 h in air. As the annealing temperature increases from 450°C to 550°C, the grain size of the films also increases from 30.8 to 39.8 nm, respectively.[56] Sr doping increases the dielectric thermal stability and it was reported that with the rise in sintering temperature, the grain size of the BST also increases.[57]

3.7.4.2 EFFECTS OF Fe DOPING

Mishra and coworkers studied about the effects of Fe on the structure, dielectric, magnetic, and electrical properties of $BaTiO_3$. They prepared $BaTi_{1-x}Fe_xO_3$ (x = 0.005, 0.01, 0.015) via solid-state route and the obtained ceramic was tetragonal with space group *P4mm*. It was observed that transition temperature of the pure barium titanate was 120°C, but after doping with iron, it shifted to 125°C. Iron ion substituted at Ti site of the barium titanate crystal.[58] Fe-doped barium titanate ceramics were prepared by modified sol–gel route by Kundu and coworkers. It was observed that dielectric constant increases by Fe doping on barium titanate as compared to pure barium titanate and the Curie temperature shifted toward lower temperature side with increase in doping concentration. Maximum dielectric constant was observed for 0.3 mol% Fe doping. The variation of dielectric constant value was described with the help of lattice parameter and the *c/a* ratio and the maximum value was reported for 0.3 mol% Fe doping. Lattice parameter value confirms the tetragonal perovskite structure for Fe-doped barium titanate. Dielectric constant was measured with the variation of temperature at different frequency.[59]

3.7.4.3 EFFECTS OF Ni DOPING

Kundu and coworkers synthesized Ni-doped barium titanate by a chemical method using polymer and polyvinyl alcohol.[59] The synthesized Ni-doped barium titanate was tetragonal in structure. It was observed that dielectric permittivity of doped specimen is enhanced as compare to pure barium titanate. The highest dielectric permittivity value was observed for 0.6 mol% of Ni ion doping. Dielectric behavior was described in terms of change in crystalline structure of the doped barium titanate. It is also seen that particle size of the doped and pure barium titanate is almost same; therefore, enhancement of dielectric constant did not result due to the particle size.

3.7.4.4 EFFECTS OF Mn DOPING

Mn ion substituted the titanium atom of barium titanate ceramic. Mn-doped barium titanate $BaTi_{1-x}Mn_xO_3$ ($0 \leq x \leq 0.01$) was prepared by Wang et al. Substitution of Ti ion by Mn ion partially traps the conduction of electrons. The dielectric constants of the doped barium titanate exhibited a peak at $T_C \sim 110°C$ upon heating, which is related to a ferroelectric–paraelectric phase transition.[60] Dielectric constant and dissipation factor of $Ba_{1.005}(Ti_{0.98}Mn_{0.02})O_{2.985}$ sintered in pure nitrogen atmosphere were 1500 and 0.01, respectively, at 1 kHz frequency from temperature −60°C to +160°C.[61] Mn doping on barium titanate changes the phase of the doped ceramic.

Hexagonal phase of Mn-doped barium titanate is reported for $BaTi_{1-x}Mn_xO_3$ (where x = 1–10%). Mn ion is also responsible for the ferromagnetism of the doped ceramic. Ferromagnetic order increases gradually with the increase in Mn-doping concentration. Decrease in dielectric constant value and dissipation factor is reported with the increase in Mn content in doped barium titanate ceramic.[62]

3.7.4.5 EFFECTS OF La DOPING

Lanthanum ion replaces the barium site in the perovskite structure of barium titanate. Dielectric properties of undoped and La-doped barium titanate were studied by M. M. Vijatovic Petrovica. Pechini process was applied for the preparation of lanthanum-doped barium titanate (0.3 and 0.5 mol% of La). It was noticed that phase transition temperature of doped barium titanate shifted to lower temperature as compared to pure barium titanate (120°C). The reported transition temperature was 111°C for 0.3 mol% lanthanum and 96°C for 5 mol% of lanthanum.[63] Mancic et al. reported that the dielectric permittivity of lanthanum-doped $BaTiO_3$ (0.3 mol% of lanthanum) was varied from 8000 to 12,000 at 1 kHz with varying temperature and dissipation factor was <1%.[64] La doping on barium titanate increases the dielectric constant of ceramic. As the La concentration increases from 0.2 to 0.8 wt %, the grain growth observed, as a result increase in dielectric constant reported.[65]

3.7.4.6 EFFECTS OF Cu DOPING

Copper doping on barium titanate affects its structural, dielectric, and electrical resistivity properties. Divalent copper replaces the A site of perovskite structure of barium titanate, which is also a divalent. The reported Curie temperature was in the range of 120–125°C for the copper-doped barium titanate ($Ba_{1-x}Cu_xTiO_3$, where x = 0.01–0.04%). Temperature dependence dielectric constant values were observed in the range of 600–911, which is less than the pure barium titanate. This might be due to conducting nature of copper ions in doped $BaTiO_3$.[66] Cu^{2+} ion inhibits grain growth of the ceramic due to the segregation at grain boundaries. Method of preparation also affects the grain size of the ceramic. Cu doping leads to the change in structure of the barium titanate. Orthorhombic phase structure was reported for $Cu_xBa_{1-x}TiO_3$ (where x = 0.01, 0.05, 0.1).[67] Partial reduction of titanium by Cu^{2+} ion induces a shifting of charges of oxygen toward titanium ion. Copper doping is responsible for the metallic behavior of the ceramic and it induces cubic–quadratic transition in barium titanate.[68]

3.7.4.7 EFFECTS OF Ca DOPING

There are a number of controversial studies on the phase transition of calcium-doped barium titanate. Mitsui et al. reported that Ca doping in barium titanate shows a sharp-phase change.[69] Berlincourt and coworkers reported that Ca doping causes a negligible change in barium titanate.[70] Possible occupancy of Ca^{2+} ion in Ba^{2+} site or in Ti^{4+} site was reported by Zhuang and coworkers.[71]

3.8 STUDY OF BARIUM STRONTIUM TITANATE

BST is a lead-free ceramic. Because of environmental friendly character of BST, it is an interesting material to substitute the lead (Pb)-based ceramics. Hence, this research topic is of great interest to scientists.

3.8.1 DIELECTRIC PROPERTY

BST is a high dielectric ceramic with low dielectric loss value; therefore, it is used in various electronic applications. Recently, BST thin films are used in place of the SiO_2 (conventional dielectric material) which cannot provide the required charge storage density.[72] Sr doping changes the dielectric constant value of the barium titanate, and the change directly depends on the concentration of the dopants.[29] Figure 3.15 shows the variation of dielectric constant value with the composition Ba/Sr in $Ba_{1-x}Sr_xTiO_3$. The Curie temperature varies with the concentration of dopants. With the increase of Sr content in BST, a linear decrease in Curie temperature was observed. Variation of dielectric constant of BST with temperature and frequency is reported.

Generally, with increase in frequency, dielectric constant value of ceramic decreases; this is because in lower frequency region, dielectric constant is affected by various types of polarization such as electronic, atomic, interfacial, and ionic polarization but in higher frequency region, it is only due to electronic polarization. Dielectric constant value of material increases gradually up to its Curie temperature and after that the value reduces, which indicates phase transition of the ceramic material at Curie temperature. Table 3.1 represents the variation in the relative permittivities, ε_r, at room temperature and Curie temperature and the dielectric loss factor, tan δ, obtained on varying the Sr concentration in $Ba_{1-x}Sr_xTiO_3$.

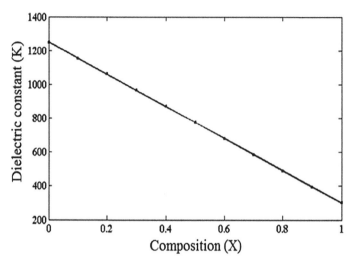

FIGURE 3.15 Variation of dielectric constant with the compositions of barium/strontium in $Ba_{1-x}Sr_xTiO_3$. (Adapted from Balachandran, R.; Ong, B. H.; Wong, H. Y.; Tan, K. B.; Rasat, M. M. *Int. J. Electrochem. Sci.* **2012**, *7*, 11895–11903. Reproduced with permission.)

TABLE 3.1 Dielectric Constants of Different Stoichiometric Ratios of Barium Strontium Titanate.

Ceramic	Phase	ε_r (RT)	$\varepsilon_r(T_C)$	T_C (°C)	tan δ	Reference
$Ba_{0.8}Sr_{0.2}TiO_3$	Orthorhombic	655	–	–	0.058	[73]
$Ba_{0.9}Sr_{0.1}TiO_3$	Tetragonal	1500	3250	79	0.012	[74]
$Ba_{0.95}Sr_{0.1}TiO_3$	Tetragonal	1700	3500	95	0.010	[74]
$Ba_{0.5}Sr_{0.5}TiO_3$	–	1164	–	–	0.063	[75]
$Ba_{0.6}Sr_{0.4}TiO_3$	Cubic	1200	–	–	0.01	[76]
$Ba_{0.9}Sr_{0.1}TiO_3$	Tetragonal	750	918	60	–	[48]
$Ba_{0.8}Sr_{0.2}TiO_3$	Tetragonal	830	980	45	–	[48]
$Ba_{0.7}Sr_{0.3}TiO_3$	Tetragonal	1056	1195	35	–	[48]
$Ba_{0.8}Sr_{0.2}TiO_3$	Tetragonal	1530	3062	80	–	[6]

Patil and coworkers reported the variation of dissipation factor (tan δ) with frequency (Fig. 3.16). It was observed that with the increase in frequency, value of tan δ decreases. At higher frequency, dipole affects the polarization and reduction in dielectric losses is reported. Similar results are reported by many workers.[48,77,78]

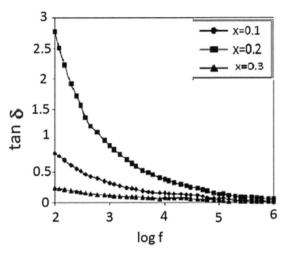

FIGURE 3.16 Variation of dielectric loss with frequency for $Ba_{1-x}Sr_xTiO_3$. (Adapted from Patil, D. R.; Lokare, S. A.; Devan, R. S.; Chougule, S. S.; Kanamadi, C. M.; Kolekar, Y. D.; Chougule, B. K. *Mater. Chem. Phys.* **2007,** *104*, 254–257. Reproduced with permission. © 2007 Elsevier.)

3.8.2 FERROELECTRIC PROPERTY

BST ($Ba_{1-x}Sr_xTiO_3$) is a solid solution of $BaTiO_3$ (BTO) and $SrTiO_3$ (STO). Strontium (Sr^{2+}) has the same valency as barium ion, its radius is almost equal to Ba^{2+}, and Sr has a high solid solubility. As explained earlier, BTO is the first perovskite compound exhibiting ferroelectric properties. An advantage of ferroelectric ceramics is the ease with which their properties can be modified by modifying the composition and the ceramic microstructure: the substitution of Sr^{2+} for Ba^{2+} in $BaTiO_3$ lowers the Curie temperature, T_C, by 4°C per atom substitution. For example, $Ba_{0.75}Sr_{0.25}TiO_3$ has a Curie temperature of about +40°C, whereas $Ba_{0.5}Sr_{0.5}TiO_3$ has a Curie temperature of about −50°C. Now, it has been established that barium titanate (BTO) has a perovskite structure and on decreasing the temperature, it undergoes successive phase transitions from cubic to tetragonal to orthorhombic and rhombohedral. Strontium titanate (STO) is a more recent ferroelectric ceramic with a large dielectric constant. It undergoes a transition from the cubic phase to an antiferrodistortive phase which involves tilting of the TiO$_6$ octahedra at 105 K. It is to be noted that at Curie temperature (Ba$_{1-x}$Sr$_x$), TiO$_3$ solid solutions have higher dielectric constant than pure BTO. BST is purely ferroelectric and exhibits spontaneous polarization below its Curie temperature. The tunability of BST is also very high in the ferroelectric phase, especially near T_C. However, the dielectric losses in this region are also very high, and hence, this phase of BST finds applications in nonvolatile memories. BST becomes paraelectric above T_C and the hysteresis effect does not prevail. This region serves well for tunable microwave device applications due to the associated high dielectric constant and low losses.[32,79,80]

3.8.3 MICROSCOPIC STUDIES OF BARIUM STRONTIUM TITANATE

The properties and morphology of the ceramic powder can be studied on the basis of its microstructure. Figure 3.17a shows the TEM images of Ba$_{0.90}$Sr$_{0.1}$TiO$_3$ powders sintered at 750°C and Figure 3.17b is the electron diffraction pattern obtained from TEM study. The electron diffraction pattern assigned to be perovskite phase of the BST ceramic. TEM study reveals the fine morphology of the ceramic.[78]

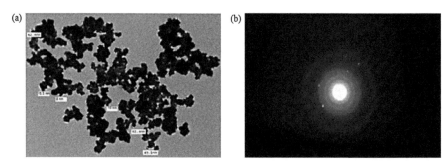

FIGURE 3.17 (a) TEM micrograph and (b) the diffraction pattern of the Ba$_{0.90}$Sr$_{0.1}$TiO$_3$ powder sample sintered at heat treatment temperature 750°C for 1 h. (Reprinted from Mahania, R. M.; Battisha, I. K.; Aly, M.; Abou-Hamad, A. B. *J. Alloys Compd.* **2010**, *508*, 354–358. Reproduced with permission. © 2010 Elsevier.)

The grain size obtained by the TEM study was compared with that of the XRD results. It is reported that the average grain size of ceramic powder was larger in TEM image than the XRD results. This can be explained due to the fact that uniform strain within the particle is not consider in the line broadening of the XRD pattern of the BST.[87] It was reported that with increase in sintering temperature of ceramic, grain size also increased. The average grain size of BST ceramic directly related to its dielectric properties.[88] Population of domains and movement of the domain walls also affect dielectric properties of BST ceramic.[89] It is reported that uniform grain size results simple and regular movement of domain wall. Grain size and its uniformity also affect the internal stress, which arises because of mechanical deformation of unit cells of ceramics. Uniform grain size of ceramic produces less internal stress which results easier domain wall motion. It significantly increases dielectric constant value of the ceramic. Average grain size of $Ba_{0.8}Sr_{0.2}TiO_3$ ceramic was estimated by using SEM analysis and high-resolution TEM image. Figure 3.18 shows the SEM micrograph of BST sintered at 1250°C and the obtained grain size of $Ba_{0.8}Sr_{0.2}TiO_3$ ceramic was in the range of 150–300 nm.[88]

FIGURE 3.18 SEM of $Ba_{0.8}Sr_{0.2}TiO_3$ ceramics ball milled for 20 h and sintered at 1250°C. (Reprinted from Mudinepalli, V. R.; Feng, L.; Lin, W. C.; Murty, B. S. *J. Adv. Ceram.* **2015**, *4* (1), 46–53. Reproduced with permission. Creative Commons license CCBY.)

SEM study showed that grain size and density of ceramic increase with increase in sintering temperature. Higher porosity and finer grain size were observed at low sintering temperature (1050°C). Change in dielectric constant and dissipation factor values was reported with decrease in grain size.[52] Grain size of the ceramic also plays an important role to study about the ferroelectric properties of BST ceramic. It was observed that with the change in grain size, ferroelectric hysteresis loops also change.[90] The coercive electric field decreases as the grain growth of barium strontium titanate takes place. It is also observed that remnant polarization is directly proportional to the grain size of ceramic, which means remnant polarization decreases with decrease in grain size.[88]

3.9 APPLICATIONS

Nowadays, BST ceramic has attracted much interest toward electronic purposes, due to its high dielectric constant and linearly adjustable Curie point with the strontium concentration over a wide range of frequency and temperature.[91–93] The electrical properties of BST ceramic powders are generally depend on various factor such as sintering temperature, grain size, porosity, and concentration of strontium.[64,94] These properties make BST a good ceramic for tunable microwave dielectric devices,[95,96] piezoelectric actuators, transducers, passive memory storage devices, thermal switches, tunable microwave devices, PTCR, MLCCs, DRAMs, etc.[91,97–100] Figure 3.19 shows a well-designed structure of the metal–insulator–metal capacitor. In this capacitor, silver was coated in the top and bottom of the dielectric material (BST). The most important properties of BST (dielectric material) ceramic capacitor for storage application in DRAM are dielectric constant value, charge storage density, and leakage current density.[101]

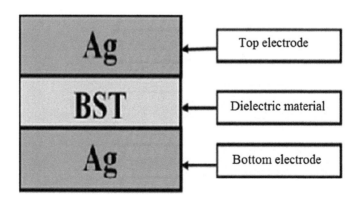

FIGURE 3.19 Structure of capacitor for DRAM cell. (Adapted from Ong, B. H.; Wong, H. Y.; Tan, K. B.; Rasat, M. M. *Int. J. Electrochem. Sci.* **2012**, *7*, 11895–11903. Reproduced with permission.)

Ferroelectric ceramic materials constructed in the form of thin films have great potential in integrated electronics as passive components and nonvolatile memory applications. The application of ceramics in devices requires that the properties of the bulk are attained in thin films. $Ba_{1-x}Sr_xTiO_3$ thin films exhibit a large dielectric constant with dc bias voltage, and hence, they are developed for microwave device applications such as filters, phase shifter, and microwave delay line system. High-quality BST thin films with good optical properties have wide applications in wireless communication. Nowadays, ceramic capacitors with ferroelectric properties are used in various applications like in medical ultrasound machines, infrared cameras, fire sensors, etc. The scope of ferroelectric ceramics continues to be wide and research in this field is driven by the demand for smarter device applications.

KEYWORDS

- **perovskite structure**
- **ferroelectric material**
- **sol–gel method**
- **barium strontium titanate**
- **microscopic studies**

REFERENCES

1. Makovec, D.; Samadmija, Z.; Drofenik, M. *J. Am. Ceram. Soc.* **2004**, *87*, 1324–1329.
2. Maga, D.; Igor, P.; Sergei, M. *J. Mater. Chem.* **2000**, *10*, 941–947.
3. Jin, Z.; Ang, C.; Yu, Z. *J. Am. Ceram. Soc.* **1999**, *82* (5), 1345–1348.
4. Jaffe, B.; Jaffe, H.; Cook, W. R. *Piezoelectric Ceramics*, 1st ed.; Academic Press: London, 1971; pp 271–279.
5. Bauger, A.; Moutin, J. C.; Niepce, J. C. *J. Mater. Sci.* **1983**, *18*, 3041–3046.
6. Li, Z.; Fan, H. *J. Phys. D: Appl. Phys.* **2009**, *42*, 075415–075422.
7. Kongtaweelert, S.; Sinclair, D. C.; Panichphant, S. *Curr. Appl. Phys.* **2006**, *6*, 474–477.
8. Roeder, R. K.; Slamovich, E. B. *J. Am. Ceram. Soc.* **1999**, *82*, 1665–1675.
9. Jeon, J. H. *J. Eur. Ceram. Soc.* **2004**, *24*, 1045–1048.
10. Hung, K. M.; Yang, W. D.; Huang, C. *J. Eur. Ceram. Soc.* **2003**, *23*, 1901–1910.
11. Sharma, K. P.; Varadan, V. V.; Varadan, V. K. *J. Eur. Ceram. Soc.* **2003**, *23*, 659–666.
12. Razak, K. A.; Asadov, A.; Yoo, J.; Haemmerle, E.; Gao, W. *J. Alloys Compd.* **2008**, *449*, 19–23.

13. Brankovic, G.; Brankovic, Z.; Goes, M. S.; Paiva, C. O.; Cilense. M.; Varela, J. A.; Longo, E. *Mater. Sci. Eng. B* **2005**, *122*, 140–144.
14. Rout, S. K.; Bera, J. *Ferroelectrics and Dielectrics*; Allied Publishers Pvt. Ltd.: New Delhi, 2004; pp 3–7.
15. Thakur, O. P.; Prakash, C.; Agarwal, D. K. *Mater. Sci. Eng. B* **2002**, *96*, 221–225.
16. Paul, S.; Kumar, D.; Manokamna, G. *J. Biosphere* **2013**, *2* (1), 55–58.
17. Zhoul, L.; Vilarinho, P.; Baptista, J. L. *J. Eur. Ceram. Soc.* **1999**, *19*, 2015–2020.
18. Joshi, P. C.; Cole, M. W. *Appl. Phys. Lett.* **2000**, *77*, 289–291.
19. Hennings, D. *Br. Ceram. Proc.* **1989**, *41*, 1–10.
20. Budd, K. D.; Payne, D. A. In *Better Ceramics Through Chemistry*; Brinker, C. J., Clark, D. E., Ulrich, D. R., Eds.; *Materials Research Society Symposia Proc.*, 1984, *Vol. 32*, Elsevier Science: North-Holland, pp 239–244.
21. Boulos, M.; Guillement-Fritsch, S.; Mathieu, F.; Durand, B.; Lebey, T.; Bley, V. *Solid State Ionics* **2005**, *176*, 1301–1309.
22. Tan, C. *Thin Solid Films* **2008**, *516*, 5545–5550.
23. Guo, L.; Luo, H.; Gao, J.; Guo, L.; Yang, J. *Mater. Lett.* **2006**, *60*, 3011–3014.
24. Mahata, M. K.; Kumar, K.; Rai, V. K. *Spectrochim. Acta: Mol. Biomol. Spectrosc.* **2014**, *124*, 285–291.
25. Yang, W.; Chang, A.; Yang, B. *J. Mater. Synth. Process* **2002**, *10* (6), 303–309.
26. Wodecka-Dus, B.; Lisinska-Czekaj, A.; Orkisz, T.; Adamcyzk, M.; Osinka, K.; Kozielski, L.; Czekaj, D. *Mater. Sci.—Poland* **2007**, *25* (3), 791–799.
27. Wu, E.; Chen, K. C.; Mackenzie, J. D. In *Better Ceramics Through Chemistry*; Brinker, C. J., Clark, D. E., Ulrich, D. R., Eds.; *Materials Research Society Symposia Proc., Vol. 32*, Elsevier Science: North-Holland, 1984; pp 169–174.
28. Izyumskaya, N.; Alivov, Y. I.; Cho, S. J.; Morkoc, H.; Lee, H.; Kang, Y. S. *Crit. Rev. Solid State Mater. Sci.* **2007**, *32, 111–202*.
29. Kao, C. F.; Yang, W. D. *Appl. Organomet. Chem.* **1999**, *13*, 383–397.
30. Atkins, P.; Overton, T.; Rourke, J.; Weller, M.; Armstrong, F. *Shriver & Atkins Inorganic Chemistry*, 4th ed.; Oxford University Press: Oxford, 2006; pp 103–106.
31. Kao, K. C. *Dielectric Phenomena in Solids*, 1st ed.; Elsevier: North-Holland, 2004.
32. Moulson, A. J.; Herbert, J. M. *Electroceramics; 2nd ed.* John Wiley & Sons: England, 2003; pp 5–94.
33. Chilibon, I.; Marat-Mendes, J. N. *J. Sol-Gel Sci. Technol.* **2012**, *64*, 571–611.
34. Hussain, S. T. *Tr. J. Chem.* **1996**, *20*, 33–37.
35. Cowley, J. M.; Moodie, A. F. *Acta Crystallogr.* **1957**, *199* (3), 609–619.
36. Froh, J. *Hyperfine Interact.* **2004**, *154*, 159–176.
37. Guo, L.; Luo, H.; Gao, J.; Guo, L.; Yang, J. *Mater Lett.* **2006**, *60*, 3011.
38. Sakabe, Y.; Wada, B.; Hamaji, Y. *J. Korean Phys. Soc.* **1998**, *32*, 260–264.
39. Valot, C. M.; Floquet, N.; Perriat, P.; Mesnier, M.; Niepce, J. C. *Ferroelectrics* **1995**, *172*, 235–241.
40. Sharma, N. C.; McCartney, E. R. *J. Aus. Ceram. Soc.* **1974**, *10*, 16–20.
41. Arlt, G.; Hennings, D.; With, G. *J. Appl. Phys.* **1985**, *58*, 1619–1625.
42. Shaikh, A. S.; Vest, R. W.; Vest, G. M. *IEEE Trans. Ultra. Ferro. Freq. Contrl.* **1989**, *36*, 407–412.
43. Chattopadhyay, S.; Ayyub, P.; Palkar, V. R.; Multani, M. *Phys. Rev. B* **1995**, *52*, 3177–13183.
44. Kim, H. T.; Han, Y. T. H. *Ceram. Int.* **2004**, *30*, 1719–1723.

45. Lokare, S. A. *Int. J. Chem. Phys. Sci.* **2015**, *4*, 155–161.
46. Pajak, Z.; Stankowski, J. *Proc. Phys. Soc.* **1958**, *72*, 1144–1146.
47. Patankar, K. K.; Kadam, S. L.; Mathe, V. L.; Kanamadi, C. M.; Kothawale, V. P.; Chougule, B. K. *Br. Ceram. Trans.* **2003**, *102*, 19–22.
48. Patil, D. R.; Lokare, S. A.; Devan, R. S.; Chougule, S. S.; Kanamadi, C. M.; Kolekar, Y. D.; Chougule, B. K. *Mater. Chem. Phys.* **2007**, *104*, 254–257.
49. Megaw, H. D. *Ferroelectricity Crystals*; Methuen & Co. Ltd.: London, 1957; p 85.
50. Hsiang, H. I.; Lin, K. Y.; Yen, F. S.; Hwang, C. Y. *J. Mater. Sci.* **2001**, *36*, 3809–3815.
51. Wang, X.; Chen, R.; Zhou, H.; Li, L.; Gui, Z. *Ceram. Int.* **2004**, *30*, 1895–1898.
52. Luan, W.; Gao, L.; Guo, J. *Ceram. Int.* **1999**, *25*, 727–729.
53. Akcay, G.; Misirlioglu, I. B.; Alpay, S. P. *J. Appl. Phys.* **2007**, *101*, 104110–104117.
54. Andrich, E. *Electron. Appl.* **1965**, *25*, 123–144.
55. Remmel, T.; Gregory, R.; Baumert, B. *JCPDS-Int. C. Diff.* **1999**, *41*, 38–45.
56. Singh, S. B.; Sharma, H. B.; Sarma, H. N. K.; Phanjoubam, S. *Phys. B* **2008**, *403*, 2678–2683.
57. Li, J.; Jin, D.; Zhou, L.; Cheng, J. *Mater. Lett.* **2012**, *76*, 100–102.
58. Mishra, A.; Mishra, N. *Int. J. Mater. Sci. Appl.* **2012**, *1* (1), 14–21.
59. Kundu, T. K.; Jana, A.; Barik, P. *Bull. Mater. Sci.* **2008**, *31*, 501–505.
60. Wang, X.; Gu, M.; Yang, B.; Zhu, S.; Cao, W. *Microelect. Eng.* **2003**, *66*, 855–859.
61. Li, Y.; Yao, X.; Wang, X.; Zhang, L. *Ferroelectrics* **2009**, *384*, 73–78.
62. Rani, A.; Kolte, J.; Gopalan, P. *Ceram. Inter.* **2015**, *41*, 14057–14063.
63. Bobic, J. D.; Ramoska, T.; Banys, J.; Stojanovic, B. D. *Mater. Char.* **2011**, *62*, 1000–1006.
64. Mancic, D.; Paunovic, V.; Vijatovic, M.; Stojanovic, B.; Zivkovic, L. *Sci. Sinter.* **2008**, *40*, 283–294.
65. Li, Y.; Yao, X.; Yao, Y.; Zhang, L.; Shen, B. *Ferroelectrics* **2007**, *356*, 102–107.
66. Rao, M. V. S.; Ramesh, K. V.; Ramesh, M. N. V.; Rao, B. S. *Adv. Mater. Phys. Chem.* **2013**, *3*, 77–82.
67. Singh, J. P.; Kumar, H.; Chandra, J. *Orient. J. Phys.* **2011**, *3* (1), 81–85.
68. Ouedraogo, A.; Palm, K.; Chanussot, G. *J. Sci. Res.* **2009**, *1* (2), 192–199.
69. Mitsui, T.; Westphal, W. B. *Phys. Rev.* **1961**, *124*, 1354–1359.
70. Berlincourt, D. A.; Kulesar, F. *J. Acoust. Soc. Am.* **1959**, *24*, 709–713.
71. Zhuang, Z. Q.; Harmer, M. P.; Smyth, D. M.; Newnham, R. E. *Mater. Res. Bull.* **1981**, *22*, 1329–1335.
72. Kingon, A. I.; Maria, J. P.; Streiffer, S. K. *Nature* **2000**, *406*, 1032–1038.
73. Cheng, J. C.; Tang, J.; Meng, X. J.; Guo, S. L.; Chu, J. H.; Wang, A. M.; Wang, H.; Wang, Z. *J. Am. Ceram. Soc.* **2001**, *84*, 887–889.
74. Mitoseriu, L.; Stoleriu, L.; Viviani, M.; Piazza, D.; Buscaglia, M. T.; Calderone, R.; Buscaglia, V.; Stancu, A.; Nanni, P.; Galassi, C. *J. Eur. Ceram. Soc.* **2006**, *26*, 2915–2921.
75. Balachandran, R.; Ong, B. H.; Wong, H. Y.; Tan, K. B.; Rasat, M. M. *Int. J. Electrochem. Sci.* **2012**, *7*, 11895–11903.
76. Zhang, H.; Zhang, L.; Yao, X. *J. Electroceram.* **2008**, *21*, 503–507.
77. Koops, C. G. *Phys. Rev. B* **1951**, *83*, 121–124.
78. Mahania, R. M.; Battishab, I. K.; Aly, M.; Abou-Hamad, A. B. *J. Alloys Compd.* **2010**, *508*, 354–358.
79. Jona, F.; Shirane, G. *Ferroelectric Crystals*; Dover Publications: New York, 1993.
80. Fleury, P. A.; Scott, J. F.; Worlock, J. M. *Phys. Rev. Lett.* **1968**, *21*, 16–19.
81. Shirane, G.; Yamada, Y. *Phys. Rev.* **1969**, *177*, 858–869.

82. Acikel, B.; Taylor, T. R.; Hansen, P. J.; Speck, J. S.; York, R. A. *IEEE Microw. Wireless Compon. Lett.* **2002**, *12*, 237–239.
83. Jain, M.; Majumder, S. B.; Katiyar, R. S.; Bhalla, A. S.; Agrawal, D. C.; Van Keuls, F. W.; Miranda, F. A.; Romanofsky, R. R.; Mueller, C. H. *Mater. Res. Soc. Symp. Proc.* **2003**, *748*, 483–488.
84. Park, B. H.; Quanxi, J. *Jpn. J. Appl. Phys.* **2002**, *41* (11b), 7222–7225.
85. Maria, J. P.; Ayguavives, F. T.; Kingon, A. I.; Tombak, A.; Mortazawi, A.; Stauf, G.; Ragaglia, C.; Roeder, J.; Brand, M. In *Material Research Society Meeting*, 2000.
86. Bellotti, J.; Akdogan, E. K.; Safari, A.; Chang, W.; Pond, J. *Ferroelectrics* **2002**, *271*, 131–136.
87. Wei, X.; Xu, G.; Ren, Z.; Wang, Y.; Shen, G.; Han, G. *Mater. Lett.* **2008**, *62*, 3666–3669.
88. Mudinepalli, V. R.; Feng, L.; Lin, W. C.; Murty, B. S. *J. Adv. Ceram.* **2015**, *4* (1), 46–53.
89. Shaw, T. M.; Trolier-McKinstry, S.; McIntrye, P. C. *Annu. Rev. Mater. Sci.* **2000**, *30*, 263–298.
90. Leu, C. C.; Chen, C. Y.; Chien, C. H. *Appl. Phys. Lett.* **2003**, *82*, 3493–3495.
91. Mao, C.; Dong, X.; Zeng, T.; Chen, H.; Cao, F. *Ceram. Int.* **2008**, *34*, 45–49.
92. Kavian, R.; Saidi, A. *J. Alloys Compd.* **2009**, *468*, 528–532.
93. Li, W.; Xu, Z.; Chu, R.; Fu, P.; Hao, J. *J. Alloys Compd.* **2009**, *482*, 137–140.
94. Kong, L. B.; Zhang, T. S.; Ma, J.; Boey, F. *Prog. Mater. Sci.* **2008**, *53* (2), 207–322.
95. Liu, W. L.; Xue, D.; Kang, H, Liu, C. *J. Alloys Compd.* **2007**, *440*, 78–83.
96. Fournaud, B.; Rossignol, S.; Tatibou, J. M.; Thollonb, S. *J. Mater. Process. Technol.* **2009**, *209*, 2515–2521.
97. Sundaresan, A.; Rao, C. N. R. *Nano Today* **2009**, *4*, 96–106.
98. Hsiang, H. I.; His, C. S.; Huang, C. C.; Fu, S. L. *Mater. Chem. Phys.* **2009**, *113*, 658–663.
99. Yu, P.; Cui, B.; Shi, Q. *Mater. Sci. Eng.* **2008**, *473*, 34–41.
100. Kavian, R.; Saidi, A. *J. Alloys Compd.* **2009**, *468*, 528–532.
101. Ong, B. H.; Wong, H. Y.; Tan, K. B.; Rasat, M. M. *Int. J. Electrochem. Sci.* **2012**, *7*, 11895–11903.

CHAPTER 4

PREPARATION AND CHARACTERIZATION OF COPPER THIN FILMS FOR ANTIMICROBIAL APPLICATIONS

UDAYA BHAT K.[1*], ARUN AUGUSTIN[1], SUMA BHAT[2], and UDUPA K. R.[1]

[1]*Department of Metallurgical and Materials Engineering NITK Surathkal, Srinivasanagar, 575025, India*

[2]*Department of Mechanical Engineering, Srinivasa School of Engineering, Mukka, Srinivasanagar, 575025, India*

Corresponding author. E-mail: udayabhatk@gmail.com

ABSTRACT

Nowadays, hospital-acquired infections is the major problem in the health sector. These infections are mainly spreading through touch surfaces in the hospitals. It is well proved that copper has an inherent property to kill the microbes while they are in contact with the microbes. Hence, the hospital touch surfaces made of copper is a feasible solution for the hospital-acquired infections. The current chapter is discusses on the preparation and characterization of the copper coatings for antimicrobial applications. The copper-coated touch surface has the advantages such as tunability of surface morphology, easiness to coat on any complex shapes, high purity, usage of less material, etc., as compared to the touch surface made by using bulk copper. This chapter deals with the mechanism of killing microbes by copper as well as the different methods of copper coating techniques, such as thermal evaporation, E-beam evaporation, sputter deposition, chemical vapor deposition, electroless deposition, and electrodeposition. The characterization techniques for thickness measurements, mechanical properties measurements, residual stress measurements, microstructural studies, and antimicrobial properties are discussed in the present chapter.

4.1 INTRODUCTION

Every day, people make physical contacts with a variety of surfaces, during their walk of life. People make contact with doors, handles, knobs, rails, and furniture in public utilities (recreation, transportation utilities, etc.) as a part of their everyday lives. Unfortunately, knowingly or unknowingly, touch of some people leaves behind dangerous microbes (microorganisms) on their surfaces, putting the next user to a higher risk level. Some of the items listed above are found to serve as reservoirs for the spread of many microbes. Some of these microbes can occupy inanimate spots in the surfaces and can survive for periods ranging from a few weeks to a few months.[1] Microbes contaminate surfaces through large volume surface soaking spills or microdroplet aspirations (like sneezes) and are subsequently transferred to others when they happen to be in touch with the contaminated surfaces (called touch events).[2] Some of these microbes can cause damage in a host. Some of them may even cause a disease in the host.[3] A recent survey indicated that the patients admitted to rooms previously occupied by patients infected or colonized with methicillin-resistant *Staphylococcus aureus* (MRSA) or vancomycin-resistant *Enterococcus* have an increased risk of acquiring the same pathogen as that of prior room occupants.[4] Because of these difficulties, emphasis is on the importance of cleaning and disinfection of the surfaces which are being in touch by people at large. Incorporation of features that will reduce or inhibit the growth of microbes with the current design methodologies offers a new ray of hope in the design that will lead to better outcomes, simultaneously reducing the overall cost.[5] Design and development of antimicrobial surfaces is an effort in that direction.

An antimicrobial surface has the ability to inhibit or reduce the growth of microbes on it.[6] Such surfaces are becoming increasingly relevant in multiple sectors, namely, clinics, food and pharmaceutical industry, tourism (airports), and even in the domestic sector. Such surfaces play an important role in improving hygienic conditions and sterile procedures in hospitals, kitchens, sanitary facilities, air conditioning and ventilation, transportation, food preparation and packaging systems, etc. In these applications, microbes like virus, bacteria, and fungi compromise the health of both consumers and patients.[7]

4.2 APPROACHES TO REDUCE GROWTH OF MICROBES

Approaches to reduce growth of microorganisms can be broadly classified as conventional and novel ones. Conventional approaches are basically intervention type. One approach is use of disinfectants. They are widely used across the globe. This approach involves removing the microbes by thorough and regular cleaning followed by the use of disinfectants. A common problem with the use of disinfectants is the lack of residual effect, which means recontamination occurs rapidly. Due to this, disinfectants need to be applied periodically. Use of detergents, organosilanes, and light-activated photosensitizers like TiO_2 belongs to this group.

This method is convenient and cheap. But many questions decide its efficiency in real life. A few of them are optimal mode of deployment, frequency of deployment, the effect of dust and other contaminants on the performance of the disinfectant, the skill of the applier, etc. Because of the requirement of routine cleaning of items and surfaces and difficulty associated with the complete removal of the microbes from contaminated surfaces, improved methods of disinfecting the surfaces have been a continuous focus. Making a surface less able to support contamination and microbial growth is also a strong route for increasing the efficiency.[8] There are other methods which are being tried. Some of them are briefed below.

Negative air ionization: Here, air ionizers are used to control the spread of infections in hospitals. It is suggested that ions in the air play an important role in preventing the transmission of infection due to some of the microbes. It is also hypothesized that the ions in the air charge medical equipments negatively and they repel airborne infectious microbes.[9] Another approach is use of nanocomposite formulations in the form of a film. These composites are loaded with the cell wall degrading enzymes or antimicrobial enzymes. A particular report says that use of carbon nanotube–lysostaphin combination as an antimicrobial film can kill 99% of MRSA within 2 h. It is also said that these films are reusable and fairly stable. When applied, these enzymes will be released slowly and they help in minimizing the growth.[10,11]

4.3 ANTIMICROBIAL METALS

Though it is realized that, in many cases, "high touch surfaces" are potential source for infections, surprisingly, only few technological improvement methods have been implemented to address the problem of contaminated surfaces.[12] Use of metals (like copper and silver) has been found to be effective to prevent surface contamination due to contacts. Historically, metals like mercury, arsenic, and antimony were also used as antimicrobial metals.[13]

Metals like copper and silver could be extremely toxic to some of the microbes even at exceptionally low concentrations. Because of their biocidal nature (biocidal means any chemical that can destroy life by poisoning), metals have been widely used as antimicrobial agents in a multitude of applications, related with food storage, healthcare, and other industries.[14] Unlike other antimicrobial agents, use of metals as antimicrobial agents has a couple of advantages.

1. Metals are stable under conditions currently found in the industry allowing their use as additives. They can be added in different forms, such as particles, ions absorbed or exchanged in a carrier, salts, hybrid structures, etc.
2. Metals can be encapsulated as nanoparticles in a variety of polymer matrices. These metal nano/polymer composites can be easily prepared and handled.[14]

Copper is cheap compared to silver. Use of copper as an antimicrobial agent has a long history. Oldest record on the use of antimicrobial ability of the copper is seen in Egyptian medical texts. It is reported that copper was used to sterilize water and healing wounds during as early as 2200–2600 BC.[15] Similarly, vessels made of Cu and Ag were used for water disinfection and food preservation by the Persian Kings.[16] The inherent antimicrobial properties of solid copper are well documented by US Environmental Protection Agency (EPA). The products made from copper are listed with EPA and the manufacturers are allowed to make public claim, "This surface continuously kills >99.9% MRSA within two hours." In 2008, 350 items were included in the list of antimicrobial materials. Minimum copper content in the antimicrobial material is 60%.[5] Also, copper has intrinsic ability to destroy a wide range of organisms.[17–22] A few of them are *Escherichia coli* O157:H7, MRSA, *Staphylococcus*, *Clostridium difficile*, *Salmonella enterica*, *Campylobacter jejuni*, influenza A virus, adenovirus, and fungi. Apart from the health industry, antimicrobial surfaces have been utilized for their ability to keep the surface cleaned. Either physical nature of the surface or the chemical "make up" is manipulated to create an environment which cannot be tolerated by microbes for a variety of reasons. Similarly, photocatalytic materials have been used for their ability to kill microorganisms, air and water cleaning, water purification, etc.[23]

The processing and manipulation of surfaces for antimicrobial applications is an issue with many questions. However, with the recent developments in chemistry and materials science, the subject is being extended to metal surfaces, coating complexes, and nanomaterials.[16] One feasible route is to develop alternative strategies such as antibacterial coatings that would make surfaces less accommodating to microbes. These surfaces are not like the antibacterial fluids that just wash away—here, the goal is to make a surface which is intrinsically deadly to harmful microbes.[24]

A common threat with antibiotic resistant bacteria is that they spread easily and rapidly through patient–patient, patient–staff, or patient–surface contacts.[25] Therefore, it is important to minimize the bacterial contamination in the health-care sector to minimize the spread of the infection. This forces the need for biocidal (any chemical that can destroy life by poisoning) surfaces to help reduce touch-acquired contamination. Copper has been identified as an effective antimicrobial surface against a broad spectrum of microbes. Also, minimum copper content in a copper alloy is about 60% copper. This gives a scope for balancing multiple properties (like improving wear resistance and strength, with sufficient antimicrobial behavior) in the copper-based material.

4.3.1 HOW DOES COPPER KILL MICROBES?

In 2008, the US EPA approved 282 copper alloys to be registered as antimicrobial agents. Copper and copper alloys are the first solid material to acquire this status. The laboratory tests have demonstrated that 99.9% of bacteria on copper and copper alloy surfaces, those with greater than 65% copper content, were killed within 2–3 h. The bactericidal action is effective against MRSA, influenza H1N1, *E. coli* 0157, and a wide range of other dangerous pathogens, including viruses. It should be stated that all the alloys listed by EPA are effective killers of germs even at room temperature and under typical indoor humidity conditions.

Copper can destroy undesirable viruses and bacteria in the water and food items. For the modern science, benefits and undesirable aspects of microbes are realized recently. In contrast to that Ayurveda, Indian medicinal system had realized the importance of copper centuries back itself. Also, ancient Egyptians had used copper extensively to keep water and food items free from microbial contamination.[26]

Though documented history on the use (some items are shown in Fig. 4.1) of copper and its alloys as antimicrobial surfaces is long, understanding the mechanism of effect of copper or (its ions) to bacteria is partial. But it is well accepted that metal atoms (or ions) have effect on the cellular (bacteria) components and hence biomolecules. Consequence of this is metal toxicity toward bacteria and other microbes.[13] Metals can exert toxic effects toward microbes by different ways, that is, by binding to or blocking functional groups in biological molecules, by displacing essential metals in enzymes, and by participating in chemical reactions that are harmful. The results of these toxic effects are damage to protein DNA and biological membranes, interference in enzyme functions and cellular processes, and oxidative stress. It may be noted that among the first row transition metals, Cu^{2+} has maximum affinity for biological molecules, and hence, it can displace other atoms (ions) from the biological molecules.[27] Another way of understanding the metal toxicity toward microbes is by considering the strength of metal ions as Lewis acids. As per this, hard Lewis acids (one with nonpolarizable electron shells) prefer ionic bonding with oxygen containing ligands. On the other hand, soft Lewis acids (one with polarizable electron shells) bond covalently with sulfur and nitrogen ligands, cysteine thiols, and nitrogen imidazoles. Metals which are toxic toward microbes are basically soft Lewis acids and they are likely to displace intermediate and hard Lewis acids from cysteine thiols due to their higher affinity for them.[13]

FIGURE 4.1 Some of the items made from copper, used as domestic tools.

Another proposed mechanism for the metal toxicity is based on the oxidative stress. Metals like Cu, Cr, and Fe are called Redox active metals. They imbibe damage by cellular oxidative stress damage. Cellular oxidative stress damage means cell damage caused by the abundance of oxidants like reactive oxygen species (ROS). Examples for such (ROS) species are oxygen ions, free radicals, and peroxide groups. For living activity, certain amount of ROS is essential. They are highly active and they carry out oxidation reactions with many organic molecules at the site of their formation. Excessive level of ROS is bad. Redox active metals generate hydrogen peroxide (H_2O_2), hydroxyl radicals ($OH^\bullet$), and superoxide (O_2^-) via Fenton-like reactions.[28] This has been accepted as the major mechanism for copper toxicity on microbes.

Some of the Fenton-like reactions[29] are written below:

$$Cu^{2+} + H_2O_2 \leftrightarrow Cu\cdots OOH^+ + H^+$$

$$Cu\cdots OOH^+ \rightarrow HO_2^\bullet + Cu^+$$

$$Cu^+ + H_2O_2 \rightarrow OH^\bullet + OH^- + Cu^{2+}$$

$$OH^\bullet + RH \rightarrow R^\bullet + H_2O$$

At the same time, recent experimental evidence showed that copper-mediated ROS generation occurs largely in the periplasm of *E. coli*. This observation has forced the ROS generation mechanism to some level of debate.[30] Now, it is being thought that the toxicity toward microbes by copper surface is due to the damage of cells caused by the very high local concentration of Cu ions generated due to the dissolution from the surface which also leads to rupture of cell membrane. It promotes further ROS generation and further cell destruction and degradation of plasmid and DNA chromosomes.[13,28]

Antimicrobial metals like copper have different and multiple cellular targets in bacteria. For bacteria, only limited options are available to nullify the effects of metal toxicity. For bacteria, most of the resistance to metals is linked to detoxification or efflux of toxic metals from the cells. This is in contrast to resistance of bacteria to organic antimicrobial compounds. Organic antimicrobial compounds are inactivated by compound breakage or inactivation by enzyme cleavage, whereas metals are immutable and bacterial immunity systems do not sufficiently discriminate to allow entry of only selected metal ions. Simultaneously, metal ion chaperones (chaperones are molecular proteins that help in folding and unfolding of molecular structures) may also be subverted to bind the toxic metals (elements like Cd and Pb) and compounds.[13,31,32]

From an atomistic approach, different antimicrobial mechanisms have been hypothesized for the copper toxicity on microbes. Some of them are as follows:

1. Copper ions can accept or donate electrons which help the ions to participate in the chemical reactions that lead to the oxidation damage to the microbes.[33]
2. Copper ions can cause cell membrane lesions that can cause the leakage of intracellular component like potassium and glutamate which causes cell apoptosis.[34]

From antimicrobial point, dry copper surfaces behave differently compared to wet copper surfaces.[35–38] Toxic effect of wet copper (generate cuprous ions) is classified as chronic (with reference to *E coli*); the effect of dry copper is called acute.[35] In wet surfaces, copper toxicity is due to the accumulation of cuprous ion at intracellular region. In microbes like *E. coli*, this accumulation causes the inactivation of hydratazes. It may be noted that the hydratazes are essential for normal cell function. Also, it is reported that cuprous ions cause the damage to Fe–S clusters in the proteins. Due to this growth, defects are developed in cells attacked by copper ions.[39] In contrast, when *E. coli* is exposed to dry copper surfaces, the situation is different. Microbes on dry surface copper surfaces are not in an environment that promotes growth. These cells face challenges that are different from those of chronically challenged cells by copper ions. When *E. coli* is exposed to wet copper surfaces, *E. coli* do not spread easily and Fe–S units in the proteins are unlikely targets for toxic effect. Santo et al.[35] observed that presence of *E. coli* increases the copper-releasing amount from the dry copper surfaces. On dry Cu surfaces, copper accumulation is rapid and extensive. Also, it is observed that the membrane integrity is lost when exposed to dry copper. They have concluded cell envelope damage as the mode of killing by dry copper surfaces. It must be reported that another possibility for antimicrobial behavior appears due to genotoxicity leading to DNA damage. Some of the damage modes due to the presence of copper ions are schematically presented in Figure 4.2.

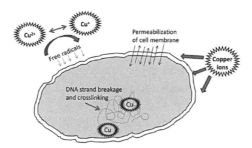

FIGURE 4.2 Some of the damage modes shown by the microbes due to the presence of copper ions near them.

4.3.2 THIN FILM COPPER COATINGS

As world population increases, the demand for copper continues to increase. In such a scenario, to serve the society's needs, use of copper in bulk form is becoming highly expensive. This also puts a limit on the consumption of copper in the health-related and other sectors where microbial contamination is a serious issue.

Copper thin films versus bulk copper: Most engineering metals (normally referred to as bulk materials) have fixed properties and when the material dimensions are reduced below a certain level, specific surface area (surface area per unit mass) changes rapidly. Behavior of these metals is entirely different compared to that exhibited by the bulk. This makes them versatile as far applications are considered. It also opens up new opportunities to use them in applications where results are visible and rapid. When the metal is used as a thin film, the quantum of the material required is very less and it reduces cost also.

The advances in coating technologies have contributed to enhance the surfaces of the tools, devices, amenities, etc. in the health and domestic fields toward more of antimicrobial or clean in nature. Though there could be competition from other techniques, thin film deposition of copper on health-care items like door handles, knobs, sanitary items, etc., is getting more and more importance. This is also true in other sectors like domestic, food processing, and transportation. Thin film copper deposition is a powerful tool to enhance the antimicrobial nature of the utilities made from stainless steel or aluminum. Copper coating increases level of reliability of human safety. They do not rely on human behavior. They serve as an extra layer of protection, constantly working in the background to fight bacteria.

4.3.3 BENEFITS OF THIN FILM COPPER COATINGS FOR ANTIMICROBIAL APPLICATIONS

Thin copper films behave differently compared to bulk materials of same chemical composition. Thin film materials have a higher surface area to volume ratio and their thickness is generally less than 1 µm. The increased surface area helps to increase the 3D contact with the microbes. Depending on the nature of the processing, the surface will have hills and valleys. They increase the possibility of microbe killing by copper ions. This is schematically presented in Figure 4.3.

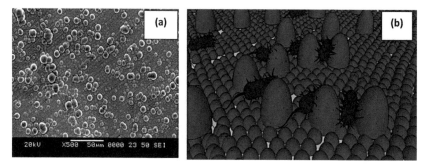

FIGURE 4.3 Topographical features of a copper coated surface: (a) a secondary electron micrograph showing topography and (b) schematic sketch presenting increased antimicrobial activity.

Another aspect which makes thin films to behave differently is that they have smaller crystallites. Grain size of a polycrystalline thin film has an important role in determining functional as well as structural properties. As the grain size reduces, the grain boundary volume increases. Grain boundaries contain a lot of loosely bounded copper atoms. These atoms could be easily released from copper surface and attached to the cell wall of the microbes. In this way, nanosized grains help to increase the antimicrobial action. Moreover, the nanograins in the thin films help to improve the mechanical properties like microhardness and scratch resistance. According to the literature,[40,41] as the grain size reduces to the range of 10–20 nm, copper exhibits Hall–Petch effect and below that reverse Hall–Petch effect is observed. The bottom-up approaches to fabrication of copper thin film coating (electroplating, sputtering, etc.) allows the easy regulation of process parameters and hence grain size.[42,43] Figure 4.4 shows the optimum grain size obtained from electrodeposition and DC magnetron sputtering, respectively.

They also have a large number of defects like vacancies, dislocations, stacking faults, twins, and grain boundaries. The fraction of the defects exposed to environment is also high. Most of the thin film deposition conditions are nonequilibrium in nature,[44] generating microstrain in the crystallite. This is schematically shown in Figure 4.5. Reduced crystallite size, high specific surface area, microstrains, and a large number of defects increase surface energy of the coating. They affect the copper ion release rate and hence antimicrobial nature of the coating. Because of these features, their properties are expected to be different than bulk material properties, and many a time, thin films are superior than the bulk.[45] If the coating has increased surface energy, the solutions (possibly containing microbes) spread on the surface more faster and rapid biocidal action is possible.

So for the antimicrobial touch surfaces like copper, hydrophilic surface is better.[46,47] The touch surfaces, like wash tabs, rails exposed to environments, and utensils, are some examples were liquid media of pathogen will be in contact with touch surfaces. In such cases, antimicrobial copper thin film with higher surface energy can help the rapid antimicrobial action.

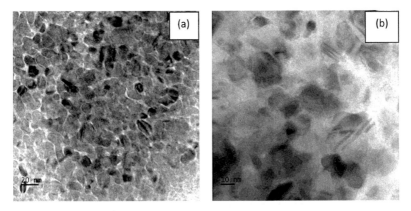

FIGURE 4.4 Typical microstructures in thin film copper coating obtained under optimized conditions: (a) electroplating and (b) sputtering.

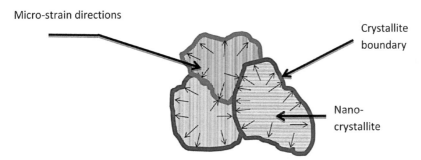

FIGURE 4.5 Schematic sketch indicating microstrain at the boundaries of small crystallites.

Another parameter which differentiates thin films from bulk copper is the possibility of developing texture in the coating. Though texturing is possible in bulk materials also, it involves a sequence of thermomechanical treatments.[48] These steps are highly energy consuming and hence costly. In the case of coatings, this is possible by controlling the coating parameters and hence less expensive. In the case of copper coatings, texture has an important role in the mechanical properties and antimicrobial activities. If the copper coating has a preferred growth along (111) direction, the mechanical properties are expected to be higher because of the close packed structure. The (111) plane has higher Young's modulus of 172 GPa compared to other planes in copper.[49] Increased mechanical properties help to increase the durability of the coating in adverse circumstances like scratch and wear. In the touch surface, applications of such coating are expected to have longer life.

4.4 METHODS OF GETTING COPPER THIN FILM COATINGS

In general, getting copper thin films depends on four aspects. They are (1) use of correct copper source, (2) transport of copper species, (3) condensation of copper species on the substrate, and (4) correct type of the substrate. The properties of the final coating depend on all four aspects. The extent may be different based on the application in view. Nature of the coating is strongly affected by the copper source (purity, shape, etc.) and the substrate used (development of residual stress, texture, etc.). Copper source can be pure copper (as in sputtering) or copper salt (as in chemical vapor deposition, CVD) or salt in an aqueous medium (as in electrodeposition). Substrate depends on the user requirements, and for the topic under consideration in this chapter, the substrate would be a structural metallic material like Al and its alloys, stainless steels, etc. Transportation of copper species and condensation of copper species may be by physical means (evaporation and condensation, in physical vapor deposition, PVD) or through the help of a current (in electrodeposition) or through a magnetic field (magnetron sputtering). Final coating also depends on other parameters like deposition temperature, deposition environment, rate of deposition, distance between the source to the substrate, etc. In following pages, some of these techniques are briefed with a focus on generating copper thin films on structural materials like stainless steels and aluminum alloys.

4.4.1 THERMAL EVAPORATION AND CONDENSATION

This is a PVD method where copper atoms are vaporized from solid copper and the vaporized copper atoms are made to deposit over the substrate without colliding with the residual gas molecules (Fig. 4.6a). In this method, a large current is passed through a filament. It heats up the copper kept in a crucible made from a refractory material. The copper evaporates and the vapor flux is directed outward and condenses on the sample kept in the line of sight.[50] All these components are enclosed in a vacuum chamber. Vacuum level decides the contamination and the vaporizing temperature. For example, if pressure is 10^{-8} Torr, the vaporizing temperature would be 727°C. If pressure is reduced to 10^{-6} Torr, the temperature needs to be raised to 857°C.[50] Though the process is simple and inexpensive, it suffers from a couple of serious limitations. First, it is a line-of-sight process and hence coating thickness would be maximum at the center line of symmetry and drops rapidly as one would go away (Fig. 4.6b). This could be partially overcome by opting for modified geometry for copper source, promoting collisions between the vaporized copper atoms by employing a higher vaporization rate, use of proper shutter plates, etc. Second, handling the evaporating sample (Cu) in a refractory crucible is difficult due to differences in the values of coefficient of thermal expansion.[51] Since it is a line of process, when the substrate is of complex shape, there will be geometrical shadowing and some portion of the object would be improperly covered.[52] Also, deposition of multielements (copper alloys) is difficult because different elements have different tendency for evaporation.[53]

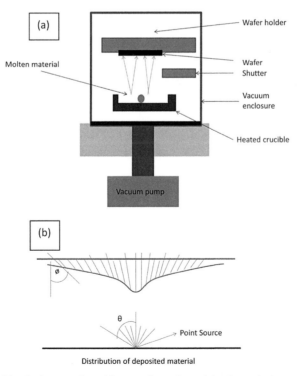

FIGURE 4.6 Physical vapor deposition configurations: (a) schematic instrumentation and (b) sketch showing distribution of vapor flux with respect to center.

4.4.2 E-BEAM EVAPORATION AND DEPOSITION

This is another type of PVD. A schematic diagram of this process is given in Figure 4.7. In e-beam evaporation, e-beam is generated in a gun which works on the principle of thermionic emission. The e-beam is accelerated, transferred, and focused on to a copper block (copper source). Copper block is water cooled to dissipate excessive heat. Due to the combination of high temperature and vacuum, the copper gets evaporated and gets condensed on to the target, which is maintained at low temperature compared to the copper source. Since the electron bean is focused, only a small region in the sample gets heated up and evaporates. There is no contamination from the crucible. The e-beam is magnetically deflected by an angle more than 180° to eliminate the contamination of the electron source due to copper deposition. The operating parameters have to be tuned to minimize the electrical arcing and discharge.[54]

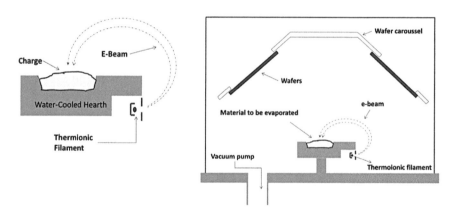

FIGURE 4.7 Schematic description of the e-beam evaporation and condensation process for the thin film deposition of copper.

Reported literature on e-beam evaporation and condensation for the copper deposition is relatively less, possibly due to the difficulties involved in melting and evaporating a high conducting metal like Cu. Recently, Komalakrishna et al.[55] successfully deposited copper thin film on aluminum using electron beam deposition technique, for the possible antimicrobial applications. Figure 4.8 shows a TEM micrograph of the copper film indicating fine crystallites. Crystallite size distribution is also presented. The deposit had fine crystallites without much defects like twins. The deposition rate employed was relatively small (about 6 nm/min). Under this condition, the copper atoms have sufficient freedom to condense and rearrange into low energy configurations. Fine crystallite size was small mainly due to large amount of nucleation sites on the substrate which became active due to high level of undercooling (temperature difference between condensing species and substrate). The copper ion release rate (in aqueous medium) measured is about 0.0015 ppm/h for an exposure of surface area =10 mm × 10 mm. This rate was observed to be constant, even up to 25 h of exposure.

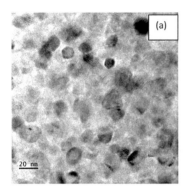
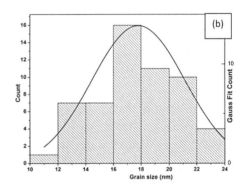

FIGURE 4.8 (a) Microstructure in copper thin film deposited using e-beam process and (b) grain size distribution.

4.4.3 SPUTTER DEPOSITION

Sputtering is a physical deposition process wherein the surface atoms are physically ejected from a target surface using momentum by bombarding with the energetic particles. Generally, bombarding species is argon ions (Ar^+). For developing thin copper films (or alloy films), pure copper (or copper alloy) is the target material (cathode). When copper surface is bombarded with energetic Ar^+ ions, they knock out neutral copper atoms, copper ions, clusters of atoms, electrons, etc. This phenomenon in total is called sputtering (Fig. 4.9). In sputter deposition, the sputtered atoms are redeposited on the anode surface (substrate).[56] Sputtering gas is converted into plasma using a high-energy source like DC voltage, radio frequency (RF) power, magnetron, etc. Gas ions in the plasma are accelerated and directed toward the copper target. Ions hit the target and eject both neutral atoms and free electrons. Some of the ions will recombine with the free electrons and will be reused in the plasma generation. Ejected atoms move straight and get deposited on the sample surface.

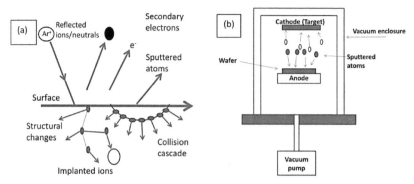

FIGURE 4.9 Sputtering for thin film deposition: (a) ejection of different species during sputtering and (b) sketch for sputter deposition unit.

There are different sputtering configurations for producing thin films.[57] In the case of DC sputtering, the plasma is created exclusively, by a DC discharge across cathode (copper target) and anode (sample to be coated). In the case of RF sputtering, plasma is created with the help of a RF source. It also uses a high voltage, but alternating at a high frequency. Finally, in the case of magnetron sputtering, strong electromagnets are placed behind the target. They generate a magnetic field that helps to trap electrons produced due to ionization. In reality, combinations like DC magnetron or RF magnetron are favored as they give a good deposition rate and minimum bombardment of the substrate by the electrons. They also reduce the extent of vacuum required. Large samples (even up to a few tens of inches) can be uniformly deposited.[58] Figure 4.10 shows morphology of the copper coating produced on aluminum using DC magnetron sputtering. Distribution of crystallite size is also given.

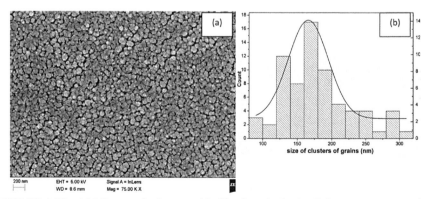

FIGURE 4.10 (a) Micrograph of copper thin film deposited using DC magnetron sputtering and (b) size distribution of clusters of grains.

4.4.4 CHEMICAL VAPOR DEPOSITION

In CVD, selected gaseous molecules are used as precursors. They are facilitated to undergo chemical reactions to produce a thin coating or powder layer on the sample surface (Fig. 4.11). In CVD, there are two distinct varieties. In first variety, reactions take place very close to the sample surface and the product phase (copper atom) is directly deposited on the sample surface. This category is called heterogeneous CVD and it is more common. In the second variety, reactions take place in gaseous phase and the product (which is in gaseous form) will get transported to the sample surface and gets deposited. This group is called homogeneous CVD.

Main steps in the CVD process could be listed as the following:

1. Transportation of reacting species from the bulk to the sample surface
2. Adsorption of the species on the surface
3. Heterogenous surface reactions on the sample surface
4. Diffusion of the product to the growth site
5. Nucleation and growth of the film
6. Desorption of the gaseous products and transport to far from the surface

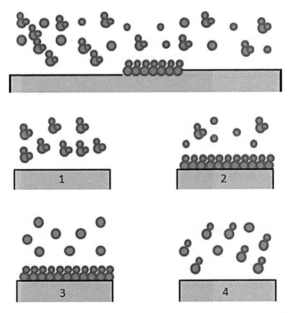

FIGURE 4.11 Schematic of CVD process. Images from top (anticlockwise) indicates successive increase in the thickness of deposited layer. Once one layer is deposited, it is indistinguishable from previous layers.

Generally, precursor gas is mixed with an inert carrier gas and it helps in controlling the reaction rate. Major parameters affecting the rate of deposition are temperature, presence of carrier gas, velocity of gas flow, distance along the direction of gas flow, and type of precursors. Common precursors for copper CVD are copper acetyl acetonate and copper hexaflouroacetyleacetonate (Cu hfac).[59,60] Deposition temperature is in the range of 200–350°C under net pressure of 1 atm.

Different types of CVD do exist. A few of them are plasma-enhanced CVD, atmospheric pressure CVD, low-pressure CVD, very low-pressure CVD, metal-organic CVD, etc.

As a group, CVD is a complex process for coating formation. CVD exhibits several advantages. Some of them are capability to produce highly pure and dense coatings, high deposition rate, etc. It is a nonline-of-sight process and hence complex-shaped components can be uniformly coated.

4.4.5 ELECTROLESS COPPER DEPOSITION

Electroless deposition is a nonelectrolytic method of deposition of metals from their solutions using a reducing agent. Plating solutions contain a metal salt and a suitable reducing agent. Actual reduction of a metal takes place only when a suitable surface is made available. Plating continues due to the catalytic nature of the already deposited metal. Hence, it is also called autocatalytic plating. Electroless deposition route is available for metals like Ni, Au, Cu, etc. Electroless deposition is slower compared to electrolytic deposition and it is much expensive due to expensive raw materials. But it has the following advantages[61]:

1. Electroless technique deposits metal uniformly over all surfaces, regardless of the shape and size.
2. Baths demonstrate 100% throwing power.
3. Metal can be deposited on conducting and nonconducting surfaces.
4. Substrates arranged in large racks can be coated simultaneously.

Common salt used for the electroless deposition of copper is cupric salt, such as copper sulfate, copper chloride, and copper nitrate. Formaldehyde is extensively used as reducing agent. Other reducing agents are dimethylamineborane, borohydride, hypophosphite, hydrazine, and sugar.

Deposition is generally done under alkaline conditions (pH > 13) and formaldehyde as reducing agent.[62] The reactions are as follows:

$Cu^{2+} + 2e^- = Cu$ $E=+0.34$ V

$HCOO^- + 2H_2 + 2e^- = HCHO + 3OH^-$ $E=-1.07$ V

Net reaction may be written as

$Cu^{2+} + 2HCHO + 4OH^- = Cu + 2HCOO^- + 2H_2O + H_2$

As deposition takes place, copper ions, formaldehyde, and alkali have to be replenished periodically.[63,64]

In the bath, complexing agents like tartrate salts, alkanol amines, ethylene diamine tetra-acetic acid, etc., are used. The baths will also have the additives like stabilizers, accelerators, etc.

Microstructure of the film developed during electroless deposition depends on three stages—namely, nucleation, growth, and coalescence of individual units to generate three dimensional films. In the nucleation stage, the copper atoms gets deposited on certain preferred locations and forms individual copper crystals of extremely fine size (nuclei).[63,65] Nuclei grow in size in all directions. Simultaneously, fresh nuclei also form. Average density of nuclei reaches a maximum. After that, favorably oriented units coalesce together and density of nuclei starts to drop. After a certain stage, coating is laterally filled and then growth is possible only along the thickness direction. Now onward, only favorably oriented grains grow. They will produce a deposition which is columnar in nature. The microstructure formation is schematically shown in Figure 4.12. For thin film coating, columnar stage may or may not be observed. For thick film coating, columnar stage will form a major fraction in the coating microstructure.[66,67] Also, within a grain, there could be multiple subgrains.

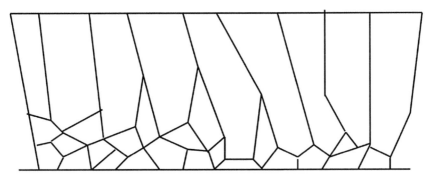

FIGURE 4.12 Schematic presentation of microstructure formation during atomistic deposition on a substrate.

4.4.6 COPPER THIN FILMS BY ELECTROPLATING

Copper is one of the commonly electroplated metal for a wide range of applications. Of the many plating systems that have been investigated, only a few have reached the stage of commercial applications. These are (1) alkaline cyanide and pyrophosphate complex ion systems and (2) acid sulfate and fluoborate simple ion systems. In recent times, some alkaline noncyanide systems have been developed for replacing cyanide.[43] It may be noted that cyanide solutions are getting discouraged because of their toxicity and waste treatment problems.

As a process, electrodeposition is deposition of a metal in the form of a coating by passing current through an electrochemical cell (Fig. 4.13). Current is sourced from an external power source. The component to be coated is cathode and copper bar is anode. Anode is a sacrificial one. Anode is getting consumed and getting deposited at the cathode according to the following reactions.

At cathode, $M^{n+} + ne^- = M$

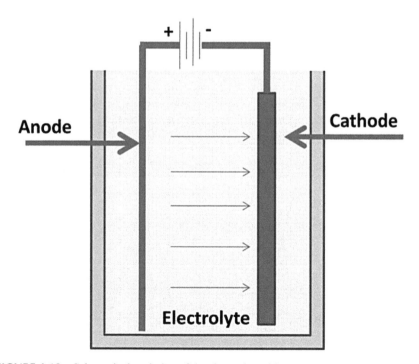

FIGURE 4.13 Schematic description of the electrodeposition setup.

At anode, M=M^{n+} + ne$^-$. At the anode, electrodissolution takes place. A copper salt will be used to make an electrolyte. Other additives are grain refiners, hardeners, brighteners, buffering agents, and leveling agents. By varying the parameters like pH, deposition time, current density, bath composition, etc., the properties of the coating, namely, composition, morphology, crystal size, crystal orientation, residual stress, etc., may be controlled. Figure 4.14 shows a cross-sectional micrograph of the copper film deposited.

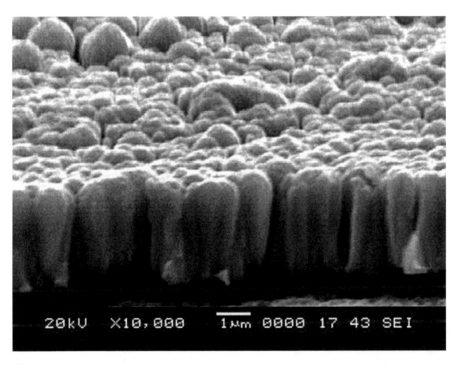

FIGURE 4.14 Cross-sectional microstructure of the electrodeposited copper layer.

Though there are multiple techniques available for the deposition of copper as thin film, the developments in these techniques were based on the requirements of the coatings industry, electronics, and microelectronics industry or for the decorative applications. It is recently that development and characterization of copper thin films is happening with an eye on the antimicrobial applications.

Benefits of electrodeposition techniques are low processing temperature, easy control of film thickness, easy deposition into complex shapes, low capital investment and possibility of alloy deposition, and sometimes deposition of nonequilibrium materials. Characteristics of the final deposit depend on the concentration of copper salts, additives, free acid level or pH, temperature, level of agitation, voltage, and current density employed.

Recent investigations have highlighted electrodepositing as an attractive approach for the preparation of nanostructured materials (Fig. 4.15). It provides a cost-effective and less equipment intensive method for the preparation of nanocrystalline coatings. The film thickness can be accurately controlled by monitoring the consumed charge. Deposition rate is also reasonably good.[46]

Continuous effort is being done to improve the property of the deposit by looking for improved technique. Couple of techniques to be mentioned here are pulse electrodeposition, and periodic reverse and asymmetric alternating current plating. In pulse electrodeposition, either potential or current is alternated between different values.[46] This results in a series of pulses of equal magnitude, duration, and polarity separated by zero current (Fig. 4.16). They favor nucleation of fresh grains and greatly increase the number of nuclei per unit surface area. They give coatings with better properties than conventional coatings. By controlling the pulse parameters, it is possible to control the composition of the deposited film and thickness of the film in atomic scale.[68,69] In periodic reverse plating, the components are plated conventionally for a selected time and then deplated by reversing the current flow, but for a shorter time. In asymmetric alternating current plating, an alternating current is used, but the mean value is not zero.[43]

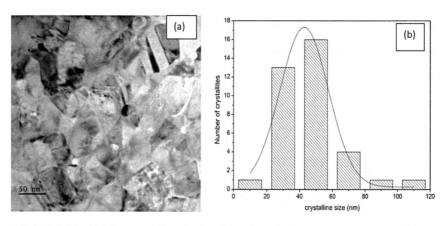

FIGURE 4.15 (a) Nanocrystallites in the electrodeposited copper and (b) crystallite size distribution.

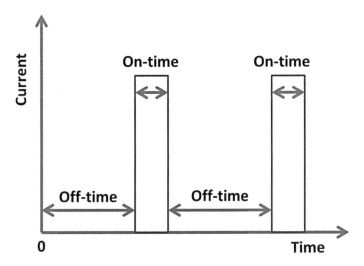

FIGURE 4.16 Important parameters affecting deposition using pulse electrodeposition technique.

4.5 CHARACTERIZATION OF COPPER THIN COATING

4.5.1 MEASUREMENT OF FILM THICKNESS

Thickness measurement techniques are grouped into direct and indirect method depending on whether the thickness is monitored directly on the sample or not. Older techniques for measuring film thickness depend on the use of a micrometer, or a mechanical stylus or a multibeam interferometer after the part is coated (or process is over). Another approach is to weigh parts before and after deposition and use the coated surface area and coating density to get the film thickness. Advantage is that measurement is done directly on the sample. Some of the drawbacks are the thickness value that is an average value and the measurement after the process has been completed. There are optical-based thickness measurement techniques. Main techniques available for film thickness measurement are single wavelength ellipsometry (SWE), reflectometry (RF), and spectroscopic ellipsometry (SE). SWE and RF are relatively inexpensive, useful especially for single film on a substrate. SWE is particularly effective for films with thickness less than a few hundred nanometers. RF is useful for thickness more than a few hundred nanometers. SE is useful for measurement with multiple films.

All three techniques are useful only if light reaches bottom of the film and interacts with bottom layer. Most oxides and semiconductors qualify these criteria, even if thickness is up to a few micrometers. For copper (or for metals), optical methods are useful only if layer is very thin.

The change in intensity of an optical ray, when it passes through a transparent/semitransparent thin film, is negligible. But as the ray passes through the material, there will be a phase change with respect to the light ray falling. This is the significant amount. The resultant ray can undergo constructive or destructive interference with the initial light ray.

Reflection of light: Light reflected from thin film–air interface and thin film–substrate interface has a difference in phase (angular) relationship. The magnitude of phase difference depends on the wavelength λ, thickness t, and optical properties of the materials. Based on this phase change, there will be destructive or constructive interference, when they come out of the surface. In reflectometry, one measures wavelength of the reflected intensity against wavelength of the light to estimate film thickness. In ellipsometry, the ratio of the wave amplitude parallel to the plane of incidence against the wave amplitude perpendicular to the plane of incidence is measured. Along with that, amplitude ratio and phase shift between the waves are also measured. The amplitude ratio and phase shifts are functions of the wavelength, thickness, optical properties of the materials, and angle of incidence.

Ellipsometry: Ellipsometry means elliptically polarized light. Here, the polarization-dependent optical techniques are used to explore the possibility of thickness measurement. The primary tools for this measurement are light source, polarizer, sample, analyzer, compensator, and detector. Polarizer and analyzers are made from materials that can polarize the light.[70] Tools used in ellipsometry are schematically presented in Figure 4.17. A light source produces unpolarized light. Then, it is passed through a polarizer and the light coming out of the polarizer has electric field oriented along a particular direction (linearly polarized light).

FIGURE 4.17 Various units of an ellipsometric thickness measuring unit.

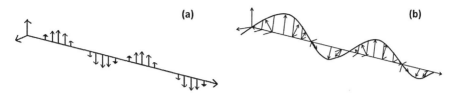

FIGURE 4.18 Vibrations in polarized light: (a) linearly polarized light and (b) elliptically polarized light.

The linearly polarized light gets reflected from the sample surface and becomes elliptically polarized. Elliptically polarized light passes through a continuously rotating polarizer (called analyzer). The amount of light passing through depends on the orientation of the polarizer with respect to the elliptical field of the wave coming from the sample. Detector converts light information to electrical signal. This is compared to known input parameters to determine the polarization change caused by the wave interacting with the sample. These parameters are Ψ and Δ.[71] They are the shifts in the vibration angles along x and y coordinates as a function of time. Using a high-quality ellipsometer, Ψ and Δ could be measured to 0.01° and 0.02° accuracy and this can easily measure a thickness range in the scale of a monolayer.[72]

Though ellipsometry is accurate, it is applicable to transparent films and for copper this is possible if coating is limited to a few nanometers in thickness. Use of the technique depends on extensive modeling and electromagnetic theory is involved if coating–substrate combination is altered.

Indirect ways of thickness measurement use the shift of interference fringes by fabricating a step on another piece of the substrate. Alternatively, fraction of the light transmitted through a microscopic slide which has a copper layer deposited along with the substrate. A third way is the use of DTM (digital thickness monitors).

In DTM method, a monitor is used in the coating chamber along with the sample. The monitor measures and continuously monitors the deposition on its surface. One such monitor is quartz crystal monitor. Quartz crystal monitor operates using the principal of piezoelectric effect. When a free running RF voltage is applied to the quartz crystal, it vibrates at its natural frequency. If temperature is fixed, the vibration frequency (or amplitude for that matter) changes only if the mass of the crystal changes. During film deposition, mass changes continuously and so is vibrational frequency. The changes are detected electronically and converted into thickness. To keep temperature fixed throughout, the crystal is cooled continuously.

4.5.2 EVALUATION OF MECHANICAL PROPERTIES OF THIN FILMS

As detailed earlier, thin films are deposited by various means. Their mechanical properties are important because sometimes high stresses are developed during deposition and entirely different scales of defects are created. They affect the life of the components especially when the components are subjected to wear and sliding conditions (bed rails, window doors, etc.).

Use of conventional indentation methods like Vickers and Brinell hardness tests is not suitable for hardness measurements of thin films. On the other hand, an improvised technique, called nanoindentation, is used. Nanoindentation is a measurement technique where many parameters like hardness, elastic properties, strength, creep behavior, etc., of the film could be evaluated. In nanoindentation, load used is very small (grams or less than that). The resulting depth is very small, often in the scale of nanometers.

In nanoindentation measurements, a hard indenter, usually an accurately ground diamond, is pressed into the film at a controlled rate. Initially, the film is deformed elastically and later plastically. Along with the load, penetration of the indenter into the film is measured, simultaneously. Upon gradual removal of the load, the elastic part is recovered and the plastic part is retained. A typical load versus elongation plot is shown in Figure 4.19. Compared to other hardness measurements, the nanoindentation gives a continuous record of both load and depth during loading and unloading. Though imaging is not essential to get indentation diameter, use of an atomic force microscope or scanning probe microscope allows one to locate the position of an indenter accurately and is very useful in the study of heterogeneous samples and topographically uneven samples.

The study of mechanical properties by nanoindentation technique has become popular recently. It is more popular in the field of thin film investigation due to various reasons. Some of them are high spatial resolution, high depth resolution, ability to get mechanical properties of surface layers adherent to substrates, etc. Nanoindentation instrumentation and techniques are extensively studied and extensive literature is available. Most commonly employed method for loading is piezoelectric or electromagnetic method. They can give a load resolution better than 0.01 µN. Extensively used indenter shape is Berkovich. It has a three-sided pyramid shape having depth-to-area ratio similar to that of a standard Vickers indenter. Extremely fine displacements are measured by employing capacitance-based methods. They can give a resolution better than 0.1 Å.

Indentation testing accompanied with instrumentations for load–displacement plots, as explained above, is useful for extracting indentation hardness of the film adherent to the substrate. Physically, indentation hardness means resistance to local deformation and is measured as the ratio of normal load to the projected indentation area at that load. $H = P_{max}/A_{proj}$, where A_{proj} is maximum projected area.

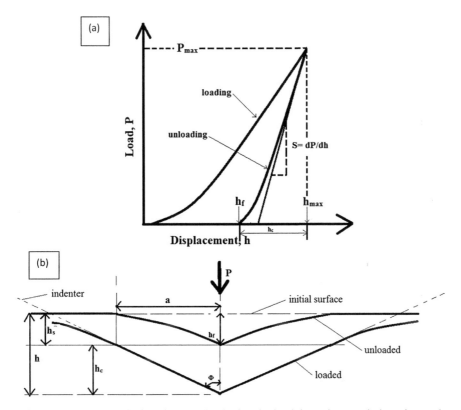

FIGURE 4.19 (a) Typical load versus depth plot obtained through a nanoindentation study and (b) meaning of various parameters are presented.

Though the hardness is independent of the applied load, during measurements certain precaution needs to be taken. One of them is deformation zone below the indenter should not reach film–substrate interface. For this, it is conventionally taken as indentation depth as less than 10% of the film thickness. Also, the hardness measurements by nanoindentation cannot be directly compared with the hardness measured through conventional route. Hardness measured through nanoindentation route is higher due to indetation size effect, strain gradient at the tip, etc. It is extremely sensitive to local inhomogeneities like change in composition, defects, etc. Hence, it is very much essential to compare hardness measured under identical conditions.

Other mechanical tests relevant for the thin film characterization are tests for estimating wear resistance, scratch resistance, and adhesion strength. Wear resistance helps to overcome the continuous wear due to the day-to-day contacts with the other objects including human body. Like wear resistance, scratch resistance is also important property because the harmful microbes can easily survive on the surface if the surface has micron-sized scratches. Scratch test can be carried out as per ASTM G 171-03. The schematic representation of scratch test set up is shown in Figure 4.20. Both scratch resistance and wear resistance depend on the microstructural parameters like crystallite size, coating texture, nanotwins, and other embedded defects.

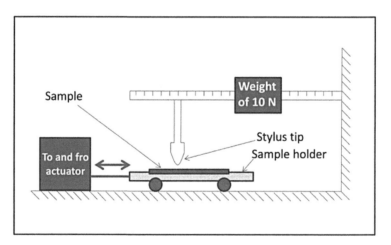

FIGURE 4.20 Schematic presentation of the scratch test geometry.

Adhesion between coating and the substrate is measured using adhesion strength. Adhesion strength can be measured in different ways, namely, cross-hatch test and pull-off adhesion test. Cross-hatch test is done using a scratch hardness tester with the help of a cross hatch cutter. The ASTM standard for this measurement is ASTM D3359. The pull-off adhesion test is done with a pull-off adhesion tester as per ASTM D4541. Adhesion strength is obtained by using a 20-mm diameter aluminum dolly. It is loaded at a rate of 0.3 MPa/s. Epoxy patch adhesive (Loctite 907 Hysol) is used to fix the dolly on the coated samples. It is fixed 4 h before the pull-off test. The schematic representation of pull-off adhesion test set up is shown in Figure 4.21.

Preparation and Characterization of Copper Thin Films

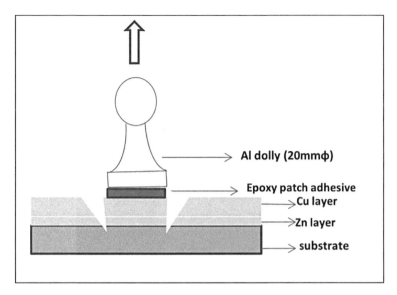

FIGURE 4.21 Schematic representation of pull-off adhesion test set up.

4.5.3 RESIDUAL STRESSES IN THE THIN FILM COATING

As explained in Section 4.4, thin film copper coatings could be produced by different routes. The deposition conditions are different in each case, and hence, residual stresses which are locked internally in the coating are different. Depending on the process used, parameters employed, film–substrate combinations used, the stresses may be either compressive or tensile. Generally, compressive stresses are beneficial. But even compressive stresses, if they are extremely high, can lead to coating failure by spallation. Similarly, when tensile stresses are high, they can lead to coating failure by cracking. Therefore, estimation of the residual stresses is important to understand the longevity of the coating. Further, residual stresses also affect the mechanical properties, optical appearance, and corrosion.

Basically, there are three types of residual stresses:

Macrostresses: These are homogeneous over macroscopic areas of the material. From applications point of view, residual macrostress is more relevant. They are tensor quantities with magnitudes varying with direction at a single point. The macrostress at a given location and along a given direction is determined by measuring the strain in that location along that direction. When macrostresses are estimated along at least three known directions, maximum and minimum normal residual stresses, the residual shear stress, their orientations with respect to a reference direction can be estimated. Macrostresses strain many crystals uniformly along the surfaces. This uniform distortion of the lattice shifts the angular position of the diffracted peak selected for residual stress measurements.

Microstresses: They are homogenous within microscopic regions but varying from one grain to another grain. Microstresses are scalar in nature. They are mainly due to defects within the crystal. Macrostresses and microstresses can be determined separately from the diffraction peak position and peak width.[73,74]

Figure 4.22 shows diffraction configuration of a monochromatic X-ray beam at a high diffraction angle (2θ), for a sample stressed along two directions (σ each). From Bragg's law, $n\lambda = 2d\sin\theta$, where n is an integer denoting order of diffraction, λ is the monochromatic X-ray wavelength, d is the lattice spacing of crystal planes, and θ is the diffraction angle. Diffraction occurs at an angle 2θ satisfying $n\lambda = 2d\sin\theta$.

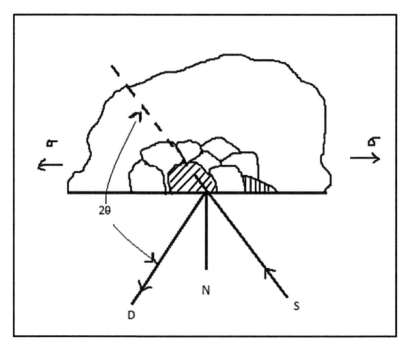

FIGURE 4.22 Diffraction configuration from a stressed surface (S: incident rays, D: diffracted rays, N: plane normal, θ: diffraction angle, do: planar spacing in unstressed condition, σ: stress applied).

Presence of residual stress in the sample alters the lattice spacing and hence alters diffraction angle θ. Measuring the change in the angular position of the diffraction peak for at least two orientations of the sample enables calculation of the stress present in the sample surface lying in the plane of diffraction, which has both incident and diffracted X-ray beams. Due to precision reasons, the diffraction peaks at higher diffraction angles ($2\theta > 120°$) are generally used.

X-ray diffraction experiments are limited to the surface of the sample. In the exposed surface layer, a condition of plane stress is assumed to exist. The plane stress is made up of two principal stresses σ_1 and σ_2 along the plane of the surface, and the stress along the perpendicular direction is zero. But the strain exists along the perpendicular direction due to two principal stresses and Poisson's ratio. The strain $\varepsilon_{\phi\Psi}$ in the direction defined by angles ϕ, Ψ is

$$\varepsilon_{\phi\Psi} = \left[\frac{1+\vartheta}{E}\left(\sigma_1\alpha_1^2 + \sigma_2\alpha_2^2\right)\right] - \left[\frac{\vartheta}{E}(\sigma_1 + \sigma_2)\right]$$

where E is the modulus of elasticity, ϑ is Poisson's ratio, and α_1 and α_2 are the angle cosines of the strain vector.

$\alpha_1 = \cos\phi \sin\Psi$

$\alpha_2 = \sin\phi \sin\Psi$

With the help of some arithmetic work principal, stresses σ_1 and σ_2 can be obtained.

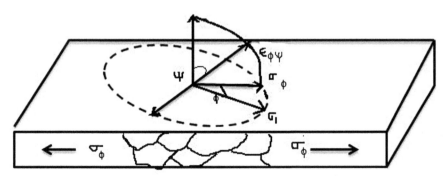

FIGURE 4.23 Various stress components giving rise to different principal stresses at the surface.

Inhomogeneous microstresses: They are inhomogenous even within a grain.

4.5.4 MICROSTRUCTURAL CHARACTERIZATION OF THIN FILMS

The performance of the copper coating (mechanical, antimicrobial, optical, etc.) depends on the structure of the films. It is also important that structure at different levels affects the performance of the coatings. For example, structure at atomic scale (atomic structure) affects the defects like grain boundaries, stacking faults, crystallite size, etc. Hence, mechanical properties like strength and hardness do get affected by them. Similarly, structure at microscale affects the properties like wetting, light reflectivity, and mechanical properties. Also, most of the coating microstructure is highly anisotropic with a large fraction of grain boundaries perpendicular to the film–substrate interface. And, in many cases, thin film structure is metastable, due to the nonequilibrium nature of the deposition process. They have a large number of low angle grain boundaries, high-angle grain boundaries, stacking faults, vacancies, twins, dislocations, etc. (Fig. 4.24). They affect the coating energy and ability of the coating to release copper atoms and ions to the aqueous medium.

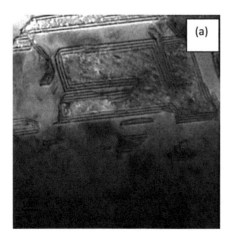
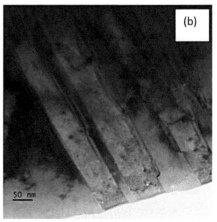

FIGURE 4.24 (a) Typical defects like stacking faults and (b) twins observed in the electrodeposited copper thin films.

Following are the important microstructural features affecting the properties of the thin film copper coating.

1. Phases and their composition
2. Shape, size, and distribution of grains
3. Defects (vacancies, grain boundaries, dislocations, stacking faults, twins, etc.)
4. Cracks, buckling, porosities, etc.
5. Anisotropy in the grain structure

The relevance of these features will be different for different properties. A common feature in the thin film copper coating is that the coating is controlled by the nucleation of the crystallites and the way they grow. Finally, the coating is observed to be made up of an array of fairly close packed pillars that are growing perpendicular to the interface. This gives rise to anisotropy in the coating structure.

Nucleation is the first stage in the development of coating microstructure. It affects crystallite size and crystallite size distribution, defect density, texture, surface morphology, etc. From both antimicrobial and mechanical properties point of view, fine crystallites with defects like twins and dislocations are beneficial, and hence, a pretreatment is given to increase the nucleation rate. This is true in processes like electrodeposition and electroless deposition.

Generally, irrespective of the coating processes employed for copper thin film deposition, the thin films have an initial layer (at the interface) consisting of randomly oriented fine grains (Figs. 4.12 and 4.14). With increase in thickness, columnar grains are established and the grain size increases. In the beginning of columnar structure, some amounts of voids are also observed. But they are closed up once the steady state growth conditions are achieved. The thickness of fine grain region and columnar region will vary depending on the type of the process and processing conditions employed. In many cases, orientation difference between nearby units is small and they cause a large number of crystallites within an outward growing grain. To accommodate continuity requirements, misfit dislocations, vacancies stacking faults, etc., are generated within the crystal. Residual stresses getting generated can also source defects like dislocations, twins, etc.

4.5.4.1 TECHNIQUES FOR MICROSTRUCTURAL CHARACTERIZATION

For characterization of thin film copper coatings, many techniques are relevant and these techniques are complimentary in nature as shown in Table 4.1. Depending on the requirements, appropriate selection has to be done. It is also essential to notice that in some techniques, volume of the material analyzed is small and hence correct sampling has to be done. Special attention has to be given to ensure that the sample is a correct representation of overall component. Also, sample has to be carefully prepared so that correct microstructure is obtained. Otherwise, even if characterization tool is selected correctly, the results obtained would be erroneous (Table 4.1).

TABLE 4.1 Characteristic Techniques and Their Applicability.

Techniques	Purpose
OM	Microstructure at low magnification, cracks, open pores
XRD	Phase composition, residual stress, crystallite size estimation, etc.
SEM	Microstructure at low and high magnification, surface topography, grain structure, inclusions, porosities, cracks, phase distribution
TEM	Microstructure at high magnification, phase and its distributions, grain structure at extremely fine scale, defects like stacking faults, dislocations, twins, etc. Grain orientation
EBSD	Grain orientations and grain size distribution, low-angle, and high-angle grain boundaries
EDS (with either SEM or TEM)	Chemical composition
WDS	Precise chemical composition
AFM	Surface topography at atomic scale
STM	Surface topography at atomic scale

AFM, atomic force microscopy; EBSD, electron backscattered diffraction; EDS, energy dispersive spectroscopy; OM, Optical microscopy; SEM, scanning electron microscopy; STM, scanning tunneling microscopy; TEM, transmission electron microscopy; WDS, wavelength dispersive spectroscopy; XRD, X-ray diffractometry.

Each technique is a subject on its own and there are standard books explaining the principle, experimental details, and analysis. A few of them are listed in references.[75–83]

4.5.5 STANDARD PROTOCOLS FOR ASSESSING THE ANTIMICROBIAL METALLIC SURFACES

Use of metals as antimicrobial surfaces is relatively new. This is in contrast to papers, textiles, and plastics, which are used for handling the food and drug items extensively. Since these are being used for fairly long time, the procedure for testing their antimicrobial effectiveness is also standardized. For metals, this is yet to take place.

4.5.5.1 TECHNIQUES FOR ANTIMICROBIAL CHARACTERIZATION

4.5.5.1.1 Zone of Inhibition Method

Antimicrobial activities of the copper surface can be tested qualitatively using the zone of inhibition (ZOI) method. It is done as following route. To check the growth, standard strains of bacteria are used. Generally, Gram-positive *Staphoycoccus epidermidis* and Gram-negative organisms *E. coli* are used. Test specimens of standard dimensions are to be used for samples and control samples (for reference). The agar plates are to be prepared using oxoid nutrient agar. A volume of 20 mL sterilized agar medium is to be dispensed into each of standard flat bottomed Petri dishes. The agar is to be allowed to solidify and it is followed by injection of a dose of 500 µL culture of the standard bacterial strains. All the precut specimens are to be kept in the Petri dish. It is essential to ensure even contact in all the samples. This set is kept in an incubator for 37°C for 2 h. After the scheduled period, the specimens are to be observed critically to check the "ZOI" (as shown in Figure 4.25). This is done visually and qualitatively on the basis of absence or presence of bacterial effects in the contact zone. The ZIO is used to determine the diffusion of antimicrobial agent from its source into the growth media. The "ZOI" formed without the microbial growth around the test specimen is generated by the antimicrobial substance which is diffused from the specimen. The ZOI basically represents the region where the antimicrobial agent released from the surface ceases the bacterial growth.[84] These images are generally collected at 5× magnification.

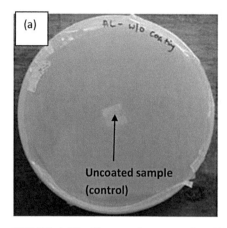 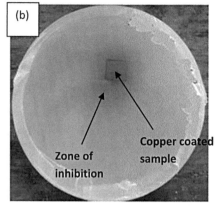

FIGURE 4.25 Photograph representing (a) zero zone of inhibition of control (uncoated aluminum) and (b) zone of inhibition corresponding to copper-coated sample. (Reprinted from Ref. 55. Open access.)

4.5.5.1.2 Colony Forming Unit Counting Technique

In this method, test (metal) surface is spread with a thin layer of bacteria culture. For this purpose, 0.75 mL of diluted inoculum is used. Inoculum is basically bacteria culture of 24-h old. Dilution is done by adding 1 mL of inoculum to 199 mL of 0.5% v/v nutrient broth. On the top of the spread inoculum film, a thin, sterile, transparent plastic film is placed to ensure that the inoculum spreads uniformly (or evenly) on the metal surface. This ensures uniform contact of the bacteria to the substrate surface. For macroscopically smooth metallic surfaces, spreading is generally uniform and in such cases use of plastic films may be skipped. Before using, the test coupons and the plastic film are to be sterilized by immersing in 70% ethyl alcohol followed by drying in a laminar flow chamber, under UV radiation. The experimental coupons are to be placed in a clean Petri dish and incubated at 30 ± 1°C for 24 h. To maintain humidity in the Petri dish, wet and sterile tissues could be used.

After 24 h of exposure to bacteria culture, the surfaces of the test specimens and the plastic film are to be washed with 8 mL of nutrient broth: soya bean casein digests broth with lecithin and polysorbate medium. The bacteria count in the solution could be determined by culturing in nutrient agar medium in a way similar to ZOI technique. As an example, photograph of *E. coli* colonies formed after 4 h of exposure on uncoated aluminum is given in Figure 4.26.

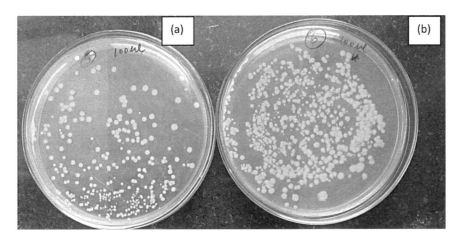

FIGURE 4.26 Photograph of *E. coli* colonies formed after 4 h of exposure on (a) copper-coated aluminum substrate and (b) noncoated aluminum substrate (control).

4.5.5.1.3 Shaking Flask Method

In this method, test coupons are suspended in 200 mL of dilute nutrient broth (0.5%v/v of 24 h grown culture). The flasks are maintained in an incubator at 30 ± 1°C for 24 h by keeping in a shaker at 85 rpm. The bacterial activity is compared with that in a control sample. The experiment is to be repeated to get concurrent values.

4.5.5.1.4 Comparison of Antibacterial Properties

Antimicrobial properties of a metal (Cu) surface is always compared with that of a control, exposed to identical environments for 24 h. The antibacterial effect is reported as antibacterial ability (R):

$$R = [\log (B/A) - \log (C/A) = \log (B/C)$$

where A is the antibacterial count in the beginning, B is the bacterial number in the control at the end of the experiment, and C is the live bacterial number in the fluid kept on the metal sample.

If R is equal or greater than 2 (two orders of magnitude or greater), the metal is considered as antimicrobial in nature (as per JIS 2801:2000 2001).

4.6 CONCLUSIONS

Amidst the worldwide concerns on increasing antibiotic resistance of microbes, the antimicrobial metals are promising. They are currently not used on large scale but their use is continuously growing. Use of metal coating provides multiple benefits like resistance to microbe growth along with protection against friction wear, moisture, and chemical exposure. Use of antimicrobial copper should be thought as a supplement method to improve the efficiency of infection control practices. Designers should always think about opportunities to upgrade surfaces to antimicrobial copper surfaces. These surfaces must not be waxed, painted, lacquered, varnished, or otherwise coated.

KEYWORDS

- **nanocrystallites touch surfaces**
- **microbes**
- **antimicrobial coatings**
- **mechanical characterization**

REFERENCES

1. Kramer, A.; Schwebke, I.; Kampf, G. *BMC Infect. Dis.* **2006**, *6* (1), 1.
2. Morgan, D. J.; Liang, S. Y.; Smith, C. L.; Johnson, J. K.; Harris, A. D.; Furuno, J. P.; Thorn, K. A.; Snyder, G. M.; Day, H. R.; Perencevich, E. N. *Infect. Ctrl. Hosp. Epidemiol.* **2010**, *31* (07), 716–721.
3. Pirofski, L.; Casadevall, A. *BMC Biol.* **2012**, *10* (1), 1.
4. Datta, R.; Platt, R.; Yokoe, D. S.; Huang, S. S. *Arch. Intern. Med.* **2011**, *171* (6), 491–494.
5. https://www3.epa.gov/ (accessed April 12, 2016).
6. Vail, D. *Dorland's Illustrated Medical Dictionary*; Elsevier: Amsterdam, 1957.
7. https://www.sciencedaily.com/releases/2014/06/140610101844.htm (accessed April 10, 2016).
8. Donskey, C. J. *Am. J. Infect. Cntrl.* **2013**, *41* (5), S12–S19.
9. Shepherd, S. J.; Beggs, C. B.; Smith, C. F.; Kerr, K. G.; Noakes, C. J.; Sleigh, P. A. *BMC Infect. Dis.* **2010**, *10* (1), 1.
10. Pangule, R. C.; Brooks, S. J.; Dinu, C. Z.; Bale, S. S.; Salmon, S. L.; Zhu, G.; Metzger, D. W.; Kane, R. S.; Dordick, J. S. *ACS Nano* **2010**, *4* (7), 3993–4000.
11. Humphreys, H. *Clin. Infect. Dis.* **2013**, cit765.

12. Weber, D. J.; Rutala, W. A. *Am. J. Infect. Ctrl.* **2013**, *41* (5), S31–S35.
13. Hobman, J. L.; Crossman, L. C. *J. Med. Microbiol.* **2015**, *64* (5), 471–497.
14. Palza, H. *Int. J. Mol. Sci.* **2015**, *16* (1), 2099–2116.
15. O'gorman, J.; Humphreys, H. *J. Hosp. Infect.* **2012**, *81* (4), 217–223.
16. Lemire, J. A.; Harrison, J. J.; Turner, R. J. *Nat. Rev. Microbiol.* **2013**, *11* (6), 371–384.
17. Yasuyuki, M.; Kunihiro, K.; Kurissery, S.; Kanavillil, N.; Sato, Y.; Kikuchi, Y. *Biofouling* **2010**, *26* (7), 851–858.
18. Hong, I.; Koo, C. H. *Mater. Sci. Eng.: A* **2005**, *393* (1), 213–222.
19. Faúndez, G.; Troncoso, M.; Navarrete, P.; Figueroa, G. *BMC Microbiol.* **2004**, *4* (1), 1.
20. Wheeldon, L.; Worthington, T.; Lambert, P. A.; Hilton, A.; Lowden, C.; Elliott, T. S. *J. Antimicrob. Chemother.* **2008**, *62* (3), 522–525.
21. Noyce, J.; Michels, H.; Keevil, C. *Appl. Environ. Microbiol.* **2007**, *73* (8), 2748–2750.
22. Weaver, L.; Michels, H.; Keevil, C. *Lett. Appl. Microbiol.* **2010**, *50* (1), 18–23.
23. Fujishima, A.; Rao, T. N.; Tryk, D. A. *J. Photochem. Photobiol. C: Photochem. Rev.* **2000**, *1* (1), 1–21.
24. Noimark, S.; Allan, E.; Parkin, I. P. *Chem. Sci.* **2014**, *5* (6), 2216–2223.
25. Dancer, S. *J. Hosp. Infect.* **2009**, *73* (4), 378–385.
26. http://www.sanskritimagazine.com/ayurveda/why-copper-was-used-by-our-ancestors/ (accessed April 11, 2015).
27. Waldron, K. J.; Robinson, N. J. *Nat. Rev. Microbiol.* **2009**, *7* (1), 25–35.
28. Grass, G.; Rensing, C.; Solioz, M. *Appl. Environ. Microbiol.* **2011**, *77* (5), 1541–1547.
29. Wang, S. *Dyes Pigm.* **2008**, *76* (3), 714–720.
30. Macomber, L.; Rensing, C.; Imlay, J. A. *J. Bacteriol.* **2007**, *189* (5), 1616–1626.
31. Harrison, M. D.; Jones, C. E.; Solioz, M.; Dameron, C. T. *Trends Biochem. Sci.* **2000**, *25* (1), 29–32.
32. Appenroth, K.-J. *Soil Heavy Metals*; Springer: Berlin, 2010; pp 19–29.
33. Rodriguez-Montelongo, L.; Lilia, C.; Farías, R. N.; Massa, E. M. *Biochim. Biophys. Acta—Bioenerget.* **1993**, *1144* (1), 77–84.
34. Borkow, G.; Okon-Levy, N.; Gabbay, J. *Wounds: A Compendium of Clinical Research and Practice* **2010**, *22* (12), 301–310.
35. Santo, C. E.; Lam, E. W.; Elowsky, C. G.; Quaranta, D.; Domaille, D. W.; Chang, C. J.; Grass, G. *Appl. Environ. Microbiol.* **2011**, *77* (3), 794–802.
36. Casey, A.; Adams, D.; Karpanen, T.; Lambert, P.; Cookson, B.; Nightingale, P.; Miruszenko, L.; Shillam, R.; Christian, P.; Elliott, T. *J. Hosp. Infect.* **2010**, *74* (1), 72–77.
37. Santo, C. E.; Morais, P. V.; Grass, G. *Appl. Environ. Microbiol.* **2010**, *76* (5), 1341–1348.
38. Noyce, J.; Michels, H.; Keevil, C. *J. Hosp. Infect.* **2006**, *63* (3), 289–297.
39. Macomber, L.; Imlay, J. A. *Proc. Natl. Acad. Sci.* **2009**, *106* (20), 8344–8349.
40. Cheng, S.; Ma, E.; Wang, Y.; Kecskes, L.; Youssef, K.; Koch, C.; Trociewitz, U.; Han, K. *Acta Mater.* **2005**, *53* (5), 1521–1533.
41. Sanders, P.; Youngdahl, C.; Weertman, J. *Mater. Sci. Eng.: A* **1997**, *234*, 77–82.
42. Dini, J. W.; Snyder, D. D. *Modern Electroplating*, 5th ed.; Wiley Online Library, 2011; pp 33–78.
43. Mech, K.; Kowalik, R.; Żabiński, P. *Arch. Metall. Mater.* **2011**, *56* (4), 903–908.
44. Bozzini, B.; Giovannelli, G.; Cavallotti, P. *J. Appl. Electrochem.* **2000**, *30* (5), 591–594.
45. Bicelli, L. P.; Bozzini, B.; Mele, C.; D'Urzo, L. *Int. J. Electrochem. Sci.* **2008**, *3* (4), 356–408.
46. Schrader, M. E. *J. Phys. Chem.* **1974**, *78* (1), 87–89.

47. Nie, Y.; Kalapos, C.; Nie, X.; Murphy, M.; Hussein, R.; Zhang, J. *Ann. Clin. Microbiol. Antimicrob.* **2010**, *9* (25), 1–10.
48. Haouaoui, M.; Hartwig, K. T.; Payzant, E. A. *Acta Mater.* **2005**, *53* (3), 801–810.
49. Xue, X.; Kozaczek, K.; Kurtz, D.; Kurtz, S. *Adv. X-Ray Anal.* **2000**, *42*, 612.
50. Mattox, D. *Vacuum Deposition, Reactive Evaporation, and Gas Evaporation*; ASM International, Member/Customer Service Center: Materials Park, OH, 1994; pp 556–572.
51. Ashmanis, I.; Kozlov, V.; Yadin, E.; Vilks, J.; Swisher, R. In *Proceedings of the Annual Technical Conference—Society of Vacuum Coaters*, 2007; Vol. 50, p 746.
52. Dobrowolski, J.; Ranger, M.; Wilkinson, R. *J. Vac. Sci. Technol. A* **1983**, *1* (3), 1403–1408.
53. Romig, Jr., A. *J. Appl. Phys.* **1987**, *62* (2), 503–508.
54. Singh, J.; Quli, F.; Wolfe, D. E.; Schriempf, J.; Singh, J. Applied Research Laboratory, Pennsylvania State University: Pennsylvania, PA, 1999.
55. Komalakrishna, H.; Augustin, A.; Bhat, U. *Am. J. Mater. Sci.* **2015**, *5* (3C), 19–24.
56. Mattox, D. *Society of Vacuum Coaters*; William Andrew Publishing, 1998.
57. McClanahan, E.; Laegreid, N. *Top. Appl. Phys.* **1991**, *64*, 339.
58. Rohde, S. ASM International, Member/Customer Service Center: Materials Park, OH, 1994; pp 573–581.
59. Shima, K.; Shimizu, H.; Momose, T.; Shimogaki, Y. *ECS J. Solid State Sci. Technol.* **2015**, *4* (8), P305–P313.
60. Chiou, J.-C.; Chen, Y.-J.; Chen, M.-C. *J. Electr. Mater.* **1994**, *23* (4), 383–390.
61. Mishra, K.; Paramguru, R. *Afr. J. Pure Appl. Chem.* **2010**, *4* (6), 87–99.
62. Silvain, J.; Chazelas, J.; Trombert, S. *Appl. Surf. Sci.* **2000**, *153* (4), 211–217.
63. Schlesinger, M.; Paunovic, M. *Modern Electroplating*; John Wiley & Sons: Hoboken, NJ, 2011; Vol. 55.
64. Bindra, P.; White, J. R. *Electroless Plating Fundamentals & Applications*; 1990; pp 289–375.
65. Paunovic, M.; Stack, C. In *Proceedings of the Electrochemical Society*, Pennington, New Jersey, 1981; pp 81–86.
66. Nakahara, S.; Okinaka, Y. *Acta Metall.* **1983**, *31* (5), 713–724.
67. Paunovic, M.; Zeblisky, R. *Plat. Surf. Finish.* **1985**, *72* (2), 52–54.
68. Chandrasekar, M.; Pushpavanam, M. *Electrochim. Acta* **2008**, *53* (8), 3313–3322.
69. Yamada, A.; Houga, T.; Ueda, Y. *J. Magn. Magn. Mater.* **2002**, *239* (1), 272–275.
70. Tompkins, H. *Handbook of Ellipsometry*; William Andrew Publishing, 2005.
71. Humlíček, J. William Andrew Publishing, 2005.
72. Irene, E. A. In *SiO₂ Films 583-636 in Handbook of Ellipsometry*; Tompkins, H. G.; Irene, E. S., Eds.; William Andrew Publishing: Norwich, NY. 2005.
73. Prevey, P. S. *ASM Int., ASM Handb.* **1986**, *10*, 380–392.
74. Prevey, P. S. *Principles of X-Ray Diffraction Stress Measurement*. Lambda Technologies. www.lambdatechs.com (accessed April 2, 2016).
75. Fultz, B.; Howe, J. M. *Transmission Electron Microscopy and Diffractometry of Materials*; Springer Science & Business Media: Berlin, 2012.
76. Schwartz, A. J.; Kumar, M.; Adams, B. L.; Field, D. P. *Electron Backscatter Diffraction in Materials Science*; Springer: Berlin, 2009; Vol. 2.
77. Egerton, R. F. *Physical Principles of Electron Microscopy*; Springer: Berlin, 2005; pp 125–153.

78. Goodhew, P. J.; Humphreys, J.; Beanland, R. *Electron Microscopy and Analysis*; CRC Press: Boca Raton, FL, 2000.
79. Flewitt, P. E. J.; Wild, R. K. *Physical Methods of Materials Characterization*, IOP Public, 1994.
80. Cullity, B. D.; Stock, S. R. *Elements of X-ray Diffraction*; Prentice Hall: Upper Saddle River, NJ, 2001; Vol. 3.
81. Brandon, D.; Kaplan, W. D. *Microstructural Characterization of Materials*; John Wiley & Sons: Hoboken, NJ, 2013.
82. Goldstein, J.; Newbury, D. E.; Echlin, P.; Joy, D. C.; Romig Jr., A. D.; Lyman, C. E.; Fiori, C.; Lifshin, E. *Scanning Electron Microscopy and X-Ray Microanalysis: A Text for Biologists, Materials Scientists, and Geologists*; Springer Science & Business Media: Berlin, 2012.
83. Seo, Y.; Jhe, W. *Rep. Progr. Phys.* **2007,** *71* (1), 016101.
84. Uzun, M. *Mater. Sci.* **2013,** *19* (3), 301–308.

CHAPTER 5

FABRICATION, SCANNING ELECTRON MICROSCOPIC, AND ELECTROCHEMICAL STUDIES OF NICKEL HYDROXIDE BATTERY ELECTRODES MODIFIED WITH ZINC OXIDE

B. J. MADHU*

Post Graduate Department of Physics, Government Science College, Chitradurga 577501, Karnataka, India

*E-mail: bjmadhu@gmail.com

ABSTRACT

Zinc oxide (ZnO) was synthesized by solution combustion technique. The influence of ZnO additive on the β-nickel hydroxide (β-Ni(OH)$_2$) electrode performance was examined. Modified β-Ni(OH)$_2$ with 5% ZnO was prepared by coprecipitation method. Structure of the modified β-Ni(OH)$_2$ with ZnO was characterized by X-ray diffraction analysis and Fourier-transform infrared analysis. Pasted-type battery electrodes were fabricated using pure β-Ni(OH)$_2$ and modified β-Ni(OH)$_2$ with 5% ZnO as active materials on a nickel sheet as a current collector. Structural and the electrochemical performance of modified β-Ni(OH)$_2$ with ZnO electrode were investigated by scanning electron microscopy, energy dispersive X-ray analysis, and cyclic voltammetry studies. Anodic and cathodic peak potential values were found to reduce after the incorporation of ZnO into the β-Ni(OH)$_2$ electrode. Furthermore, incorporation of ZnO is found to increase the reversibility of an electrode reaction and also enhance the separation of an oxidation current peak from the oxygen evolution current. These results suggest that the modified β-Ni(OH)$_2$ with 5% ZnO electrode possess better electrochemical properties and hence, can be recognized as a promising candidate for battery electrode applications.

5.1 INTRODUCTION

During recent decades, considerable attention has concentrated on the design of novel electrode materials for modern energy conversion and energy storage components. Particularly, batteries based on nickel–metal hydride (Ni/MH) materials are treated as the most favorable candidates because of their superiority in terms of capacity, output power, capacity, cost, and reliability.[1–5] In the present chapter, fabrication, scanning electron microscopic and electrochemical properties of nickel hydroxide battery electrodes modified with zinc oxide are discussed. Understanding and interpreting morphological and electrochemical measurements of battery electrode materials are critical to develop efficient battery electrodes.

Presently, significant interest is concentrated on the improvement of alkaline batteries with greater specific energies due to the increasing demand for electronic and telecommunication devices.[6–11] Battery chemistry is known to play a crucial role. Secondary batteries with greater specific energy can be produced with the development of Ni–MH battery technology. MH-negative electrode possesses a greater capacity per unit volume and lengthier cycle life than the cadmium (Cd)-negative electrode in nickel–cadmium batteries.[12–14] The performance of alkaline batteries is strongly influenced by the positive nickel electrode. The cell capacity is restricted by the positive electrode. This is because the capacity of a negative electrode is superior to that of the positive electrode. For Ni–MH batteries, the performance of the Ni electrode is to be adequately enhanced to match the superior characteristics of the negative electrode, namely, the hydrogen storage alloy electrode. Thus, improvement in the nickel hydroxide (Ni(OH)$_2$) electrode is very essential for alkaline storage batteries.

5.2 NICKEL HYDROXIDE BATTERY ELECTRODES

Among the existing battery electrodes, the Ni(OH)$_2$ is widely used in rechargeable Ni-based batteries as a positive electrode material.[15,16] Generally, Ni(OH)$_2$ occurs in two different phases, specifically α-Ni(OH)$_2$ and β-Ni(OH)$_2$. During charging, the α-Ni(OH)$_2$ and β-Ni(OH)$_2$ are transformed into γ-NiOOH and β-NiOOH, respectively.[17–19] The β-phase is found to be more stable as compared to α-Ni(OH)$_2$ phase. The α-Ni(OH)$_2$ is known to be unstable in the alkaline media. Hence, β-Ni(OH)$_2$ is employed as an active material in the alkaline batteries. In a charging process, β-Ni(OH)$_2$ is typically oxidized to β-NiOOH. It has been established that the addition of compounds containing transition metal atom into the Ni electrode is an efficient method to improve the utilization and cycle life of active material.[20–24] Further, addition of metallic compounds into the Ni electrode is found to inhibit the swelling of the nickel electrode during charging process. Thus, addition of metallic compounds into the Ni electrode is found to extend the cycle life of rechargeable batteries.[25–27] Various studies have been carried out on different compositions of Ni(OH)$_2$ to achieve best performance.[28–32]

5.3 SYNTHESIS OF NICKEL HYDROXIDE

In the present work, chemical precipitation technique has been used for the synthesis of β-nickel hydroxide. Analar grade reagents such as potassium hydroxide (KOH) and nickel sulfate (NiSO$_4$) were used. For the preparation of solution and washing of the precipitate, the triple distilled water was used. Solution of 1 M KOH was added to 1 M NiSO$_4$ solution by dripping at a flow rate of 10 mL min^{-1} with constant stirring. When the pH of the suspension reaches 13, the addition of the reagent was terminated. Then, the mixture was permitted to stand for 24 h for the complete digestion of precipitate. Precipitate from the excess reagent was separated by centrifugation at 1500 rpm for 1 h. In addition, precipitate was washed carefully with triple distilled water. To the wash water, excess barium chloride (1 M) was added, which leads to the precipitation of barium sulfate. Washing of the precipitate was finished when the white precipitate of BaSO$_4$ was no more found in the wash water. Finally, nickel hydroxide precipitate was dried at 60°C in an oven for 48-h duration.

5.4 PREPARATION OF MODIFIED NICKEL HYDROXIDE WITH ZINC OXIDE

5.4.1 SYNTHESIS OF ZINC OXIDE

Various techniques are available for the synthesis of zinc oxide. Among those techniques, the solution combustion technique is known to be simple, inexpensive, and fast. This technique leads to the synthesis of fine and homogeneous powders. In the present studies, zinc oxide was synthesized by combustion technique using stoichiometric mixture of zinc nitrate as oxidizer and urea as a fuel. The redox mixture in the aqueous form was taken in a Pyrex vessel and heated in a muffle furnace kept at 500 ± 10°C temperature. The redox mixture finally produces porous powder.

Structure of the ZnO was determined by X-ray diffraction (XRD) analysis using Bruker AXS D8 Advance diffractometer. Figure 5.1 represents the XRD pattern of ZnO. Analysis of XRD pattern shown the existence of hexagonal system of ZnO with primitive lattice (JCPDS 79-2205), without any signals from other compounds, which specifies the creation of single phase ZnO powder.

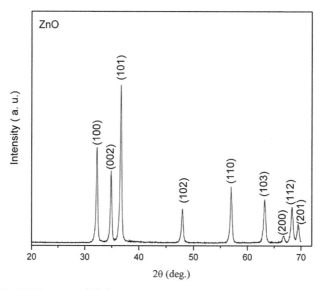

FIGURE 5.1 XRD pattern of ZnO.

5.4.2 SYNTHESIS OF MODIFIED β-NICKEL HYDROXIDE ELECTRODES WITH ZnO

Modified β-nickel hydroxide electrode material with ZnO was synthesized using coprecipitation method. In this method, the incorporation of ZnO to active β-nickel hydroxide is achieved by addition of ZnO during the synthesis of β-nickel hydroxide. Modified β-nickel hydroxide electrode materials with 5% ZnO have been prepared.

5.5 FABRICATION OF ELECTRODES

Nickel hydroxide powder alone cannot be utilized for the fabrication of electrode owing to its poor electrical conductivity. Definite amount of conductive substance, for example, graphite powder, is added to nickel hydroxide electrode. Graphite powder possesses beneficial features such as chain structure, low density, and good water absorption features. Thus, graphite powder can raise the void and electrode surface per unit volume and may improve the contact of active sample with electrolyte solution. Consequently, enhancing the content of graphite powder will improve the utilization of the active material. On the other hand, it is well known that when the content of nickel hydroxide is too high, the electrode resistance becomes excessive; the utilization of the active sample decreases and that will in turn reduce the capacity of the electrode material.[27,33] Therefore, an increase in $Ni(OH)_2$ content in the electrode sample will not increase the utilization efficiency. Further, it is difficult to fabricate the electrodes without polytetrafluoroethylene (PTFE) solution, since a certain amount of binder is essential to bind the particles efficiently. Utilization of the active material can be improved by increasing the content of the PTFE solution. Though it increases the specific capacity of the electrode, further rise in PTFE content beyond certain level will reduce the capacity due to an increase in the internal resistance. Electrode with fibrous structure that holds the active material powder effectively can be obtained by using an optimum amount of PTFE solution.[19]

Based on the above factors, the two compositions of electrodes were attained, namely, (1) pure electrode without additives: 85% $Ni(OH)_2$ powder + 10% graphite powder + 5% PTFE and (2) modified β-$Ni(OH)_2$ electrode with ZnO: 85% of modified β-$Ni(OH)_2$ with ZnO + 10% graphite powder + 5% PTFE. Electrodes were fabricated by mixing modified $Ni(OH)_2$ powder with graphite and the PTFE solution to form slurry. Obtained slurry was pasted on a Ni sheet. Fabricated electrode was dried at around 80°C temperature for 1 h. Electrode dimension was kept 1 cm × 1 cm covering the rest with insulating Teflon tape.

5.6 X-RAY DIFFRACTION ANALYSIS

The structure of modified β-nickel hydroxide (β-$Ni(OH)_2$) with 5% ZnO was investigated by XRD analysis using Bruker AXS D8 Advance diffractometer. Figure 5.2 displays XRD pattern of modified β-$Ni(OH)_2$ with 5% ZnO. XRD pattern has shown the occurrence of amorphous phase of β-nickel hydroxide with the absence of sharp peaks from nickel hydroxide. In addition, the XRD pattern revealed the existence of hexagonal system of zinc oxide with primitive lattice (JCPDS 036-1451) and orthorhombic system of nickel oxalate hydrate with end centered lattice (JCPDS 025-0582) within the electrode material.

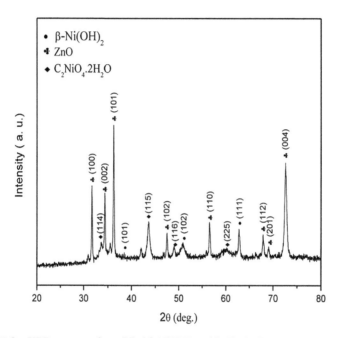

FIGURE 5.2 XRD pattern of modified β-Ni(OH)$_2$ with 5% ZnO.

5.7 FOURIER TRANSFORM INFRARED ANALYSIS

The FT-IR spectrum of the modified Ni(OH)$_2$ was recorded using the Bruker Alpha spectrophotometer in KBr pellets. Figure 5.3 shows the FT-IR spectrum of modified β-Ni(OH)$_2$ with 5% ZnO. FT-IR spectrum indicates that synthesized nickel hydroxide can be regarded as β-form, due to presence of (1) strong and narrow band around 3640 cm^{-1} relating to stretching vibration of υ(OH), which reveals the presence of hydroxyl groups in free alignment; (2) a band around 510 cm^{-1} related to the lattice vibration of OH; and (3) a weak band at 492 cm^{-1} appearing due to lattice vibration of Ni–O, υ(Ni–O).[3,27,34]

Broad and dominant band placed at 3421 cm^{-1} is allocated to the O–H stretching vibration of H$_2$O molecules and of an H-bound OH group. Furthermore, other peak noticed at 1638 cm^{-1} is ascribed to bending vibration of H$_2$O molecules.[35] Peaks situated between 800 and 1800 cm^{-1} could be attributed to the existence of anions, which have not possibly been entirely removed during washing. Detected peaks at 1476 and 1360 cm^{-1} are allotted to different vibrational modes of the carbonate groups appeared due to adsorption of an atmospheric carbon dioxide.[36] Band at 1052 cm^{-1} is related to vibration of SO$_4^{2-}$.[37]

The band observed around 2920 cm^{-1} is associated with infrared active vibrations of H$_2$O which is adsorbed on the sample.[38] The sharp peak observed around 450 cm^{-1} is ascribed to ZnO translation mode. Noticed absorption bands in 500–1000 cm^{-1} region are attributed to ZnO.[39]

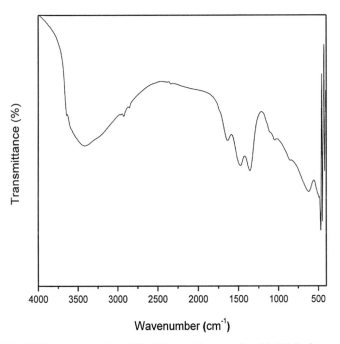

FIGURE 5.3 FT-IR spectrum of modified β-nickel hydroxide with 5% ZnO.

5.8 SCANNING ELECTRON MICROSCOPY STUDIES

Microscopic observations have been undertaken on modified β-nickel hydroxide electrode with 5% ZnO using SEM technique. Morphological studies were carried out by scanning electron microscope (SEM) (Model: Leica S440i INCA X-sight). Figure 5.4 displays SEM image of representative modified β-nickel hydroxide with 5% ZnO. SEM image displays that modified β-nickel hydroxide electrode sample is scaly and it possess a fibrous pattern with mixtures of asymmetrical shapes. The fibrous electrode is known to possess high specific surface area and comprises large number of active sites for the redox reactions. Further, porous configuration of an electrode sample provides additional channels for an electrolyte. This easily relieves the internal stress formed during charging and discharging processes, which in turn protect the electrode from physical damage.[32]

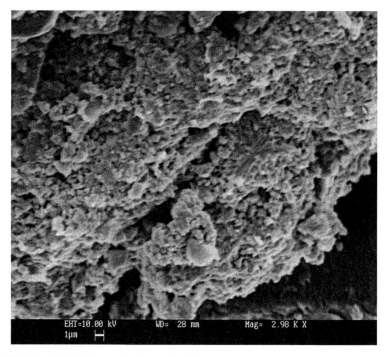

FIGURE 5.4 SEM micrograph of modified β-nickel hydroxide with 5% ZnO.

5.9 ENERGY DISPERSIVE X-RAY STUDIES

Composition studies of electrode materials were carried out using SEM attached with energy dispersive X-ray (EDX) analyzer (Model: Leica S440i INCA X-sight). Figure 5.5 displays the EDX spectrum of modified β-nickel hydroxide electrode with 5% ZnO. EDX spectrum displays the existence of

Fabrication, Scanning Electron Microscopic, and Electrochemical Studies   141

20.59 wt% of Ni, 29.43 wt% of Zn, and 31.56 wt% of O inside the modified β-nickel hydroxide electrode with 5% ZnO.

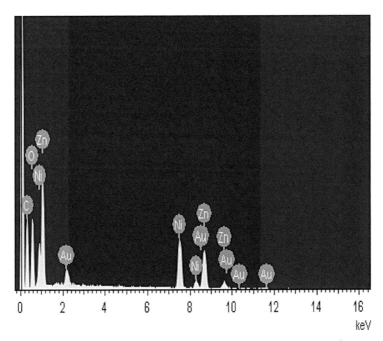

FIGURE 5.5 EDX spectrum of modified β-nickel hydroxide with 5% ZnO.

5.10 CYCLIC VOLTAMMETRY STUDIES

Cyclic voltammetry (CV) studies were undertaken using electrochemical workstation (Model: CHI604D). The fabricated electrode was used as a working electrode. Platinum foil was used as a counter electrode. Further, Hg/HgO electrode was used as a reference electrode and 6 M potassium hydroxide (KOH) solution was used as an electrolyte. The electrodes were activated in 6 M KOH solution before performing CV studies. Subsequently resting for 30 min, the cyclic voltammograms were obtained. All CV measurements were undertaken at room temperature.

Figure 5.6 displays CV curves of pure β-Ni(OH)$_2$ and modified β-Ni(OH)$_2$ with 5% ZnO in 6 M KOH electrolyte at a scanning rate of 25 mV s^{-1} at potential window of 0–0.7 V versus Hg/HgO. Observed pair of strong redox peaks in the CV curve is owed to Faradaic reactions of β-nickel hydroxide. To compare the electrochemical properties of electrodes, CV data in Figure 5.6

consisting of anodic and cathodic peak potentials are tabulated in Table 5.1. Variation in anodic peak potentials, cathodic peak potentials, and oxygen evolution reactions (OERs) with the addition of ZnO by coprecipitation method are shown in Figure 5.6 and tabulated in Table 5.1.

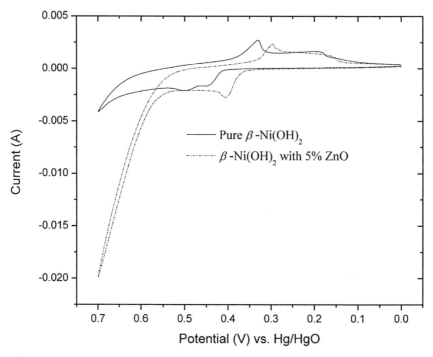

FIGURE 5.6 Cyclic voltammograms of the pure and modified β-Ni(OH)$_2$ electrodes.

Generally, the average of an anodic and cathodic peak potentials (E_{rev}) is considered as an evaluation of the reversible potential for Ni(OH)$_2$ electrodes. The potential difference (ΔE_{ac}) between anodic (E_a) and cathodic (E_c) peak potentials is considered as a measure of reversibility of the redox reaction.[20] As compared to pure β-Ni(OH)$_2$, the E_{rev} and ΔE_{ac} values of modified β-nickel hydroxide with 5% ZnO are reduced, which specifies that insertion of ZnO is found to increase the reversibility of electrode reaction.

When battery electrode is being charged, OER is considered to be a parasitic side reaction. The OER is found to possess negative effects on charge efficiency and structure of an electrode. It is very much beneficial to decrease an anodic oxidation peak potential (E_a) of β-Ni(OH)$_2$ electrode and increase the oxygen evolution potential (E_{OE}) so as to isolate these two processes and increase the charging efficiency of the β-Ni(OH)$_2$ electrode. In the present studies, as compared to pure β-Ni(OH)$_2$, the ($E_{OE} - E_a$) value of modified β-nickel hydroxide with 5% ZnO was increased, which specifies that the incorporation of ZnO is found to enhance the separation of anodic peak from an oxygen evolution current. Large value of ($E_{OE} - E_a$) permits electrode to be charged fully before the occurrence of oxygen evolution. Hence, modified β-Ni(OH)$_2$ with 5% ZnO electrode material can effectively restrain the OER and improve the charging efficiency.

TABLE 5.1 CV Characteristics for Pure and Modified β-Ni(OH)$_2$ Electrodes.

Electrode	E_a (mV)	E_c (mV)	E_{OE} (mV)	ΔE_{ac} (mV)	E_{rev} (mV)	$E_{OE} - E_a$ (mV)
Pure β-Ni(OH)$_2$	500	331	550	169	415.5	50
Modified β-Ni(OH)$_2$ electrode with 5% ZnO	407	297	537	110	352	130

5.11 CONCLUSIONS

Zinc oxide was synthesized using solution combustion technique using zinc nitrate as oxidizer and urea as a fuel. The effect of ZnO additive on the β-Ni(OH)$_2$ electrode performance was examined. Modified β-nickel hydroxide with 5% ZnO was synthesized by coprecipitation technique. Structure of the modified β-nickel hydroxide with ZnO was characterized by XRD analysis and Fourier transform infrared analysis. In the present study, a pasted type battery electrode is fabricated using β-nickel hydroxide as an active material on a nickel sheet as a current collector. Structural and the electrochemical performance of modified β-Ni(OH)$_2$ with ZnO electrode were investigated by SEM, EDX analysis, and CV studies. Anodic and cathodic peak potential values were found to reduce after the incorporation of ZnO into the β-nickel hydroxide electrode by coprecipitation technique. Furthermore, incorporation of ZnO is found to increase the reversibility of an electrode reaction and also enhance the separation of an oxidation current peak from the oxygen evolution current. These findings suggest that the modified β-Ni(OH)$_2$ with 5% ZnO electrode possess better electrochemical properties and hence can be recognized as a favorable candidate for battery electrode applications.

KEYWORDS

- nickel hydroxide
- novel battery electrode materials
- chemical precipitation technique
- scanning electron microscopy
- electrochemical properties

REFERENCES

1. Bode, H.; Dehmelt, K.; Witte, J. Zur Kenntnis der Nickelhydroxidelektrode-I. Über das Nickel(II)-Hydroxidhydrat. *Electrochim. Acta* **1966,** *11*, 1079–1087.
2. Falk, S. U.; Salkind, A. J. *Alkaline Storage Batteries*; John Wiley & Sons: New York, 1969.
3. Oliva, P.; Leonardi, J.; Laurent, J. F.; Delmas, C.; Braconnier, J. J.; Figlarz, M.; Fievet, F.; de Guibert, A. Review of the Structure and the Electrochemistry of Nickel Hydroxides and Oxyhydroxides. *J. Power Sour.* **1982,** *8*, 229–255.
4. Lim, H. S.; Verzwyvelt, S. A. Effects of Electrode Thickness on Power Capability of a Sintered-Type Nickel Electrode. *J. Power Sour.* **1996,** *62*, 41–44.
5. Li, L.; Wu, Z.; Yuan, S.; Zhang, X.-B. Advances and Challenges for Flexible Energy Storage and Conversion Devices and Systems. *Energy Environ. Sci.* **2014,** *7*, 2101–2122.
6. Wu, Z.; Huang, X.-L.; Wang, Z.-L.; Xu, J.-J.; Wang, H.-G.; Zhang, X.-B. Electrostatic Induced Stretch Growth of Homogeneous β-Ni(OH)$_2$ on Graphene with Enhanced High-Rate Cycling for Supercapacitors. *Sci. Rep.* **2014,** *4*, 36691–36698.
7. Wang, Z.-L.; Xu, D.; Xu, J.-J.; Zhang, X.-B. Oxygen Electrocatalysts in Metal–Air Batteries: From Aqueous to Nonaqueous Electrolytes. *Chem. Soc. Rev.* **2014,** *43*, 7746–7786.
8. Huang, X.-L.; Chai, J.; Jiang, T.; Wei, Y.-J.; Chen, G.; Liu, W.-Q.; Han, D.; Niu, L.; Wang, L.; Zhang, X.-B. Self-Assembled Large-Area Co(OH)$_2$ Nanosheets/Ionic Liquid Modified Graphene Heterostructures Toward Enhanced Energy Storage. *J. Mater. Chem.* **2012,** *22*, 3404–3410.
9. Zhang, L.; Zhang, X.-B.; Wang, Z.; Xu, J.; Xu, D.; Wang, L. High Aspect Ratio γ-MnOOH Nanowires for High Performance Rechargeable Nonaqueous Lithium–Oxygen Batteries. *Chem. Commun.* **2012,** *48*, 7598–7600.
10. Gifford, P.; Adams, J.; Corrigan, D.; Venkatesan, S. Development of Advanced Nickel/Metal Hydride Batteries for Electric and Hybrid Vehicles. *J. Power Sour.* **1999,** *80*, 157–163.
11. Chan, C. C. The State of the Art of Electric and Hybrid Vehicles. *Proc. IEEE* **2002,** *90*, 247–275.
12. Deabate, S.; Henn, F.; Devautour, S.; Giuntini, J. C. Conductivity and Dielectric Relaxation in Various Ni(OH)$_2$ Samples. *J. Electrochem. Soc.* **2003,** *150*, 23–31.

13. McBreen, J. In *Modern Aspects of Electrochemistry*; White, R. E.; Bockris, J. O'. M.; Conway, B. E.; Eds.; Plenum Press: New York, 1990; No. 21.
14. Armstrong, R. D.; Briggs, G. W. D.; Charles, E. A. Some Effects of the Addition of Cobalt to the Nickel Hydroxide Electrode. *J. Appl. Electrochem.* **1988**, *18*, 215–219.
15. Kohler, U.; Antonius, C.; Bauerlein, P. Advances in Alkaline Batteries. *J. Power Sour.* **2004**, *127*, 45–52.
16. Fetcenko, M. A.; Ovshinsky, S. R.; Reichman, B.; Young, K.; Fierro, C.; Kock, J.; Zallen, A.; Mays, W.; Ouchi, T. Recent Advances in NiMH Battery Technology. *J. Power Sources* **2007**, *165*, 544–551.
17. Xu, P.; Han, X. J.; Zhang, B.; Lv, Z. S.; Liu, X. R. Characterization of an Ultrafine β-Nickel Hydroxide from Supersonic Co-precipitation Method. *J. Alloys Cmpd.* **2007**, *436*, 369–374.
18. Vidottil, M.; Salvadora, R. P.; Torresi, S. I. C. Synthesis and Characterization of Stable Co and Cd Doped Nickel Hydroxide Nanoparticles for Electrochemical Applications. *Ultrason. Sonochem.* **2009**, *16*, 35–40.
19. Chang, Z.; Zhao, Y.; Ding, Y.; Chang, Z.; Zhao, Y.; Ding, Y. Effects of Different Methods of Cobalt Addition on the Performance of Nickel Electrodes. *J. Power Sour.* **1999**, *77*, 69–73.
20. Shruthi, B.; Bheema Raju, V.; Madhu, B. J. Synthesis, Spectroscopic and Electrochemical Performance of Pasted β-Nickel Hydroxide Electrode in Alkaline Electrolyte. *Spectrochim. Acta A* **2015**, *135*, 683–689.
21. Watanabe, K.; Koseki, M.; Kumagai, N. Effect of Cobalt Addition to Nickel Hydroxide as a Positive Material for Rechargeable Alkaline Batteries. *J. Power Sources* **1996**, *58*, 23–28.
22. Guerlou-Demourgues, L.; Delmas, C. Electrochemical Behavior of the Manganese-Substituted Nickel Hydroxides. *J. Electrochem. Soc.* **1996**, *143*, 561–566.
23. Ding, Y.; Yuan, J.; Chang, Z. Cyclic Voltammetry Response of Coprecipitated $Ni(OH)_2$ Electrode in 5 M KOH Solution. *J. Power Sources* **1997**, *69*, 47–54.
24. Sood, A. K. Studies on the Effect of Cobalt Addition to the Nickel Hydroxide Electrode. *J. Appl. Electrochem.* **1986**, *16*, 274–280.
25. Chakkaravarthy, C.; Periasamy, P.; Jegannathan, S.; Vasu, K. I. The Nickel/Iron Battery. *J. Power Sour.* **1991**, *35*, 21–35.
26. Li, J.; Li, R.; Wu, J.; Su, H. Effect of Cupric Oxide Addition on the Performance of Nickel Electrode. *J. Power Sour.* **1999**, *79*, 86–90.
27. Nathira Begum, S.; Muralidharan, V. S.; Ahmed Basha, C. The Influences of Some Additives on Electrochemical Behaviour of Nickel Electrodes. *Int. J. Hydrogen Energy* **2009**, *34*, 1548–1555.
28. Watanabe, K.; Kumagai, N. Thermodynamic Studies of Cobalt and Cadmium Additions to Nickel Hydroxide as Material for Positive Electrodes. *J. Power Sour.* **1998**, *76*, 167–174.
29. Constantin, D. M.; Rus, E. M., Oniciu, L.; Ghergari, L. The Influence of Some Additives on the Electrochemical Behaviour of Sintered Nickel Electrodes in Alkaline Electrolyte. *J. Power Sour.* **1998**, *74*, 188–197.
30. Pralong, V.; Delahaye-Vidal, A.; Beaudoin, B.; Leriche, J. B.; Tarascon, J. M. Electrochemical Behavior of Cobalt Hydroxide Used as Additive in the Nickel Hydroxide Electrode. *J. Electrochem. Soc.* **2000**, *147* (4), 1306–1313.
31. Elumalai, P.; Vasan, H. N.; Munichandraiah, N. Electrochemical Studies of Cobalt Hydroxide—An Additive for Nickel Electrodes. *J. Power Sources* **2001**, *93*, 201–208.

32. Shruthi, B.; Madhu, B. J.; Bheema Raju, V. Influence of TiO_2 on the Electrochemical Performance of Pasted Type β-Nickel Hydroxide Electrode in Alkaline Electrolyte. *J. Energy Chem.* **2016,** *25,* 41–48.
33. Ding, Y.; Yuan, J. L.; Li, H.; Chang, Z.; Wang, Z. A Study of the Performance of a Paste-Type Nickel Cathode. *J. Power Sour.* **1995,** *56,* 201–204.
34. Acharya, R.; Subbaiah, T.; Anand, S.; Das, R. P. Preparation, Characterization and Electrolytic Behavior of β-Nickel Hydroxide. *J. Power Sour.* **2002,** *109,* 494–499.
35. Aghazadeh, M.; Golikand, A. N.; Ghaemi, M. Synthesis, Characterization, and Electrochemical Properties of Ultrafine β-Ni(OH)$_2$ Nanoparticles. *Int. J. Hydrogen Energy* **2011,** *36,* 8674–8679.
36. Jeevanandam, P.; Koltypin, Y.; Gedanken, A. Synthesis of Nanosized α-Nickel Hydroxide by a Sonochemical Method. *Nano Lett.* **2001,** *1,* 263–266.
37. Kosova, N. V.; Devyatkina, E. T.; Kaichev, V. V. Mixed layered Ni–Mn–Co Hydroxides: Crystal Structure, Electronic State of Ions, and Thermal Decomposition. *J. Power Sour.* **2007,** *174,* 735–740.
38. Li, H. B.; Yu, M. H.; Wang, F. X.; Liu, P.; Liang, Y.; Xiao, J.; Wang, C. X.; Tong, Y. X.; Yang, G. W. Amorphous Nickel Hydroxide Nanospheres with Ultrahigh Capacitance and Energy Density as Electrochemical Pseudocapacitor Materials. *Nat. Commun.* **2013,** *4,* 1894, DOI:10.1038/ncomms2932.
39. Theo Kloprogge, J.; Hickey, L.; Frost, R. L. FT-Raman and FT-IR Spectroscopic Study of Synthetic Mg/Zn/Al-Hydrotalcites. *J. Raman Spectrosc.* **2004,** *35,* 967–974.

CHAPTER 6

COMPARATIVE SPECTROSCOPIC STUDY OF POLYINDOLE/POLY(VINYL ACETATE) COMPOSITES

DEEPAK J. BHAGAT[1*] and GOPAL R. DHOKANE[2]

[1]*Department of Physics, Nehru Mahavidyalaya, Nerparsopant 445102, Maharashtra, India*

[2]*Department of Physics, Arts, Science and Commerce College, Chikhaldara 444807, Maharashtra, India*

*Corresponding author. E-mail: bhagatd@rediffmail.com

ABSTRACT

This chapter reports the comparative investigation of polyindole/poly(vinyl acetate) composites synthesized chemically using different oxidants. The nickel nitrate and cupric chloride were used as oxidants. The prepared samples were analyzed through Fourier-transform infrared spectroscopy, photoluminescence spectroscopy and ultraviolet–visible (UV–Vis) spectroscopy, field emission scanning electron microscopy (FE-SEM), and X-ray diffraction pattern (XRD). The optical bandgap energy values for composites were determined to be 4.26 and 4.79 eV for different oxidants. The determined optical bandgap energy has conventional value for the photovoltaic activities and also indicates that the composites have semiconducting properties. As-synthesized composites have application potential in optical devices and polymeric solar cells.

6.1 INTRODUCTION

In recent years, development of renewable energy sources is getting more awareness in the world as natural reserves will after all finish in future.

Though the consumption of sunlight seems to be the most useful and popular approach for recognition of renewable energy,[1] solar cells can directly transfer solar energy into electrical energy. Solar energy is giant energy reservoir that could be trapped and used efficiently by developing efficient solar cells such as organic polymer solar cells.[2] Conducting polymers have tremendous research interest due to their flexibility in industrial applications.[3]

Conducting polymer composites are novel category in material science synthesized through several suitable composition of conducting polymer with insulating polymeric material. Hence, attractive properties of both materials are united effectively.[4,5] These composite materials reveal diverse properties like good mechanical properties and process ability owing to the insulating polymer while electrical conductivity represented by conducting polymer. Recently, polymer composites have got tremendous attention in materials science due to their ease of processing, tunable properties, and broad variety of applications. The conducting polymer composites are one of the most important and fascinating areas in research of polymer composite. The polyindole is an emergent conducting polymer which has combination of optical and mechanical properties, also its electrical conductivity can be enhanced; hence, there is remarkable growth in its requirement as polymeric materials.[6–10]

Some research work on the synthesis of PIN, its derivatives, copolymers, and composites via different polymerization methods; furthermore, its purpose has been reported, for example, Gupta et al. synthesized polyindole chemically and studied its morphological and thermal analysis.[11] Giribabu et al. reported synthesis of PIN nanowires and its electrochemical properties.[12] Koiry et al. prepared long PIN fibers via interfacial polymerization and study its electro-activity.[13] Gopi et al. prepared poly(indole-co-thiopene) copolymer coatings electrochemically on low nickel stainless steel.[14] Zhijiang et al. investigated electrochemical properties of zinc/polyindole secondary battery.[15] Tuken et al. investigated corrosion performance of polyidole on nickel-coated mild steel.[16] Altindal et al. explored electrical characteristics of Al/polyindole schottky barrier diodes.[17] Joshi et al. reported nanoscale electrical properties of polyindole–Au nanocomposites prepared via in situ polymerization.[18] Handore et al. synthesized polyindole–ZnO nanocomposites and studied its ionic conductivity.[19] Kumar et al. investigated electro-catalytic applications of polyindole/SnO_2 nanocomposites.[20] Ramesan et al. explored DC conductivity of polyindole and nanomagnetite (Fe_3O_4) particles.[21] Kumar et al. coated oxidize polyindole over graphene in aqueous media and studied their perm selective behavior toward dopamine hydrochloride.[22] Mudila et al. synthesized polyindole/graphene oxide nanocomposites for electro-chemical energy storage.[23] Ryu et al. focused on the physiochemical properties of the polyindole/thiol composites.[24] Urkmez et al. synthesized and characterized the poly(dimethylsiloxane)/polyindole composites.[25] Pandey et al. developed polyindole-modified potassium ion sensor using dibenzo-18-crown-6 mediated poly(vinyl chloride) matrix membrane.[26] Sari et al. studied electrorheological and creep behavior of polyindole and polyindole/polyethylene.[27] Arslan et al. studied dielectric and electrorheological properties of kaolinite, polyindole, and polyindole/kaolinite composite suspensions.[28]

The PIN has several interesting properties, but it suffers from some limitations like poor mechanical strength and thermal stability. The synthesis of PIN composites needs clever approach to overcome these limitations and enhanced the properties of this conducting polymer to our preferred levels. In conducting polymer composites, properties of conducting polymer were marvelously blended with those of the counter polymer. The intention behind the synthesis of conducting polymers with thermoplastic polymers is to increase their processibility.[29,30] Moreover, mechanical properties of conducting polymer can be enhanced by synthesizing composite in the presence of host matrix.[26,31] As mentioned in literature review, several matrix polymers have been used in combination with a PIN.

The PVAc was chosen here as the best candidate to perform as stabilizer or matrix polymer. The PVAc has various favorable characteristics such as harmless, easily producible, nonpoisonous, cost effective, environment friendly, highly processable, convenient in application; additionally, it has good adhesiveness and mechanical properties.[33] The PVAc can serve as polymeric plasticizer or steric stabilizer for PIN that helps in enhancement of processibility due to its low T_g.[34]

The present chapter reports on the comparative spectroscopic study of the polyindole/poly(vinyl acetate) (PIN/PVAc) composites prepared via chemical polymerization technique using different oxidants. The PIN/PVAc composites synthesized using oxidants nickel nitrate and cupric chloride. The prepared composites were analyzed through Fourier transform infrared spectroscopy (FTIR), photoluminescence (PL), X-ray diffraction (XRD), and ultraviolet–visible (UV–Vis) spectroscopy, and field emission scanning electron microscopy (FE-SEM).

6.2 EXPERIMENTAL

6.2.1 PRECURSORS

All chemicals were analytical grade and purchased from SD Fine Chemicals, India, such as monomer indole, nickel nitrate, and cupric chloride used as oxidizing agents, methanol used as organic solvent. The poly(vinyl acetate) (PVAc) used as counter polymer procured from Himedia Chemicals, India.

6.2.2 SYNTHESIS OF COMPOSITES

PIN/PVAc composites were synthesized via chemical oxidative polymerization method at room temperature. The nickel nitrate ($Ni(NO_3)_2$) cupric and chloride ($CuCl_2$) were used as oxidant. Mixed solutions of PVAc and methanol were prepared by dissolving polyvinyl acetate (1 g) in methanol (9 mL) and stirred about 1 h and kept it for 24 h to make homogenous solution. Then, monomer indole (0.5 g) was added into PVAc solutions and continuously stirred for 1 h. The indole was polymerized by adding oxidants $CuCl_2$ and $Ni(NO_3)_2$ (each 0.08823 g) (15 wt%), respectively, and continuously stirred about 2 h to complete reaction of polymerization. PIN/PVAc solutions were kept for 1 h to get settle down. Then, these solutions were poured on chemically cleaned optically plane glass plates to prepared films.

The films were dried by an isothermal loss of solvent. Hence, entire assembly was placed in a dust-free chamber maintained at constant temperature. After loss of solvent isothermally, films were washed with hot distilled water and remove from glass plates. Then, films were again dried for 6 h at room temperature. In this way, we prepared PIN/PVAc composites with different oxidant.

The flowchart clearly represents the synthesis method of PIN/PVAc composites with different oxidant. It reveals stepwise presentation of adopted technique (Fig. 6.1).

6.2.3 CHARACTERIZATIONS

The as-synthesized composites are analyzed through various characterizations techniques, for example, FTIR of sample was taken on SHIMADZU FTIR-8400 to know the chemical bonding resent in the prepared composite materials. The PL spectra of both samples were taken on FL

Spectrophotometer, Hitachi, F-7000 to study the luminescent properties of composites using emission spectra.

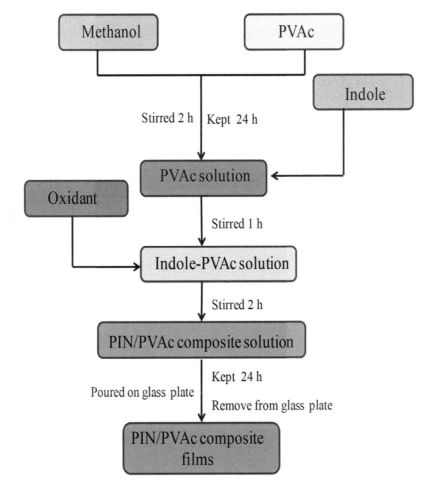

FIGURE 6.1 Flowchart of synthesis of PIN/PVAc composites.

The UV–Vis spectroscopy was taken on Agilent Technologies, Cary 60 UV–Vis, for the evaluation of optical bandgap and studied the %absorption of material. The FE-SEM analysis performed through S-4800, Hitachi, Japan to investigate the surface morphology as well as their nature of composite materials XRD (Rigaku, Miniflex-II, Japan). The XRD patterns were calculated using Cu$K\alpha$ (λ = 1.5406 Å) radiation. The intensity was measured as a function of 2θ in the range 10–70° at room temperature.

6.3 RESULTS AND DISCUSSIONS

On the basis of various characterizations, the analysis of PIN/PVAc composites is discussed below in detail. The comparison of both samples is also explained through the characterization techniques.

6.3.1 FTIR ANALYSIS

The FTIR spectra of PIN/PVAc composites are displayed in Figure 6.2. The FTIR spectra show some major peaks that clear indication of various bond formation and that are discussed below.

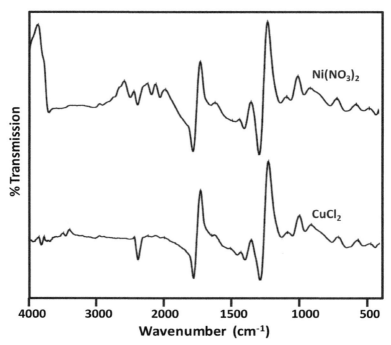

FIGURE 6.2 FTIR spectrum of PIN/PVAc composites.

The FTIR spectra of both samples show minor variation in peaks which indicate the successful synthesis of the PIN/PVAc composites. The slight shifting of major as well as minor peaks in both spectra is clearly observed from the figure. As shown in the figure for both samples, the peaks at 2467 cm^{-1} differ from each other. The difference is that the noisy peaks observed in sample Ni(NO$_3$)$_2$ and single major absorption peaks in sample CuCl$_2$. Moreover, the peaks present in both samples are listed and assigned in Table 6.1.

TABLE 6.1 Wavenumbers and Their Assignments of Peaks for PIN/PVAc Composites.

Wavenumbers (cm^{-1}) for samples		Assignments	References
Using Ni(NO$_3$)$_2$	Using CuCl$_2$		
654 cm^{-1}	659 cm^{-1}	C=H bond	[29,30]
772 cm^{-1}	778 cm^{-1}	C–H bond in benzene	[33,34]
1132 cm^{-1}	1131 cm^{-1}	C–N stretch	[35]
1286 cm^{-1}	1287, 2960 cm^{-1}	Indicates presence of PVAc	[35,36]
1395 cm^{-1}	1400 cm^{-1}	N–H stretch is present; it means nitrogen is not present in the polymerization	[35,36]
1639 cm^{-1}	1641 cm^{-1}	–C=C– vibration is only present in the case of indole	[34–36]
2467 cm^{-1}	2477 cm^{-1}	Amine group	[36]

6.3.2 PHOTOLUMINESCENCE ANALYSIS

The PL is new property in the investigation of conducting polymer composites as optical material. The PL spectrum of both samples with excitation of 310 nm displayed in Figure 6.3A ranges over 400–600 nm. Figure 6.3B displays PL spectrum of both samples with excitation of 340 nm in the range of 400–600 nm.

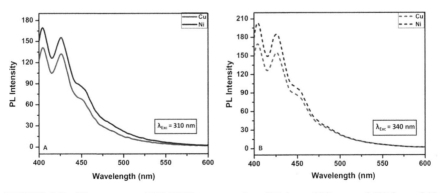

FIGURE 6.3 PL spectrum of PIN/PVAc composites: (A) λ_{Exc} = 310 nm and (B) λ_{Exc} = 340 nm.

The wavelengths of excitation chosen for the PIN/PVAc composites are 310 and 340 nm. This is due to π–π* transition of benzenoid group that responsible for PL in PIN.[37] It clearly observed from both the figures that PL spectra of both samples represent two major peaks at 404, 425, and tailed to around 600 nm. These peaks are lying in ultraviolet and visible region, respectively. The possibility of existence of numerous electronic states contributing in photo-excitation process is due to existence of two major peaks in PL spectrum. The heteroatomic conducting polymers reflect the superior intensity in PL spectra; it may be due to systematic arrangement of quinoid and benzoid group. It favors growth of excitons and raises in delocalization length of singlet exciton.[38,39]

The PL spectrum of both samples represents the emission at same wavelength for both major peaks with different intensity. The PL intensity of samples is differing from each other. It is clearly observed from Figure 6.3 that the increase in excitation wavelength results in the intensity of emission; it also increases in both samples.

6.3.3 UV–VIS ANALYSIS

Figure 6.4 shows UV–Vis spectrum of the PIN/PVAc composites in the range 200–350 nm. The figure clearly displays that the lower wavelength side has higher percentage absorption. It was clearly seen that the strongest peak of the absorption appears at wavelength 230 nm due to the optical transition from valence band to the conduction band. The information relating to the

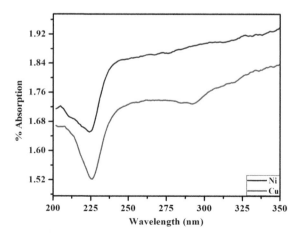

FIGURE 6.4 UV–Vis spectrums.

electronic transitions in the solid studied via absorption of the energy in UV–Vis region of the electromagnetic spectrum. The absorption of photons causes excitation of electrons to the empty band from the filled band. This results in rapid raise in absorption coefficient. The start of this rapid difference in absorption coefficient is called fundamental absorption edge. In broad system, resultant energy is recognized as optical energy gap or bandgap energy.[40,41]

6.3.4 OPTICAL BAND GAP

From Figure 6.5, optical band gaps of composite films were determined. The composites have several applications that depend on its optical bandgap energy and it has been determined using given relation,[41]

$$\alpha = \frac{A(h\nu - E_g)^n}{h\nu} \quad (6.1)$$

where α is absorption coefficient, $h\nu$ is incident photon energy, A is constant, and E_g is optical bandgap of material. The figure directly shows that PIN/PVAc composite synthesized via oxidant nickel nitrate has lower bandgap energy value than cupric chloride. The optical bandgap energy values of films have been determined as 4.26 and 4.79 eV. These values obviously show that these materials have application potential in the photo catalytic activities, optical devices, and polymeric solar cells.

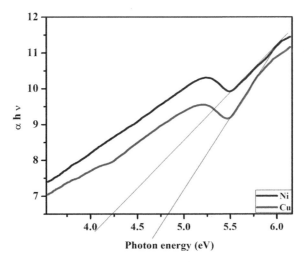

FIGURE 6.5 Plot of $\alpha h\nu$ versus photon energy (eV).

6.3.5 REFRACTIVE INDEX

The refractive index is mainly elemental property of material because of its direct relationship with the electronic polarizability of ions and the local field in the optical material. The refractive index of optical materials is crucial property in designing integrated optical devices like switches, modulators and filters.[6] The plot of refractive index versus photon energy (eV) displays in Figure 6.6. The refractive index (n) of composite materials has been calculated via given equation,[40]

$$n = \frac{1}{T_s} + \left(\frac{1}{T_s} - 1\right)^{1/2} \tag{6.2}$$

where T_s is % transmission coefficient. The refractive index of samples constantly increases up to 5.20 eV and then speedily increases up to 5.75 eV; after that, it decreases gradually with increase in photon energy, it can be owing to drop in transmittance with enhance in absorption coefficients. The regular raise in refractive index of samples represents the normal dispersion behavior. The figure clearly represents that the PIN/PVAc composite prepared using cupric chloride have greater value of refractive index than that of oxidant nickel nitrate.

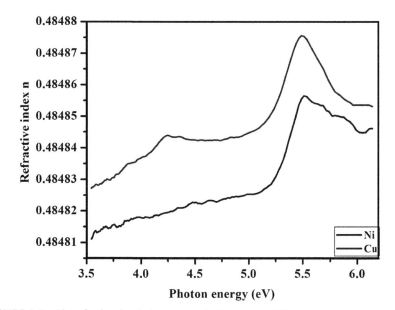

FIGURE 6.6 Plot of refractive index versus photon energy (eV).

6.3.6 EXTINCTION COEFFICIENT

Figure 6.7 displays plot of the extinction coefficient versus photon energy of composites. The extinction coefficients have been determined through the given expressions,[6,40]

$$k = \frac{\alpha \lambda}{4\pi} \qquad (6.3)$$

where k is extinction coefficient, α is percentage absorption coefficient, and λ is wavelength. The extinction coefficient is the measure of fraction of light lost because of scattering and absorption per unit distance of penetration medium. The extinction coefficients of composites are reduces sharply up to 5.5 eV and small raise up to 6.0 eV. Subsequently, it drops off steadily with raise in photon energy. The extinction coefficients ease with enhancement in photon energy; it possibly related to drop in absorption coefficients and improvement in transmittance clearly observed from figure. The raise in absorbance and scattering outcomes in fraction of light lost. Furthermore, the loss factor is influence by photon energy; it drops off with raise in photon energy. Additionally, optical absorption edge can examined directly for composites under investigation. The steady reduce in scattering from 3.5 to 5.5 eV for constant distance of penetration medium, it seen from curve of extinction coefficients.

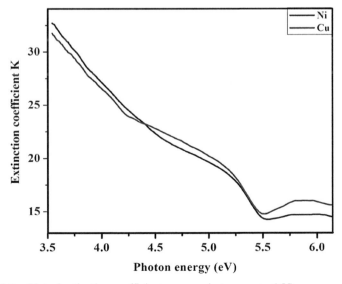

FIGURE 6.7 Plot of extinction coefficients versus photon energy (eV).

6.3.7 OPTICAL CONDUCTIVITY

The optical conductivity (σ_{opt}) of composites is determined to know the optical response of materials. It has been calculated from following equation,[42]

$$\sigma_{opt} = \frac{\alpha n c}{4\pi} \tag{6.4}$$

where σ_{opt} is an optical conductivity, c is velocity of light, n is refractive index and α is absorption coefficient. Figure 6.8 represents the plot of optical conductivity versus photon energy (eV) of PIN/PVAc composites. The optical conductivity directly depends on refractive index and absorption coefficient of composites, it clearly observed from the figure. The reduction in absorption coefficient is due to constant increase in optical conductivity.

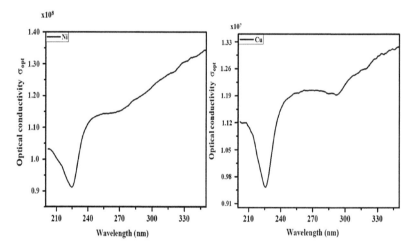

FIGURE 6.8 Plot of optical conductivity (σ_{opt}) versus photon energy (eV).

The optical conductivity values of PIN/PVAc composites established for oxidant $Ni(NO_3)_2$ and oxidant $CuCl_2$ are 1.335×10^8 and 1.319×10^7 S^{-1}, respectively. The values of optical conductivity of composites are given in Table 6.2.

TABLE 6.2 Optical Band Gap, Optical Conductivity.

Sample	Optical band gap	Optical conductivity
Cu	4.26	1.35×10^8
Ni	4.79	1.319×10^7

6.3.8 DIELECTRIC CONSTANTS

The complex dielectric constant reveals the basic intrinsic property of materials. The real part of dielectric constant is representing how much it will slow down velocity of light in material, although imaginary part of dielectric constant represents how dielectric material absorbs energy from electric field due to dipole motion. The real dielectric constant (ε_r) as a function of wavelength is displayed in Figure 6.8, whereas imaginary dielectric constant (ε_i) as a function of wavelength represents in Figure 6.9 of all composites. The real and imaginary parts of dielectric constant have been calculated by using following expressions.[6,7,41]

$$\varepsilon_r = n^2 - k^2 \tag{6.5}$$

$$\varepsilon_i = 2nk \tag{6.6}$$

where ε_r is real part of dielectric constant, ε_i is imaginary part of dielectric constant, n is refractive index, and k is extinction coefficient. Figures 6.9 and 6.10 reflect that real and imaginary part of dielectric constant increase with raise in wavelength. The data of real and imaginary part of dielectric constant provide knowledge concerning loss factor which is ratio of imaginary-to-real dielectric constant. This conclusion reveals that loss factor in composites increases with raise in wavelength. The real part of dielectric constant enhances regularly with raise in wavelength in minor region up to 280 nm; later, it increases quickly with raise in wavelength. Furthermore, imaginary part of dielectric constant frequently increases with raise in wavelength.

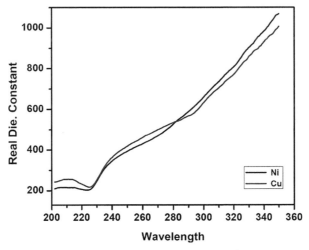

FIGURE 6.9 Plot of real dielectric constant versus wavelength.

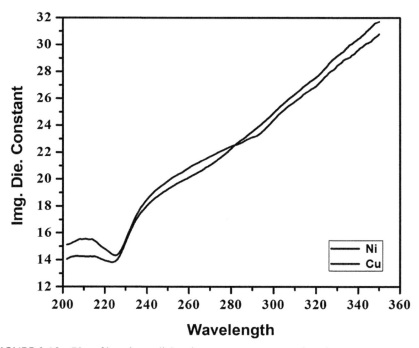

FIGURE 6.10 Plot of imaginary dielectric constant versus wavelength.

6.3.9 SURFACE MORPHOLOGICAL STUDY

The FE-SEM micrographs of PIN/PVAc composites are displayed in Figure 6.11. The micrographs of both samples display the agglomeration of particles. The spongy amorphous nature was observed from the figure. The micrographs clearly represent that the particles are asymmetrical in size and shape. It is directly seen from the figure that micrographs of both samples differ from each other. The PIN/PVAc composite prepared using oxidant nickel nitrate has spongy, globular structure and less agglomeration as compared to other micrograph, whereas PIN/PVAc composite synthesized by using the oxidant cupric chloride shows spongier and "cauliflower"-like structure. Moreover, there is formation of some large knife as well as large hemispherical structures. Also, the large amount of agglomeration is observed in this sample. It can be seen that both samples have little bit different morphology only due to change in oxidant, which was used to synthesize the PIN/PVAc composites at the time of polymerization.

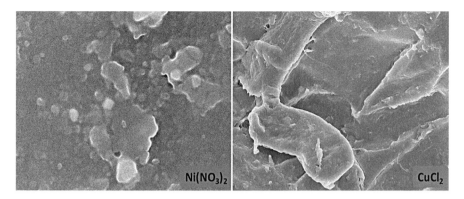

FIGURE 6.11 FE-SEM micrographs.

6.3.10 XRD ANALYSIS

In general, Figure 6.12 represents the XRD patterns for the PIN/PVAc composites. The XRD pattern of PIN/PVAc composites synthesized via Ni(NO$_3$)$_2$ represents no characteristic peaks, which correspond to any crystalline phase. The broad hump appears at nearly same 2-theta scale in XRD patterns for all samples of PIN/PVAc composites. The broad hollow in XRD patterns suggests an amorphous structure, that is, the feature of polymer. Also, the amorphous nature of the PIN/PVAc composites has evident from the noisy peaks present in the XRD patterns. The PIN/PVAc composites prepared through oxidant CuCl$_2$ display noisy peaks with some major peaks, that is, the indication of semicrystalline nature. The broad hollow 2-theta scale and polymer chain separation values are determined from the following equation[43]:

$$R = \frac{5\lambda}{8\sin\theta} \quad (6.7)$$

where λ is wavelength of X-ray source and θ is diffraction position. The polymer chain separation of samples was estimated using peak value of amorphous hollow.

The d-values observed for sharp peaks in XRD patterns of all samples are matched with standard d-values of CuCl$_2$. The average crystalline size from sharp peaks is estimated through the Scherer's formula.[44]

$$D = K\lambda/\beta\cos\theta \quad (6.8)$$

where D is the crystalline size, K is the shape factor, which can be assigned a value 0.89 if the shape is unknown, θ is the diffraction angle at maximum peak intensity, and β is the full width at half maximum of diffraction angle in radians. Also, XRD patterns agree with ICDD file number 01-078-7708.[45]

This can be clearly observed from the figure as well as the table that the XRD pattern of both samples has big difference in appearance. The PIN/PVAc composites prepared through the oxidant nickel nitrate have the high intensity amorphous hollow that is character of conducting polymer which is amorphous in nature, whereas the PIN/PVAc composites prepared through the oxidant cupric chloride have amorphous hollow with some major peaks which is the characteristic of semicrystalline nature of conducting polymer. It clearly represents that the oxidant performed key role in synthesis process at the time of polymerization. The nature of prepared conducting polymer or conducting polymer composites directly depends upon the oxidizing agent used for polymerization for synthesis polymeric materials. By comparing XRD pattern of both samples, it can be clearly observed that the PIN/PVAc composites prepared through oxidant nickel nitrate represent more noisy peaks in XRD pattern as compared to oxidant cupric chloride.[46–48]

Figure 6.10 and the table show the difference between two samples. The amorphous hollow appears at 2-theta scale/degree for both are displayed in the table. The polymer chain separation values are calculated and given in Table 6.3.

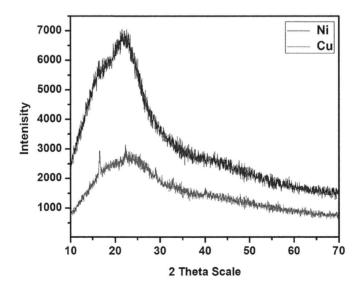

FIGURE 6.12 XRD pattern.

TABLE 6.3 Broad Hump, Average Polymer Chain, and Average Crystalline Size.

Sample	Broad hump at 2θ position	Average polymer chain separation R (Å)	Average crystalline size D (nm)
Cu	22.1402	5.007	16.021
Ni	21.05	4.7851	–

The synthesis method is most important in the property of materials. In present study, the polymerization method was similar to synthesize the PIN/PVAc composites. But the two different oxidants were used for polymerization of indole that affects the chemical and physical properties of composites. The above discussed results of various characterizations show the differences in both samples of PIN/PVAc composites. It confirmed that the oxidant plays important role in chemical structure, surface morphology as well as their band gap.

6.4 CONCLUSIONS

The PIN/PVAc composites were successfully synthesized via chemical polymerization technique using different oxidants. The photoluminescence spectrum reflects the higher intensity for PIN/PVAc composite synthesized using oxidant nickel nitrate. The optical conductivity values of PIN/PVAc composites established for oxidant nickel nitrate and cupric chloride are 1.335×10^8 and 1.319×10^7 S^{-1}, respectively.

The optical bandgap energy values of the composites were calculated to be 4.26 and 4.79 eV for oxidants nickel nitrate and cupric chloride, respectively. The calculated optical bandgap energy has conventional value for the photovoltaic activities. These composites have potential of application in the renewable energy sources such as polymeric solar cells and optical devices.

ACKNOWLEDGMENTS

Authors are very much thankful to Principal, Arts Science and Commerce College, Chikhaldara for providing necessary facilities. Also, they are thankful to Head Department of Physics, Sant Gadge Baba Amravati University, Amravati for providing X-ray diffraction and photoluminescence facility. Also, author is thankful to Head Department of Chemistry, Sant Gadge Baba Amravati University, Amravati for providing ultraviolet–visible facility.

One of the authors D. J. Bhagat dedicated this chapter to his parents.

KEYWORDS

- photoluminescence
- polyindole/poly(vinyl acetate)
- spectroscopy
- microscopy
- composites

REFERENCES

1. Tsur, Y.; Zemel, D. A. Long-Term Perspective on the Development of Solar Energy. *Sol. Energy* **2000**, *68*, 379–392.
2. Spanggaard, H.; Krebs, F. C. A Brief History of the Development of Organic and Polymeric Photovoltaics. *Sol. Energy Mater. Sol. Cells* **2004**, *83*, 125–146.
3. Mahmud, H. N. M. E.; Kassim, A.; Zainal, Z.; Yunus, W. M. M. Fourier Transform Infrared Study of Polypyrrole-Poly(Vinyl Alcohol) Conducting Polymer Composite Films: Evidence of Film Formation and Characterization. *J. Appl. Polym. Sci.* **2006**, *100*, 4107–4113.
4. Mallikarjuna, N. N.; Venkataraman, A.; Aminabhavi, T. M. A Study on γ-Fe$_2$O$_3$ Loaded Poly(Methyl Methacrylate) Nanocomposites. *J. Appl. Poly. Sci.* **2004**, *94*, 2551–2557.
5. Murugendrappa, M. V.; Ambika-Prasad, M. V. N. Dielectric Spectroscopy of Polypyrrole-[Gamma]-Fe$_2$O$_3$ Composites. *Mater. Res. Bull.* **2006**, *41*, 1364–1371.
6. Bhagat, D. J.; Dhokane, G. R. Studies on Thermal Analysis and Optical Parameters of Cu Doped Poly(Vinyl Acetate)/Polyindole Composites. *Appl. Sur. Sci.* **2015**, *351*, 1440–1445.
7. Bhagat, D. J.; Dhokane, G. R. Electro-optical Properties of One Pot Synthesized Polyindole in the Presence of Poly(Vinyl Acetate). *Electron. Mater. Lett.* **2015**, *11*, 346–351.
8. Winey, K. I.; Kashiwagi, T.; Mu, M. Improving Electrical Conductivity and Thermal Properties of Polymers by the Addition of Carbon Nanotubes as Fillers. *Mater. Res. Bull.* **2007**, *2*, 348–353.
9. Bauhofer, W.; Kovacs, J. Z. A Review and Analysis of Electrical Percolation in Carbon Nanotube Polymer Composites. *Compos. Sci. Technol.* **2009**, *69*, 1486–1498.
10. Alig, I.; Poetschke, P.; Lellinger, D.; Skipa, T.; Pegel, S.; Kasaliwal, G. R. Establishment, Morphology and Properties of Carbon Nanotube Networks in Polymer Melts. *Polymer* **2012**, *53*, 4–28.
11. Gupta, B.; Chauhan, D. S.; Prakash, R. Controlled Morphology of Conducting Polymers: Formation of Nanorods and Microspheres of Polyindole. *Mater. Chem. Phys.* **2010**, *120*, 625–630.
12. Giribabu, K.; Manigandan, R.; Suresh, R.; Vijayalakshmi, L.; Stephen, A.; Narayanan, V. Polyindole Nanowires: Synthesis, Characterization and Electrochemical Sensing Property. *Chem. Sci. Trans.* **2013**, *2*, S13–S16.

13. Koiry, S. P.; Saxena, V.; Sutar, D.; Bhattacharya, S.; Aswal, D. K.; Gupta, S. K.; Yakhmi, J. V. Interfacial Synthesis of Long Polyindole Fibers. *J. Appl. Polym. Sci.* **2007**, *103*, 595–599.
14. Gopi, D.; Saraswathy, R.; Kavitha, L.; Kim, D. K. Electrochemical Synthesis of Poly(Indole-*co*-Thiophene) on Low-Nickel Stainless Steel and Its Anticorrosive Performance in 0.5 mol L^{-1} H$_2$SO$_4$. *Polym. Int.* **2013**, *63*, 280–289.
15. Zhijiang, C.; Chengwei, H. Study on the Electrochemical Properties of Zinc/Polyindole Secondary Battery. *J. Power Sources* **2011**, *196*, 10731–10736.
16. Tuken, T.; Yazici, B.; Erbil, M. The Use of Polyindole for Mild Steel Protection. *Surf. Coat. Technol.* **2006**, *200*, 4802–4809.
17. Altindal, S.; Sari, B.; Unal, H. I.; Yavas, N. Electrical Characteristics of Al/Polyindole Schottky Barrier Diodes. I. Temperature Dependence. *J. Appl. Polym. Sci.* **2009**, *113*, 2955–2961.
18. Joshi, L.; Prakash, R. Polyindole-Au Nanocomposite Produced at the Liquid/Liquid Interface. *Mater. Lett.* **2011**, *65*, 3016–3019.
19. Handore, K. N.; Bhavsar, S. V.; Pande, N.; Chhattise, P. K.; Sharma, S. B.; Dallavalle, S.; Gaikwad, V.; Mohite, K. C.; Chabukswar, V. V. Polyindole–ZnO Nanocomposite: Synthesis, Characterization and Heterogeneous Catalyst for the 3,4-Dihydropyrimidinone Synthesis Under Solvent-Free Conditions. *Polym. Plast. Technol. Eng.* **2014**, *53*, 734–741.
20. Kumar, A.; Pandey, A. C.; Prakash, R. Electro-oxidation of Formic Acid Using Polyindole-SnO$_2$ Nanocomposite. *Catal. Sci. Technol.* **2012**, *2*, 2533–2538.
21. Ramesan, M. T. Synthesis and Characterization of Magnetoelectric Nanomaterial Composed of Fe$_3$O$_4$ and Polyindole. *Adv. Polym. Technol.* **2013**, *32* (19), 21362.
22. Kumar, A.; Prakash, R. Graphene Sheets Modified with Polyindole for Electro-chemical Detection of Dopamine. *J. Nanosci. Nanotechnol.* **2014**, *14*, 2501–2506.
23. Mudila, H.; Rana, S.; Zaidi, M. G. H.; Alam, S. Polyindole/Graphene Oxide Nanocomposites: The Novel Material for Electrochemical Energy Storage, Fullerenes. *Nanotubes Carbon Nanostruct.* **2014**, *23*, 20–26.
24. Ryu, K. S.; Park, N. G.; Kim, K. M.; Lee, Y. G.; Park, Y. J.; Lee, S. J.; Jeong, C. K.; Joo, J.; Chang, S. H. The Physiochemical Properties of Polyindole/Thiol Composites. *Synth. Met.* **2003**, *135–136*, 397–398.
25. Urkmez, G.; Sari, B.; Unal, H. I. Synthesis and Characterization of Novel Poly(Dimethylsiloxane)/Polyindole Composites. *J. Appl. Polym. Sci.* **2011**, *121*, 1600–1609.
26. Pandey, P. C.; Prakash, R. Polyindole Modified Potassium Ion-Sensor Using Dibenzo-18-Crown-6 Mediated PVC Matrix Membrane. *Sens. Actuators, B* **1998**, *46*, 61–65.
27. Sari, B.; Yavas, N.; Makulogullari, M.; Erol, O.; Unal, H. I. Synthesis, Electrorheology and Creep Behavior of Polyindole/Polyethylene Composites. *React. Funct. Polym.* **2009**, *69*, 808–815.
28. Arslan, Y.; Unal, H. I.; Yilmaz, H.; Sari, B. Electrorheological Properties of Kaolinite, Polyindole, and Polyindole/Kaolinite Composite Suspensions. *J. Appl. Polym. Sci.* **2007**, *104*, 3484–3493.
29. Kalasad, M. N.; Gadyal, M. A.; Hiremath, R. K.; Ikram, I. M.; Mulimani, B. G.; Khazi, I. M.; Anantha Krishnan, S. K.; Rabinal, M. K. Synthesis and Characterization of Polyaniline Rubber Composites. *Compos. Sci. Technol.* **2008**, *68*, 1787–1793.
30. Aydinli, B.; Toppare, L.; Tincer, T. A Conducting Composite of Polypyrrole with Ultrahigh Molecular Weight Polyethylene Foam. *J. Appl. Polym. Sci.* **1999**, *72*, 1843–1850.

31. Tsocheva, D.; Tsanov, T.; Terlemezyan, L. Ageing of Conductive Polyaniline/Poly(Ethylene-*co*-Vinylacetate) Composites Studied by Thermal Methods. *J. Therm. Anal. Calorim.* **2002**, *68*, 159–168.
32. Cui, H. W.; Du, G. B. Preparation and Characterization of Exfoliated Nanocomposite of Polyvinyl Acetate and Organic Montmorillonite. *Adv. Polym. Technol.* **2012**, *31*, 130–140.
33. Goh, S. H.; Chan, H. S. O.; Ong, C. H. Miscibility of Polyaniline/Poly(Vinyl Acetate) Blends. *Polymer* **1996**, *37*, 2675–2679.
34. Talbi, H.; Ghanbaja, J.; Billaud, D.; Humbert, B. Vibrational Properties and Structural Studies of Doped and Dedoped Polyindole by FTIR, Raman and EEL Spectroscopies. *Polymer* **1997**, *38*, 2099–2106.
35. Ramesan, M. T. Fabrication and Characterization of Conducting Nanomaterials Composed of Copper Sulphide and Polyindole. *Polym. Compos.* **2012**, *33*, 2169–2176.
36. Bhagat, D. J.; Dhokane, G. R. Novel Synthesis and DC Electrical Studies of Polyindole/Poly(Vinylacetate) Composite Films. *Chem. Phys. Lett.* **2015**, *619*, 27–31.
37. Shimano, J. Y.; MacDiarmid, A. G. Polyaniline, a Dyanamic Block Copolymer: Key to Attaining Its Intrinsic Conductivity. *Synth. Met.* **2001**, *123*, 251–262.
38. Wohlgenannt, M.; Vardeny, Z. V. Spin Dependent Exciton Formation Rates in Conjugated Materials. *J. Phys. Condens. Matter* **2003**, *15*, R83–R107.
39. Abthagir, P. S.; Saraswathi, R. Charge Transport and Thermal Properties of Polyindole, Polycarbazole and Their Derivatives. *Thermochim. Acta* **2004**, *424*, 25–35.
40. Bhagat, D. J.; Dhokane, G. R. UV–Vis Spectroscopic Studies of One Pot Chemically Synthesized Polyindole/Poly(Vinyl Acetate) Composite Films. *Mater. Lett.* **2014**, *136*, 251–253.
41. Nemade, K. R.; Waghuley, S. A. UV–Vis Spectroscopic Study of One Pot Synthesized Strontium Oxide Quantum Dots. *Results Phys.* **2013**, *3*, 52–54.
42. Sharma, P.; Katyal, S. C. Determination of Optical Parameters of a-$(As_2Se_3)_{90}Ge_{10}$ Thin Film. *J. Phys. D: Appl. Phys.* **2007**, *40*, 2115–2120.
43. Waghuley, S. A.; Yenorkar, S. M.; Yawale, S. S.; Yawale, S. P. Application of Chemically Synthesized Conducting Polymer–Polypyrrole as a Carbon Dioxide Gas Sensor. *Sens. Actuators B: Chem.* **2008**, *128*, 366–373.
44. Cullity, B. D. *Elements of X-ray Diffraction*; Addison–Wesley Publishing Company Inc.: London, 1978.
45. Malcherek, T.; Schluter, J. Structures of the Pseudo-Trigonal Polymorphs of $Cu_2(OH)_3Cl$. Corrigendum, *Acta Cryst.* **2009**, *65*, 334–341.
46. Bhagat, D. J.; Dhokane, G. R. Frequency Dependent Conductivity and Dielectric Behaviour Studies of Cu Doped Polyindole in Presence of Poly(Vinyl Acetate). *J. Inorg. Organomet. Polym.* **2017**, *27*, 46–52.
47. Bhagat, D. J.; Dhokane, G. R. Study of Chemically Synthesized Polyindole/Poly(Vinyl Acetate) Conducting Composite Film. *Macromol. Symp.* **2016**, *362*, 145–148.
48. Bhagat, D. J.; Dhokane, G. R.; Electro-Optical Properties of Poly(Vinyl Acetate)/Polyindole Composite Film. *AIP Conf. Proc.* **2016**, *1728*, 020171.

CHAPTER 7

WELDING OF ALLOY C-276

MANIKANDAN M.[*], ARIVAZHAGAN N., and NAGESWARA RAO M.

School of Mechanical Engineering, VIT University, Vellore, India

[*]Corresponding author. E-mail: mano.manikandan@gmail.com

ABSTRACT

Superalloy C-276 is susceptible to hot cracking during gas tungsten arc welding (GTAW) technique. The occurring of microsegregation during solidification leads to the formation of P and µ phases. These phases have been accountable for the hot cracking. The chapter investigates the improvement of weldability of alloy C-276 by switching over from gas tungsten arc (GTA) to pulsed current gas tungsten arc (PCGTA) welding. Pulsed and continuous current GTAW was carried out both by autogenous mode and using different filler wires (ERNiCrMo-3 and ERNiCrMo-4). The weld joints were investigated on microstructure, microsegregation, and mechanical properties. The optical and scanning electron microscopy revealed that current pulsing improved the overall mechanical properties with refined microstructure in the fusion zone of all the weldments. Energy dispersive X-ray spectroscopy results bring out that joints fabricated by PCGTA autogenous and ERNiCrMo-4 shows reduced microsegregation compared to similar GTA weld joints. Joints fabricated by ERNiCrMo-3 indicate higher segregation with lower mechanical properties compared to other weldments. Bend test results in the absence of cracking irrespective of the type of welding are adopted in the current chapter.

7.1 INTRODUCTION

Alloy C-276 is a highly corrosion resistant nickel-based alloy derived from the Ni–Cr–Mo ternary system. Further, this alloy C-276 exhibits better corrosion resistance compared to other common materials such as alloy 625, AISI 316, and Monel 400 especially under marine environment involving crevice corrosion conditions particularly.[1] The alloy C-276 is having a broad range of application in the nuclear industry.[2] The welding of alloy C-276 is possible with commercial welding techniques like gas metal arc welding (GMAW) and gas tungsten arc welding (GTAW). Many researchers have reported that the precipitation of intermetallic phases μ and P during solidification of alloy C-276.[3,4] Cieslak et al.[5] investigated the arc welding on alloys C-4, C-22, and C-276. The authors observed that elemental segregation during solidification leading to the formation of topologically closed phases (TCP) P and μ in alloys C-22 and C-276. The authors also reported that alloy C-276 shows the highest susceptibility to hot cracking due to these phases. These phases are possible equilibrium structures in the Ni–Cr–Mo ternary system.[4] The presence of these phase reduced the mechanical, metallurgical and corrosion behavior of the alloy.[6] In general, the P and μ phases in some nickel-based alloys are unavoidable due to the higher volume content of Mo (24%). For example, Perricone and DuPont[7] reported the absence of these phases in the weldment containing 12% Mo in Ni–Cr–Mo alloy.

The core issue is brought down by the formation of secondary P and μ phases in the alloy C-276 during the industrial welding process. This is possible by reducing the severity on the microsegregation during the fusion-welding process. In this context, the pulsed current gas tungsten arc welding (PCGTAW) has shown some real hands in the welding of alloy C-276. PCGTAW is a variation of constant current GTAW. In contrast to GTAW, during pulsed current gas tungsten arc (PCGTA) the material will be melt during peak current pulses for brief intervals of time allows the heat dissipate into the base material.[8]

Many kinds of literature reported that the faster cooling rate and shorter solidification time could lead to improving the metallurgical and mechanical properties by reducing the deleterious secondary phases in the weld joints. Farahani et al.[9] compared the gas tungsten arc (GTA) and PCGTA-welded fusion zone microstructure of superalloy 617 and concluded that PCGTAW obtained the refined microstructure and improved mechanical properties compared to GTAW. Radhakrishna et al.[10] compared the GTAW and electron beam welding (EBW) in superalloy 718 and concluded that faster cooling achieved in EBW obtained reduced Nb segregation in the fusion zone. Madhusudhan Reddy et al.[11] Also attempted the comparative studies on GTAW and EBW in dissimilar austenitic and ferritic stainless steel. The author concluded that rapid solidification in EBW leads to improved metallurgical and mechanical properties compared to GTAW joints. Manikandan et al.[12] attempted comparative studies on weldment produced by GTAW and PCGTA in alloy 718 and concluded that current pulsing achieved a faster cooling rate with a consequent reduction in Nb-rich Laves phase in the weld zone. Kumara and Shahi[13] studied the influence of welding technique on the metallurgical and mechanical performance of stainless steel weldment. Also, the authors indicated that low heat input with faster solidification resulted in higher hardness and impact toughness as compared to welding techniques with higher heat input. Manikandan et al.[14] studied the continuous Nd:YAG laser welding and concluded that finer grain size, superior mechanical properties, and also acceptable limits of microsegregation compared to the arc-welding process. Guangyi et al.[15] reported that pulsed laser beam welding shows relatively less microsegregation compared to the arc welding process. Guangyi et al.[15] also indicated that microsegregation in C-276 joints produced by pulsed laser beam welding is relatively less compared to the arc welding process.

There is no published literature available with particular reference to alloy C-276 by GTAW and PCGTAW welding process by autogenous; ERNiCrMo-3 and ERNiCrMo-4 filler wires. Since the weld microstructure and mechanical properties strongly depend on the welding process and the filler wire, this research is undertaken to investigate the effect of pulsed current on metallurgical and mechanical properties of alloy C-276 weld by GTAW. The authors published[8,16] the detailed microstructure and microsegregation analysis on weld-interface regions. The present chapter focused on the weld center of the fusion zone. Mechanical properties of the weld joints fabricated by different welding process are correlated with microstructure and scanning electron microscopy/energy dispersive X-ray spectroscopy (EDS) analysis. The aim of the present study is to compare the weldability of six different welding process of alloy C-276. The outcome of the study shall be highly beneficial to the industries operating with the alloy C-276.

7.2 MATERIALS AND METHODS

7.2.1 MATERIAL AND WELDING PROCEDURE

The base material was alloy C-276 with a thickness of 4 mm. The chemical composition of as received base metal and filler wires such as ERNiCrMo-3 and ERNiCrMo-4 are listed in Table 7.1. The process parameters of the weld joints are given in Table 7.2. The weld joint was carried out both GTA and PCGTA mode with (ERNiCrMo-3 and ERNiCrMo-4) and without filler wire (autogenous). To ensure freedom from weld deficiencies, radiography examination were carried out on the weld joints.

TABLE 7.1 The Chemical Composition of As-received Base Metal Alloy C-276 and Filler Wires.

Base/filler metal	Chemical composition (% wt.)								
	Ni	Mo	Cr	W	Co	Mn	Fe	Nb	Others
Alloy C-276	Bal	16.5	15.5	3.5	0.05	0.47	6.00	–	0.18 (V), 0.006 (P), 0.001 (S), 0.03 (Si), 0.004 (C)
ERNiCrMo-3	Bal	11.0	22.5	–	–	0.6	1.5	5.0	0.016 (P), 0.014 (S), 0.4 (Si), 0.6 (Cu), 0.3 (Al), 0.3 (Ti), 0.2 (C)
ERNiCrMo-4	Bal	16.5	16.0	3.5	2.0	1.0	6.0	–	0.05 (P), 0.04 (S), 0.6 (Cu), 0.03 (C)

TABLE 7.2 Process Parameters Used for Welding.

Welding parameters Current (A)	GTA auto-genous 160	PCGTA autog-enous 120	GTA ERNi CrMo-3 130	PCGTA ERNiCr Mo-3 120	GTAW ERNi CrMo-4 120	PCGTA ERNiCr Mo-4 120
Peak current (A)	–	240	–	240	–	240
Voltage (V)	15	17	11	13	13.2	11.7
Pulse frequency (Hz)	–	6	–	6	–	6
Pulse on time	–	50%	–	50%	–	50%

7.2.2 METALLURGICAL CHARACTERIZATION

Microstructure examination was conducted by both optical and scanning electron microscope. The samples were extracted in the weld joints plane transverse to the weld bead. The standard metallographic procedure was adapted to polish the specimens. To reveal the microstructure, samples are etched with 80 mL HCl, 4 mL HNO_3, 1 g $CuCl_2$, and 20 mL glycerol solution. EDS elemental analysis was carried out to study about the extent of microsegregation taking place during solidification. Emphasis was given matrix metal Ni and alloying elements Cr, Mo, W, and Fe.

7.2.3 MECHANICAL TESTING

Microhardness test was carried out in the composite region which consists of base metal, weld zone, and heat-affected zone (HAZ). The test was performed with a load of 500 gf with a dwell period of 10 s at regular intervals of 0.25 mm. Tensile test specimens were prepared as per ASTM: E8M-13a. The test specimens were extracted with an axis perpendicular to weld bead. The test was done in universal testing machine (Instron make Model 8801) with a strain rate of 2 mm/min. Charpy V-notch impact test evaluated the toughness of the weld joints. Specimens were prepared, and the test was conducted as per ASTM: E23-12C. The energy observed in the fracture were recorded. Tensile and impact test were done in triplicate to check the reproducibility of the test results. To determine the mode of failure, SEM fractographic analysis was carried out on the tensile and impact failure samples. Bend test was performed as per ASTM E190-92(2008). The test was done using the universal testing machine. The load was applied to the root of the weld doing up to 180° maximum bending angle.

7.3 RESULTS

7.3.1 MICROSTRUCTURE EXAMINATION

The micrograph of as-received base metal which was treated in solution showed the fine equiaxed grains with clear grain boundaries (Fig. 7.1). Also, annealing twins can be seen in several grains.

FIGURE 7.1 As-received base metal alloy C-276 microstructure.

Figure 7.2a represented optical microstructure of autogenous welding fusion zone closest to GTA weld center. The fusion zone weld center is showing cellular structures. The corresponding regions microstructure of autogenous PCGTA welding process shows in Figure 7.2b. The weld center dominates the fine equiaxed dendrites.

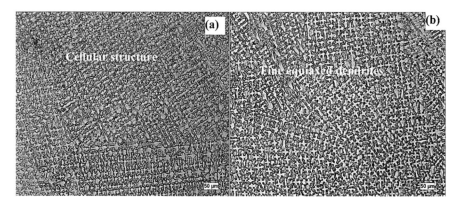

FIGURE 7.2 Fusion zone microstructure of autogenous weldments (a) GTAW and (b) PCGTAW.

The optical microstructure of ERNiCrMo-3 weld center of the GTA weldment is dominating the coarse cellular structure (Fig. 7.3a), whereas the equivalent region of the PCGTA weldment is found to be the fine equiaxed dendrite structure (Fig. 7.3b).

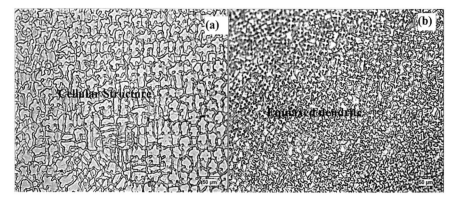

FIGURE 7.3 Fusion zone microstructure of joint produced by ERNiCrMo-3. (a) GTAW and (b) PCGTAW.

The optical micrograph of the ERNiCrMo-4 weld center of the GTA weldment is shown in Figure 7.4a. The coarse cellular structure and columnar grains can be seen running nearly diagonally through the micrograph, whereas the fusion zone of the PCGTA weldment predominated with fine equiaxed dendrite structures (Fig. 7.4b).

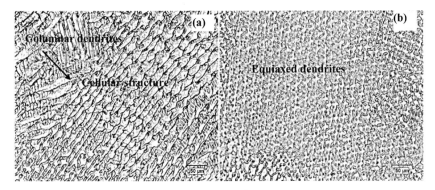

FIGURE 7.4 Fusion zone microstructure of joint produced by ERNiCrMo-4. (a) GTAW and (b) PCGTAW.

7.3.2 SEM/EDS ANALYSIS

SEM micrograph of autogenous GTA weld is dominating the cellular structure with columnar dendrites (Fig. 7.5a). The EDS results obtained from fusion zones of the autogenous GTA weldments are shown in Fig. 7.5(i–ii). Table 7.3 is a compilation of levels of different elements—Ni, Cr, Mo, W, and Fe—at subgrain boundary and subgrain body (matrix) in the fusion zone of the weld joint. It is observed that there is segregation of Mo and W in subgrain boundaries as compared to subgrain interiors, whereas the fine equiaxed dendrite structures were predominated in the fusion zone employing PCGTA autogenous welding process (Fig. 7.6a). The individual EDS results are represented in Figure 7.6(i–ii) and Table 7.4. It is observed that the element levels between boundary and matrix are less differentiated as compared to the GTA autogenous weld.

The SEM/EDS results of GTA ERNiCrMo-3 weld is represented in Figure 7.7(i–ii) and Table 7.5. It is witnessed that the Mo segregates aggressively in the subgrain boundary as compared to the subgrain body and similarly W also segregates little higher in subgrain boundary. The individual EDS results of PCGTAW are shown in Figure 7.8(i–ii) and Table 7.6. It is observed from the subgrain boundary segregation of Mo and W was noticed.

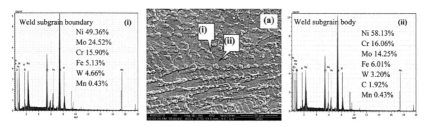

FIGURE 7.5 SEM/EDS analysis of weld joint produced by autogenous GTAW. (a) SEM-fusion zone: (i) EDAX of weld subgrain boundary and (ii) EDAX of weld subgrain body.

TABLE 7.3 EDS Analysis of Autogenous GTA Weldment.

Type of welding	Zone	Ni	Cr	Mo	W	Fe
Autogenous GTA	Weld subgrain boundary	49.36	15.90	24.52	4.66	5.13
	Weld subgrain body	58.13	16.06	14.25	3.20	6.01

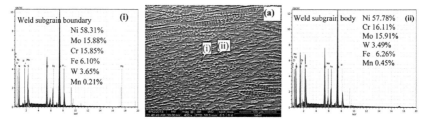

FIGURE 7.6 SEM/EDS analysis of weld joint produced by autogenous PCGTAW. (a) SEM-fusion zone: (i) EDAX of weld subgrain boundary and (ii) EDAX of weld subgrain body.

TABLE 7.4 EDS Analysis of Autogenous PCGTAW Element.

Type of welding	Zone	Ni	Cr	Mo	W	Fe
Autogenous PCGTA	Weld subgrain boundary	58.31	15.85	15.88	3.65	6.10
	Weld subgrain body	57.78	16.11	15.91	3.49	6.26

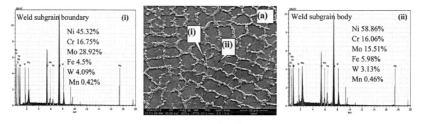

FIGURE 7.7 SEM/EDS analysis of weld joint produced by GTAW with ERNiCrMo-3. (a) SEM-fusion zone: (i) EDAX of weld subgrain boundary and (ii) EDAX of weld subgrain body.

Welding of Alloy C-276

TABLE 7.5 EDS Analysis of GTAW Fabricated with ERNiCrMo-3.

Type of welding	Zone	Ni	Cr	Mo	W	Fe
ERNiCrMo-3	Weld subgrain boundary	45.32	16.75	28.92	4.09	4.5
	Weld subgrain body	58.86	16.06	15.51	3.13	5.98

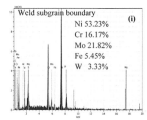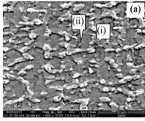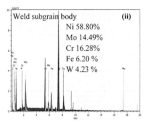

FIGURE 7.8 SEM/EDS analysis of weld joint produced by PCGTAW with ERNiCrMo-3 (a) SEM-fusion zone: (i) EDAX of weld subgrain boundary and (ii) EDAX of weld subgrain body.

TABLE 7.6 EDS Analysis of PCGTAW Fabricated with ERNiCrMo-3.

Type of welding	Zone	Ni	Cr	Mo	W	Fe
PCGTA ERNiCrMo-3	Weld subgrain boundary	53.23	16.17	21.82	3.33	5.45
	Weld subgrain body	58.80	16.28	14.49	4.23	6.20

The subgrain boundary of ERNiCrMo-4 weld made by GTAW process is enriched in Mo and W and impoverished in Ni compared to subgrain body. It is well established by SEM/EDS analysis and is represented in Figure 7.9(i–ii) and Table 7.7, whereas the effect of segregation is absent in the case of PCGTA welding process. It is confirmed by the SEM/EDS analysis and is shown in Figure 7.10(i–ii) and Table 7.8.

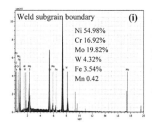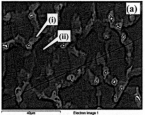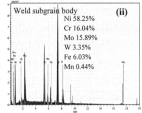

FIGURE 7.9 SEM/EDS analysis of weld joint produced by GTAW with ERNiCrMo-4 (a) SEM-fusion zone: (i) EDAX of weld subgrain boundary and (ii) EDAX of weld subgrain body.

TABLE 7.7 EDS Analysis of GTAW Fabricated with ERNiCrMo-4.

Type of welding	Zone	Ni	Cr	Mo	W	Fe
GTAW	Weld subgrain boundary	54.98	16.92	19.82	4.32	3.54
ERNiCrMo-4	Weld subgrain body	58.25	16.04	15.89	3.35	6.03

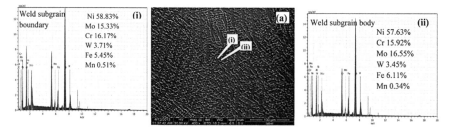

FIGURE 7.10 SEM/EDS analysis of weld joint produced by PCGTAW with ERNiCrMo-4. (a) SEM-fusion zone: (i) EDAX of weld subgrain boundary and (ii) EDAX of weld subgrain body.

TABLE 7.8 EDS Analysis of PCGTAW Fabricated with ERNiCrMo-4.

Type of welding	Zone	Ni	Cr	Mo	W	Fe
PCGTAW	Weld subgrain boundary	58.33	16.17	15.33	3.71	5.45
ERNiCrMo-4	Weld subgrain body	57.63	15.92	16.55	3.45	6.11

7.3.3 MECHANICAL TESTING

Microhardness test measurements were carried out across the weldments. The hardness values obtained in the different weldments are represented in the bar chart are shown in Figure 7.11. The results indicate that average hardness of the fusion zone is higher than the HAZ and base metal. Percent increase in average hardness of fusion zone concerning base metal is provided in Table 7.9. The average hardness was found to be in the range of 224–256 HV in the fusion zone of autogenous, ERNiCrMo-3, and ERNiCrMo-4 welds employing both GTAW and PCGTA-welding process.

Figure 7.12(a)–(g) give photographs of the broken tensile test specimen for all six different weldments. It can be seen from the illustration fracture occurred in the weld joint in all cases. Figure 7.13 shows bar chart representation of cumulative tensile properties of weld joints fabricated by GTA and PCGTA autogenous, different filler wires welding. It is observed from the results weld joint exhibited good ductility value ranging from 34% to 66%. As compared to the GTAW, PCGTA gave better strength and higher ductility values. The average UTS value of PCGTA weldments (autogenous, ERNiCrMo-3, and ERNiCrMo-4) distinguished themselves by way of exhibiting higher by 2%, 3%, and 6% compared to that of GTA weldments. Table 7.10 compared the UTS of base metal with the weld joints. It is observed that the average weldment of GTA and PCGTA fabricated by ERNiCrMo-3 shows less than the as-received base metal (750 MPa) (Fig. 7.13).

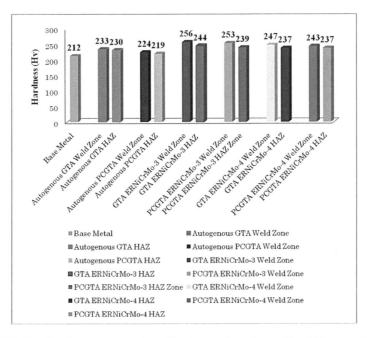

FIGURE 7.11 Bar chart representation of hardness values obtained by different weldments.

TABLE 7.9 Percent Improvement of Average Fusion Zone Hardness Concerning Base Metal.

Type of welding	% Increase
GTA autogenous	9
PCGTA autogenous	6
GTA ERNiCrMo-3	21
PCGTA ERNiCrMo-3	19
GTA ERNiCrMo-4	17
PCGTA ERNiCrMo-4	15

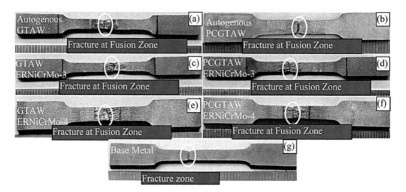

FIGURE 7.12 Photographs of tensile failure samples (a) GTA autogenous, (b) PCGTA autogenous, (c) GTA with ERNiCrMo-3, (d) PCGTA with ERNiCrMo-3, (e) GTA with ERNiCrMo-4, (f) PCGTA with ERNiCrMo-4, and (g) base metal.

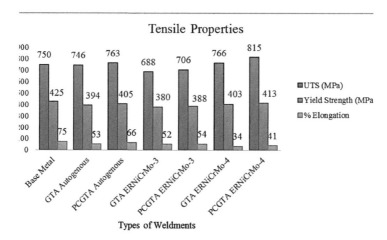

FIGURE 7.13 Bar chart representation of average tensile properties obtained by different weldments.

TABLE 7.10 Extent of Increase or Decrease in UTS of Weldment Compared to UTS of Base Metal.

Type of welding	% Increase (>)/decrease (<)
GTA autogenous	<0.5
PCGTA autogenous	>2
GTA ERNiCrMo-3	<8
PCGTA ERNiCrMo-3	<6
GTA ERNiCrMo-4	>2
PCGTA ERNiCrMo-4	>7

Welding of Alloy C-276

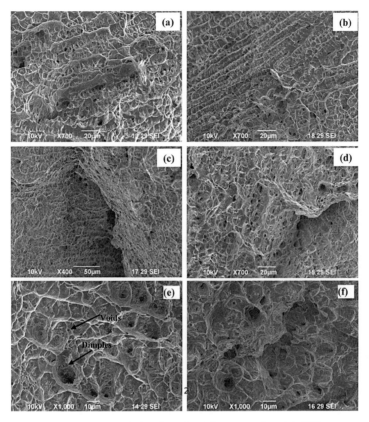

FIGURE 7.14 Fractograph analysis of tensile failure samples. (a) GTA autogenous, (b) PCGTA autogenous, (c) GTA with ERNiCrMo-3, (d) PCGTA with ERNiCrMo-3, (e) GTA with ERNiCrMo-4, and (f) PCGTA with ERNiCrMo-4.

Figure 7.15 shows the photograph of impact test samples. It can be seen that the weldments are not completely undergone rupture. Figure 7.16 shows the SEM fractographic analysis. Fractography revealed the presence of dimples and microvoids. Figure 7.17 represents the impact toughness values of GTA and PCGTA weldments fabricated by autogenous mode and ERNiCrMo-3 and ERNiCrMo-4 filler wires. It is observed from the data that the average toughness of PCGTA weldments is higher than the GTAW in all the cases. The PCGTAW average toughness was 4% higher by autogenous and ENiCrMo-3 and 6% in ERNiCrMo-4 compared to similar GTA weldment. The average toughness obtained with PCGTAW autogenous and ERNiCrMo-4 (56 J) is very close to that of base metal (57 J). The impact toughness of ERNiCrMo-3 is less compared to other weldments and base metal.

Deformation at notch (No breakage)

FIGURE 7.15 Photograph of impact test samples. (a) GTA autogenous, (b) PCGTA autogenous, (c) GTA with ERNiCrMo-3, (d) PCGTA with ERNiCrMo-3, (e) GTA with ERNiCrMo-4, and (f) PCGTA with ERNiCrMo-4.

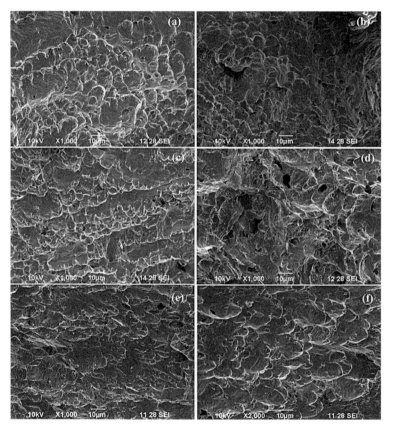

FIGURE 7.16 Factograph of impact test samples. (a) GTA autogenous, (b) PCGTA autogenous, (c) GTA with ERNiCrMo-3, (d) PCGTA with ERNiCrMo-3, (e) GTA with ERNiCrMo-4, and (f) PCGTA with ERNiCrMo-4.

The bend test evaluates the ductility and soundness of the weld joint. Figure 7.18 shows the root bend samples of all the six cases. No cracks or fissures were noticed in the all six cases.

Welding of Alloy C-276

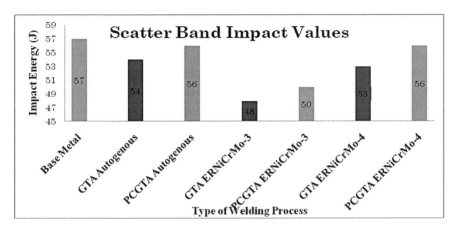

FIGURE 7.17 Bar chart representation of impact values obtained by different weldments.

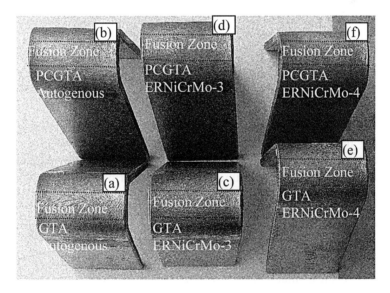

FIGURE 7.18 Photographs of bend test specimens. (a) GTA autogenous, (b) PCGTA autogenous, (c) GTA with ERNiCrMo-3, (d) PCGTA with ERNiCrMo-3, (e) GTA with ERNiCrMo-4, and (f) PCGTA with ERNiCrMo-4.

7.4 DISCUSSION

Significant refinement observed in the microstructure in both optical and scanning electron when one switches over from GTAW to PCGTAW. Weld metal solidification influences many ways in the current pulsing. Current pulsing cycling welded current from a low level to high level at a selected frequency. The cycling of current resulted in (1) variation cyclic temperature, (2) higher weld metal cooling rates, (3) higher fluid flow, and (4) abridged thermal gradients in the weld pool.[17] This contribution leads to refined microstructure in the PCGTA weldments (Figs. 7.2b, 7.3a, and 7.4a).

In all the three cases (autogenous, filler wires ERNiCrMo-3, and ERNiCrMo-4), GTA weldments (Figs. 7.2a, 7.3a, and 7.4a) show coarse columnar dendrites/coarse cellular structures. The slower cooling in GTAW tends to have coarse microstructure compared to the faster cooling achieved in the PCGTAW with fine microstructure.[18] A similar observation is reported in the fusion of alloy 617 on switching over from GTAW to PCGTAW.[9]

EDS analysis was carried out to evaluate the microsegregation of alloying elements Ni, Cr, Mo, W, and Fe. The exception of Cr, all of them shows a tendency to microsegregation. The P and μ phases are composed of above alloying elements. These phases occur during solidification. Solidification of alloy C-276 begins with the crystallization of γ, but end with the final solidification of P and μ phases. Solidification starts by a primary liquid → γ, and proceeds to cause enrichment of the interdendritic liquid in Mo and W until a eutectic reaction L → γ + P + μ occurs terminating the solidification process.[5]

The tendency of microsegregation of alloying elements can be quantity by using the concept of distribution coefficient (k). Previous researcher Perricone and DuPont[7] used the concept to quantify the microsegregation of various Ni–Cr–Mo based superalloys. The distribution coefficient (k) is calculated using the equation:

$$k = \frac{C_{core}}{C_o} \qquad (7.1)$$

where C_{core} is the elemental level measured in the subgrain body and C_o is the elemental level in the nominal composition of the alloy.

Table 7.11 displayed the k values of different alloying elements for both GTAW and PCGTAW calculated using eq 7.1. The k values were observed closer to 1 for alloying elements Mo and W on moving from GTAW to PCGTAW. The severity of microsegregation is reduced in the PCGTAW case. A similar reduction in microsegregation was observed in the superalloy 718 by a current pulsing technique in GTAW.[12,19]

Filler wire employed by ERNiCrMo-3 both GTAW and PCGTAW show higher segregation of Mo, W, and Fe compared to ERNiCrMo-4 and autogenous weldments. The larger variation in chemical composition between the base metal and ERNiCrMo-3 filler wires leads to higher microsegregation. The nominal chemical composition of alloy C-276 has Ni–16Cr–16Mo–5Fe–4W, whereas ERNiCrMo-3 is Ni–22Cr–10Mo–1Fe. In the case of ERNiCrMo-4 filler wire, no such microsegregation is observed. The chemical compositions are at the same level as in the base metal.

TABLE 7.11 Distribution Coefficient (k) Values of Alloying Elements in GTAW and PCGTAW.

Elements	Ni	Cr	Mo	W	Fe
GTA autogenous	1.00	1.01	0.87	0.92	0.99
PCGTA autogenous	1.00	1.01	0.97	1.01	1.03
GTA ERNiCrMo-3	1.02	1.01	0.94	0.90	0.98
PCGTA ERNiCrMo-3	1.02	1.02	0.88	1.22	1.02
GTA ERNiCrMo-4	1.01	1.01	0.97	0.97	0.99
PCGTA ERNiCrMo-4	1.00	1.00	1.01	1.00	1.00

The segregation of alloying elements Mo and W can be understood regarding mismatch of atomic radii between matrix element Ni and alloying elements. The mismatch of atomic radii is high as 9% for Mo and 10% for W compared to Ni. Alloying elements Cr and Fe did not show any microsegregation in both GTA and PCGTAW. The atomic radii between Cr and Fe are less than 1% compared to matrix Ni element. These observed level have no tendency for microsegregation during solidification.[8,16]

TCP phases (μ and P) influenced the weld microstructure during solidification. It is observed from the present study the microsegregation taken place in the GTAW; PCGTAW has resulted in the absence from microsegregation in ERNiCrMo-4 and autogenous case. The results obtained from the PCGTA clearly show that the occurrences of P and μ phases are largely reduced. The faster cooling rate achieved in the PCGTA that lead to a lower microsegregation.[17]

The extent of segregation governs the following inequality in GTAW:

GTA ERNiCrMo-3 > GTA ERNiCrMo-4 > GTA autogenous   (7.2)

The extent of segregation governs the following inequality in PCGTAW. The observed trends are similar to a later case:

PCGTA ERNiCrMo-3 > PCGTA ERNiCrMo-4 > PCGTA autogenous   (7.3)

The average fusion zone hardness values improved from the base metal is represented in Table 7.9. Figure 7.11 is a bar chart representation of the corresponding hardness values. It can be seen that in all the three cases autogenous and filler wires (ERNiCrMo-3 and ERNiCrMo-4), GTAW resulted in 2–3% higher hardness compared to the PCGTAW. Segregation of Mo and W in the fusion zone of GTA resulted in higher hardness compared to PCGTAW mode as clearly brought out by SEM/EDS analysis (Fig. 7.5–7.10). Dislocation theory states that the hardness or the strength of an alloy is controlled by the density and mobility of dislocations in the alloy. In general, alloys and metal can be strengthened by two ways: (1) to reduce the density of dislocation and (2) to increase the resistance to against the mobility of dislocations in high-density dislocations materials. Many methods can obstruct dislocation of the movement. Hardness can be achieved by precipitation, solid solution hardening, and refinement in microstructure and work hardening. In the present study, it is observed by EDS analysis that the movement of the dislocation can be obstructed by the Mo and W-rich eutectoids in the case of GTAW. The lower cooling rate and the minor temperature gradient between the fusion boundary and center line for continuous current GTA welding with argon shielding gas leads to segregation as reported by Manikandan et al.[12] It leads to increased significantly higher hardness in the case of GTA weldments.

The following inequality is governed by the average microhardness in GTAW:

GTA ERNiCrMo-3 > GTA ERNiCrMo-4 > GTA autogenous   (7.4)

The following inequality is governed by the average microhardness in PCGTAW. The observed trends are similar to a later case.

PCGTA ERNiCrMo-3 > PCGTA ERNiCrMo-4 > PCGTA autogenous   (7.5)

These inequalities are similar to those presented in SEM/EDS analysis. The hardness trends are in line with the extent of segregation. The present study clearly shows that the segregation leads to higher hardness in both GTA and PCGTA weldments.

Figure 7.12 shows the tensile failure samples of GTA and PCGTAW. Figure 7.13 represents the bar chart comparison of tensile property results obtained in the present study. It is observed from the result in all the cases that PCGTA shows 2–6% improvement in strength (UTS) compared to GTAW. It is evident from the microstructure analysis and EDS analysis that the refined microstructure with reduced microsegregation obtained in PCGTAW is widely responsible for the higher strength and ductility compared to GTAW. Table 7.10 compared the strength of different weldments compared to that of base metal. It is seen from Table 7.10 and Figure 7.13 that the UTS values obtained from autogenous GTAW and ERNiCrMo-3 (GTAW and PCGTAW) are less compared to that of base metal. The higher segregation observed in the GTAW is responsible for the less strength compared to PCGTAW and ERNiCrMo-4 filler wire.

Figure 7.13 is a bar chart representation of the comparative tensile results obtained in the present study. It is observed from Figure 7.13 that ultimate tensile strengths of weld joints produced by autogenous GTA and ERNiCrMo-3 (GTA and PCGTA) are less compared to that of base metal. Table 7.10 shows the change in UTS of various weldments compared to that of base metal. In all the three cases, PCGTAW joints exhibit a higher strength and higher ductility as compared to GTAW. It is observed from the tensile results that PCGTAW gives 2–6% improvement in UTS compared to GTAW. The refined of fusion zone microstructure with a reduction in microsegregation occurring on switching over from GTAW to PCGTAW are believed to be largely responsible for the superior mechanical properties of the latter type of joints.

It is observed from the present study that for both GTAW and PCGTAW, use of filler wire ERNiCrMo-3 resulted in less strength as compared to the autogenous mode and use of ERNiCrMo-4 filler wire. This is believed to be related to the relatively high degree of segregation in ERNiCrMo-3 weldments, as brought out by the EDS analysis.

It is noticed from Figure 7.12 in all the cases that failure occurred in the fusion zone. Fractographic analysis (Fig. 7.14) was carried out to evaluate the mode of failure. SEM fractography analysis revealed that in both GTAW and PCGTAW, the fracture occurs preferentially along the interdendritic regions. It is observed that PCGTA shows little marginally improved ductility compared to GTAW. The present study believed that PCGTAW joints are expected to be free from P and μ phases. The presence of TCP phase in the interdendritic regions initiate the favorable cite for the crack initiation and propagation. Previous researchers[6] also witnessed that TCP phases adversely affect the several mechanical properties which include the strength and ductility. The presence of TCP phases not only depletes the matrix of the alloying elements but also provides a favorable site for easy crack initiation and propagation.

Impact toughness obtained in the present studies are compared in the bar chart representation (Fig. 7.17). Similar to the tensile results PCGTWA had higher toughness compared to GTAW. The toughness was improved ~4% in autogenous and ERNiCrMo-3 filler wire, whereas in the case of ERNiCrMo-4, ~6% was observed. It is believed that the refined microstructure with reduced microsegregation seen in current pulsing is to be largely responsible for the higher toughness. It is relevant to note in this context that Farahani et al.[9] reported that current pulsing improved the strength and toughness of the weld joint compared to GTAW in the alloy 617. The toughness value observed from autogenous and PCGTA ERNiCrMo-4 shows near to that base metal; this is believed that relatively reduced segregation compared to the GTA case. The joint fabricated by PCGTA ERNiCrMo-3 filler shows lesser toughness. The higher segregation noticed in the EDS analysis is widely responsible for the reduced toughness values. It is observed from the present study (Fig. 7.15(a–f)) the impact test specimens did not break completely in all the weld joint employed. This indicates that the failure occurs in ductile mode. The SEM factograph (Fig. 7.16) confirms the ductile mode of failure with the support of dimples present on the fracture surface.

The photograph of the bend test samples fabricated by autogenous and filler wires (ERNiCrMo-3 and ERNiCrMo-4) both GTA and PCGTA are presented in Figure 7.18. It can be seen that no cracks were found in any of the weld joints even up to the maximum bending angle of 180°. This confirms the weld joints are a good degree of formability and freedom from defects.

7.5 CONCLUSION

This chapter reports the joints fabricated by GTA and PCGTA welding using autogenous and filler wires (ERNiCrMo-3 and ERNiCrM0-4). The main findings observed in this chapter are listed below.

1. Weld joints fabricated with GTAW mode, both autogenous and with filler wires, show relatively coarse cellular microstructure Welds made with PCGTAW, in contrast, show relatively fine equiaxed dendrite microstructure.
2. Weld joints fabricated with GTAW mode, both autogenous and ERNiCrMo-4, shows enrichment of Mo (segregation) in subgrain boundaries and impoverished in Ni compared to the matrix. PCGTAW, in contrast, shows no such segregation were observed

in the fusion zone. It is understood that the susceptibility of hot cracking of this alloy also comes down in the PCGTAW mode.
3. Weld joints fabricated by ERNiCrMo-3 both GTA and PCGTA show severe segregation. The larger difference in the chemical composition between the base metal and filler wire causes the severe segregation.
4. The extension of microsegregation is interconnected with the hardness values observed. In both GTA and PCGTA weldments, higher segregation leads to higher hardness.
The following inequality observed in the average microhardness in GTAW.
The average microhardness in GTA welds is governed by the following inequality.
The following microhardness in PCGTA weld is regulated by the following inequality. The observed trends are similar to above case.
PCGTA ERNiCrMo-3 > PCGTA ERNiCrMo-4 > PCGTA Autogenous
5. Weld joints fabricated by PCGTAW gave superior tensile strength with good ductility and good impact toughness compared to GTAW. The refined microstructure with reduced microsegregation is believed to be largely responsible for the improvement in the mechanical behavior.
6. Weld joints produced by ERNiCrMo-3 filler wire, both GTA and PCGTA, show less strength compared to base metal and other weldments. The higher segregation noticed in ERNiCrMo-3 is widely responsible for the reduced strength.
7. Bend test result did not show cracking in the weldments produced in all the cases. This indicates that the ductility and soundness of the weld joints were good.

ACKNOWLEDGMENTS

This chapter was supported by the Defence Research Development organization (DRDO) (Grant No. ERIP/ER/1103952/M/01/1403). We also thank Department of Science and Technology for the funding received from them under the FIST program; Instron makes universal testing machine used in the present study was procured under this program.

KEYWORDS

- superalloy C-276
- pulsed current gas tungsten arc welding
- microstructure
- mechanical properties

REFERENCES

1. Inconel Alloy C-276, Technical Data Sheet: Special Metals Corporation. http://www.specialmetals.com/documents/Inconel%20alloy%20C-276.pdf (accessed Oct 23, 2013).
2. Wu, D. J.; Ma, G. Y.; Niu, F. Y.; Guo, D. M. Pulsed Laser Welding of Hastelloy C-276: High-Temperature Mechanical Properties and Microstructure. *Mater. Manuf. Process.* **2013**, *28* (5), 524–528.
3. Mattews, S. J. Thermal Stability of Solid Solution Strengthened High-Performance Alloys. *Miner., Met., Mater. Soc.* **1976**, 215–226.
4. Raghavan, M.; Berkowitz, B. J.; Scanlon, J. C. Electron Microscopic Analysis of Heterogeneous Precipitates in Hastelloy C-276. *Metall. Trans. A* **1982**, *13A*, 979–984.
5. Cieslak, M. J.; Headley, T. J.; Romig, A. D. The Welding Metallurgy of Hastelloy Alloys C-4, C-22, and C-276. *Metall. Trans. A* **1986**, *17A*, 2035–2047.
6. Sims, C. T.; Hagel, W. C. *The Superalloys*; John Wiley & Sons: Hoboken, NJ, 1972.
7. Perricone, M. J.; DuPont, J. N. Effect of Composition on the Solidification Behavior of Several Ni–Cr–Mo and Fe–Ni–Cr–Mo Alloys. *Mater. Trans.* **2006**, *37A*, 1267–1280.
8. Manikandan, M.; Arivazhagan, N.; Nageswara Rao, M.; Reddy, G. M. Improvement of Microstructure and Mechanical Behavior of Gas Tungsten Arc Weldments of Alloy C-276 by Current Pulsing. *Acta Metall. Sin. (Engl. Lett.)* **2015**, *28* (2), 208–215.
9. Farahani, E.; Shamanian, M.; Ashrafizadeh, F. A Comparative Study on Direct and Pulsed Current Gas Tungsten Arc Welding of Alloy 617. *AMAE Int. J. Manuf. Mater. Sci.* **2012**, *02* (01), 1–6.
10. Radhakrishna, C. H.; Prasad Rao, K.; Srinivas, S. Laves Phase in Superalloy 718 Weld Metals. *J. Mater. Sci. Lett.* **1995**, *14*, 1810–1812.
11. Madhusudhan Reddy, G.; Mohandas, T.; Sambasiva Rao, A.; Satyanarayana, V. V. Influence of Welding Processes on Microstructure and Mechanical Properties of Dissimilar Austenitic-Ferritic Stainless Steel Welds. *Mater. Manuf. Process.* **2005**, *20* (2), 147–173.
12. Manikandan, S. G. K.; Siva Kumar, D.; Prasad Rao, K.; Kamaraj, M. Effect of Weld Cooling Rate on Laves Phase Formation in Inconel 718 Fusion Zone. *J. Mater. Process. Technol.* **2014**, *214*, 358–364.
13. Kumar, S.; Shahi, A. S. On the Influence of Welding Stainless Steel on Microstructural Development and Mechanical Performance. *Mater. Manuf. Process.* **2014**, *29* (8), 894–902.

14. Manikandan, M.; Hari, P. R.; Vishnu, G.; Arivarasu, M.; Devendranath Ramkumar, K.; Arivazhagan, N.; Nageswara Rao, M.; Reddy, G. M. Investigation of Microstructure and Mechanical Properties of Super Alloy C-276 by Continuous Nd:YAG Laser Welding. *Proc. Mater. Sci.* **2014,** *5*, 2233–2241.
15. Guangyi, M. A.; Dongjiang, W. U.; Dongming, G. U. O. Segregation Characteristics of Pulsed Laser Butt Welding of Hastelloy C-276. *Metall. Mater. Trans. A* **2011,** *A42*, 3853–3857.
16. Manikandan, M.; Arivazhangan, N.; Nageswara Rao, M.; Reddy, G. M. Microstructure and Mechanical Properties of Alloy C-276 Weldments Fabricated by Continuous and Pulsed Current Gas Tungsten Arc Welding Techniques. *J. Manuf. Process.* **2014,** *16*, 563–572.
17. Janaki Ram, G. D.; Venugopal Reddy, A, Prasad Rao, K, Madhusudhan Reddy, G. Control of Laves Phase in Inconel 718 GTA Welds with Current Pulsing. *Sci. Technol. Weld. Join.* **2004,** *9* (5), 390–398.
18. Van der Voort, G. F. *Metallography and Microstructures ASM Handbook*; ASM International: Novelty, OH, 2004.
19. Janaki Ram, G. D.; Venugopal Reddy, A.; Prasad Rao, K.; Reddy, G. M.; Sarin Sundar, J. K. Microstructure and Tensile Properties of Inconel 718 Pulsed Nd-YAG Laser Welds. *J. Mater. Process. Technol.* **2005,** *167* (1), 73–82.
20. John Dupoint, N.; John Lippold, C.; Samuel Kiser, D. *Welding Metallurgy and Weldability of Nickel-Base Alloys*; John Wiley & Sons Publication: Hoboken, NJ, 1999.
21. Verhoven, J. D. *Fundamentals of Physical Metallurgy*; Wiley: New York, 1975.

CHAPTER 8

SURFACE PLASMON RESONANCE-BASED SENSORS

RAJ KUMAR GUPTA*, DEVANARAYANAN V. P., and V. MANJULADEVI

Department of Physics, Birla Institute of Technology and Science, Pilani 333031, Rajasthan, India

*Corresponding author. E-mail: raj@pilani.bits-pilani.ac.in

ABSTRACT

The sensors based on optical phenomenon, surface plasmon resonance (SPR) is gaining tremendous scientific attention due its label-free nature, very large sensitivity, real-time detection, and analysis capabilities. The resonance condition is very sensitive to a minute change of the refractive index of the dielectric material deposited over the metal surface. This provides opportunity to develop a sensor by proper functionalization of the metal surface. In this chapter, we discuss the fundamentals for excitation of surface plasmon wave and SPR phenomenon. We review few important techniques for observing the SPR phenomenon which have led to the development of SPR-based sensors. We discuss the potential of SPR phenomenon for gas, chemical, and biological sensing applications with examples. In addition, we discuss briefly some nontraditional applications of SPR phenomenon.

8.1 INTRODUCTION

A surface plasmon (SP) wave is charge density oscillation at metal-dielectric interface which can be excited by a polarized light wave incident onto the metallic layer (e.g., gold film) via a coupling high refractive index (RI) medium (e.g., glass prism).[1,2] At surface plasmon resonance (SPR), the component of wavevector of incident light planar to the interface matches to that of SP wave resulting in the absorption of energy of the reflected beam.[3] SPs are first observed in 1902, when a metallic grating was illuminated with polychromatic light and a narrow dark band in the spectrum of diffracted light was observed.[4] In 1958, Turbadar observed a large drop in reflectivity when illuminating thin metal film deposited on a substrate.[5] In 1968, Otto explained Turbadar's results and demonstrated that the drop in the reflectivity in the attenuated total reflection method is due to the excitation of SPs.[6] In the same year, Kretschmann and Reather reported excitation of SPs in another configuration of the attenuated total reflection method. In the late 1970s, SPs were first employed for the characterization of thin films.[7] Liedberg demonstrated first SPR biosensor in 1983.[8]

SPs occur as a result of collective oscillations of the free electrons of a metal. They can be excited on metallic surfaces with a p-polarized light by using a coupling medium such as a prism. To excite the SPs in a thin metallic film on the prism surface, the light incident on the metal–prism interface has to undergo a total internal reflection in the prism. If the component of the incident wavevector parallel to the metal surface (say k_x) is changed, the SPR will occur when k_x is equal to the wavevector of the SP wave (k_p). Let θ_i be the angle of incidence (greater than critical angle), n_{pri} is RI of the prism, and λ is the wavelength of the light source, then

$$k_x = \frac{2\pi}{\lambda} n_{pri} \sin(\theta_i)$$

where k_x can be changed by changing either angle of incidence (θ_i) or the wavelength of the source (λ). In normal SPR device, reflected intensity is recorded as a function of θ_i.

The wavevector of the SP wave is given by

$$k_p = \frac{2\pi}{\lambda} \sqrt{\frac{\varepsilon_1 \varepsilon_2}{\varepsilon_1 + \varepsilon_2}}$$

where ε_1 and ε_2 are dielectric constants of the metal film and organic film deposited on the metal surface.

At SPR,

$$k_x = k_p$$

At resonance, the intensity of the reflected light reduces to a minimum value. The minimum in the reflected intensity at resonance can be thought of as being due to destructive interference between the totally reflected light and the light emitted by SPs. Hence, the resonance can be detected by detecting the minimum intensity of the reflected light. Due to molecular-specific interaction, the RI of the film deposited on the metal surface changes leading to a change in k_p. Consequently, the resonance will occur at a different angle of incidence. Recoding the shift in the resonance angle, change in RI due to molecular-specific interaction can be evaluated. Such physical phenomenon will be employed to understand the molecular interactions and can be exploited for sensor applications.

In the SPR instrument, if the angle of incidence is fixed and a polychromatic light is used, then the reflected light of particular wavelengths will be absorbed at resonance. Such absorption band in the reflected light is dependent on the RI of the organic film on the metal surface. Such phenomenon is effectively used for sensing applications.

The wavevector (k_p) of the SP having energy $\hbar\omega$ is always larger than the wavevector of light in free space. The momentum of SP is large due to the strong coupling between light and surface charges. The electromagnetic field has to drag the surface charges along the metal surface. This shows that the light propagating in free space cannot excite SPs. Thus, the wavevector of light has to be increased over its free space value. Figure 8.1 shows that the addition of an interface (evanescent wave) is a solution to increase the wavevector value of the exciting light above its free space value.[3]

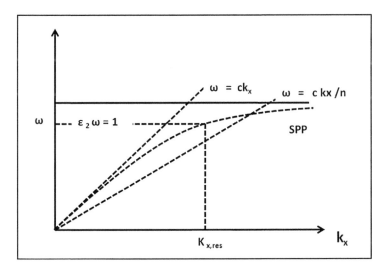

FIGURE 8.1 Dispersion relation in the free space ($\omega = ck_x$) and that of in coupling medium ($\omega = ck_x/n$). Here, ω is angular frequency, c is velocity of light in free space, and n is RI of a medium.

The excitation of SP with the evanescent wave is possible in two configurations: (1) Kretschmann–Reather geometry and (2) Otto geometry. In Kretschmann–Reather geometry, a high RI waveguide (n_w) is interfaced with a metal-dielectric waveguide having thin metal film with permittivity ε_m and thickness (t), and a semi-infinite dielectric with a RI n_d ($n_d < n_w$). The high RI waveguide is generally glass prism. Here, the light propagating through the prism is made incident on the metal film. Above the critical angle of the waveguide, the light confines inside the prism due to total internal reflection. In this condition, a part of the electromagnetic energy propagates as evanescent wave. Evanescent wave couples with the SPs on the metal surface when the metal film is sufficiently thin ($t < 100$ nm). The wavevector of SPs along the thin metal film is influenced by the presence of dielectric on other side of metal film. The change in the dielectric constant on the metal surface leads to the change in wavevector of SP. This condition alters the wave-matching condition and thereby shifts the resonance condition. The geometry is as shown in Figure 8.2.

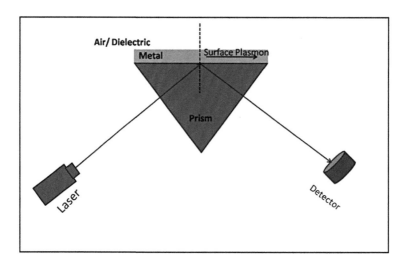

FIGURE 8.2 Schematic diagram for SPR sensor in Kretschmann–Reather geometry.

In Otto geometry, the high RI waveguide is interfaced with dielectric-metal waveguide consisting of a thin dielectric film with RI n_d ($n_d < n_w$) and thickness (t) and semi-infinite metal with permittivity, ε_m. Here, the thickness of dielectric layer chosen to be appropriate for the coupling of evanescent wave to the SP. The geometry is as shown in Figure 8.3.

Surface Plasmon Resonance-Based Sensors 195

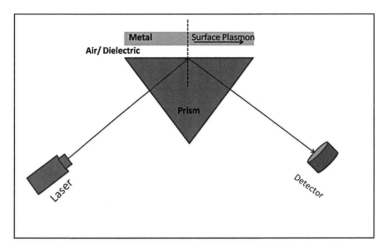

FIGURE 8.3 Schematic diagram for SPR sensor in Otto configuration.

The SPR instrument can work in different modes of scanning—(a) wavelength interrogation and (b) angular interrogation.

8.2 WAVELENGTH INTERROGATION-BASED SPR INSTRUMENT

Wavelength interrogation-based SPR instrument records the drop in intensity in the wavelength spectrum of the total internally reflected beam from the sensor assembly (waveguide with metal-coated surface) when it is illuminated by collimated polychromatic light beam. The incident optical wave and the SP wave in the metal couple at a particular wavelength with respect to the coupling condition as discussed earlier. The wavelength at which the SP at the interface between metal and the sensing medium couples to evanescent wave causes a dip in its intensity in the wavelength spectrum. The wavelength, at which the minimum intensity is recorded, is very sensitive to variations in the RI of the dielectric adjacent to the metal layer. Therefore, the variations in the RI of the sensing medium can be detected by measuring the resonant wavelength.[9] The sensitivity of such a system is defined as the ratio of shift in resonance wavelength on analyte adsorption to RI variation of medium containing the analyte.[10] The schematic of such a sensor is as shown in Figure 8.4.

In the wavelength interrogation method, white LED is used as the light source and a monochromator is used to record wavelength spectrum. The sensitivity of the instrument is 5.83×10^{-4} RI unit (RIU).[11] In an interesting

SPR setup in the wavelength interrogation method, the source was used as LEDs of five different wavelengths. The angle of incidence was altered by rotating mirror and was fixed at the resonant angle for the central wavelength in the spectrum. The CCD camera was used as the detector to record the intensity profile for each wavelength. The interference filters with a full-width half maximum of 10 nm were used to decrease the noise in central wavelength. The setup provides a resolution of 3×10^{-6} RIU. Such a system may become popular because of compactness and economic feasibility and sensitivity similar to a monochromator-assisted SPR spectral sensor.[12] The earlier discussed spectral SPR sensor uses prism as its waveguide.

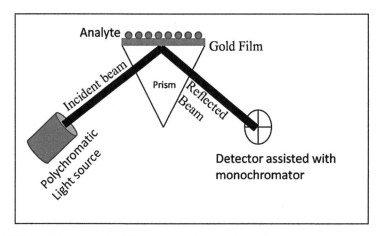

FIGURE 8.4 Schematic of a SPR instrument in wavelength interrogation mode.

The miniaturization of the sensor is further possible with the introduction of optical fiber. In the optical fiber-based SPR sensor, the fiber acts as a coupling medium. The fiber optic-based SPR sensor utilizes a standard single mode optical fiber with gold coating on the cladding stripped region. The guided mode of the light beam excites the SP in the gold layer. SPR sensors using single mode fibers utilize wavelength interrogation mode yielding broad SPR dips caused by the variation on SPR condition along the sensing region leading to low sensitivity to RI variation on the metallic surface.[13] The sensitivity is improved using the splicing of single mode fiber to multimode fiber. The spliced region of single mode fiber has gold deposition of 50 nm. The splicing junction acts like prism in Kretschmann–Reather configuration. This technology makes the fundamental mode to propagate inside the fiber to sense leading to improved sensitivity to RI measurement.[14] The schematic of a fiber-optic SPR sensor is shown in Figure 8.5.

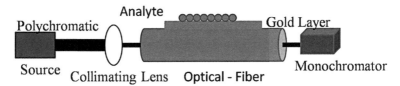

FIGURE 8.5 Schematic of fiber optic-based SPR instrument.

8.3 ANGULAR INTERROGATION-BASED SPR INSTRUMENT

This is the most popular mode of SPR sensor because it provides very high sensitivity. In angular interrogation method, the angle of incidence of a tightly focused monochromatic light onto the sensing area via the coupling prism is changed and reflected intensity is collected simultaneously. The dip in intensity indicates the resonance. Any adsorption on the sensing element causes a shift in the resonance angle. The angular interrogation can be achieved either by rotating prism assembly or by rotating source and detector simultaneously. Such a mechanical driven scanning system makes the instrument bulky, costly, and complex in operation. In general, the basic SPR instrumentation involves the optical components to be mounted on the arms of a goniometer. In this configuration, the angle of incidence and reflection are changed equally, and the reflected intensity is measured simultaneously. In another configuration, the prism assembly and the detector are rotated and the light source is kept fixed.[15] Goniometric SPR with a fixed photo-detector is achieved by employing two prism configuration where the illuminated prism face is in contact with another prism with a wedge like separation.[16] This configuration never keeps the interrogation spot constant at the surface of illuminated prism. Constant interrogation spot is obtained at the surface of the prism by fixing the reflecting surface of the prism at the axis of rotation of the arms of goniometer. The inclusion of optical components always causes the alignment problems and also leads to fall of intensity after each refraction/reflection in the optical components. Two-prism SPR sensor with an extra prism results in lower intensity of the reflected beam onto the detector as compared to the single prism configuration and, therefore, degrades the sensitivity of instrument. Recently, Devenarayanan et al. have adopted a new opto-mechanical scanning mechanism to obtain the SPR spectra in the high sensitive angular interrogation mode without employing a goniometer. Here, the laser as well as the detector is kept stationary. As the angle of incidence changes due to rotation of the scanning mirror, the reference spot position on the gold surface shifts and as a consequence the spot of reflected beam on the quadrant photodiode shifts. To retain the spot position fixed onto the gold surface, the prism assembly is translated vertically till original reference spot position is regained. Under this condition, the total reflected intensity and the angle of incidence are recorded. Figure 8.6 shows the schematic of the instrument.[17,18] The proposed scheme employs less number of optical components with low mass, and therefore, the overall optical complexity, size and mass of the instrument reduces, and the portability increases. The developed SPR instrument is employed to study the dielectric layers of various materials, for example, water, liquid crystal, stearic acid, cadmium stearate, and aqueous solution of sucrose. The dielectric layers were formed onto the gold surface by drop cast, spin coating, and Langmuir–Blodgett (LB) techniques. The angular resolution obtained in this configuration is 22 µdeg which is comparable to many commercial SPR instruments. Using the developed SPR instrument, the lowest detection limit of the sucrose in aqueous solution was found to be 100 fM. The resolution of the instrument was thus found to be 1.92 µRIU. The instrument is calibrated using sucrose solution. The sensitivity of the instrument was found to be 52.6°/RIU. The developed instrument is very low cost, portable, and sensitive.[19]

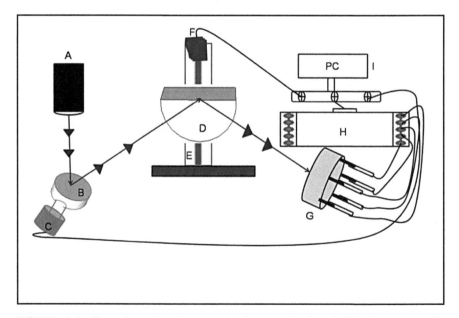

FIGURE 8.6 The schematic of an angular interrogation-based SPR instrument. (A) Laser (B) mirror (C) piezomotor, (D) prism assembly, (E) translation stage, (F) motor, (G) segmented photodiode, (H) data acquisition hardware, and (I) PC. (Reprinted with permission from Ref. 45. © 2016 Springer.)

8.4 SPR MICROSCOPY

In general, the real samples under investigation consist of a large number of components. The presence of other components can affect the interaction of an analyte with that of sensing element of the sensor. Simultaneous sensing of multi-analytes in a given sample is not possible employing the conventional single channel SPR instruments. Therefore, multichannel SPR instruments should be fabricated to address such issues. In 1989, Hickel et al. have demonstrated SPR microscopy wherein the different phases of a lipid monolayer onto gold substrate yield different contrast.[20] The basic principle of SPR microscopy is shown in Figure 8.7. An array of sensing area will be created on the 50-nm thick gold film by immobilizing different ligands. Interaction of the analytes from the sample with the immobilized ligands changes the SPR condition differently, and therefore, the intensity of reflected light collected from different sensing area can be different depending on the nature and amount of the specific analytes. This provides an avenue to assess the multi-analytes present in a real sample in terms of their nature and concentration. The SPR imaging is extremely popular in the bio-sensing and biomedical application.[21,22]

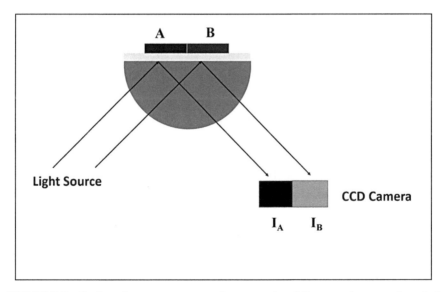

FIGURE 8.7 Surface plasmon resonance microscope. A and B are sensing areas deposited with different ligands on the surface of gold film. The difference in SPR condition from these areas gives rise to a difference in reflected intensity as recorded by a charged coupled detector (CCD) camera.

8.5 SPR FOR SENSING APPLICATIONS

8.5.1 GAS SENSING

The nitrogen dioxide (NO_2) is one of the harmful gases for living being. Short-term exposure itself may lead to inflammation in respiratory organs. This is one of the common air pollutants and has to be monitored regularly. Sensor based on SPR is used for the detection of NO_2 in air. The reported method utilizes spin-coated phthalocyanine over the gold surface as the functional layer for the detection of NO_2 gas. The lowest detectable concentration of NO_2 was found to be 1 ppm.[23] The SPR sensor possessing one active layer for the detection of a particular gas molecule fails to detect a combination of gas molecules. This is lack of selectivity of the sensor and one of the challenges in the field. The tin oxide layer on gold surface improves not only the sensitivity but also the selectivity of gas sensor. The lowest detection limit of 100 ppm for nitrogen oxide–nitrogen dioxide mixture was achieved.[24] The SPR sensor is developed with polyethylene glycohol ($n = 600$) and is employed for sensing hydrocarbons, aldehydes, and alcohols. The calibration curves for each of these species showed different sensitivity and therefore indicated the possibility of selective detection of such gas molecules.[25] In an interesting work by Bingham et al.,[26] the sensing of inert gas molecules, for example, He, Ar, and N_2, was studied using the localized SPR phenomenon in silver and gold nanoparticles. A change in RI ~2.45×10^{-4} RIU was measured reliably and reproducibly. The sensitivity was observed as ~200 nm/RIU. This indicates that the localized SPR technique is highly sensitive and reliable even for the inert analytes during the detection process. The SPR sensors which are based on angular or wavelength scanning modes employ prism as a wave guide. Therefore, such instruments are neither portable nor modular to perform geometrically confined localized as well as remote experiments. The SPR sensor based on optical-fiber technology can be portable and field deployable. The fiber-optic SPR sensors are used for the detection of gases in remote areas. The reported ammonia sensor utilizing the fiber-optic SPR technique possesses polyaniline film on ITO as the sensing layer. The sensor performs in wavelength interrogation mode. The lowest detection limit for ammonia with such a sensor is 10 ppm.[27] The doping of nickel oxide with ITO for the deposition of sensing layer on the fiber improves the sensitivity of fiber-optic sensor for the hydrogen sulfide gas.[28]

8.5.2 CHEMICAL SENSING

Nonylphenol is an organic material that is generally used in oil, laundry, and detergent industries. It is nonbiodegradable and can pollute the aquatic resources. Human can consume nonylphenol through fish and other edible water resources. The pollutants can lead to imbalance in hormonal secretion and may lead to cancer. It can present in river water ~4 µg/L. SPR technology is used for the detection of nonylphenol in shell fish.[29] The functional layer on the gold layer was created by immobilizing 9-(*p*-hydroxyphenyl) nonanoic acid on dextran matrix using amine coupling chemistry. The lowest detectable concentration of nonylphenol was 0.2 µg/L. In the fish samples, nonylphenol was detected as low as 10 ng/g. SPR sensor can be utilized for the detection of glucose level in diabetic patients. The detection of glucose was performed in wavelength mode with prism as the wave guide. The detection limit of the sensor is 8.67×10^{-6} RIU which is equivalent to 6.23 mg/dL of glucose in water.[30] Quality of the milk available in market may have contaminants that are harmful to human body. Such contaminants can be easily detected using the SPR sensors. The reported method detects staphylococcal enterotoxin B contaminant in milk using the SPR technique. The sensor works in wavelength mode with a prism as the optical waveguide. The detection sensitivity is found out to be 5 ng/mL without the functionalization of sensing element. The functionalization provides a lowest detection limit of 0.5 ng/mL.[31] To improve the milk yield, the cows are treated with recombinant bovine somatotropin (rBST). However, the studies showed that the milk from the treated cows can possess insulin-like growth factor (IGF-1) with rBST. IGF-1 can lead to cancer and is a potential threat to human health. Guidi et al.[32] have developed a solid phase enzyme-linked immunoassay and characterized the hyperimmune polyclonal-anti-IFG-1 antibodies with respect to its specific binding to IGF-1. Such antibodies were used as functional material and a lowest detectable level of IGF-1 in milk was detected as 1 ng/L. The detection is automated and real time. Heavy metals in potable water pose serious threat on human health. Potable water gets contaminated with heavy metal due to industrial discharge and other form of pollutions. The detection of heavy metal in water can be achieved using SPR technology. The squarylium dye (SQ) changes RI of a copolymer on interaction with copper ion (Cu^{2+}). A thin film consisting of copolymer, polyvinyl chloride–polyvinyl acetate–polyvinyl alcohol, SQ, and others was deposited onto the gold surface. Such functional layer was employed for the detection of different alkali metals, alkaline earth metals, and transition metal ions. The SPR sensor showed highest selectivity for the Cu^{2+} ions. The lowest detectable concentration of the ion was reported to be 1 pM.[33] In another interesting work, the functional layer for the detection of heavy metal was created using metallothinein protein.[34] Such sensor was employed for the detection of Cd, Zn, and Ni in buffer at concentration down to 100 ng/mL. The online (in vitro) detection of chemicals in fluids is possible with the help of optic-fiber SPR probe. The fiber-optic probe for the detection of urea is such an example for online chemical sensor. The sensor works in wavelength mode wherein the sensing element is functionalized with specific enzyme, urease onto silver layer. The sensitivity of the sensor is improved by the addition of silica layer between silver and the enzyme. This silica layer will protect silver from oxidation and improves the enzyme reactivity. The sensor is able to perform well for the urea concentration range of 0–160 mM.[35] From these examples, we found that SPR platform is powerful to detect minute amounts of chemicals in liquids and found application in medical field.

8.5.3 BIOSENSOR

SPR is very popular for biosensing application because it provides not only label-free measurement but also very sensitive to the targeted analytes. Some of the interesting biosensing application using SPR technology is discussed below. The sensitivity of the SPR biosensor with bare metal surface is less due to the inertness of noble (gold) metals toward any other chemical species. To establish the molecular specific interaction, the metal surface should be treated chemically to yield specific functionalization. The self-assembled monolayer (SAM) of organothiols is stable in nature over the noble metal (gold) surface. The SAM of such organothiol can act as a platform for immobilization of any ligands. This is the essence of any biosensor with appropriate functionalization of the gold surface.

The SPR sensor for the detection of cardiac muscle death in less than 10 min was achieved. Myoglobin and cardiac troponin I are known to be the markers for cardiac muscle injury. The antibodies of the specific antigens were developed and were immobilized to the carboxymethylated dextran layer on a gold surface. The troponin level lies in the range of 1–3 ng/mL in the human blood after the myocardial damage. Using the SPR sensor, the lower detection limit of myoglobin and troponin was found to be 2.9 and 1.4 ng/mL.[36]

Prostate-specific antigen (PSA) is a marker for prostate cancer. The healthy level of PSA in serum is <4 ng/mL. Monoclonal antibodies against PSA were immobilized onto the sensing surface of SPR instrument and it was used for the detection of PSA. The lowest detectable limit was found to be 0.15 ng/mL.[37]

The graphene layer is one of the candidates which can enhance the sensitivity of the SPR sensor.[38] The graphene sheets having carboxylic acid group can bind with the biomolecules through amine linking chemistry. This stable adsorption of biomolecules over sensing layer enhances the sensitivity of SPR biosensors.[39] The utility of SPR biosensing application can be enhanced using the optic-fiber sensor. A novel smartphone-based optic-fiber SPR sensor was developed. Here, the flash light present in the phone acts as the light source and the camera of the phone acts as the detector. The sensor utilizes the relative intensity change as the signature of binding between analyte and ligand. This sensor has a detection limit of 47 nM for the IgG protein.[40]

8.5.4 OTHERS

The SPR sensors are used for the detection of change in dielectric properties of molecular fluids. A study was conducted for the detection of cation and anion in solution using SPR sensors. The authors have used the electric field-assisted method for the detection of cations and anions. They recorded change in intensity of reflected light as the measure of adsorption of cations or anions absorbed on the metallic surface.[41] The side polished fiber-based RI wavelength sensor provides the detection of RI of liquids having RIU 1.32–1.40 with sensitivity up to 4365.5 nm/RIU and a FOM of 51.61 RIU^{-1}.[42] Prism-based SPR sensor is used to measure the optical anisotropy in the thin films of molecules. The method utilizes the anisotropic molecular layers obtained through various thin film deposition techniques like spin coating and LB technique.[43] The RI measurements in two orthogonal directions yield remarkably different values.

Wang et al. have shown that the SPR phenomenon can be employed for voltage-driven tunable optical filter. In their interesting work, the RI of the gold/liquid crystal interfaces was altered by the application of electric field. For a given applied voltage, due to resonance condition, a band of wavelength was found to be absorbed in the reflected beam leading to the appearance of complimentary color in the reflected light. Such technique can be equally applied for the development of tunable infrared filter.[44]

The SPR phenomenon is widely used in the modern sensing technology. The selectivity of analytes is still a big challenge in the field. Major scientific attentions are now devoted toward the development of specific SPR sensor which can yield reliable and reproducible data. Some of the latest development in the field are sensing through SPR imaging, SPR fluorescence, and electrochemical SPR.

ACKNOWLEDGMENTS

We are thankful to Department of Science and Technology, India for funding the project (IDP/IND/12/2010). We also thank BITS Pilani for providing fellowship to Devanarayanan.

KEYWORDS

- **surface plasmon resonance**
- **metal–dielectric interface**
- **resonance angle**
- **functionalized layer**
- **sensing**

REFERENCES

1. Raether, H. *Surface Plasmon on Smooth and Rough Surfaces and Grating*; Springer: Berlin, 1988.
2. Kretschmann, E.; Raether, H. *Z. Natureforsch.* **1968**, *23A*, 2153.
3. Novotny, L.; Hecht, B. *Principles of Nano-optics*; Cambridge University Press: New York, 2008.
4. Wood, R. W. *Philos. Mag.* **1902**, *4*, 396.
5. Turbadar, T. *Proc. Phys. Soc.* **1959**, *73*, 40.
6. Otto, A. *Zeits Phys.* **1968**, *216*, 398.
7. Pockrand, I.; Swalen, J. D.; Gordon, J. G.; Philpot, M. R. *Surf. Sci.* **1978**, *74*, 237.
8. Liedberg, B.; Nylander, C.; Lundstrom, I. *Sens. Act.* **1983**, *4*, 299.
9. Homola, J. *Sens. Actuators B* **1997**, *41*, 207.
10. Shalabney, A.; Khare, C.; Rauschenbach, B.; Abdulhalim, I. *Sens. Actuators B* **2011**, *159*, 201.
11. Ho, H. P.; Wu, S. Y.; Yang, M.; Cheung, A. C. *Sens. Actuators B* **2001**, *80*, 89.

12. Sereda, A.; Moreau, J.; Boulade, M.; Olivero, A.; Canva, M.; Maillart, E. *Sens. Actuators B* **2015**, *209*, 208.
13. Slavik, R.; Homola, J.; Ctyroky, J. *Sens. Actuators B* **1999**, *54*, 74.
14. Liu, Z.; Wei, Y.; Zhang, Y.; Sun, B.; Zhao, E.; Zhang, Y.; Yang, J.; Yuan, L. *Sens. Actuators B* **2015**, *221*, 1330.
15. Tiwari, K.; Sharma, S. C. *Sens. Actuators A* **2014**, *261*, 128.
16. Mohanty, B. C.; Kasiviswanathan, S. *Rev. Sci. Instr.* **2005**, 76, 033103.
17. Liang, H.; Miranto, H. *Sens. Actuators B* **2010**, *149*, 212.
18. Liu, Y.; Xu, S.; Tang, B.; Wang, Y.; Zhou, J.; Zheng, X.; Zhao, B.; Xu, W. *Rev. Sci. Instr.* **2010**, *81*, 036105.
19. Gupta, R. K.; Devanarayanan, V. P.; Manjuladevi, V. *Sens. Actuators B* **2016**, *227*, 643.
20. Hickel, W.; Kamp, D.; Knoll, W. *Nature* **1989**, *339*, 186.
21. Brockman, J. M.; Nelson, B. P.; Corn, R. M. *Annu. Rev. Phys. Chem.* **2000**, *51*, 41.
22. Wong, C. L.; Olivo, M. *Plasmonics* **2014**, *9*, 809.
23. Wright, J. D.; Cado, A.; Peacock, S. J.; Rivalle, V.; Smith, A. M. *Sens. Actuators B* **1995**, *29*, 108.
24. Yang, D.; Lu, H. H.; Chen, B.; Lin, C. W. *Sens. Actuators B* **2010**, *145*, 832.
25. Miwa, M.; Arakawa, T. *Thin Solid Films* **1996**, *281–282*, 466.
26. Bingham, J. M.; Anker, J. N.; Kreno, L. E.; Duyne, R. P. V. *J. Am. Chem. Soc.* **2010**, *132*, 17358.
27. Mishra, S. K.; Kumari, D.; Gupta, B. D. *Sens. Actuators B* **2012**, *171*, 976.
28. Mishra, S. K.; Rani, S.; Gupta, B. D. *Sens. Actuators B* **2014**, *195*, 215.
29. Samsonova, J. V.; Uskova, N. A.; Andresyuk, A. N.; Franek, M.; Elliott, C. T. *Chemosphere* **2004**, *57*, 975.
30. Lam, W. W.; Chu, L. H.; Wong, C. L.; Zhang, Y. T. *Sens. Actuators B* **2005**, *105*, 138.
31. Homola, J.; Dostalek, J.; Chen, S.; Rasooly, A.; Jian, S.; Yee, S. S. *Int. J. Food Microbiol.* **2002**, *75*, 61.
32. Guidi, A.; Robbio, L. L.; Gianfaldoni, D.; Revoltella, R.; Bono, G. D. *Biosens. Bioelectr.* **2001**, *16*, 971.
33. Ock, K.; Jang, G.; Roh, Y.; Kim, S.; Kim, J.; Koh, K. *Microchem. J.* **2001**, *70*, 301.
34. Wu, L. P.; Li, Y. F.; Huang, C. Z.; Zhang, Q. *Anal. Chem.* **2006**, *78*, 5570.
35. Bhatia, P.; Gupta, B. D. *Sens. Actuators B* **2012**, *161*, 434.
36. Masson, J.-F.; Obando, L.; Beaudoin, S.; Booksh, K. *Talanta* **2004**, *62*, 865.
37. Besselink, G. A. J.; Kooyman, R. P. H.; van Os, P. J. H. J.; Engbers, G. H. M.; Schasfoort, R. B. M. *Anal. Biochem.* **2004**, *333*, 165.
38. Verma, R.; Gupta, B. D.; Jha, R. *Sens. Actuators B* **2011**, 160, 623.
39. Chiu, N.-F.; Huang, T.-Y. *Sens. Actuators B* **2014**, *197*, 35.
40. Liu, Y.; Liu, Q.; Chen, S.; Cheng, F.; Wang, H.; Peng, W. *Sci. Rep.* **2015**, *5*, 12864.
41. Ko, H.; Kameoka, J.; Su, C. B. *Sens. Actuators B* **2009**, *143*, 381.
42. Zhao, J.; Cao, S.; Liao, C.; Wang, Y.; Wang, G.; Xu, X.; Fu, C.; Xu, G.; Lian, J.; Wang, Y. *Sens. Actuators B* **2016**, *230*, 206.
43. Devanarayanan, V. P.; Manjuladevi, V.; Poonia, M.; Gupta, R. K.; Gupta, S. K.; Akhtar, J. *J. Mol. Struct.* **2016**, *1103*, 281.
44. Wang, Y. *Appl. Phys. Lett.* **1995**, *67*, 2759.
45. Gupta, R. K. (2017). Sensing Through Surface Plasmon Resonance Technique. In: Reviews in Plasmonics 2016. Geddes, C. Ed.; Springer, pp 39–53.

CHAPTER 9

DESIGN ISSUES IN HIGH STRAIN RATE DYNAMIC COMPRESSIVE FAILURE OF STRUCTURAL CERAMICS, POLYMERS, AND COMPOSITES

SAIKAT ACHARYA[1*], K. S. GHOSH[2], D. K. MONDAL[2], and A. K. MUKHOPADHYAY[1]

[1]*Advanced Mechanical and Materials Characterization Division, CSIR-Central Glass and Ceramic Research Institute, Kolkata 700032, India*

[2]*Department of Metallurgical and Materials Engineering, National Institute of Technology Durgapur, Durgapur 713209, India*

*Corresponding author. E-mail: saikat@cgcri.res.in

ABSTRACT

Evaluation of dynamic compressive fracture for advanced structural materials like ceramics and polymers is essential for developing high strain rate-resistant structures. Performances of the individual components present in a composite structure control the overall efficiency during its field application. That is why, it is indeed necessary to take care of the characteristically brittle microstructure of the ceramics in general and alumina in particular. Recently, polycrystalline alumina is found to be used extensively in dynamic load-bearing applications. Hence, an idea for developing a smarter design by introducing a weak interface between the two monolithic alumina ceramic disks was evolved to deflect and/or arrest the cracks to enhance their failure times. Thus, specimens of monolithic ceramics and ceramic–polymer-layered composites were developed from commercial alumina and an appropriate polymer. They were subsequently exposed to high strain rate (e.g., 10^3 s^{-1}) impact tests using a split Hopkinson pressure bar (SHPB) apparatus in association with in-situ high-speed real-time videography. The maximum compressive strength for monolithic alumina was ~3 GPa and that for layered structure was slightly higher, for example, ~3.2 GPa. Time to reach the maximum compressive strength from impact for monolithic alumina and the layered structures were 35 and 55 µs, respectively. In addition, to understand the deformation behavior of brittle polymers and reinforced polymers, the polymethyl-methacrylate and glass fiber-reinforced polymer samples were also examined in high strain rate SHPB experiments. The yield strengths of these two polymers were highly sensitive to variations in strain rate. Postmortem examinations of the fragmented specimens were performed by field emission scanning electron microscopy and transmission electron microscopy. Based on these experimental data, some new failure mechanisms were proposed for these materials.

9.1 INTRODUCTION

Characteristically ceramics are strong in compression and weak in tension. The compressive yield strength (σ_{yc}) of ceramics has been observed to be strain rate sensitive.[1-17] On the other hand, the compressive yield strength (σ_{yc}) of polymers is also known to be strain rate and temperature sensitive.[18-29] But there is hardly any study reported on comparative high strain rate dynamic failure behavior of brittle ceramics, for example, alumina, ceramic–polymer-layered composites, for example, alumina–epoxy, brittle polymers, for example, poly(methyl methacrylate) (PMMA), and fiber-reinforced polymers, for example, glass fiber-reinforced polymer (GFRP). This fact acted as a major motivation for the present work.

As a classical brittle ceramic, alumina is susceptible to strain-rate-dependent brittle failure.[1] The fracture process shows evidences of basal twins,[2] dislocations, grain boundary microcracking due to dislocation pileups, and formation of subgrain structures.[3,4] The deformation is also associated with the occurrence of tensile stress field around damage-induced defects.[4] Density and grain size of alumina also play an important role in the high strain rate deformation and failure processes. For high-purity fine-grain alumina under high strain rate failure, the critical resolved shear stress that operates on a slip plane has higher magnitude than those which occur in low purity coarse grain alumina.[4] During high strain rate experiments, a strain-rate-dependent drastic change in the facture pattern of alumina occurs from a primarily elastically deformed state to a predominantly inelastically deformed state.[5]

In monolithic alumina, the process of microcrack development has been observed to be influenced by the rate of application of load and/or strain rate. The range from microcrack development to dislocation assisted crack formation was proposed as the primary failure mechanism in the conventional dynamic deformation under compressive fracture at low strain rate. However, the nucleation of microcracks, their growth, and coalescence have been suggested as the major failure mechanism in high strain rate fracture.[1–8] But their roles in the global failure process under high strain rate compression are far from well understood.[1–8]

Concurrent attempts with a mixed bag of success or failure were also made by introduction of weak interface and/or interphase engineering with a view to deflect or arrest cracks.[6–13] Preliminary studies on high strain rate dynamic compressive failure behavior of ceramic–polymer-layered composites highlighted a possibility of stress cushioning to a degree that was higher than that attainable with the monolithic ceramic.[14] However, further experiments are necessary for development of better understanding of the scientific issues related to failure at high strain rate.[14]

It is established that the rate of load application and, in turn, the induced strain thus generated influences the microcrack development process. At relatively lower strain rates, for example, $10^{-5} \leq (d\varepsilon/dt) \leq 10^2$ s^{-1} for monolithic alumina, the strain rate sensitivity of the compressive strength depends essentially on subcritical growth of axial microcracks.[15] At relatively higher loading rates, for example, $10^2 \leq (d\varepsilon/dt) \leq 10^4$ s^{-1}, a dependency on crack inertia to control the strength has been often suggested.[16] In addition, both the grain size and processing methodology are found to be essential parameters in determining the dynamic compressive strength.[15–17] However, at higher loading rates, the experimental conditions play an important role as well. In spite of extensive research, the experimental data available for the polycrystalline alumina do not cover the wide spectrum as desired. For instance, mostly the grain sizes referred to are either smaller than 4 μm (fine grain) or larger than 17 μm, and the loading rates remain either $\leq 10^0$ s^{-1} or $\geq 10^3$ s^{-1}.[1–6,8–17] Thus, the present work has been performed on a dense, for example, ρ ~97% ρth polycrystalline alumina of 5-μm grain size employing a split Hopkinson pressure bar (SHPB) apparatus in the strain rate regime $10^2 \leq (d\varepsilon/dt) \leq 10^3$ s^{-1}.

A lot of investigation has been performed in polymer and composite materials to understand their dynamic deformation and failure. In polymers, high rates of loading cause a rise in temperature and hence a thermal decomposition at the affected region.[18–21] Further, restricted shear deformation owing to excessive stress concentration at the crack tips region also occurs.[21–23] In addition, failure mode in high strain rates is associated with amount of energy absorption that is much more than that occurs at low strain rates.[23] The compressive strength for unidirectional fiber-reinforced composites has been observed to be nearly half of their tensile strength.

Thus, compressive strength of composites is becoming a limiting factor in designing the structures.[24–30] Notwithstanding the wealth of literature,[18–29] relatively little is reported concerning high strain rate behavior of brittle and reinforced polymers, for example, PMMA and GFRP.

In the present study, a modified SHPB is used to perform controlled high strain rate deformation and failure analysis of alumina, alumina–epoxy-layered composites, PMMA, and GFRP materials. The aim of the work is to understand their dynamic behavior over a wide range of strain rates. The intention is to have an idea about the individual performance as well as combinatorial performance of ceramics and polymers in mitigating high strain rate dynamic impact as well. Such inputs are considered important for futuristic design of high strain rate damage resistant structures.

9.2 MATERIALS AND METHODS

Specimens of monolithic and layered structure from commercial alumina, PMMA, and GFRP were developed for impact experiment under uniaxial stress condition using an SHPB test setup.

9.2.1 LAYERED AND MONOLITHIC ALUMINA SPECIMENS

Pure alumina powder (99.7%) of commercial variety (CT3000SDP, ALMATIS, Germany) was pressureless sintered for 2 h at a temperature of 1600°C in air to prepare cylindrical disks[17] for the SHPB tests. Bulk density (ρ) of the sintered sample was estimated by the water immersion technique through Archimedes' principle and observed to be of ~97.2% of the theoretical (ρ_{th} ~3.98 g cm^{-3}). The monolithic alumina disks had final sintered diameter (d) and length (l) of 6.00 ± 0.05 and 3.00 ± 0.05 mm, respectively. For developing the alumina–polymer-layered structure, the diameter (d) and length (l) of the sintered alumina specimens were also kept at 6.00 ± 0.05 and 1.50 ± 0.05 mm, respectively. To make an alumina–polymer-layered specimen, two of these thinner alumina disks were joined together by using a thin layer of commercially available polymeric adhesive. The specimen size for testing is determined mainly based on the capacity of the test system which is primarily dependent on the dimensions of the bars and the striking capability of the launcher.[14,17,21] In the cases of both the monolithic alumina and the alumina–polymer-layered composite, the slenderness ratio (l/d) was 0.5, which matched with those typically used for ceramics.[1–7]

9.2.2 PMMA SPECIMENS

Commercial PMMA rods (Anulon-120) of diameter 25 mm were procured from M/S Plastic Abhiyanta, Kolkata, India for preparing PMMA test samples. It has been reported that cylindrical polymer samples with length (l)-to-diameter (d) ratio in the range 0.5–2.0 are considered most appropriate for SHPB experiments.[18] Thus, for the present study, diameter (d) and length (l) of the cylindrical PMMA samples were kept in the range of 10.09 ± 0.04 and 5.09 ± 0.04 mm, respectively. Here, the purpose was to maintain the slenderness ratio (l/d) of about 0.5 to attain stress equilibrium during deformation.[19–23]

9.2.3 GFRP SPECIMENS

The resin content of the GFRP sample was about 57 wt%. It had about 43 wt% of glass fiber reinforcement. GFRP specimens of diameter (d) ~10 ± 0.5 mm and length (l) ~5 ± 0.5 mm were used so that length (l)-to-diameter (d) ratio (l/d) remained about 0.5, which would help to attain dynamic stress equilibrium during deformation.[19–23] These cylindrical axially reinforced unidirectional GFRP samples were used in the SHPB experiments.

9.2.4 HIGH STRAIN RATE EXPERIMENTS

The SHPB technique is widely used for characterizing materials at high rates in the strain rate regime 10^2–10^4 s^{-1}.[14–23] A schematic diagram of modified SHPB setup used in the present study is shown in Figure 9.1.

The SHPB setup comprises the two maraging steel bars—incident and transmitter with a set of striker bars of same materials. The launching system for the striker bar is set with desired gas pressure using a fill valve. The materials of the bars along with the striker one are so chosen that they remain elastic when the specimen itself experiences a considerable strain. The elastic strain pulses which are produced due to impact of the striker bar at one end of the incident bar travel down the other end of the transmitter bar through the specimen which is usually kept as sandwiched between the two bars, Figure 9.1. Due to impedance mismatch, a part of the incident strain pulse gets reflected back from the bar–specimen interface and the rest is transmitted through the transmitter bar. The amplitude of the elastic pulses depends on the velocity of the striker bar. It is well known[7] that the measure of the transmitted pulse provides the compressive stress (σ_c) which is experienced by the specimen while that of the reflected pulse provides the strain rate ($\dot{\varepsilon}$) of the specimen. The strain rate is integrated with respect to time t to get the axial strain (ε) of the specimen in test. Poison's ratio mismatch during high strain rate compression may cause erroneous results. Thus, to minimize frictional effects, molybdenum di-sulfide grease was used between the bars and the specimen in the experiment. To prevent damage at the bar ends from the hard alumina fragments impedance, matched tungsten carbide platens were used between the bars and the specimen. Maraging steel collars were used as shrunk fit jackets to the platens to prevent their frequent failure.

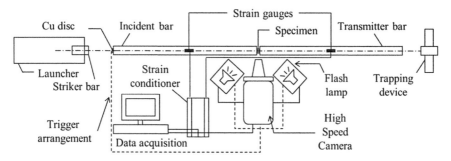

FIGURE 9.1 A schematic diagram of SHPB set up with a high-speed camera for real time video images during the experiment. (Reprinted with permission from Ref. 17. © 2015 Elsevier.)

High-speed real-time video images during the experiment often help to analyze the details of failure process. So, SHPB setup was properly synchronized with the high-speed camera by a laboratory made trigger arrangement. To facilitate attainment of constant strain rates in the samples during the tests, tiny copper disk was used as pulse shapers. The disk was kept between the striker and incident bars. The copper disk was attached to the incident bar by applying a thin layer of grease.

It is important that the specimen should not encounter multiple loading pulses. Thus, failure of the specimen must be ensured during the first pass of the loading pulse. SHPB apparatus should be reasonably friction-free. To ensure that, bar-to-bar tests without the specimen were carried out to check the equilibrium condition of the apparatus. After the impact experiments, the fractured, deformed, or pulverized specimens were collected for post-mortem examinations. The examinations were carried out by using the field emission scanning electron microscopy (FESEM) and transmission electron microscopy (TEM) techniques. Further details about the examinations are given elsewhere.[14,17,21]

9.3 RESULTS

9.3.1 LAYERED AND MONOLITHIC ALUMINA

A typical illustrative experimental data comprising strain rate and stress history of layered and monolithic samples are shown in Figure 9.2a,b for the layered and monolithic alumina samples, respectively. The average compressive yield strengths (σ_{yc}) were, for example, 3.21 ± 0.28 and 3.02 ± 0.34 GPa for the layered alumina–polymer composite and the monolithic alumina samples. Thus, the magnitude of (σ_{yc}) of layered alumina samples was about 70% higher than that of the monolithic alumina. The average time to compressive failure (t_f) was, for example, ~50 and ~36 μs for the layered and monolithic alumina samples. Hence, failure time for the layered sample was enhanced by ~40%. Both the specimens started to lose their load-carrying capacity soon after reaching their maximum compressive strength.

In Figure 9.2a,b, layered specimen attended the peak stress, for example, 3.21 GPa at 48 μs, whereas monolithic specimen attended the peak compressive stress, for example, 3.02 GPa at 36 μs. However, maximum strain rates achieved by both the samples were nearly same, for example, ~10^3 s^{-1}.

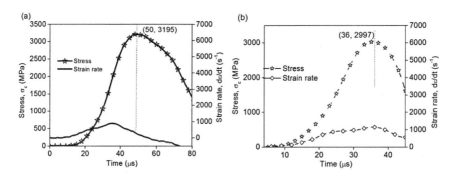

FIGURE 9.2 Plot of stress and strain rate as a function of time in the SHPB experiment: (a) alumina-polymer layered structure; and (b) monolithic alumina specimens (a: Reprinted/modified with permission from Ref. 14. © 2014 Elsevier; b: Reprinted/modified with permission from Ref. 17. © 2015 Elsevier.)

As mentioned before, after the SHPB experiments, the fragments of samples were collected for postmortem analysis using FESEM and TEM techniques. Layered alumina fragments after the tests depicted shear plane (indicated by black arrow) with grain localized plasticity (indicated by white arrow), Figure 9.3a. Characteristically curved slip bands (indicated by black arrow) and stepped microfracture (indicated by white arrow) on different facets of neighboring grains were also frequently observed (Fig. 9.3b) in the layered alumina sample.

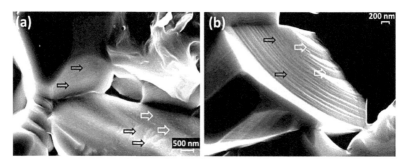

FIGURE 9.3 FESEM photomicrographs of fragments from SHPB tests of the alumina-polymer layered specimen: (a) shear plane with grain localized plasticity; and (b) curved slip bands and stepped microfracture.

Postmortem FESEM examination of the monolithic alumina fragments showed intragranular microfracture, for example, stepped shear induced deformation (indicated by black arrow) on the same grain, Figure 9.4a. Extensive shear-induced slip bands (indicated by white arrow) on different facets of neighboring grains with microcleavages were also frequently observed in the monolithic alumina sample, Figure 9.4b.

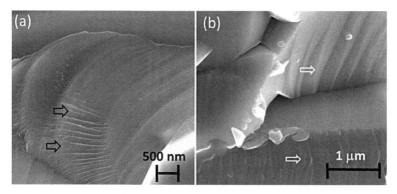

FIGURE 9.4 FESEM photomicrographs of fragments from SHPB tests of the monolithic alumina specimen: (a) grain localized curved shear-induced stepped microfracture with microplasticity and (b) grain localized slip bands with microcleavages.

Postmortem TEM examination of the layered alumina fragments showed dislocations entanglement (Fig. 9.5a) and that of monolithic alumina fragments exhibited dislocation array in a single alumina grain, Figure 9.5b.

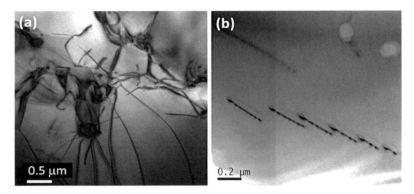

FIGURE 9.5 TEM photomicrographs: (a) dislocation entanglement in alumina-polymer layered specimen and (b) dislocation array in monolithic alumina specimen.

Additional deformation features observed in the layered alumina sample included bifurcation of dislocations (Fig. 9.6a) and the corresponding SAD pattern, Figure 9.6b. This kind of bifurcation, however, was not noticed in the case of the monolithic alumina specimen (Fig. 9.7). These results have been discussed further later in the present work.

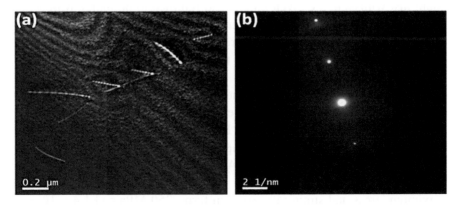

FIGURE 9.6 TEM photomicrographs of weak beam dark field image in alumina-polymer layered specimen: (a) bifurcation of dislocations and (b) corresponding SAD pattern.

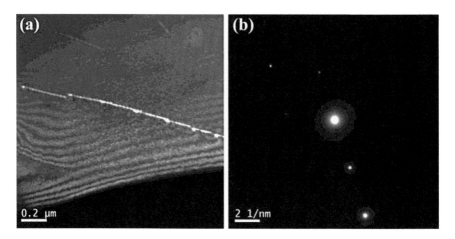

FIGURE 9.7 TEM photomicrographs of weak beam dark field image in monolithic alumina specimen: (a) no bifurcation of dislocations observed and (b) corresponding SAD pattern.

9.3.2 POLY(METHYLMETHACRYLATE)

The high strain rate experiments on PMMA samples were performed in a wide strain rate range, for example, strain rate varied from 600 to 1400 s^{-1}. Obviously, pressure varied in the range of 0.4–1.4 kg cm^{-2} such that not all the samples were broken. Two sets (e.g., P and Q) of samples were broken at two different pressures to understand the impact of strain rate. Third set of samples (R) was exposed to the threshold impact so that they could just withstand the given impact and returned to normalcy with some residual strain state. Similarly, the fourth set of samples (S) was exposed to a relatively low strain rate so that the residual stress became reasonably less; however, the time required to unload the specimen was nearly same to that of the third set of samples.

Typical average data of strain rates as a function of time are plotted in Figure 9.8a for the PMMA samples P, Q, R, and, S. In each case, four samples, for example, P1, P2, P3, and P4, were tested. Similarly, the average stress–strain diagrams of those samples are shown in Figure 9.8b. The compressive stress (σ_c) of the PMMA samples changed from ~280 to 350 MPa, reflecting a 25% increase with the enhancement of strain rates. The maximum stress and strain rates were in turn ~350 MPa and 1.4×10^3 s^{-1}, while the critical strain rate was ~7.7×10^2 s^{-1}.

Figure 9.9a showed FESEM photomicrograph of highly localized shear deformation along with an underneath crack while Figure 9.9b provided

Design Issues in High Strain Rate Dynamic Compressive Failure 217

evidence for localized heating from possibly a rise in local temperature due to high strain rate compression. These evidences may suggest that elastic strain energy was released at the crack tip and localized melting of PMMA chains had taken place during deformation.

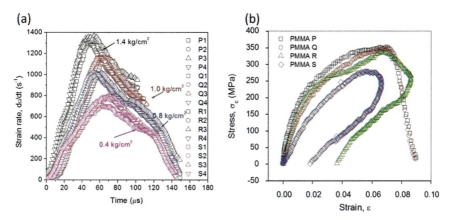

FIGURE 9.8 Plot of PMMA samples after the present SHPB experiment: (a) strain rate and time history at different impact; and (b) average stress and strain (Reprinted/modified with permission from Ref. 21. © 2014 Springer.)

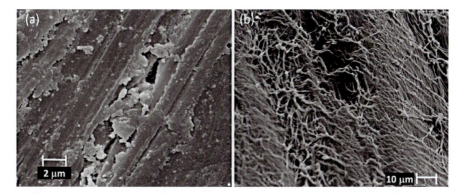

FIGURE 9.9 FESEM photomicrographs of fragmented PMMA specimen after SHPB experiment: (a) shear plane along with crack propagation and (b) localized rise of temperature. (Reprinted/modified with permission from Ref. 21. © 2014 Springer.)

It is well known that in brittle polymers, for example, PMMA, the initiation of bulk plastic flow is followed by initiation and propagation of cracks. The final failure process is however dominated by random cracking and localized shear flow due to high stress concentration at existing crack tips.[18–20] Simplistically speaking, the elastic strain energy stored in the PMMA samples gets released during high strain rate deformation. As a result, the local rise of temperature occurs at the crack tip. Due to dynamic deformation, the maximum temperature at the crack tip region was estimated to be around 530°C.[19] Based on the experimental evidences (Figs. 9.7 and 9.8), therefore, it seems plausible to argue that the samples like "P" and "Q" which suffered most catastrophic failure might have experienced mostly the effect of both high strain rate and the induced thermal decomposition due to rise in local temperature. On the other hand, at relatively lower strain rates, for example, as those experienced by "R" and "S," the samples were deformed without fracture since the amount of localized rise in temperature was far from significant.

9.3.3 GLASS FIBER-REINFORCED POLYMER

Typical average data of stress as a function of strain with parametric variation of strain rates are plotted in Figure 9.10a for the GFRP samples. Further, the data of Figure 9.9a proved that there was typically a peak stress, a downslide, and then a second rise of the stress, Figure 9.10a. It reflected the first cracking of the epoxy matrix followed by interfacial debonding followed by the fiber taking up the load up to its fracture.

Similarly, the average data on strain rate evolution as a function of time with parametric variation of applied impact pressures on the striker bars are shown for the GFRP samples in Figure 9.10b. With increase in impact pressure from about 0.6 to 0.9 bar, the average rise of strain rate had occurred from ~550 to 800 s^{-1}, Figure 9.10b. Further increase in impact pressure up to 1.2 bar enhanced the strain rate to an average of 1250 s^{-1}, Figure 9.9b. These data showed that the higher was the impact pressure, the larger was the magnitude of maximum strain rate that could be achieved. The data presented in Figure 9.10a,b confirmed further that in the case of GFRP, the maximum compressive strength (σ_c) increased by about 55% (i.e., from ~250 to 385 MPa) with the enhancement of the corresponding strain rate from ~5.50 × 10^2 to 1.25 ×10^3 s^{-1}. Failure of the present GFRP material however occurred due to interfacial separation at matrix–fiber interface as shown by the photomicrographs presented in Figure 9.11a, followed by localized delamination, debonding, and finally both localized temperature rise-induced shear flow of the matrix as well as fiber fracture, especially at high strain rates, Figure 9.11b.

Design Issues in High Strain Rate Dynamic Compressive Failure 219

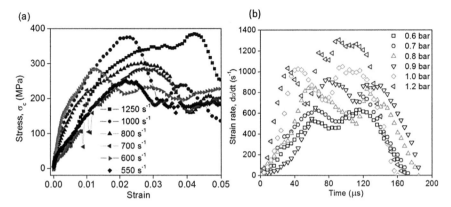

FIGURE 9.10 Plot of present GFRP samples : (a) stress versus strain as a function of various strain rates and (b) strain rate and time history at different impact. (Reprinted with permission from Ref. 31. © 2017 IOP Publishing Ltd.)

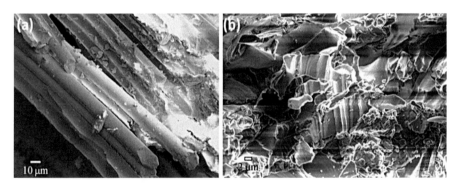

FIGURE 9.11 FESEM photomicrographs of fractured surfaces from the broken GFRP specimens: (a) fibre fracture by delamination, debonding and fracture and (b) disintegration of matrix through brittle fracture with and without shear dominated failure.

Thus, the FESEM evidences confirmed that especially at high strain rates, failure of the present GFRP material happened due to brittle fracture of the matrix with and without extensive shear deformation, interfacial separation at matrix–fiber interface, localized fiber delamination, debonding, and finally fiber fracture. This is a consequence of energy absorption by the polymeric matrix at high strain rates.

There was a general empirical power law dependence of the maximum compressive yield strength (σ_c) and the average strain rate ($d\varepsilon/dt$) of alumina, PMMA, and GFRP which could be expressed as $(\sigma_c) \sim A \, (d\varepsilon/dt)^n$, where A is an empirical constant and n is the strain rate sensitivity. The individual relationships for alumina, PMMA[21] and GFRP[31] are respectively shown along with the relevant literature data in Figure 9.11a–c, while the overall comparative picture for the data from the present work is shown in Figure 9.11d. It can be noticed that the value of n for PMMA was 0.09. It was higher than those of GFRP ($n = 0.06$) and alumina ($n = 0.02$) in the strain rate regime 10^{-4}–550 s^{-1}. However, at $(d\varepsilon/dt) > 550$ s^{-1}, the sensitivity index of GFRP drastically changed to the highest value of 0.86 which was much higher than those evaluated in the present work for PMMA or alumina, Figure 9.12d.[21, 31]

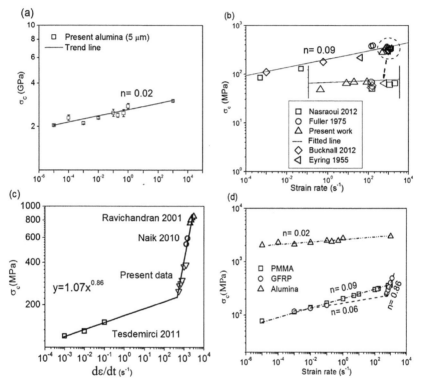

FIGURE 9.12 Plot of compressive strength as a function of strain rate: (a) alumina, (b) PMMA, (c) GFRP and composite, and (d) alumina, PMMA and GFRP. (a: Reprinted/modified with permission from Ref. 21. © 2014 Springer; c: Reprinted/modified with permission of Ref. 31. © 2017 IOP.)

9.4 DISCUSSION

9.4.1 LAYERED AND MONOLITHIC ALUMINA

The XRD data analysis of the present alumina ceramic has confirmed the major phase as α-phase (JCPDS43-1484). The grain size of sintered alumina has been observed to vary from about 1.5 to 12 μm with an average of about 5 μm.[17] The SHPB experimental results (Fig. 9.2) showed that the dynamic compressive strength of the layered alumina structure was higher than that of the monolithic alumina. The dynamic compressive strength of the layered structure was 70% more than that of the monolithic one. The signature of both deformation and failure was reported and observed to be stochastic in nature.[14-17] However, these signatures appeared to be related to the several factors. For instance, the experimental condition, that is, with or without confinement, is an important factor to be considered for testing alumina ceramic in a strain rate regime $10^0 \leq \dot{\varepsilon} \leq 10^3$ s^{-1}. An equally important aspect is the processing history, for example, whether the alumina material is pressureless sintered[4] or hot-pressed[15] and/or hot-isostatically pressed.[15]

The photomicrographs presented in Figure 9.3 indicate that under the present experimental condition, the ceramic–polymer-layered specimen had undergone complex failure process in a given grain or assemblage of grains in multiple planes. The simultaneous formations of shear planes with grain localized plasticity and curved stepped microfracture strongly suggested the presence of both in-plane and out-of-plane shear deformation modes which were active during the failure process. The characteristic features of the deformation that had occurred at the respective grains are dependent on their relative orientations. Since steep stress gradients occurred in the layered specimen during failure, the failure morphology typically represented by the photomicrographs in Figure 9.3 was obviously complex, as expected.

Postmortem examination of the alumina fracture fragments as depicted by the FESEM photomicrographs presented in Figure 9.4 suggests that the sample suffered not only conventional global brittle fracture but grain localized plasticity, microcleavages, and intragranular microcracking occurred as well. These other failure processes can be considered as additional means of energy dissipation. Further, there are evidences of failure that had occurred at a single and/or multiple planes, Figure 9.4. These evidences strongly implied that the high strain rate compressive failure of a grain or an assemblage of grains is dependent on their relative orientations with respect to the surrounding microstructural constituents as well as their relative orientations with respect to the direction of propagation of the stress wave.

The FESEM-based photomicrographic evidences shown in Figure 9.4 have strongly suggested that during intragranular deformation and fracture processes, the in-plane and out-of-plane shear stress components became active. Therefore, the possible existence of a three-dimensional stress state during passage of the stress wave through the microstructure of the present coarse grain monolithic alumina cannot be totally ruled out. Formation of subgrain structure during high strain rate-induced fracture in pressureless sintered commercial alumina[17] strongly supports this argument. However, the exact genesis of subgrain formation could not be revealed and was beyond the scope of the present work. But considering the present wealth of literature, it appears that this observation[17] is of very significant importance so far as high strain rate deformation and failure of brittle materials, for example, ceramics, are concerned in the strain rate regime $10^0 \leq \dot{\varepsilon} \leq 10^3$ s^{-1}. The reason is that such a process may help further dissipation of energy even in a commercial grade coarse grain size alumina subjected to high strain rate compressive deformation. To the best of our knowledge other than those reported in Ref. [17], probably there is no other literature evidence available in support of the formation of subgrain structure in alumina.

9.4.2 POLY(METHYLMETHACRYLATE)

The compressive stress endurance of amorphous polymers, for example, PMMA, is highly sensitive to strain rate which is dependent on the impact energy. A part of the impact energy is converted to heat which plays an important role in the deformation process. Whether the heating will be locally confined at the crack tips region or would be dissipated generally throughout the bulk of the material is decided by the material under the deformation process. However, due to rapidity of the process, the complete homogeneous transfer of heat can seldom take place. The net deformation energy imparted to the sample gets dissipated either as fracture energy or gets absorbed as elastic strain energy or by both of these processes.[19,20] The enhancement of chain mobility is reported to occur in the condition of surface depression following the onset of local plasticity.[19]

Further, it is well known[22–24] that molecular mobility of polymer chains is dependent on its stiffness. The higher the stiffness, the lower becomes the molecular mobility. Strain rate plays an important role in decreasing the mobility of the polymer chains by making the chain stiffer. Recent efforts[25] have demonstrated that if PMMA specimens are stretched beyond the yield point, the relaxation rates are enhanced considerably. Obviously, an increase of strain rate decreases the molecular mobility. Many molecular theories have been proposed[23–26] to determine the yield stress of amorphous polymers subjected to dynamic loading under tension or compression. These models have tried to explain the yield behavior of the present PMMA. It appears that the deformation and failure of PMMA due to dynamic compression is a process that combines both mechanical and thermal activation processes.[20–26]

9.4.3 GLASS FIBER-REINFORCED POLYMER

The impact response of a polymeric material is its property to absorb and dissipate energy during high strain rate loading. In general, the deformation of reinforced polymers, for example, GFRP, at high strain rates primarily leads to conversion of the impact energy to heat energy. As explained before, whether the heat energy would be locally confined at the crack tip region or gets transferred throughout the bulk of the material would depend on the matrix material the deformation process.[19] It can be noticed from Figure 9.9b that with the increase of strain rate, the elastic strain energy of the matrix material increases. Thus, the specimen absorbs the total energy received from the high strain rate compressive impact imparted either as fracture energy or elastic strain energy or both when it is exposed to dynamic deformation. When the strain energy exceeds the fracture energy of the reinforced matrix material, for example, GFRP, the sample breaks into pieces. To regain its initial condition, material releases the energy that it accumulates during deformation. At high strain rate, material gets shorter span of time to release a part of the energy it absorbs under constraints. Over and above, the rate of energy accumulation becomes faster. Therefore, periodic time gap between accumulation and release of energy becomes progressively smaller. As a result, cracks between the matrix material and reinforcing agents start developing which progress across the matrix material. The failure happens when the strain rate reaches a critical value that leads to the accelerated growth and rapid diverging of the existing cracks until fragmentation of the specimen occurs under the constraints of high strain rate compressive fracture. [31].

9.5 SUMMARY AND CONCLUSION

The major conclusions of the present SHPB experiments followed by corroborative postmortem examination by FESEM and TEM were as follows:

a) The maximum compressive strength of layered alumina–polymer composite was 70% higher than that of the monolithic alumina.
b) The time to compressive failure of layered alumina–polymer composite was about 40% higher than that of the monolithic alumina, thereby suggesting an important means to cushion stress wave propagation.
c) The maximum strain rate of layered alumina–polymer composite, for example, 9×10^2 s^{-1}, was about 10% smaller than that, for example,

1×10^3 s^{-1}, achieved in the monolithic alumina, thereby supporting the stress cushioning role of the thin polymer layer. The high strain rate dynamic failure of monolithic alumina and layered alumina–polymer composite was associated with brittle fracture, extensive grain-localized plasticity and shear deformation, microcracking, and dislocation entanglement.

d) The maximum compressive strength of PMMA was 0.35 GPa, that is, about an order of magnitude smaller than that of the layered alumina–polymer composite. The yield strength increased by 25% from 0.28 GPa as $\dot{\varepsilon}$ was enhanced from about 4.8×10^2 to 9.3×10^2 s^{-1}. The critical $\dot{\varepsilon}$ was ~7.7×10^2 s^{-1}. Thus, the critical strain rate of PMMA was also much smaller than those of monolithic and layered alumina–polymer composite, as expected. The high strain rate failure of PMMA was associated with the thermal decomposition that happened most probably due to localized rise in temperature. Localized shear flow occurred due to high concentration of the compressive elastic pulses at the crack tips.

e) The maximum compressive strength of GFRP was 0.39 GPa, that is, about an order of magnitude smaller than that of the layered alumina–polymer composite but slightly higher than that of PMMA. The yield strength increased by 55% from 0.25 GPa as $\dot{\varepsilon}$ was enhanced from about 5.50×10^2 to about 1.25×10^3 s^{-1}. Thus, GFRP exhibited the highest strain rate among all the four materials while the highest (σ_c) was achieved by the alumina–polymer-layered composite. The high strain rate dynamic failure of GFRP was associated with brittle fracture of the matrix with and without extensive shear deformation, interfacial separation at matrix–fiber interface, localized fiber delamination, debonding, and finally fiber fracture.

ACKNOWLEDGMENTS

The authors thank Dr. K. Muraleedharan, Director, CSIR-CGCRI for encouragement and support to carry out work in this field. The authors gratefully acknowledge the part of the experiment carried out at high-speed experimental mechanics laboratory, IIT Kanpur under the supervision of Prof P. Venkitanarayanan. Mr. M. K. Gautam of the same laboratory is thanked for rendering his help in conducting the experiments. CSIR is acknowledged for the financial support (Project NWP 0027 and ESC0104).

KEYWORDS

- alumina
- deformation
- strain rate
- compressive strength
- polymer

REFERENCES

1. Lankford, J. Mechanisms Responsible for Strain-Rate-Dependent Compressive Strength in Ceramic Materials. *J. Am. Ceram. Soc.* **1981**, *64*, C-33–C-34.
2. Wang, Y.; Mikkola, D. E. {0 0 0 1}<1 0 1 0> Slip and Basal Twining in Sapphire Single Crystals Shock-Loaded at Room Temperature. *J. Am. Ceram. Soc.* **1992**, *75*, 3252–3256.
3. Grady, D. E. Shock-Wave Compression of Brittle Solids. *Mech. Mater.* **1998**, *29*, 181–203.
4. Lankford, J.; Predebon, W. W.; Staehler, J. M.; Subhash, G.; Pletka, B. J.; Anderson, C. E. The Role of Plasticity as a Limiting Factor in the Compressive Failure of High Strength Ceramics. *Mech. Mater.* **1998**, *29*, 205–218.
5. Chen, M. W.; Mccauley, J. W.; Dandekar, D. P.; Bourne, N. K. Dynamic Plasticity of Failure of High Purity Alumina Under Shock Loading. *Nat. Mater.* **2006**, *5*, 614–618.
6. Tasdemirci, A.; Hall, I. W. Numerical and Experimental Studies of Damage Generation in Multi-layer Composite Materials at High Strain Rates. *Int. J. Impact Eng.* **2007**, *34*, 189–204.
7. Walley, S. M. Historical Review of High Strain Rate and Shock Properties of Ceramics Relevant to Their Application in Armour. *Adv. Appl. Ceram. Struct., Funct. Bioceram.* **2010**, *109*, 446–466.
8. Pavlacka, R.; Bermejo, R.; Chang, Y.; Green, D. J.; Messing, G. L. Fracture Behaviour of Layered Alumina Microstructural Composites with Highly Textured Layer. *J. Am. Ceram. Soc.* **2013**, *96*, 1577–1585.
9. Pender, D. C.; Thompson, S. C.; Padture, N. P.; Giannakopoulos, A. E.; Suresh, S. Gradients in Elastic Modulus for Improved Contact-Damaged Resistance. Part II: The Silicon Nitride–Silicon Carbide System. *Acta Mater.* **2001**, *49*, 3263–3268.
10. Lee, W.; Howard, S. J.; Clegg, W. J. Growth of Interface Defect and Its Effect on Crack Deflection and Toughening Criteria. *Acta Mater.* **1996**, *44*, 3905–3922.
11. Gooch, W. A.; Chen, B. H. C.; Burkins, M. S.; Palicka, R.; Rubin, J. J.; Ravichandran, R. Development and Ballistic Testing of a Functionally Gradient Ceramic/Metal Applique. *Mater. Sci. Forum* **1999**, *308–311*, 614–621.
12. Liu, L. S. et al., In *Functionally Graded Materials VII*: Impact Characteristic Analysis of Ceramic/Metal FGM; Materials Science Forum, 2003; Vols. 423–425, pp 641–644.

13. Liu, L. S.; Zhang, Q. J.; Zhai, P. C. The Optimization Design on Metal/Ceramic FGM Armor with Neural Net and Conjugate Gradient Method. In *Functionally Graded Materials VII*; Materials Science Forum, 2003; Vols. 423–425, pp 791–796.
14. Acharya, S.; Mukhopadhyay, A. K. Dynamic Compressive Fracture for Ceramic Polymer Layered Composites. *Proc. Eng.* **2014**, *86*, 281–286.
15. Staehler, J. M.; Predebon, W. W.; Pletka, B. J.; Subhash, G. Micromechanisms of Deformation in High-Purity Hot-Pressed Alumina. *Mater. Sci. Eng. A* **2000**, *291*, 37–45.
16. Lankford Jr. J. The Role of Dynamic Material Properties in the Performance of Ceramic Armor. *Int. J. Appl. Ceram. Technol.* **2004**, *1*, 205–210.
17. Acharya, S.; Mukhopadhyay, A. K.; et al. Deformation and Failure of Alumina Under High Strain Rate Compressive Loading. *Ceram Int.* **2015**, *41*, 6793–6801.
18. Woldesenbet, E.; Vinson Jack, R. Specimen Geometry Effects on High-Strain-Rate Testing of Graphite/Epoxy Composites. *AIAA J.* **1999**, *37* (9), 1102–1106.
19. Moy, P.; Weerasooriya, T.; Chen, W.; Hsieh, A. Dynamic Stress–Strain Response and Failure Behaviour of PMMA. In *IMECE2003-43371*; pp 105–109.
20. Swallowe, G. M.; Field, J. E.; Horn, L. A. Measurements of Transient High Temperatures during the Deformation of Polymers. *J. Mater. Sci.* **1986**, *21*, 4089–4096.
21. Acharya, S.; Mukhopadhyay, A. K. High Strain Rate Compressive Behaviour of PMMA. *Polym. Bull.* **2014**, *71*, 133–149.
22. Swallowe, G. M.; Lee, S. F. A Study of the Mechanical Properties of PMMA and PS at Strain Rates of 10^{-4} to 10^3 over the Temperature Range 293–363 K. *J. Phys. IV France* **2003**, *110*, 33–38.
23. Fuller, K. N. G.; Fox, P. G.; Field, J. E. The Temperature Rise at the Tip of Fast-Moving Cracks in Glassy Polymers. *Proc. R. Soc. Lond. A* **1975**, *341*, 537557.
24. Li, Z.; Lambros, J. Strain Rate Effects on the Thermomechanical Behaviour of Polymers. *Int. J. Solid Struct.* **2001**, *38*, 3549–3562.
25. Richeton, J.; Ahzi, S.; Vecchio, K. S.; Jiang, F. C.; Adharapurapu, R. R. Influence of Temperature and Strain Rate on the Mechanical Behaviour of Three Amorphous Polymers: Characterization and Modelling of the Compressive Yield Stress. *Int. J. Solid Struct.* **2006**, *43*, 2318–2335.
26. Nasraoui, M.; Forquin, P.; Siad, L.; Rusinek, A. Influence of Strain Rate, Temperature and Adiabatic Heating on the Mechanical Behaviour of Poly-Methyl-Methacrylate: Experimental and Modelling Analyses. *Mater. Des.* **2012**, *37*, 500–509.
27. Blumenthal, W. R.; Caddy, C. M.; Lopez, M. F.; Gary, III, G. T.; Idar, D. J. Influence of Temperature and Strain Rate on the Compressive Behaviour of PMMA and Polycarbonate Polymers. *AIP Conf. Proc.* **2001**, *620*, 665–668.
28. Tesdemirci, A.; Hall, I. W.; et al. Experimental and Numerical Investigation of High Strain Rate Mechanical Behaviour of a [0/45/90/-45] Quadriaxial E-Glass/Polyester Composite. *Proc. Eng.* **2011**, *10*, 3068–3073.
29. Naik, N. K.; Venkateswara Rao, K.; Veerraju Ch.; Ravikumar, G.; Stress–Strain Behaviour of Composites under High Strain Rate Compression along Thickness Direction: Effect of Loading Condition. *Mater. Des.* **2010**, *31*, 396–401.
30. Oguni, K.; Ravichandran, G. Dynamic Compressive Behaviour of Unidirectional E-glass/Vinyl Ester Composites. *J. Mater. Sci.* **2001**, *36*, 831–838.
31. Acharya, S.; Mondal, D. K.; Ghosh, K. S.; Mukhopadhyay, A. K. Mechanical Behaviour of Glass Fibre Reinforced Composite at Varying Strain Rates. *Mater Res Express 4*, 035303, **2017**.

CHAPTER 10

APPLICATIONS OF ELECTRON BACKSCATTER DIFFRACTION IN MATERIALS SCIENCE

RAVI CHANDRA GUNDAKARAM[*]

Centre for Materials Characterization and Testing, International Advanced Research Centre for Powder Metallurgy and New Materials (ARCI), Balapur P.O., Hyderabad 500005, Telangana, India

[*]E-mail: ravi.gundakaram@arci.res.in

ABSTRACT

In the synthesis and study of materials, microstructure is an important factor. Control of the microstructure is very essential in industrial processes for reproducibility in components. In addition, it is now recognized that by suitable microstructural tuning, many properties can indeed be enhanced. Among the experimental techniques available for the study of microstructure, electron backscatter diffraction (EBSD) is both a suitable and a powerful tool. The technique conveniently probes the lengthscales between those studied by optical microscopy and transmission electron microscopy. This chapter provides an introduction to EBSD and discusses experimental details on obtaining and analyzing EBSD patterns. Traditionally, EBSD has been used to study the crystallographic orientation of each grain in a metal or an alloy but in recent years, it has also been used for microstructural studies of ceramic materials. By suitable examples, the development of the technique is illustrated in this chapter, and applications such as the determination of grain boundary character distribution and microtexture are described. In conjunction with energy dispersive spectroscopy, EBSD can be used to analyze the phases present in a material and also for the study of new phases. Finally, EBSD in the transmission mode is touched upon. The chapter ends with an outlook for the future of the technique.

10.1 INTRODUCTION

Advances in materials science in the past few decades have been tremendous. It is now possible to tailor-make materials for demanding design criteria. The other aspect in the progress of the field is the realization of the importance of microstructure. The same material can be used for improved performance, or new functionalities can be built in, by tuning the microstructure. Thus, in addition to the range of novel materials, the materials scientist can now utilize the power of microstructural engineering.

A very important aspect in the study of materials is the ability to characterize them. Based on the length scale of the feature to be studied, a suitable experimental tool can be chosen. The technique of electron backscatter diffraction (EBSD), also known as orientation imaging microscopy (OIM), has come to fill a niche area and this chapter aims to provide a brief introduction to EBSD.

At the very outset, a word or two (or maybe many more) for the justification of this chapter may be in order. Its writeup is an expansion of the talk given by the author at the World Congress on Microscopy (WCM 2015), held at the Mahatma Gandhi University, Kottayam, India, in October 2015. There are several illuminating articles in the literature[1-5] to state a few, written by experts, some of whom have been in the field from the very beginning. The current writeup is in no way a match to these in terms of scope, clarity, and depth of information. However, it was noticed that even in a conference on microscopy, where many of the presentations were on electron microscopy in particular, there was no other talk covering the important discipline of EBSD. It is also our experience that several researches in materials science and engineering are still not well versed with what EBSD can do. Thus, it was felt that it might be more helpful for an interdisciplinary audience, like the researchers in the Conference, several of whom are from the life sciences, to present the field with examples but with not too much technical detail.

10.1.1 THE ROLE OF EBSD IN MATERIALS CHARACTERIZATION

Several parameters of importance microstructurally have been studied for long. EBSD has helped in providing additional information about these parameters, while at the same time helping in enhancing the accuracy of the measurement. As an example, the grain size distribution in a sample has traditionally been studied by etching the sample and making measurements under a microscope, or by using X-ray diffraction as an experimental technique. Since the measurements are made locally in EBSD, the accuracy can be improved by covering the area of interest in greater detail. The further value addition brought in is to provide the information about the orientation of individual grains. With this, many other parameters about the nature of the grain boundaries can be studied. Strains can be mapped over sufficiently large areas, again with good accuracy. It is not an exaggeration to say that the discipline of grain boundary engineering has received a big boost due to EBSD.

10.1.2 THE TECHNIQUE

In the EBSD measurement, the sample is tilted typically to an angle of 70° to the horizontal (Fig. 10.1). This ensures that the path length of the electrons that have been backscattered is reduced, thus allowing greater numbers of these electrons to undergo diffraction and escape from the specimen.[6] The incident electron beam impinges on the sample, diffraction occurs between the backscattered electrons, and the signal is collected on the phosphor screen.

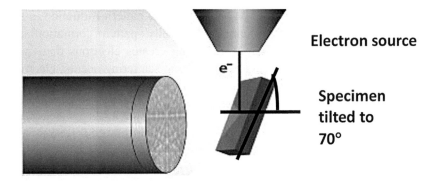

FIGURE 10.1 Schematic of the EBSD experiment. (Adapted from the OIM data analysis manual, EDAX-TSL, with permission.)

An example of the bands is given in Figure 10.2a. The bands are indexed as shown in Figure 10.2b, and the crystallographic orientation at the point where the incident beam falls on the sample is determined. Each of these steps is described briefly in what follows.

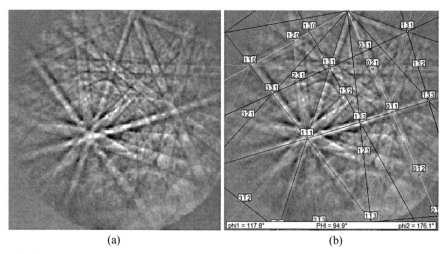

FIGURE 10.2 (a) Example of EBSD bands on the phosphor screen (left). (b) Indexing of the bands shown in the figure (right).

10.1.3 PHYSICS BEHIND BAND FORMATION

It is generally accepted that the physics of EBSD pattern formation is still not completely understood. Phenomenologically, pattern formation is a two-step process. First, incoherent scattering creates electrons that travel in different directions. Following this, coherent scattering occurs; electrons undergo diffraction moving in pairs of cones known as Kossel cones. When these intersect the phosphor screen (nearly) parallel pairs of lines are seen, which are the EBSD bands.[7]

10.1.4 ANALYSIS OF EBSD PATTERNS

The purpose of "indexing" the bands is to determine the crystallographic orientation at the point of study. But before indexing, it is necessary to detect the bands that appear on the phosphor screen. Since software finds it easier to detect a bright spot as compared to a band, each band is first converted to a spot in the Hough space using a Hough transform,[8] which is basically a mathematical operation to convert a line into a point while retaining a one-to-one correspondence between the band and the location of the spot in the Hough space.

Once the bands are identified, there are two possible ways to determine the crystallographic orientation. In the first method, the indices associated with the zone axis are identified, from which the orientation can be determined.[9] It may be mentioned here that a zone axis is a straight line that is common to two or more bands, running perpendicular to the plane of the bands and occurring at the intersection of the bands.

In the method of indexing being followed more recently, the angles between the different bands are determined and compared to a theoretical look-up table. All possible sets of three bands are formed from the detected bands. The angles between the bands for each triplet are compared to the look-up table.[10] When there is more than one possible solution, as is often if not usually the case, a voting scheme is used to identify the most possible indexing of the pattern and thus the orientation is determined.

Two commonly used parameters in EBSD data analysis are the image quality (IQ) and the confidence index (CI). The IQ is used to describe the quality of an EBSD pattern, which is dependent on the material and its condition such as strains in the sample. The CI is a measure of the reliability of the indexing solution. Once the indexing is accomplished correctly, the system moves to the next point and the indexing process is repeated until the entire area of interest is covered.

In an EBSD scan, an area of interest is first decided upon. A step size is also chosen, which depends on the feature size that needs to be studied. For example, if the grain size is x, a step size of $x/5$ is usually adequate. It is important to sample the feature adequately. While a large step size leads to undersampling, oversampling needs to be avoided too since it does not add to useful information but causes large data sets which take up computer time. When starting to work with a new sample, a quick scan can be set up with a step size of about a micron to obtain a rough idea. The step size can then be refined as needed.

Background subtraction: The band strength on the phosphor screen is usually not too strong. To enhance the contrast, the background is collected. The camera parameters can usually be adjusted so that there nearly uniform illumination on the phosphor screen and this background is subtracted so that the bands appear more clearly. Figure 10.3a,b show images of the phosphor screen, before and after background subtraction.

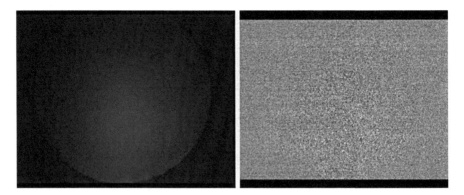

FIGURE 10.3 Images of the phosphor screen, before (left) and after background subtraction.

10.1.5 MATERIALS THAT CAN BE STUDIED

As the name signifies, EBSD hinges on diffraction occurring; hence, the prime requirement is that the specimen under study has to be crystalline. Partly crystalline materials may also be studied if the bands, even if diffuse, have sufficient intensity for indexing. Much of the early research was on metals and metallic alloys since these are amenable to good polishing and also gave good patterns. Ceramic materials can also be studied by providing a thin (≈2 nm) coating of gold or other conducting material on the surface to minimize charging effects. The use of the variable pressure mode in a scanning electron microscope obviates the need for the conductive coating. This has the additional advantage that after the EBSD study; the sample can be used for further studies in, say, nanoindentation. The sample can be in bulk form or as a film, although in the latter case, it may be more difficult to obtain good patterns due to the strains present. Furthermore, the information depth in EBSD is accepted to be 50–200 nm, so in the case of thin films, possible interference from the substrate must be kept in mind.

10.1.6 PARAMETERS THAT CAN BE OBTAINED FROM EBSD

The most basic parameter that EBSD delivers is the crystallographic orientation of each grain. Calculation of the misorientations between neighboring grains helps us in understanding the nature of the grain boundaries. The grain size can be measured, and more accurately as compared to other techniques such as the line intercept method using an optical or electron microscope. Microtexture studies are conveniently carried out by EBSD. Strains in the sample can be measured and this is an active area of work at present.

It is to be kept in mind that EBSD is basically a microstructural tool. The input to an EBSD measurement is crystallographic data and the output is microstructural information about the sample. That said, combining EDS and EBSD measurements helps us in carrying out phase analysis. Each measurement begins with obtaining an image of the sample in the scanning electron microscope. Thus, unlike in routine X-ray measurements, the area of study can be controlled finely by the experimentalist and since the spot size of the electron beam is much finer than the X-ray pencil, spatial resolution is that much higher.

It is hoped that this chapter gives a flavor to some of the possibilities of EBSD given above.

10.2 EXPERIMENTAL DETAILS

10.2.1 SPECIMEN PREPARATION FOR EBSD

If the specimen surface is not smooth, more of scattering can be expected instead of diffraction. Thus, the prime requirement is that the surface has a mirror finish. It is also desirable that the strains are kept to a minimum so that the bands are well defined.

The most common method to obtain a mirror finish is to mount the sample in Bakelite or cold-setting resin and polish the surface on abrasive paper with progressively increasing grit. Once a reasonably good surface is obtained (periodic observation in an optical microscope would do), the polishing is continued with diamond pastes of decreasing particle size. A final polishing step using colloidal silica or alumina with particle size of 0.05 μm (or 0.04 μm in some cases) completes the polishing sequence. Colloidal silica may also provide a gentle etch that is in some cases actually beneficial to obtain good EBSD patterns.

Other methods of specimen preparation are electropolishing, vibratory polishing, and ion beam milling. The former is suitable only for samples that are electrically conducting since the sample is used as one of the electrodes. Wrong choice of the electrolyte and other experimental parameters can lead to a very poor surface totally unsuitable for EBSD. Figure 10.4a shows the SEM image of a poorly electropolished copper surface. Figure 10.4b,c show the IQ map and the crystal orientation map (COM), respectively. It is hoped that these figures illustrate the importance of the proper conditions for polishing to obtain a smooth surface for EBSD.

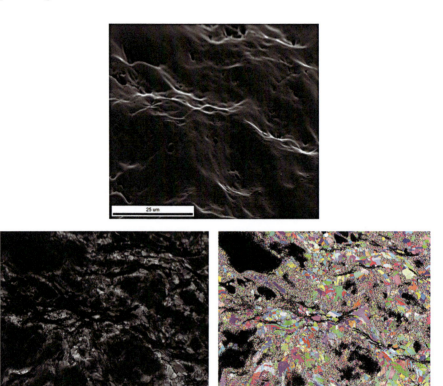

FIGURE 10.4 (a) SEM image of a poorly electropolished copper surface, showing the surface to be not suitable for a good EBSD scan. (b and c) Image quality map (left) and crystal orientation map for the surface shown in (a).

The latter two techniques mentioned above, namely vibratory polishing and ion beam milling, are more suitable for the finer steps of polishing. In vibratory polishing, a rotating platen is made to vibrate at a certain frequency. The platen is lubricated by a polishing medium, that is, colloidal silica and the samples to be polished are left upside down on the platen, often kept pressed down by a weight. The polishing takes place over an extended period of time but the effect is gentle and a good surface finish is obtained for most samples.

Applications of Electron Backscatter Diffraction 235

In the case of gentle ion beam milling, argon gas (Ar) is ionized and the low energy stream of ions is made to fall on the rotating sample surface through guns facing each other. A schematic of the gentle ion beam mill used in this work is shown in Figure 10.5.

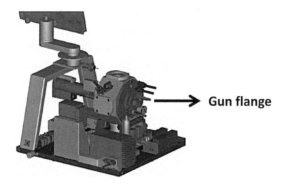

FIGURE 10.5 Schematic of the Ar ion mill used in the present study (Bal-Tec Res101).

Depending on the energy of the ions, the angle of incidence, and the nature of the interaction between the Ar ions and the material, the surface gets polished to a good finish. Figure 10.6a,b shows the bands (collected from different locations in the sample) before and after ion beam milling in a ZnO sample. It can be seen that the pattern strength has improved considerably after ion beam milling.

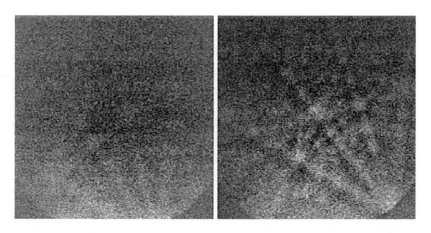

FIGURE 10.6 EBSD pattern quality before (image on the left) and after ion beam milling. The images are from different locations in the sample. It can be seen that the EBSD pattern strength is enhanced after the ion beam milling.

10.2.2 CARRYING OUT EBSD MEASUREMENTS

In the actual measurement, an area of interest and the step size are chosen. The electron beam is made to impinge on the sample, the bands are observed and the orientation is determined and stored in the computer memory. The electron beam is moved to the next data point based on the preset step size and the process is repeated until the entire range is covered. In EBSD, a grain is defined as a three-dimensional collection of contiguous points with the same crystallographic orientation. The procedure described above collects information in two dimensions. Points within the grain have the same orientation and after the scan is completed, it is straightforward to reconstruct the extent of the grains. Since the orientation of each point is known, the difference in orientation (misorientation) between neighboring grains can be calculated and the grain boundary character distribution (GBCD) can be determined at once.

10.2.3 ANALYSIS OF EBSD DATA

In EBSD measurements, what is basically measured are the Euler angles of the point under observation. Once the scan is completed, the data are usually subjected to a process of cleanup. There are several parameters that can be used for the cleanup, the most common being the Confidence Index. If neighboring points surrounding a point have the same CI and the point at the center has a different CI, the value of one of the surrounding points is assigned to the data point at the center, thus "cleaning up" that point (Fig. 10.7). The process is repeated for all the points in the scan. While cleanup is an accepted part of EBSD analysis, it should be used with care, placing special emphasis on observing if only "bad" data points are cleaned of if there are systematic changes in orientation across the area of the scan.

Some of the parameters that can be obtained from EBSD analysis are the grain size, nature of the boundaries (as defined by the misorientation between adjacent grains), crystallographic orientation of each grain, microtexture, etc. EBSD also lends itself to the study of twins in a very convenient way.

Grain-size measurements are often carried out on etched samples using the line intercept method. However, measurements using EBSD are more accurate as each and every grain is covered in pixels and a sum up gives a much better value even for complex shapes. An etched sample shows the extent of grains (Fig. 10.8) but orientation information cannot be obtained from such an image. By measuring the Euler angles at every point, the orientation of each grain can be determined.

Applications of Electron Backscatter Diffraction 237

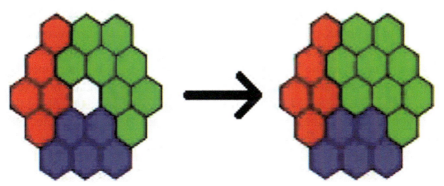

FIGURE 10.7 Schematic showing data cleanup. The point shown in white has been "cleaned up." (Taken from the OIM Analysis manual, V6.1.3, © EDAX (2011), with permission)

FIGURE 10.8 SEM image of an etched sample showing grains. The orientation of each grain is not known from such an image.

The misorientation between adjacent grains can be measured quantitatively by determining what is known as the axis–angle pair. This basically tells the angle of misorientation between the grains about a particular axis. The software allows the misorientation to be calculated between a grain and its neighbors. It is also possible to study the average misorientation between all the grains in the scan.

EBSD is a convenient and powerful tool for the study of microtexture. Normally, all crystallographic orientations have an equal probability of occurring in a sample. However, in certain cases, usually due to special processing conditions, there is a marked preference for one or more orientations to occur. This is known as texture and is an important parameter that can be used to tailor materials properties. Traditionally, texture measurements have been carried out using X-rays and this continues to be the mainstay even today. The strength of EBSD, however, is in carrying out studies locally and thus the term "microtexture" as opposed to macrotexture that is determined by XRD. Recent advances in X-ray sources and detectors facilitate microtexture studies but the added advantage of EBSD is that information about the location of grains in the microstructure is preserved, whereas X-ray methods do not provide this information.

More recently, strain measurements are being carried out using EBSD. A careful study of the shift in band positions in strained regions of a sample as compared to unstrained regions provides an estimate of strain. Further information is available in the literature.[11,12]

10.3 SELECTED APPLICATIONS

10.3.1 DEVELOPMENT OF THE TECHNIQUE

In this section, we briefly trace the development of EBSD from its initial stages to the present day, with illustrative examples. Processing details that may be needed to understand the context are also provided briefly. Early work focused on orientation measurements, which we illustrate using the example of Ta coatings. The next logical step would be to carry out microtexture measurements, and we show the same with the example of Ni coatings. In a multiphasic sample, the location of the different phases in the microstructure can be determined, as we show using the example of zinc oxide varistor materials. Combining the power of EDS and EBSD gives us the facility to distinguish phases with the same crystal structure. Compositional gradients that may occur due to certain processing conditions can be studied, which we show in Ti–N coatings. Finally, phase determination of total unknowns can be attempted by carrying out quantitative measurements on EBSD patterns.

10.3.2 ORIENTATION MEASUREMENTS: THE EXAMPLE OF Ta COATINGS

Ta coatings are especially suitable for applications where high-temperature strength, electrical conductivity at very high temperatures, and corrosion resistance to many acids and other chemicals are required. The coatings in the present study were deposited by the cold gas dynamic spray (CGDS) technique. Basically, powder particles that are carried by a gas stream of high velocity impinge on the substrate, where they get flattened into what are known as splats. By depositing successive splats in an array, the entire area of the substrate can be covered. Coating thickness can be built by multiple passes over the same area. To improve intersplat bonding, the as-deposited coatings are heat treated at elevated temperatures. In the present study, the coatings were heated to 750, 1000, and 1250°C. EBSD data were collected on all samples, in addition to the as-deposited coatings.

It is of interest to study the microstructure, both on the surface and in cross section. Figure 10.9 shows the IQ maps recorded on the surface for the Ta coating in the as-deposited condition and for those heat treated at 750, 1000, and 1250°C. The corresponding COMs are shown in Figure 10.10. The areas that are black in the COMs are from bad data points that usually occur at intersplat boundaries and are excluded from the analysis in these and all other COMs. The choice of black for these points is due to the fact that all other colors are present in the inverse pole figure (IPF).

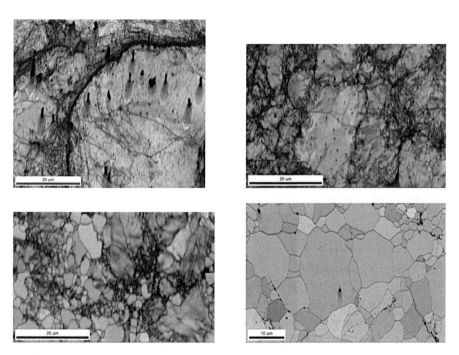

FIGURE 10.9 Image quality maps for the as-coated Ta specimen (top left) and for those heat treated at 750°C, 1000°C, and 1250°C (top right, bottom left, and bottom right, respectively).

The colors in Figure 10.10 are read in conjunction with the IPF, Figure 10.11, and each color refers to a unique orientation. Thus, contiguous points (areas) of the same color are grains in the sample. It is interesting to note that a single splat at the bottom right of the COM for the as-deposited coating (Fig. 10.9, map on the top left) contains several grains. As the heat treatment temperature increases, the bonding between the splats increases and the coating becomes dense.

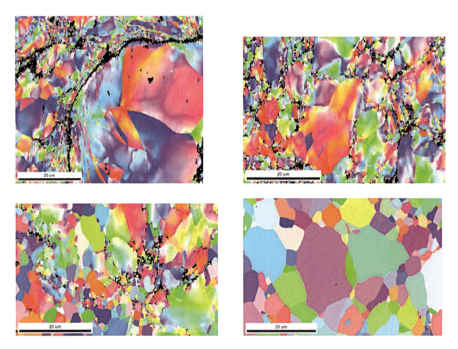

FIGURE 10.10 Crystal orientation maps (COMs) for the as-coated Ta specimen (top left) and for those heat treated at 750, 1000, and 1250°C (top right, bottom left, and bottom right, respectively). The IQ maps are in the previous figure.

To illustrate some of the microstructural features that can be determined by EBSD, we show the grain boundary map (Fig. 10.12a) and the grain size chart (Fig. 10.12b) for the sample heat treated at 1000°C. Figure 10.12a shows the misorientation between neighboring grains. Misorientation of 1°–5°, termed as subgrain boundaries, are shown colored in red, 5°–15° misorientation, low angle grain boundaries, are shown in green, and 15°–180° misorientation are the high angle grain boundaries shown in blue. The total length of each type of boundary is also shown in the legend. Of course, the shape and size of each grain can also be read from the figure. Thus, the information in Figure 10.12a is called the Grain Boundary Character Distribution (GBCD). Strains (or dislocations) in the sample manifest as subgrain boundaries, and thus, a high fraction of the subgrain boundaries indicates a high dislocation density. The study of dislocations is beyond the scope of EBSD at the present time and transmission electron microscopy is a suitable technique.

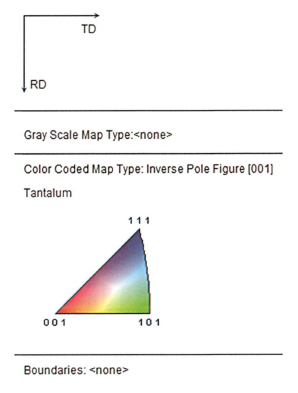

FIGURE 10.11 Color-coded inverse pole figure (IPF) for tantalum. The orientations in the maps of Figure 10.10 are read in conjunction with this IPF. Each color represents a unique crystallographic orientation.

Figure 10.12b shows the grain size distribution for the data in Figure 10.12a. Such charts are helpful in understanding the variation in grain size. The *y*-axis of this plot shows the area fraction. However, other parameters such as the number and number fractions as well as the grain diameter can also be depicted, depending on the need on hand.

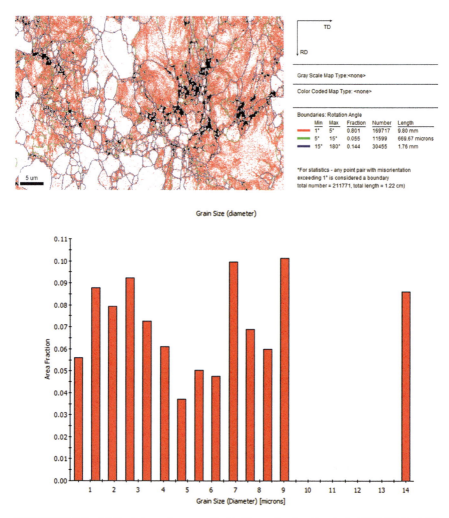

FIGURE 10.12 (a) Grain boundary map for the Ta coating heat treated at 1000°C (top). The legend to the right of the figure gives information about the grain boundaries. (b) Grain size chart for the data in (a).

As mentioned above, a single splat can contain several grains. While the example of Ta-coatings is being used to illustrate the ability of EBSD to quantitatively determine the different crystallographic orientations, Figure 10.10 also shows the evolution of the microstructure. It can be seen that with increasing temperature of the heat treatment, the intersplat bonding improves and the spread in the grain size decreases. To analyze the data further, we show the grain orientation spread (GOS) in Figure 10.13a–d

Applications of Electron Backscatter Diffraction 243

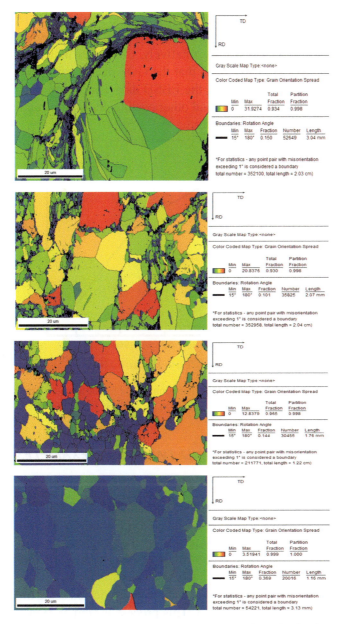

FIGURE 10.13 Top to bottom (a) Grain orientation spread (GOS) for the as-coated Ta specimen. The legend for the colors is given on the right. (b) Grain orientation spread (GOS) for the Ta coating heat treated at 750°C. The legend for the colors is given on the right. (c) Grain orientation spread (GOS) for the Ta coating heat treated at 1000°C. The legend for the colors is given on the right. (d) Grain orientation spread (GOS) for the Ta coating heat treated at 1250°C. The legend for the colors is given on the right.

for the data in Figure 10.10. The color coding in these figures is from blue (lowest spread) to red (highest spread). The black lines in these maps are the high angle grain boundaries. Qualitatively, the colors indicate the spread, and quantitatively, it can be seen that the GOS decreases monotonously as shown in Figure 10.14. Overall, the EBSD data were helpful in optimizing the heat treatment temperature and other coating deposition conditions.

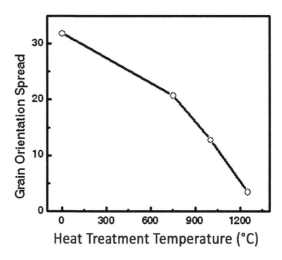

FIGURE 10.14 Variation of the grain orientation spread for the as-coated sample and those heat treated at 750, 1000, and 1250°C. Digit 0 on the *x*-axis is for the as-coated sample.

10.3.3 MICROTEXTURE AND TWINNING: EXAMPLES OF Ni AND Cu

As mentioned earlier, texture is the occurrence of one or more orientations in preference to others. Processing conditions can be tuned so as to create texturing in some materials so that the properties/performance can be enhanced. We illustrate the power of EBSD to study microtexture with nickel coatings as an example.

The nickel coatings used in this study were fabricated using the technique of pulse reverse electrodeposition. This process is similar to electrodeposition, with the addition of a reverse pulse so that a small amount of material deposited during the forward pulse sequence can be removed. By optimizing the forward and reverse pulse conditions (intensity of the signal and time), a desired microstructure, including texture, can be obtained.[13]

Figure 10.15a shows the COM for a nickel coating textured along the (0 0 1) and (1 1 1). Figure 10.15b shows the pole figures and the texture numbers given alongside. The color coding for the texture number is from blue (low) to red (high). Thus, a red dot at the center of the respective pole figure indicates texturing in that direction.

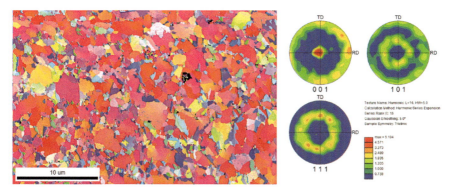

FIGURE 10.15 (a) COM of a nickel coating textured along (0 0 1) (left). (b) Pole figures showing the texturing (right). The legend on the bottom right shows the color coding in the pole figures along with the texture number.

A short recap on the comparison of texture measurements using X-ray diffraction and EBSD may be in order. As mentioned earlier, texture can be studied using both these techniques. However, the X-ray beam on the sample is often 1 cm^2 or more in area, whereas the spot size of the electron beam on the sample in the SEM is just a few nm^2. Further, a large area EBSD scan may be no more than a few hundred square micron in area and still larger areas will need stage control in the SEM and data stitching facilities, not to mention special software to correct for image drifts that may have occurred. Thus, texture measurements using X-rays provide global information whereas EBSD provides local ("microtexture") information. Thus, the two techniques are complimentary. Recently, it has been reported that information from both techniques is comparable, provided the EBSD scan contains 10,000 grains or more.[14]

Another aspect that can be studied conveniently is twinning. Primary twins are defined as having a 60° misorientation about <1 1 1> while secondary twins have a 38.9° misorientation about <1 1 0>. Twin boundaries are special boundaries that are desirable in certain cases. In the case of the copper coatings of the present study, the high fraction twin boundaries helps in enhancing electrical conductivity. Figure 10.16a shows the COM of a spark plasma sintered copper sample showing twinning. Figure 10.16b shows the high angle grain boundaries in blue and the twin boundaries in red; the extent of twinning can be seen very clearly from this figure. To illustrate further the extent of twinning, Figure 10.16c shows the coincident site lattice (CSL) boundary chart for the data set in Figure 10.16a. In the CSL notation, Σ3 denotes twins. It can be seen clearly that the fraction of twinning in this sample is high, which is the result of carefully controlling the processing conditions.

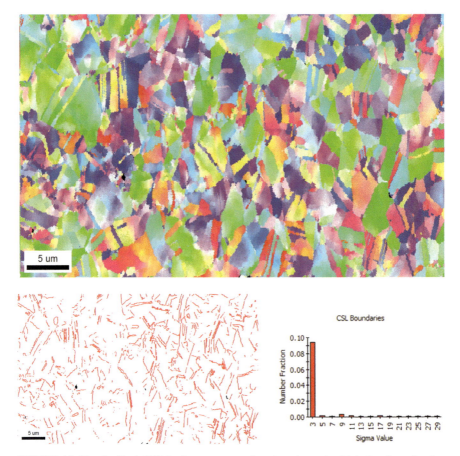

FIGURE 10.16 (a, Top) COM of a copper coating that showed a high fraction of twins. (b, bottom left) Twin boundaries (red) for the data set shown in (a) (left). (c, bottom right) Coincident site lattice boundary chart for the same data set (right).

10.3.4 STUDY OF MULTIPHASIC SAMPLES: THE EXAMPLE OF ZINC OXIDE VARISTOR MATERIAL

Many materials used in practical applications contain more than one phase. Often, the secondary phases form as impurities during processing and need to be studied so as to understand their nature and then suitably modify the processing conditions to eliminate the undesirable phases where possible.

However, in certain cases, the secondary phases are desirable and indeed necessary for the enhancement of some properties. Precipitation hardening is a case in point where fine precipitates actually lead to enhanced hardness. In the case of oxide dispersed strengthened steels, the fine oxide dispersoids provide resistance to creep and bring about other improvements. In the present case, we provide the example of zinc oxide varistor material where the secondary phases are necessary for the varistor property. In addition, it has been seen that the location of the secondary phases in the microstructure is also important, and EBSD has helped in understanding the same.

A varistor is a device that provides protection from surges in power. Taking the example of electrical circuitry, under normal conditions, the current passes through the rest of the circuit for routine performance. However, in the event of a surge, the varistor comes into the circuit and protects the devise from damage. Further details are available in Ref. [15]. The electrical behavior of grains and grain boundaries is important for varistor behavior and EBSD is a very suitable technique study the GBCD.

It is important to note that EBSD is sensitive to the crystal system. In the case where more than one phase is present in the specimen with the same crystal structure, EBSD in general cannot distinguish between the two, especially if the lattice parameters are comparable. For example, two phases with the cubic structure will be treated by the technique as a single phase and indexing will be carried out accordingly. Simultaneous EDS and EBSD measurements (next section) can sometimes be used to identify each phase when one or more elements are not common to either phase.

In the present study, ZnO is the base material and varistor activity is realized when doped with bismuth and antimony. These elements combine to form the pyrochlore phase ($Bi_3Sb_3Zn_2O_{14}$) while the reaction of Zn with Sb produces a small fraction (3% or less, as determined by X-ray diffraction) of the spinel phase ($Zn_2Sb_2O_7$). Bi_2O_3 is also present (less than 1.5%).

As it happens, all the four phases in the present case have different crystal structures: ZnO is hexagonal, pyrochlore is cubic, spinel is orthorhombic, and Bi_2O_3 has the monoclinic structure. Thus, EBSD can be used to detect these without ambiguity. For the present discussion, we concentrate on

ZnO and pyrochlore. Figure 10.17a shows the COM for these two phases, with the high angle grain boundaries marked as black lines, in a sample showing good varistor behavior. The color-coded IPFs given to the right of Figure 10.17b reflect the crystal geometries for the phases.

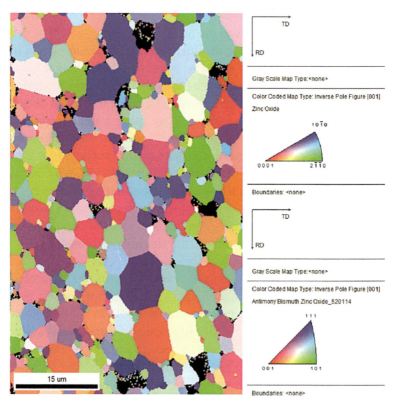

FIGURE 10.17 (a) COM of the ZnO varistor material (left). (b) IPFs for the hexagonal phase (ZnO, on the top) and the cubic phase (pyrochlore) (right).

The software provides the facility to create partitions showing each phase separately. Thus, Figure 10.18a,b shows the ZnO (left) and pyrochlore phases using the partition facility. Figure 10.17 is naturally obtained by the superposition of Figure 10.18a,b. Figure 10.19 shows the phase map with ZnO colored white and the pyrochlore phase shown in green. The black lines are the high angle grain boundaries. It can be seen that in this sample, due to the optimum processing conditions, the pyrochlore phase occurs preferentially at the triple points (common points to three neighboring grains), and when this happens, the varistor behavior was found to be good.

Applications of Electron Backscatter Diffraction 249

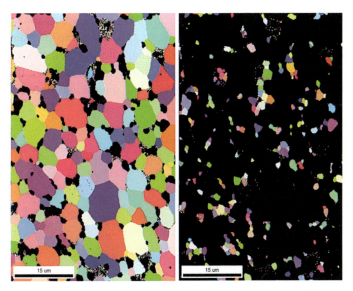

FIGURE 10.18 COMs for ZnO (left) and pyrochlore, shown partitioned. Superposition of the two images leads to Figure 10.17.

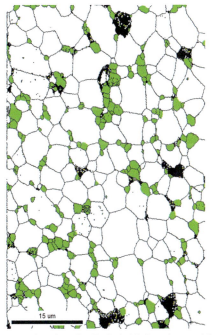

FIGURE 10.19 Phase map showing the ZnO (white) and pyrochlore (green) phases. The latter preferentially segregates at the triple points and at grain boundaries. The black lines are the high angle grain boundaries.

Figure 10.20a,b on the other hand shows the phase maps for the sample that had shown poor varistor behavior. In this case, the prochlore phase was also found inside the ZnO grains, which alters the electrical property of the grain. Thus, EBSD was helpful not only in deciding on the optimal processing conditions but also in understanding the reasons between optimal and nonoptimal performance.

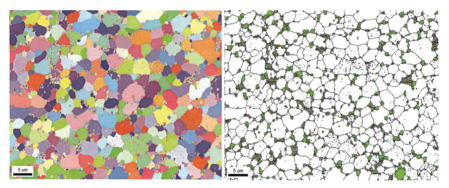

FIGURE 10.20 (a) COM (left) for a sample showing poor varistor behavior. (b) Phase map for the data shown in (a). The green areas are the pyrochlore phase.

10.3.5 COMBINING EDS AND EBSD MEASUREMENTS: COMPOSITIONAL GRADIENTS IN Ti–C–N COATINGS

As mentioned in the earlier section, EDS measurements when carried out in conjunction with EBSD help in identifying different phases with the same crystal structure, with the requirement that one or more elements are different between the phases. In this section, we show the utility of simultaneous EDS and EBSD studies with the example of Ti–C–N coatings with compositional gradients.

One aspect needs attention that is the data acquisition rates of EDS and EBSD. About a decade ago, it was possible to collect about 100 EBSD data points per second with optimal settings. With present day detectors, about 1400 patterns can be collected and indexed per second. However, EDS detectors can hardly match these speeds. Thus, it is important to adjust the speeds for simultaneous data collection. In other words, EBSD data collection should be "slowed down" so as to give sufficient time for EDS data to be collected.

Another aspect is the spatial resolution of EDS, which is about 1 μm under normal conditions. However, the sample is tilted to 70° in the EBSD

Applications of Electron Backscatter Diffraction 251

configuration, due to which the interaction volume decreases and the resolution improves to about 200 nm. However, quantification of elements needs to be carried out with care in the EBSD configuration since the EDS counts are a function of the tilt angle.

Ti–C–N coatings were prepared by detonation spray. A charge is ignited so that the shock wave carries the powder particles at supersonic velocities to the substrate and dense coatings are generated by repeated passes. As opposed to the CGDS method discussed earlier, detonation spray is a high temperature, high velocity process and is especially suited for the coating of materials with high melting points. When the powder particle impacts the substrate, it is flattened and forms a splat (similar to CGDS). It has been noticed that there is a slight difference in the elemental composition between the edge of the splat and its core due to the processing conditions; often, nonequilibrium phases are formed that are not easy to study.

Figure 10.21 shows the COM from an area of the Ti–C–N coating. We focus on the set of large grains at the center of the scan. EDS maps were collected for Ti, Co, Mo, Ni, N, and C and are shown in Figure 10.22a–f. The map for each element is shown in a different color. Although qualitative, the intensity of the color is an indication of the concentration of that element, with white indicating its absence.

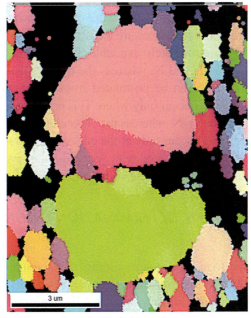

FIGURE 10.21 COM for a region of a Ti–C–N coating.

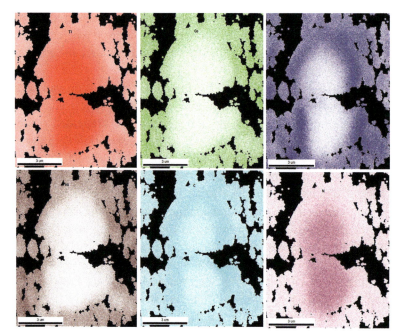

FIGURE 10.22 EDS maps for Ti, Co, Mo, Ni, C, and F, respectively, for the region shown in Figure 10.21.

As mentioned above, owing to the nature of the deposition process in the present case, nonequilibrium phases are expected to form, for which crystallographic data are not immediately available. From the figure, it can be seen that one phase contains Ti and N, whereas the other is likely to be a carbide of Co, Mo, and Ni. This can be postulated from the observation that areas containing Co, Mo, and Ni occur only where Ti is not present. Further, the core of each grain seems to be Ti–N, whereas the other elements (added for different reasons in the processing) occur only at the edges of the grains. The enhanced spatial resolution of EDS, combined with orientation information, can be used to obtain a better understanding where compositional gradients are present.

10.3.6 TOWARD THE DETERMINATION OF UNKNOWN PHASES

Early EBSD work focused on determining the symmetry elements of a previously unknown phase. Since the arrangement of EBSD bands reflects the crystallography of the underlying phase, it is possible to carry out an analysis of the bands and the zone axes and arrive at a possible solution to the phase problem.[16] Elemental information collected using EDS can be used as an extra filter to identify the unknowns. An approach has been detailed in the article by Dingley and Wright.[17]

10.3.7 EBSD IN TRANSMISSION MODE

Traditionally, EBSD measurements have been carried out in reflection geometry, that is, both the electron source and the phosphor screen are on the same side of the sample. The recent exciting development is to carry out measurements in transmission geometry (t-EBSD).[18] In this configuration, shown in Figure 10.23, the sample is tilted to an angle between 10° and 40° but in the opposite direction as compared to the usual measurements (compare with Fig. 10.1).

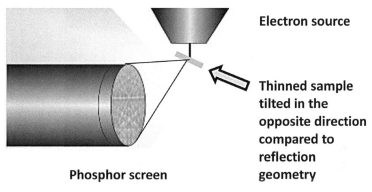

FIGURE 10.23 Schematic of the EBSD measurement in transmission geometry. (Adapted from the OIM data analysis manual, EDAX-TSL, with permission)

Thus, the signal is now through the bulk of the sample (or rather, from the lower portion from which the electron beam exits the sample). This places the requirement that the sample thickness should be sufficiently low to allow the electron beam to pass through. However, this is not much of a concern since specimens prepared for transmission electron microscopy can be used directly. Analysis of the collected data is, however, similar in the case of transmission and reflection. The advantage of t-EBSD is that spatial resolution can be improved and grain sizes of the order of 20 nm or lower can be measured. Several studies are being reported in the literature[19,20] and the fine-grained systems that were hitherto unavailable for study by EBSD are now being brought into the fold. In Figure 10.24a–c, we show a thinned sample for the t-EBSD study, the EBSD pattern recorded at the spot marked by the green cross and the indexing of the pattern. Figure 10.25 shows the COM for a t-EBSD scan.

FIGURE 10.24 (a–c) Sample used for the measurement in transmission geometry (left); EBSD pattern recorded at the spot marked by the green cross in (a) (middle), and the indexing of the bands (right).

FIGURE 10.25 COM for a t-EBSD scan.

10.4 CONCLUSIONS AND OUTLOOK FOR THE FUTURE

It is possible to say with confidence that EBSD as it is being carried out today has come a long way from the initial days when everything had to be done by hand; patterns had to be recorded on photographic film which then had to be developed, and analysis of each pattern had to be carried out manually. Advances in phosphor screens and in data analysis software, coupled with enhanced computing power, help in the recording and indexing of about 1400 points per second. Speeds of about 100 patterns per second, which was the state of the art a decade ago, seem now to be from a distant era.

One of the problems faced is with samples that have a bimodal distribution of grain sizes, especially when fine grains are interspersed between grains that are much larger. In such a situation, the choice of step size for the scan can be problematic. Ideally, at least five data points in a line and a couple of rows are desirable for grain size determination. However, this ends up oversampling the larger grains, increasing the data collection time manifold.

Software that senses the changes in orientation intelligently and adjusts the step size automatically is being developed, and this is an important advance.

Combining EBSD with focused ion beam milling[21] has opened up the possibility of studying microstructures in three dimensions. The other development, as discussed in Section 10.3.7, is to collect data in transmission. This has made it possible to study "real" nanomaterials where the feature size(s) (such as grain size) necessary for the property under consideration is/are less than 100 nm.

Much progress has been made in the measurement of strains. Elastic deformation and plastic deformation, the study of which is essential in varied disciplines and is of industrial importance, have been measured using EBSD. We are closer today than ever before toward the identification of new phases and indeed the study of total unknowns can be attempted.

All in all, it is an exciting field of work and it may not be an exaggeration to say that almost every discipline of materials science can now be studied using EBSD.

ACKNOWLEDGMENTS

The author is grateful to his Team Members Mr. M. Ramakrishna, Mr. L. Venkatesh, Dr. P. Suresh Babu, and Dr. Suresh Koppoju for valuable discussions. Sincere thanks are due to collaborators in different areas of work, Dr. S. Kumar, Dr. S. B. Chandrasehkar, Dr. K. Hembram, Dr. G. Siva Kumar, and many other colleagues at the International Advanced Research Centre for Powder Metallurgy and New Materials (ARCI). The author also thanks students Mr. P. S. V. Pramod Kumar and Mr. Paul Gaurav.

KEYWORDS

- materials science
- EBSD
- microstructure
- microstructural engineering
- crystalline materials
- charging effects

REFERENCES

1. Dingley, D. J.; Randle, V. *J. Mater. Sci.* **1992**, *27*, 4545.
2. Schwarzer, R. A. *Phys. Met. Metallogr.* **2003**, *96*(Suppl.), S104.
3. Randle, V. *Int. Mater. Rev.* **2004**, *49*, 1.
4. Randle, V. *Adv. Imag. Electron Phys.* **2012**, *151*, 363.
5. Wilkinson, A. J.; Britton, T. B. *Mater. Today* **2012**, *15*, 366.
6. Randle, V.; Engler, O. *Introduction to Texture Analysis*; Overseas Publishers Association: Amsterdam, 2000; pp 157.
7. Zaefferer, S. *Ultramicroscopy* **2007**, *107*, 254.
8. Hough, P. V. C. U.S. Patent 3,069,654, Method and Means for Recognizing Complex Patterns, 1962. March 25, 1960.
9. Dingley, D. J.; Baba-Kishi, K.; Randle, V. *An Atlas of Backscatter Kikuchi Diffraction Patterns*; Institute of Physics Publishing: Bristol, 1994.
10. Dingley, D. J.; Field, D. P. *Mater. Sci. Technol.* **1997**, *13*, 69.
11. Wilkinson, A. J.; Dingley, D. J.; Meaden, G. *Electron Backscatter Diffraction in Materials Science*; Schwartz, A. J.; et al., Eds.; Springer Science + Business Media, LLC: Berlin, 2009; pp 231–249.
12. Lehockey, E. M.; Lin, Y.-P.; Lepik, O. E. *Electron Backscatter Diffraction in Materials Science*; Schwartz, A. J.; et al., Eds.; Kluwer Academic/Plenum Publishers: Dordrecht, 2000; pp 247–264.
13. Chandrasekar, M. S.; Pushpavanam, M. *Electrochim. Acta* **2008**, *53*, 3313.
14. Wright, S. I.; Nowell, M. M.; Bingert, J. F. *Metall. Mater. Trans. A* **2007**, *38*, 1845.
15. Peiteado, M. *Bol. Soc. Esp. Ceram. Vidrio* **2005**, *44*, 77.
16. Dingley, D. J.; Baba-Kishi, K.; Randle, V. *Atlas of Backscatter Kikuchi Diffraction Patterns*; Institute of Physics Publishing: Bristol, UK, 1994.
17. Dingley, D. J.; Wright, S. I. *Electron Backscatter Diffraction in Materials Science*; Schwartz, A. J.; et al., Eds.; Springer Science + Business Media, LLC: Berlin, 2009; pp 97–107.
18. Keller, R. R.; Geiss, R. H. *Microscopy* **2012**, *245*, 245.
19. Saleh, A. A.; et al. *Mater. Charac.* **2016**, *114*, 146.
20. Babinsky, K.; et al. *Ultramicroscopy* **2015**, *159*, 445.
21. Zaefferer, S.; Wright, S. I.; Raabe, D. *Metall. Met. Trans.* **2008**, *39A*, 374.

CHAPTER 11

LIGHT MICROSCOPY IN STUDYING OF FORMATION AND DECOMPOSITION OF WOOD

GALINA F. ANTONOVA* and VICTORIA V. STASOVA*

V. N. Sukachev Institute of Forest, SB RAS Krasnoyarsk, 660036, Russian Federation

*Corresponding authors. E-mail: antonova_cell@mail.ru; vistasova@mail.ru

ABSTRACT

The changes in wood structure in the course of wood formation or any impact of mechanical/chemical treatments on mature wood should be taken into account to understand the processes themselves and/or to control of them. With this reason, we used the light microscopy during the study of wood formation in *Pinus syvestris* L. and *Larix sibirica* Ltb. and the impact of acoustic waves on oak (*Quercus robitr* L.) wood. Annual wood ring formation was studied in some directions: (1) seasonal distribution of the processes of cell wall production by cambium, the development of primary and secondary walls of early- and latewood tracheids during xylem formation in the season and under the effect of external factors on morphological characteristics of tracheids; (2) the presence of starch granules as the index of carbon store in favorable conditions or its expenditure in unfavorable external conditions during phloem development in pine and larch; (3) the biochemical changes in mono-, di-, and polymeric compounds (carbohydrates, organic, and phenolic acids, hemicelluloses, cellulose and lignin) included in biosynthesis and the formation of wall tracheid structure during wood development. Light microscopy was necessary tool to check of cell wall morphogenesis and lignification in pine (*Pinus sylvestris* L.) callus grown in the different conditions of cultivation to compare lignification in vivo and in vitro. Light microscopy was also used to study the structure of oak (*Quercus robur*) wood from different habitats and to estimate the changes in oak wood structure under ultrasound waves. The examples of light microscopy applying in those fields are given in the chapter.

11.1 INTRODUCTION

During the study of wood formation or of any impact of mechanical/chemical treatments on mature wood, it is important to evaluate the changes in wood structure for understanding current processes.

The annual ring wood formation in coniferous depends on three morphological processes: production xylem cells by cambium, radial diameter growth of primary walls (expansion growth), and deposition of structural components in secondary walls (secondary thickening) of tracheids. Special features of cell xylem development within each of these zones are resulted in the existence of two types (early and late) of tracheids differed in radial diameters and cell wall thickness. Environmental factors influence the number of cells and the relationship of early and late tracheids in annual wood layer. Such variations are resulted from physiological and biochemical events in the cells during cell development within the zones under external conditions.[1,2] One of decisive factors, affecting cell development transition from early to late type, is internal water stress.[3-6] The changes along the chain of metabolic reactions, evoked by low water potential, lead to the development of cells produced by the cambium to be as late tracheids. Low water potential (internal water stress) in developing xylem can arise because of the lack of rainfalls (external stress), the exhaustion of water stores in soil, and the strengthening of evapotranspiration at high air temperature.[7] All together reduce the internal water stores into stem tissues. Another reason of that may be a flooding of root system ("physiological drought"). Physiological drought provoked by oxygen deficiency in soil decreases water absorption.[8] Understanding the interrelation between the degree and stage of cell development and the biochemical changes in the course of tracheid development can help to interpret physiological events in cell walls of forming tissue, leading to the differences in the morphological structure of the cells themselves and of formed wood ring.

Besides xylem cells, the cambium produces phloem cells which provide the development of xylem cells by assimilates coming from photosynthetic apparatus. The development of phloem cells and the accumulation of reserve substances (starch granules, fatty acids, tannins) within them also depend on external factors and development degree of xylem cells.

Applying light microscopy to study wood formation in Scots pine (*Pinus sylvestris* L.) and larch (*Larix sibirica* Ldb.) trees was used to understand the correspondence between morphological and biochemical changes in the cells during of their development.[2,3,9]

Another trend in the study of wood formation was to investigate cell development in tissue culture. The cultures of tissues and cells are often used to estimate an influence of some substances on cell development, to study the pathway of biosynthesis of hydroxycinnamic acids, being the precursors of lignin, as well as the enzymes taking part in its biosynthesis, the changes during somatic embryogenesis, etc.[10–15] To study of lignification suspension and callus cultures are used as modal systems. Callus cultures help to estimate secondary cell wall development and lignification in tissue.[16–18] So, the system of *Pinus radiata* D. Don cell culture, suggested by authors, showed the possibility to induce differentiated tracheid elements with lignified cell walls. The changes in cultivation conditions of this system affect at key points that in turn influences cell wall development.[16]

To check the changes in morphological state of the cells in pine (*P. sylvestris* L.) callus, in their biochemical composition, and, in particular, in lignin deposited under different conditions of the cultivation, we also used light microscopy.[19] The study was carried out to compare lignification in vivo and in vitro[19] because the conditions of cell development in artificial (culture) and natural (vital tree) are different by both water ensuring and contact cell to cell. This must influence cell metabolism and, in particular, a composition of lignin and its precursors.

Change in cell wall structure occurs not only during development of forming wood cells. Any physical or biochemical impact on mature wood is resulted in the structural changes as well. The determination of destruction degree in mature wood and its components can help to control and/or optimize the production of substances in the process of such impact. One of the physical factors affecting wood structure is ultrasound waves. Such influence depends on the medium of their dissemination and of ultrasound power. The changes in the structure of oak (*Quercus robur*) wood under ultrasound in different mediums with and without the pretreatment by temperature were studied.[20,21] The affect of acoustic waves on oak wood at air, water, and water–alcohol solution with different expositions was studied by anatomical and biochemical methods, gel-filtration, X-ray diffraction, and IR-spectroscopy.

The changes in wood structure were recorded in the combination of light microscopy with such metachromatic dye as cresyl violet. Under metachromatic dyes, the staining of tissues changes due to both the alteration in the accessibility of chemical groups in substrate and of dye molecules aggregation level. Preliminary study of the mechanism of interaction between cresyl violet and the components of developing woody tissue showed that the sorption of the dye occurs due to wood carboxyl groups.[22] The changes in the density of spatial structure of wood modify the staining of the tissue under cresyl violet. Consecutive deposition of hemicelluloses, cellulose, and lignin in the course of formation and development of xylem cells leads to the reorganization in spatial structure of the cell walls and the consolidation of the tissue itself especially during lignification. From other side, the destruction of wood under ultrasound waves leads to the disintegration of wood components and, consequently, influences the distribution of the dye in tissue and changes of its staining.

The examples of the light microcopy application during the study of wood formation in conifers (morphological and biochemical aspects) and the effect of ultrasound waves with various powers on oak wood are presented in the chapter.

11.2 THE EXAMPLE OF APPLYING OF LIGHT MICROSCOPY

11.2.1 WOOD FORMATION

Wood formation in Scots pine (*P. sylvestris* L.) and larch (*L. sibirica* Ltb.) stems were studied by some ways. By one of them, the annual wood layer growth and the influence of external factors on morphological characteristics of main structural elements in conifers tissues (tracheids) in the course of the season were estimated. As mentioned above, annual wood rings of pine and larch consist of early- and latewood tracheids differed by radial diameters and thickness of cell walls where principal biomass of xylem accumulated. To find the reasons of such differences, the cores were extracted from the living tree stems through certain temporal intervals (10–11 days) in the season and fixed with formalin–acetic acid–ethanol (5:5:90). At the core cross-sections, the number of cells, radial and tangential sizes of tracheids, and their lumens were estimated after the staining with cresyl violet to inspect the number of cells in cambial zone and in the zones of growth expansion and secondary thickening (maturation) during the season and to test the beginning of lignin deposition.[1,2,9] The deposition of lignin was also checked with phloroglucinol–hydrochloric acid (Wiesner reaction). Both of the dyes showed equal staining of the beginning of lignification. The data obtained were used to calculate the cambial activity, the rate, and the duration of the development of radial diameters and secondary walls of tracheids in the zones of growth and secondary wall thickening as well as of cell wall biomass increment in certain periods of the growth season. All these data and temperature–precipitation data in the each temporal period were used to determine the optimal conditions for the development of tracheid parameters. Such approach permitted to establish different influence of external factors on radial diameter growth and secondary cell wall thickening (Table 11.1) and, as the result, to understand the reasons of the discrepancies during the formation of early- and latewood tracheids under changing weather conditions in the season.[1,2]

TABLE 11.1 Optimal Values of Temperature and Precipitation for Annual Wood Ring Formation in *Pinus sylvestris* L. Stems.

Characteristics of annual wood layer	T_{mean} daily (°C)	T_{max} of day (°C)	T_{min} of night (°C)	Precipitation (mm)
The number of cells in radial row	20	24	9	3–4[a]
Radial diameter	21–23	27–28	8–9	15–20[b]
Cross-sectional area of cell wall	16	20–21	9	1.9–2.3[b]

[a]Precipitations in a 24 h period.
[b]Total precipitations during development period of tracheid in the zone. (Reprinted with permission from Ref. 9.)

Reprinted from Antonova, G. F.; Stasova, V. V. Seasonal Distribution of Processes Responsible for Radial Diameter and Wall Thickness of Scots Pine Tracheids // Sibirskij Lesnoj Zurnal. *Siberian J. Forest* **2015**, *2*, 33–40.

Practically, similar optimum temperature was observed for the processes, taking part in annual xylem formation in larch (*L. sibirica* Ldb.) trees except T_{min} (11–12°C) to produce tracheids.[2]

The radial diameter of tracheids has been found to depend on the development rate in growth expansion zone, whereas cell wall thickness depends on the duration in secondary wall thickening zone.[9] Different physical causes are the basis of these events. The influence of temperature on biosynthesis rate of primary wall components is the base of the dependence of radial diameters on the growth rate. With rising of temperature, the synthesis rate of substances increases but the development duration of tracheids in growth expansion zone declines.[9] The dependence of cell wall thickness on development duration has to do with the diffusion of assimilates along radial row to the cells where structural wall components are synthesized that in turn depends on the number of cells in secondary wall thickening zone.[9]

In parallel with wood formation in larch and pine stems, the development of phloem was studied to estimate the production of phloem cells by cambium, their development duration and the changes in chemical composition of cells compounds. Besides, the presence of starch granules as the index of carbon store or expenditure in favorable and unfavorable external conditions was estimated.[23] It is well known that, under favorable conditions, an excess of photo-assimilates can be stored within parenchyma cells of phloem and xylem as starch grains to be then consumed in cellular syntheses with a lack of substrates in the worst conditions.

The changes in starch granules in the course estimated as scores are presented at Figure 11.1. At the beginning of annual ring wood formation, the starch granules have been shown to be exported from xylem ray parenchyma cells, then from gradually xylem resin cannels, and after that from rays and axial parenchyma of phloem (Fig. 11.1a,b). The last events were especially observed in July when photosynthesis was suppressed by high temperatures and the lack in substrates for synthetic and energetic processes appeared.

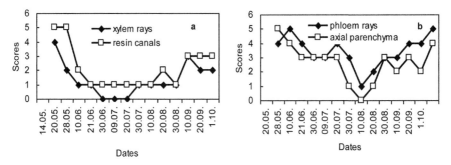

FIGURE 11.1 The content of starch (in scores) within the cells of the rays and resin canals of xylem (a) and of the rays and axial parenchyma of phloem and (b) ethanol solution iodine.

11.2.2 BIOCHEMICAL CHANGES IN CELLS

To understand biochemical reasons of the differences in morphological parameters of early and late tracheids in the course of wood formation, the changes in mono-, di-, and polymeric compounds (carbohydrates, organic, and phenolic acids, hemicelluloses, cellulose and lignin), included in biosynthesis and the formation of cell wall structure, were analyzed. With this purpose, the cell layers with different developmental degree were sampled consistently layer by layer from cuttings of tree stem and the development state of the cells in the each layer was tested with cresyl violet (Table 11.2). This permitted to check thoroughly cell developmental stages during their sampling.

TABLE 11.2 Cell Layers with Different Development Degree During Early- and Latewood Formation in Pine (*Pinus sylvestris* L.) Stem.

Cell layers within zones of		Earlywood	Latewood
Cambium			
Expansion growth			Absent
Secondary wall thickening	D1		
	D2a		
	D2b		
	D2c		

The tissues sampled were immediately fixed with ethanol at the final concentration not exceeding 30%, weighed and kept in refrigerator until the analyses. Careful sampling permitted to find the specific peculiarities in the deposition of pectin, arabinogalactans, the fractions of A and B hemiceluloses as well as cellulose at developmental stages of primary and secondary walls of the tracheids during early and latewood formation in larch (*L. sibirica* Ldb.) and to compare the data with morphological structure of forming cells.[24]

Thin walls of early tracheids are often considered to be the result from substrate lack. According to our data, there was not any limitation by substratum (carbohydrates) for wall thickening of early tracheids (Fig. 11.2). At each step of early tracheid maturation (zone D), the amount of carbohydrates was always more than that during latewood cell formation.

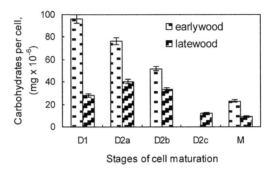

FIGURE 11.2 The content of carbohydrates within the cells of secondary cell wall thickening zone before (D1) and after (D2a, D2b, and D2c) the beginning of lignification, and mature xylem (M) during early- and latewood formation. Adapted from G. F. Antonova and I. A. Chapligina. The cell wall structure formation of earlywood and latewood in larch (*Larix sibirica* Ldb.). In *Proc. of IUFRO Intern. Symp. Working Group S2.02.07 "Larix-2007"*, Québec, Canada Sep., pp. 77-81, 2007.

The reason for the differences in the thickness of cell walls of two wood types was the various durations of cell development in secondary wall thickening zone as noted above.[1,2] So, mean duration of early tracheid development in secondary thickening zone was 10–20 days, whereas of late tracheids ranged 30–55 days (Fig. 11.3).

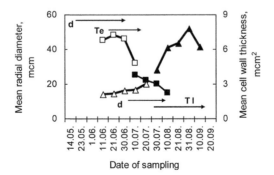

FIGURE 11.3 Characteristics of radial diameter (de) and cell wall thickness (Te) of early- and latewood (d1 and T1, correspondingly) tracheids and the duration of their development in the season. Adapted from "Development of annual layer in stem wood of Pinus sylvestris and Larix sibirica L." Lesovedenie, 1992,5, part b from fig. in paper.

The differences in the deposition of pectin substances, in the content of uronic, phenolic, and ascorbic acids during cell wall structure formation of early- and latewood tracheids in larch have also been found.[24,25] Additionally, lignin deposition in early- and latewood tracheids occurred with different intensities in both pine and larch.[26,27] The rate of lignin deposition increased gradually toward the last stage of earlywood tracheid maturation, whereas it was the highest at the outset stage of lignification and decreased toward the end of latewood tracheids maturation (Fig. 11.4).

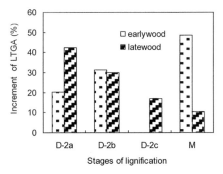

FIGURE 11.4 The dynamics of lignin (LTGA) deposition within the cells of secondary cell wall thickening zone (D2a, D2b, and D2c) and mature xylem (M) during early- and latewood formation. Adapted from Galina F. Antonova, Tamara N. Varaksina, Tatiana V. Zheleznichenko, Victoria V. Stasova Lignin deposition during earlywood and latewood formation in Scots pine stems Wood Sci Technol 2014, DOI 10.1007/s00226-014-0650-3.

This coincided with the changes in the ratio of ascorbic/dehydroascorbic acids that influences the level of redox reactions of lignin precursors (Fig. 11.5).

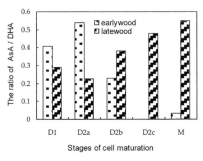

FIGURE 11.5 The ratio of ascorbic (AsA) and dehydrooascorbic (DHA) acids within the cells of secondary cell wall thickening zone before (D1) and after (D2a, D2b, and D2c) the beginning of lignification and in mature xylem (M) during early- and latewood formation. Adapted from Galina Antonova The Role of Ascorbate in Growth and Development of Cells During.... In *"Oxidative Stress in Plants: Causes, Consequences and Tolerance" (Ed.: Naser A.Anjum, Shahid Umar, Altaf Ahmad)*, I.K. International Publishing House Pvt. Ltd., New Delhi – Bangalore, ch. 15, pp. 443-466, 2012.

Taken together, morphological and biochemical data can serve for understanding the morphological differences in early- and latewood tracheids.[1,2]

11.2.3 STUDY OF PINE CALLUS CULTURE

To initiate morphogenesis of cells and lignification in the callus culture of pine (*P. sylvestris* L.), such components as sucrose, polyvinylpyrrolidone (PVP), ferulic, and ascorbic acids with different content in nutrient medium were used. Furthermore, the affect of luminance level and of cultivation duration on cell morphology and lignification in pine callus culture was investigated.[19]

Squashed preparations of callus cells with different thickening of cell walls are presented in Figure 11.6.

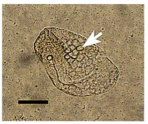  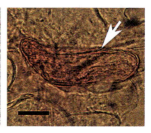

Cell with bordered pores (the beginning of wall thickening)   Cells with spiral thickening   Cell with overall thickening

FIGURE 11.6 The examples of callus culture cells on Scots pine with different degree of cell wall development.

After subcultivation of the callus with the variation of nutrient medium composition, of light–darkness combination and with different periods of the cultivation, the content of lignin, its molecular weight (by gel-filtration method), and the composition of its structural units were evaluated in callus cells.

The content of sucrose (5%) in nutrient medium, the cultivation in the conditions of light–darkness, and the increase the duration of cultivation from 21 to 60 days have been found to intensify lignification and to increase the content of the high-molecular fraction in lignin. PVP with the cultivation in the darkness during 21 days suppressed the lignification and led to the appearance of syringyl units in the composition of lignin. With increasing of cultivation duration, PVP provoked the synthesis of the high-molecular fraction in lignin. Ferulic acid promoted the development of secondary thickening in callus cell walls, increased low-molecular, and declined high-molecular fractions in lignin. Ascorbic acid intensified the growth of callus and impeded its lignification. The data obtained showed that lignin produced during the callus cultivation was more similar to lignin of earlywood but not of latewood.

11.3 THE IMPACT OF ULTRASOUND WAVES ON OAK WOOD

Destruction of wood under a physical impact and, in particular, ultrasound waves was also tested by light microscopy. We studied the changes in the structure of oak (*Q. robur* L.) wood under ultrasound in different mediums with and without the pretreatment by temperature.[20,21] The work was carried out together with Institute of Solid State Physics RAS (Russia) where the treatment of oak wood by supersonic waves, X-ray diffraction, and IR-spectroscopy were realized. The morphological and biochemical analyses were carried out in our institute. The oak wood of the trees from two stands with different wet growth conditions (valley and hill sites) was used.

The affect of acoustic waves on oak wood at air, water, and water–alcohol solution with different expositions was studied by anatomical and biochemical methods and gel filtration. The variations in morphological characteristics of wood cells as well as the changes in the chemical composition of wood and its components, in particular, in lignin were investigated to estimate the degree of wood destruction.[20]

External signs of wood destruction were fixed by the changes in the staining of oak wood cross-sections with cresyl violet and by the appearance the splits in cell walls (Fig. 11.7).

Before US   After US   Splits after US

FIGURE 11.7 Changes in the staining of oak wood and the appearance of the splits in cell wall structure after the impact of ultrasound (US).

Specific reagents (Wiesner and Mäule) showed what guaiacyl structures located within libriform fiber walls were destroyed under ultrasound more than syringyl structures located in middle lamella (Fig. 11.8).

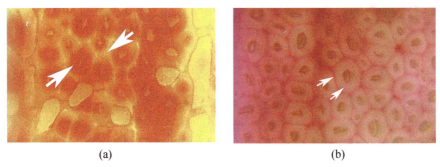

(a)   (b)

FIGURE 11.8 Specific reactions of syringyl (a) (by Mäule) and guaiacyl (b) (by Wiesner) subunits in lignin of oak wood after ultrasonic impact.

After supersonic treatment, the principal changes in anatomical wood structure were observed in cell walls of libriform and its porosity (Table 11.3).

The changes in anatomical parameters of the wood treated by ultrasound have been found to depend on the structure of wood formed under different site conditions. The width of annual ring and cross-section area of large vessels in the wood formed in wet conditions (valley site) increased, whereas in the wood formed in dry site (hill site) decreased after ultrasound treatment. Specific volumes of large vessels were enlarged in both cases whereas that of small vessel in the wood from valley site increased and, in contrary, declined in wood from hill site. Cell walls thickness of libriform did not practically change in the wood from valley site and decreased in wood from hill site. At the same time, the cross-section area of libriform lumen has expanded in both cases but in wood from valley site more than from hill site.

Under the impact of ultrasound with frequency 19–21 kHz and a low power range, the spatial organization of oak wood structure changed in dependence on the medium (air, 65% water–alcohol solution, water) of the treatment and the exposition time (15, 35 min, and 1 h, correspondingly).[20] The effect of acoustic waves in any medium during 15 and 35 min on wood structure was practically similar but the treatment in water medium caused structural changes in middle lamella mostly.

The changes in the chemical composition of wood and its components, in particular, in lignin were observed as well. Biochemical analysis showed that ultrasound treatment affected also the amounts and composition of the substances extracted by 65% water–alcohol and by water and, in particular,

TABLE 11.3 Anatomical Characteristics of Oak (*Quercus robur* L.) Wood Before and After Ultrasound Treatment.

Characteristics	Valley site Control	Valley site Treated	Hill site Control	Hill site Treated
The width of annual ring	1443.06 ± 6.66	1682.29 ± 8.57	1061.3 ± 49.84	997.3 ± 55.4
Large vessels: cross-section area (mcm^2)	48,671.1 ± 3027.3	58,538.1 ± 602.9	72,367.8 ± 4747	65,133.8 ± 1407.5
Small vessels: cross-section area (mcm^2)	882.0 ± 44.1	949.52 ± 45.21	1108.5 ± 47.6	1365.45 ± 76.12
Specific volume of vessels (%) large small)	16.37 ± 2.35 3.57 ± 0.58	28.93 ± 2.53 5.64 ± 0,49	30.82 ± 1.05 5.17 ± 0.38	33.62 ± 1.33 3.67 ± 0.33
Porosity of libriform (%)	3.10 ± 1.33	9.79 ± 0.52	5.76 ± 0.36	9.75 ± 1.22
Thickness of libriform cell walls (mcm)	5.65 ± 0.44	5.60 ± 0.13	6.43 ± 0.22	5.49 ± 0.30
Cross-section area of libriform lumen (mcm^2)	3.76 ± 1.07	18.90 ± 0.95	10.70 ± 1.40	16.82 ± 0.83

Adapted from Antonova, A. V. Bazenov, T. N. Varaksina, S.I.Evgrafova, T.N. Konovalov,N.T. Konovalova,N. V. Pashenova and V. V. Stasova. The effect of acoustic waves of the high power on oak wood. Khimija Rastitel'nogo Syr'ja (Chemistry of raw material) no. 3, pp. 1-10, 2009.

carbohydrate and soluble phenols.[20] Acoustic waves stimulated the structural changes in oak wood and provoked the increase in alcohol-soluble fraction in lignin. According to gel-filtration analysis, the molecular weight of lignin and mostly its alcohol-insoluble fraction changed. Histochemical reactions and IR spectroscopy showed that it is syringylpropane moiety of oak wood lignin that is mainly destroyed by acoustic waves. The data of IR spectroscopy also showed the instability of syringyl subunits of oak wood lignin under the impact of supersonic waves.[20] According to the data of X-ray diffractometry, crystal part and the size of elementary cell in cellulose microfibrills of oak wood increased after the treatment by supersonic waves at air.[20]

The rise of ultrasound power up to 80 W/cm^2 with short-term high power (30 s) also led to the enlargement of libriform porosity and the cross-section area of small vessels, and increase of the amount of water–alcohol-soluble substances (Table 11.4), in particular, carbohydrates, phenolic compounds, and of free phenolic acids.[21]

TABLE 11.4 Chemical Composition of Oak Wood Under the Treatment of High Power Ultrasound (US).

Characteristics	Valley site	Hilly site
Substances soluble in 65% ethanol (% of dry weight)		
Before US	3.18 ± 0.08	4.62 ± 0.10
After US	11.31 ± 0.12	7.21 ± 0.05
Substances soluble in water (% of dry weight)		
Before US	7.95 ± 0.06	8.37 ± 0 08
After US	18.25 ± 0.09	14.12 ± 0.04
Lignin (% of dry weight)		
Before US	25.6 ± 0.05	31.8 ± 0.05
After US	32.4 ± 0.06	33.5 ± 0.07
Lignin not dissolved in 96% ethanol (% of total amount)		
Before US	10.65 ± 0.12	7.15 ± 0.07
After US	5.06 ± 0.11	4.30 ± 0.10

The treatment of oak wood by supersonic waves induced the increase in substances soluble in 65% ethanol and in water, the decrease in lignin not dissolved in 96% ethanol. At the same time, high energy ultrasound did not influence lignin localization but affected sub-molecular structure that changed the degree of lignin solubility in alcohol. The increase in the alcohol-soluble fraction was due to the destruction of lignin itself and mainly its alcohol-insoluble fraction. As the result, molecular weight distribution of lignin and its fractions, according to gel-filtration technique, changed.[21] Apparently, the degree of interaction of lignin and cell wall polysaccharides also changed.

11.4 SUMMARY

Thus, the application of light microscopy for the registration of morphological changes in wood structure in the combination with biochemical and physical analyses gives very valuable information about wood formation in different weather conditions and understanding of the reasons in morphological differences of two wood types in coniferous. On the other hand, light microscopy helps to test the decomposition of mature wood under physical impacts and, in particular, under ultrasound waves.

KEYWORDS

- **annual wood ring formation in coniferous**
- **biochemical changes in cell wall compounds**
- **control of (*Pinus sylvestris* L.) callus growth**
- **destruction of oak (*Quercus robur*) wood under ultrasound waves**
- **light microscopy**
- **metachromatic dye cresyl violet**

REFERENCES

1. Antonova, G. F.; Stasova, V. V. Effect of Environmental Factors on Wood Formation in Scots Pine Stems. *Trees* **1993**, *7*(4), 214–219. DOI: 10.1007/BF00202076.
2. Antonova, G. F.; Stasova, V. V. Effect of Environmental Factors on Wood Formation in Larch (*Larix sibirica* Ldb.) Stems. *Trees* **1997**, *11*(8), 462–468. DOI: 10.1007/PL00009687.
3. Zahner, R. Internal Moisture Stress and Wood Formation in Conifers. *Forest Prod.* **1963**, 13, 240–247.
4. Zahner, R.; Lotan, J. E.; Baughman, W. D. Earlywood–Latewood Features of Red Pine Grown Under Simulated Drought and Irrigation. *For. Sci.* **1964**, *10*(3), 361–370.

5. Nonami, H.; Boyer, J. S. Primary Events Regulating Growth at Low Water Potential. *Plant Physiol.* **1990**, *93*, 1600–1609.
6. Nonami, H.; Boyer, J. S. Wall Extensibility and Cell Hydraulic Conductivity Decrease in Enlarching Stem tissue at Low Water Potential. *Plant Physiol.* **1990**, *93*, 1610–1619.
7. Gregg, B. M.; Dougherty, P. M.; Hennessey, T. C. Growth and Wood Quality of Young Loblolly Pine Trees in Relation to Stand Density and Climatic Factors. *Can. J. For. Res.* **1988**, *18*, 851–858.
8. Lyr, H.; Polster, H.; Fiedler, H.-J. *The Physiology of Woody Plants*; Gustav Fischer Verlag: Jena, 1967; pp 444.
9. Antonova, G. F.; Stasova, V. V. Seasonal Distribution of Processes Responsible for Radial Diameter and Wall Thickness of Scots Pine Tracheids. *Siber. J. For. Sci.* **2015**, *2*, 33–40. DOI: 10.15372/SJFS20150203.
10. Venverloo, C. J. The Lignin of *Populus nigra* L. cv. 'Italica'. A Comparative Study of the Lignified Structures in Tissue Cultures and the Tissues of the Tree. *Acta Bot. Neerl.* **1969**, *18*, 241–314.
11. Brunow, G.; Kilpeläinen, I.; Lapierre, C.; Lundquist, K.; Simola, L. K.; Lemmetyinen, J. The Chemical Structure of Extracellular Lignin Released by Cultures of *Picea abies*. *Phytochemistry* **1993**, *32*, 845–850.
12. Rogers, L. A.; Dubos, S.; Surman, C.; Willment, G.; Mansfielt, S. D.; Campbell, M. M. Light, the Circadian Clock, and Sugar Perception in the Control of Lignin Biosynthesis. *J. Exp. Bot.* **2005**, *56*, 1651–1663.
13. Stasolla, C.; Kong, L.; Yeung, E. C.; Thorpe, T. A. Maturation of Somatic Embryos in Conifers: Morphogenesis, Physiology, Biochemistry and Molecular Biology. *In Vitro Cell. Dev. Biol. Plant* **2002**, *38*, 93–105.
14. Stasolla, C.; Egerstdotter, U.; Scott, J.; Kadla, J.; Sederoff, R.; van Zyl, L. Lignin Composition in Wild Type and Cinnamyl Alcohol Dehydrogenase Deficient Mutant Cultured Cells of *Pinus taeda*. *Plant Physiol. Biochem.* **2003**, *41*, 439–445.
15. Kärkönen, A.; Koutaniemi, S. Lignin Biosynthesis Studies in Plant Tissue Cultures. *J. Integr. Plant Biol.* **2010**, *52*(2), 176–185.
16. Möller, R.; McDonald, A. G.; Walter, C.; Harris, P. J. Cell Differentiation, Secondary Cell-wall Formation and Transformation of Callus Tissue of *Pinus radiata* D. Don. *Planta* **2003**, *217*, 736–747.
17. Möller R.; McDonald G. A.; Walter C.; Harris P. G. Cell Differentiation, Secondary Cell-wall Formation and Transformation of Callus Tissue of Pinus Radiata D. Don. *Planta* **2003**, *217*, 736–747.
18. Möller, R.; Koch, G.; Nanayakkara, B.; Schmitt, U. Lignification in Cell Cultures of *Pinus Radiata*: Activities of Enzymes and Lignin Topochemistry. *Tree Physiol.* **2005**, *26*, 201–210.
19. Antonova, G. F.; Zheleznichenko, T. V.; Stasova, V. V. Lignification in Scots Pine Callus as Reaction on Cultivation Conditions and Nutrient Medium. *Siber. J. For. Sci.* **2014**, 6, pp 46–59.
20. Konovalova, N. T.; Konovalov, T. N.; Stasova, V. V.; Varaksina, T. N.; Antonova, G. F. The Effects of Supersonic Treatment on the Structure and Components of Oak Wood. In *Proc. The 5th Int. Symp.: Wood Structure and Properties*, Sept 3–6, 2006, Sliac-Sielnica, Slovakia. Tech. Univ.: Zvolen, Slovakia, 2006; pp 59–65.
21. Antonova, G. F.; Bajenov, A. V.; Varaksina, T. N.; Evgrafova, S. I.; Konovalov, T. N.; Konovalova, N. T.; Pashenova, N. V.; Stasova, V. V. The Effect of Acoustic Waves of the High Power on Oak Wood. *Chem. Raw Mater.* **2009**, *3*, 1–10.

22. Antonova, G. F.; Shebeko, V. V. Application of Cresyl-violet for Studying Wood Formation. *Chem. Wood (USSR)* **1981**, *4*, 102–105.
23. Antonova, G. F.; Stasova, V. V. *Seasonal Development of Phloem in Siberian Larch Stems*. Springer-Verlag: Berlin, 2008. DOI: 10.1134/S1062360408040024.
24. Antonova, G. F.; Chapligina, I. A. The Cell Wall Structure Formation of Earlywood and Latewood in Larch (*Larix sibirica* Ldb.). In *Proc. IUFRO Int. Symp. Working Group S2.02.07: Larix-2007*, Québec, Canada, 2007; pp 77–81.
25. Antonova, G. F. The Role of Ascorbate in Growth and Development of Cells During the Formation of Annual Rings in Coniferous Trees, Chapter 15. In *Oxidative Stress in Plants: Causes, Consequences and Tolerance*; Anjum, N. A., Umar, S., Ahmad A. Eds.; I. K. International Publishing House Pvt. Ltd.: New Delhi, Bangalore, 2012; pp 443–466.
26. Antonova, G. F.; Varaksina, T. N.; Zheleznichenko, T. V.; Stasova, V. V. Lignin Deposition During Earlywood and Latewood Formation in Scots Pine Stems. *Wood Sci. Technol.* **2014**, *48*(5), 919–936. DOI: 10.1007/s00226-014-0650-3.
27. Antonova, G. F.; Varaksina, T. N.; Stasova, V. V. Differences in the Lignification of Earlywood and Latewood in Larch (*Larix sibirica* Ldb.). *Euras. J. For. Res.* **2007**, *10*(2), 149–161.

CHAPTER 12

MICROSTRUCTURAL ANALYSIS OF POLYMER BLENDS, COMPOSITES, AND NANOCOMPOSITES

M. FATHIMA RIGANA, SIMI ANNIE THARAKAN,
P. THIRUKUMARAN, A. SHAKILA PARVEEN,
R. BALASUBRAMANIAN, S. BALAJI, C. P. SAKTHI DHARAN,
and M. SAROJADEVI*

Department of Chemistry, Anna University, Chennai 600025, Tamil Nadu, India

*Corresponding author. E-mail: msrde2000@yahoo.com

ABSTRACT

Microstructural analysis is a tool to determine grain size, phases present, chemical homogeneity, and distribution of phases. In this chapter, we are discussing about the blends (Schiff-base-functionalized dicyanate esters/epoxy blends, azomethine functionalized dicyanate esters/epoxy blends, polyurethane (PU)/polyacrylonitrile (PAN)/GPR blends), composites (CDPNM/epoxy/glass composites, epoxy–PU IPN/Woven glass composites, modified PMR PI/carbon fiber composites, PBz/jute fiber composites), nanocomposites (optically active—PI/OAPS nanocomposites, epoxy/cyanate ester functionalized POSS nanocomposites, eugenol-based PBz/Epoxy/OAPS nanocomposites, PBz-functionalized POSS nanocomposites, Unsaturated polyster (UPR)/silica nanocomposites, PBz/Silica nanocomposites, UPR/alumina nanocomposites, UPR/nano-ZnO nanocomposites, UPR/CaCO$_3$ nanocomposites, cyanate ester/CNT nanocomposites, polytriazoleimide/GO nanocomposites). The dispersion of the fibers/nanomaterials in the polymer matrix was studied using scanning electron microscopy, transmission electron microscopy, high-resolution transmission electron microscopy, atomic force microscopy, and energy dispersive X-ray analysis, and their effect on the thermal, mechanical, and electrical properties were discussed.

12.1 INTRODUCTION

In today's fast-developing industrialized world where hundreds of novel products are being introduced to the market space, microstructure analysis got prime importance in evaluating and assessing the performance and the failure mechanisms of new materials. Some of the areas at which the microstructural analysis is being widely used are at the study of crack development, porosity, topography, phase transitions, internal composition of the developed products, and the materials. Scanning electron microscope (SEM) and transmission electron microscope (TEM) are two important tools which are widely used for the microstructure analysis.[1–5]

While SEM is used for topographical, compositional, and morphological characterization, TEM is used for studying the internal structure and the composition. SEM gives elemental composition and the texture of the surface with scattered electrons, whereas TEM uses transmitted electrons which can give the image of internal structure—which explains morphology, crystallography, magnetic domains, etc. SEM is used for powders, polished and etched surfaces, IC chips, whereas TEM is used for imaging of dislocations, tiny precipitates, grain boundaries, and other defect structures in solids including polymers.

SEM is helpful to study the surface features and is best utilized for fracture surface studies to distinguish between ductile and brittle fracture of polymers, polymer blends, and composites, that is, morphological and compositional alteration occurred to the structure of pure polymers when they are fabricated into different kinds of blends and composites with the addition of other polymeric resins and fibers, respectively. Generally, blends and composites show enhanced thermal, mechanical, and electrical properties depending on the resins added or the filling materials used for fabrication, which is due to the physical and chemical changes in the microstructure of the product.[6–11]

12.2 BLENDS

12.2.1 BLENDS OF DICYANATE ESTERS WITH EPOXY RESINS

12.2.1.1 SCHIFF-BASE-FUNCTIONALIZED DICYANATE ESTERS/ EPOXY BLENDS

Sakthidharan et al. in 2015 prepared Schiff-base-functionalized dicyanate ester polymers and blended with epoxy resins at various compositions.[12]

SEM analysis was conducted to investigate the dispersion of cyanate ester in the epoxy resin. SEM images of cured 1,4-bis[4(2-cyanatophenylazomethyl) phenoxy]butane (but-CE), 1,5-bis[4(2-cyanatophenylazomethyl)phenoxy] pentane (pen-CE), 1,6-bis[4(2-cyanatophenylazomethyl)phenoxy] hexane (hex-CE), 5% but-CE blend (5% cyanate ester + 95% epoxy), and 10% but-CE blends (10% cyanate ester + 90% epoxy) are shown in Figure 12.1

The fractured surface of the but-CE (Fig. 12.1a) shows rose-like layered structure and its higher magnification (Fig. 12.1a') shows rose-petals like morphology. Fractured surface of cured cyanate esters pen-CE (Fig. 12.1b and b') and hex-CE (Fig. 12.1c and c') display rock and layered structure, respectively. The SEM images of neat cyanate ester (hex-CE) resemble that of 2,2'-bis(4-cyanatophenyl)isopropylidene reported by Hu et al.[13] These observations indicate a brittle behavior for all the three cured neat cyanate ester resins as expected for thermosetting resins. SEM images (Fig. 12.1d,e) illustrate uniform distribution of the cyanate ester in the epoxy resin without any void.

12.2.1.2 AZOMETHINE FUNCTIONALIZED DICYANATE ESTERS/ EPOXY BLENDS

The morphology of azomethine functionalized dicyanate ester/epoxy blends was studied by Sakthidharan et al. in 2015. Figure 12.2a–e displays the fracture surface SEM images of the cured neat cyanate esters 1,4-bis[4(4-cyanatophenylazomethyl)phenoxy]butane (CE1), 1,5-bis[4(4-cyanatophenylazomethyl)phenoxy]pentane (CE2), 1,6-bis[4(4-cyanatophenylazomethyl) phenoxy]hexane (CE3), 1% cyanate ester/epoxy, and 5% cyanate ester/epoxy blends, respectively. CE1 shows (Fig. 12.2a and a') curved hollow structure, whereas CE2 and CE3 show layered structure [on both lower (Fig. 12.2d and e) and higher magnification (Fig. 12.2d' and e')]. SEM images of cyanate ester/epoxy blends (Fig. 12.2b and c) illustrate the uniform distribution of the cyanate ester in the epoxy blends without any voids.[14]

12.2.1.3 POLYURETHANE/POLYACRYLONITRILE/GPR BLENDS

Guhanathan et al. in 2003 prepared tricomponent interpenetrating polymer network (IPN) using castor oil, toluenediisocyanate, acrylonitrile (AN), ethylene glycol diacrylate, and general-purpose unsaturated polyester resin (GPR) with various compositions.[15] The morphology of the cured

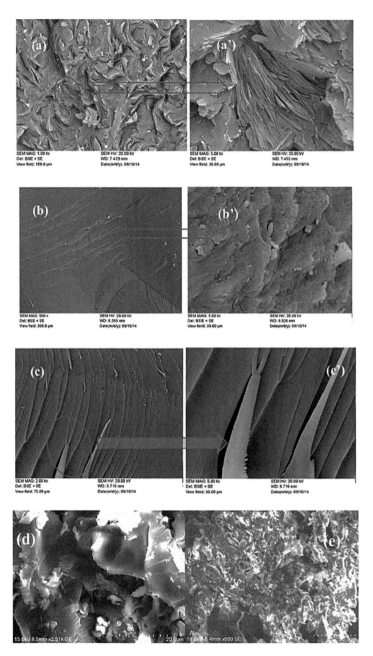

FIGURE 12.1 SEM images of cured (a) neat but-CE (a′) neat but-CE-magnified; (b) neat pen-CE and (b′); neat pen-CE-magnified; (c) neat hex-CE (c′) neat hex-CE-magnified; (d) 5% but-CE + 95% epoxy system, and (e) 10% but-CE + 90% epoxy system. (Reproduced from Ref. 12 by permission of The Royal Society of Chemistry.)

Microstructural Analysis of Polymer Blends, Composites 279

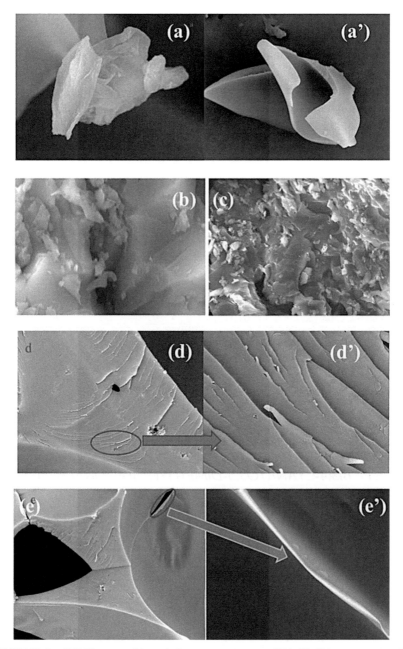

FIGURE 12.2 SEM images of (a and a') neat cyanate ester CE1, (b) 1% cyanate ester CE1 + 99% epoxy system, (c) 5% cyanate ester CE1 + 95% epoxy system, (d and d') neat cyanate ester CE2, and (e and e') neat cyanate ester CE3. (Reproduced from Ref. 14 by permission of The Royal Society of Chemistry.)

unmodified and IPN-modified polyester matrix was investigated by SEM. The tensile fractured surface of the unmodified resin (Fig. 12.3a) shows a brittle failure. The tensile fractured surface of IPN-modified GPR (Fig. 12.3b) with the particular composition (IPN 10% + GPR 90%) is smooth at the magnification of 500, thereby showing perfect compatibility between the polyurethane (PU), polyacrylonitrile (PAN), and polyester indicating a compositional homogeneity on a microscopic scale with respect to the IPN-modified system.

 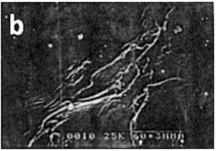

FIGURE 12.3 SEM micrograph of (a) pure GPR and (b) PU/PAN/GPR IPN (IPN 10). (Reproduced with permission from Ref. 15. © 2004 John Wiley.)

12.3 COMPOSITES

12.3.1 EPOXY/GLASS FIBER COMPOSITES

12.3.1.1 BIS(4-CYANATO3,5-DIMETHYLPHENYL)NAPHTHYL METHANE (CDPNM)/EPOXY/GLASS COMPOSITES

In 2008, Jayakumari et al. prepared dicyanate ester resin containing naphthalene moiety and blended with epoxy resin in different loading levels[16] (3%, 6%, and 9% 3CDPNM, 6CDPNM, and 9CDPNM). Composite laminates were fabricated with glass fiber and epoxy–cyanate ester blend. The phase separations brought about by the inclusion of cyanate ester into the epoxy system were studied by SEM analysis. The SEM micrographs of the fractured surface of pure epoxy and cyanate modified epoxy at 1000 magnification are shown in Figure 12.4a–d. The bright lines that emanate from the crack with the cut edges were observed in pure epoxy system (Fig. 12.4a). The polymer ligament tear appears as straight cuts with sharp edges. This shows that the cured unmodified resin is brittle, whereas for the modified resins, the fractured surfaces (Fig. 12.4b–d) appear as local shear deformation. Although few definite cut edges were seen at scattered points, the local yielding and shearing seem to be prevalent to some extent.

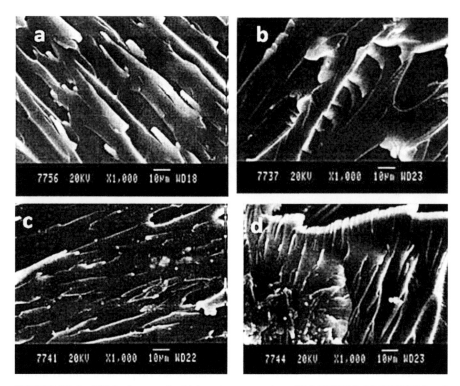

FIGURE 12.4 SEM micrograph of (a) neat epoxy system, (b) 3CDPNM, (c) 6CDPNM, and (d) 9CDPNM. (Reproduced with permission from Ref. 16. © 2008 John Wiley.)

12.3.1.2 WOVEN GLASS/EPOXY–PU IPN

Sampath et al. in 2008 prepared glass fiber composites with neat epoxy as well as PU/epoxy IPNs as matrix resin and studied the effect of this combination on the interlaminar fracture toughness using SEM.[17] The toughening mechanism believed to be in operation in this system is largely a particle cavitation mechanism and shear yielding of the matrix as shown in Figure 12.5a–c. The cured neat epoxy/glass fiber composite was found to undergo a brittle fracture (Fig. 12.5a). In the case of 5% PU–epoxy/glass fiber composite, a more ductile tearing of the matrix resin was observed (Fig. 12.5b). With increase in PU content (10%), there is evidence of increasing sheer yielding (Fig. 12.5c). At 15% PU content, more number of cavitated particles are seen indicating (Fig. 12.5d) the involvement of other toughening mechanisms.

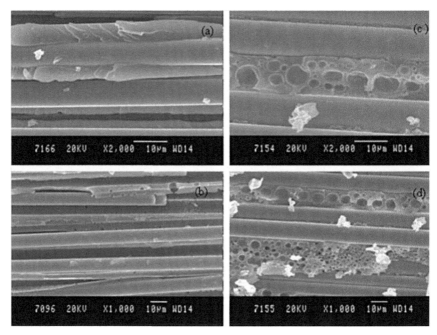

FIGURE 12.5 Scanning electron micrograph showing (a) neat epoxy–glass fiber composites, (b) 5% PU in epoxy–glass fiber composites, (c) 10% PU in epoxy–glass fiber composites, and (d) 15% PU in epoxy–glass fiber composites. (Reproduced with permission from Ref. 17. © 2008 John Wiley.)

12.3.1.3 MODIFIED PMR PI/CARBON FIBER COMPOSITES

Selladurai in 2012 fabricated carbon fiber-reinforced MPMRs PI (modified polymerization of monomeric reactants) composites.[18] The MPMRs prepolymer was prepared using the dimethyl ester of 3,3′,4,4′-benzophenonetetracarboxylic acid, bis(3,5-dimethyl-4-amino phenyl)-4′-methyl phenyl methane, and monomethyl ester of 5-norbornene-endo-2,3-dicarboxylic acid (NE). The SEM images of the postcured carbon fiber/MPMRs composite with two different magnifications are shown in Figure 12.6a,b. Some pores/voids are observed in the composites. But these are very few in number. These voids may be due to the smooth surface of the carbon fiber which resists the adherence of the resin in spite of the consolidation during the prepreg preparation and curing process. Void content and the C-scan analysis also lead to similar conclusions.

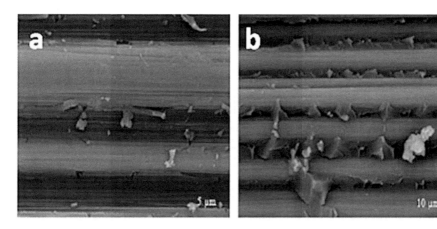

FIGURE 12.6 SEM micrographs of the carbon fiber (T300; 8 harness satin weave-3K)/MPMRs composite at different magnifications. (Reproduced with permission from Ref. 18. © 2012 Elsevier.)

12.3.1.4 PBZ/JUTE FIBER COMPOSITES

Thirukumaran et al. in 2014 fabricated bio-based composites using polybenzoxazines (PBz) resins based on renewable natural resources [viz. cardanol (C), eugenol (E), furfurylamine (F), and stearylamine (S)] and jute fiber. The SEM images of the fracture surface of the EF-Pbz/jute fiber and CF-Pbz/jute fiber composites (Fig. 12.7a,b) show less number of voids than that of the ES-Pbz/jute fiber and CS-Pbz/jute fiber composites (Fig. 12.8a,b) indicating that the compatibility between the resin and the fiber is better in the former than in the latter.[19,20]

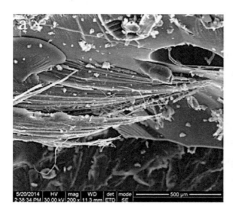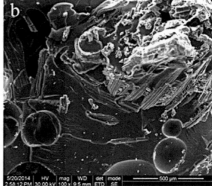

FIGURE 12.7 SEM images of (a) CF-Pbzo and (b) CS-Pbzo/jute fiber composites. (Reproduced with permission from Ref. 19. © 2016 Wiley.)

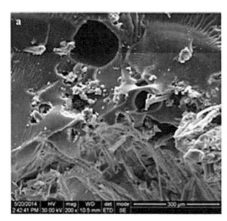 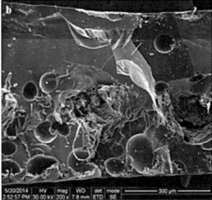

FIGURE 12.8 SEM images of (a) EF-Pbzo and (b) ES-Pbzo/jute fiber composites. (Reproduced with permission from Ref. 20. © 2014 John Wiley.)

12.4 NANOCOMPOSITES

12.4.1 POSS NANOCOMPOSITES

12.4.1.1 OPTICALLY ACTIVE—PI/OAPS NANOCOMPOSITES

In 2015, Balaji et al. synthesized optically active polyimide DAPIM-BTDA (or PMDA)-PI/OAPS (DAPIM-3,5-diaminophenyl)((R)-3,4-dihydro-1-phenylisoquinoline-2 (1H)-yl)methanone, BTDA—3,3′,4,4′-benzophenone tetracarboxylic dianhydride, PMDA—pyromelltic dianhydride nanocomposites.[21] The SEM images of neat PI and PI/OAPS (octa(aminophenyl) silsesquioxane) nanocomposites are used to study the morphology and phase separation. The fracture surface of both the neat PIs is found to be smooth, while that of PI/OAPS nanocomposites is rough. It can be seen that the fracture surface of the OAPS-incorporated PIs has many nanosized particles that are dispersed uniformly in the PI matrix. The uniform dispersion of OAPS in the PI can be attributed to covalent bond between OAPS and PI. The average size of the silica particles is in the range of 200–410 nm (0.20–0.41 mm). The fractured surface of the nanocomposites indicates that PI matrix has fine interconnected phase or cocontinuous phase, demonstrating good miscibility between the polymer and the silica phase.

12.4.1.2 POSS-CYANATE ESTER/EPOXY

In 2012, Rakesh et al. synthesized cyanate ester functionalized POSS (POSS-Cy) by reacting hydroxyl functionalized (polyhedral oligomeric silsesquioxane) POSS-OH with cyanogen bromide (CNBr) in the presence of triethylamine (Et$_3$N).[22] Epoxy/POSS-Cy nanocomposites were prepared by curing epoxy resin with POSS-Cy (5, 10, 15%) and diaminophenylmethane as the curing agent. The SEM micrographs of fractured surface of neat epoxy, 90% Ep/10% POSS-Cy with different magnifications indicate that there are some regions within the sample with well-dispersed POSS-Cy particles. POSS-Cy particles are highly hydrophobic and tend to bind together to form clusters due to agglomeration during processing. Higher magnification was applied to 10% POSS-Cy nanocomposite to closely monitor the dispersion of the POSS-Cy particles. From the results, it can be seen that the particles exist in small clusters due to agglomeration. The fracture surface of 10% POSS-CY shows that the POSS was dispersed in the epoxy system and phase separation also occurred to some extent.

12.4.1.3 EUGENOL-BASED PBZ/EPOXY/OAPS NANOCOMPOSITES

In 2014, Thirukumaran et al. synthesized eugenol-based polybenzoxazines/epoxy/OAPS nanocomposites.[23] Figure 12.9 displays the SEM images of PBz/Epoxy/OAPS hybrids with various OAPS contents (0, 1, 3, and 5 wt%). The fracture surface of neat PBz was found to be smooth (Fig. 12.9a). The SEM images of the PBz/OAPS hybrids (Figure 12.9b–d) at all compositions indicate even distribution of OAPS within the PBz matrix. The degree of roughness of the PBz/epoxy/OAPS hybrids increases upon increasing the OAPS content. The surface appears to be free of visible defects.

The AFM images of the PBz/epoxy/OAPS nanocomposites (Fig. 12.10) indicate that the size of the nodules formed by the OAPS particles is of uniform dimension, showing uniform distribution as evidenced by the SEM images.

The morphology of the PBz/epoxy/OAPS hybrid nanocomposites was further examined by TEM. Figure 12.11 shows TEM image of the hybrid material containing 5 wt% OAPS. Several dark spherical particles with varying size from 25 to 35 nm appear which are quite homogeneous. These dark particles are attributed to OAPS nanoparticles because of high electron density of OAPS cages. Moreover, there is no aggregation of OAPS moieties

286   Microscopy Applied to Materials Sciences and Life Sciences

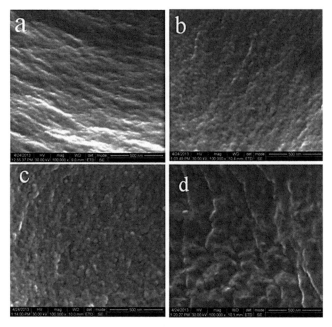

FIGURE 12.9 SEM images of the PBz/epoxy/OAPS composites: (a) 0 wt% of OAPS, (b) 1 wt% of OAPS, (c) 3 wt% of OAPS, and (d) 5 wt% of OAPS. (Reproduced with permission from Ref. 23. © 2014 John Wiley.)

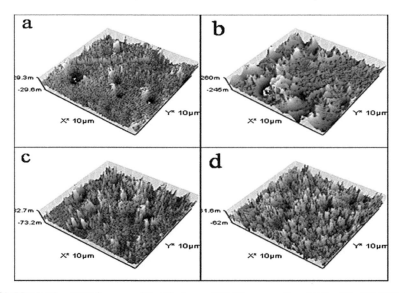

FIGURE 12.10 AFM images of the Pbz/epoxy/OAPS composites: (a) 0 wt% of OAPS, (b) 1 wt% of OAPS, (c) 3 wt% of OAPS, and (d) 5 wt% of OAPS. (Reproduced with permission from Ref. 23. © 2014 John Wiley.)

FIGURE 12.11 TEM image of the Pbz/epoxy/OAPS composites containing 5 wt% of OAPS. (Reproduced by permission from Ref. 23. © 2014 John Wiley.)

which shows uniform dispersion of OAPS in PBz/epoxy/OAPS networks. Surely, this morphology is relevant in determining the ultimate properties of the nanocomposites.

12.4.1.4 PBZ-FUNCTIONALIZED POSS NANOCOMPOSITES

In 2015, Thirukumaran et al. synthesized polybenzoxazine-tethered polyhedral oligomeric silsesquioxane nanocomposites by reacting OAPS with eugenol/guaiacol/vanillin and *p*-formaldehyde.[24] SEM micrographs (Fig. 12.12) of these hybrid nanocomposites [POSS–EPbz, POSS–GPbz, and POSS–VPbz] indicate that the POSS moieties remained evenly dispersed within the polybenzoxazine matrix. The extent of dispersion of the nanofiller plays an important role on the properties of the resulting nanocomposites. The presence of carbon, nitrogen, oxygen, and silicon and their composition in the hybrids analyzed using energy dispersive X-ray spectroscopy (EDX), integrated with scanning electron microscopy (Fig. 12.13), indicates the presence of POSS and polybenzoxazine in the matrix.

The AFM images of POSS–EPbz, POSS–GPbz, and POSS–VPbz nanocomposites are given in Figure 12.14. These heterogeneous materials show the presence of two different domains, in which the rough areas correspond to POSS rich domains, whereas the smooth areas correspond

288   Microscopy Applied to Materials Sciences and Life Sciences

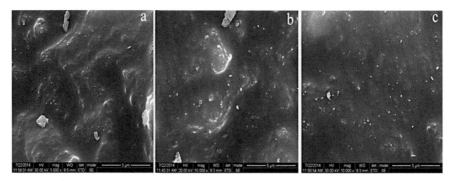

FIGURE 12.12 SEM images of the POSS–Pbz nanocomposites: (a) POSS–EPbz, (b) POSS–GPbz, and (c) POSS–VPbz. (Reproduced by permission from ref. 24. © The Royal Society of Chemistry.)

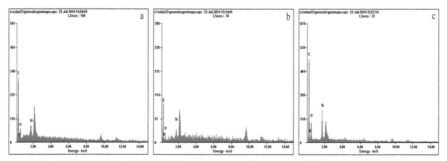

FIGURE 12.13 EDX of POSS–Pbz nanocomposites: (a) POSS–EPbz, (b) POSS–GPbz, and (c) POSS–VPbz. (Reproduced with permission from Ref. 24. © The Royal Society of Chemistry.)

to polybenzoxazine regions. The POSS seems to be distributed uniformly in the PBz matrix. This may be due to the fact that the POSS is functionalized with benzoxazine and during the curing process, the benzoxazine rings on adjacent POSS molecules open up and get covalently bonded, thereby preventing the formation of aggregates in the matrix.

High-resolution transmission electron microscopy (HR-TEM) analysis was performed to analyze the microstructure of POSS–Pbz nanocomposites (Fig. 12.15). The dark portions seen in the HRTEM images indicate that the POSS particles are spherical, about 5–10 nm in size, and uniformly distributed in the network structure. Moreover, there is no aggregation of POSS moieties. Such a uniform morphology is important for imparting high ultimate properties in the nanocomposites.

Microstructural Analysis of Polymer Blends, Composites 289

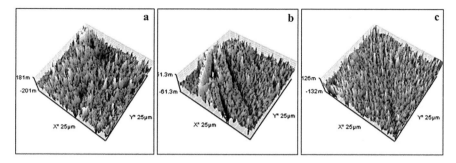

FIGURE 12.14 AFM images of the POSS–Pbz nanocomposites: (a) POSS–EPbz, (b) POSS–GPbz, and (c) POSS–VPbz. (Reproduced by permission from Ref. 24. © The Royal Society of Chemistry.)

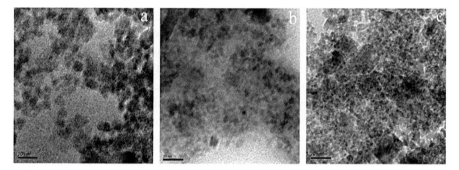

FIGURE 12.15 HRTEM images of the POSS–Pbz nanocomposites: (a) POSS–EPbz, (b) POSS–GPbz, and (c) POSS–VPbz. (Reproduced with permission from Ref. 24. © The Royal Society of Chemistry.)

12.4.2 SILICA NANOCOMPOSITES

12.4.2.1 UNSATURATED POLYSTER/SILICA NANOCOMPOSITES

Baskaran et al. in 2010 prepared nanocomposites of unsaturated polyester (UPR) with silica by casting technique. Nanosilica was synthesized by a sol–gel technique.[25] UPR/silica nanocomposites containing different proportions of nanosilica (1, 3, 5, 7, and 9 wt%) were prepared using methyl ethyl ketone peroxide and cobalt naphthenate as initiator and accelerator, respectively. The size of the particles was found to be 59 nm and the particles did not show any agglomeration up to 5% silica content as seen from the TEM images of the synthesized silica and UPR/

silica nanocomposites containing 5 and 7 wt% of nanosilica. It is obvious that the presence of 7 wt% of nanosilica in UPR resin matrix leads to aggregates.

12.4.2.2 PBZ/SILICA NANOCOMPOSITES

Shakila Parveen et al. in 2015 fabricated polybenzoxazine/silica nanocomposites using nanosized silica particles. The surface of the nanocomposites was proved to be hydrophobic.[26] Figure 12.16(a–d) shows the top-view SEM images of the neat PBZ and PBZ/SiO$_2$ hybrids. As seen from Figure 12.16a, the surface of the neat polybenzoxazine is very smooth. Compared to the neat polymer surface, the roughness of the PBz–SiO$_2$ surface (Fig. 12.16b–d) increased with the addition of SiO$_2$ nanoparticles. This rough surface was found to contain both micro- and nanoscale binary structure in which each microisland of the polybenzoxazine surface is covered with branch-like nanostructures. Such hierarchical morphology increases the surface roughness, thereby inducing hydrophobicity.

The elemental composition of polybenzoxazines silica hybrids analyzed using EDX spectroscopy, integrated with scanning electron microscopy which confirms the presence of carbon, nitrogen, oxygen, and silicon (Fig. 12.17).

The AFM images of polybenzoxazine and its nanohybrids (Fig. 12.18) indicate that the size of the nodules formed by the SiO$_2$ particles in the PBz matrix is of uniform dimensions showing uniform distribution as evidenced from SEM images. The surface roughness of the PBz/SiO$_2$ nanocomposites was calculated from AFM measurements by using the equation given below:

$$R_t = R_p + R_v$$

where R_t is the total roughness of the sample measured, R_p is the maximum profile peak height, and R_v is the maximum profile valley depth.

The total roughness was found to be 7, 81, 108, and 134 nm for PBZ:S0, PBZ:S1, PBZ:S3, and PBZ:S5, respectively. The roughness value increases on increasing the silica content from 1 to 5 wt%, which is in agreement with the AFM images.

HR-TEM was performed to analyze the microstructure of POSS–PBZ nanocomposites (Fig. 12.19). Dark regions corresponding to spherical silica nanoparticles are observed in the HR-TEM images of the POSS–PBz nanocomposites. These spherical particles are dispersed uniformly in nanometer

Microstructural Analysis of Polymer Blends, Composites 291

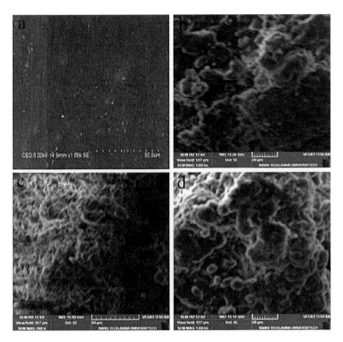

FIGURE 12.16 SEM images of PBZ–SiO$_2$ hybrids: (a) PBZ:S0, (b) PBZ:S1, (c) PBZ:S3, and (d) PBZ:S5. (Reproduced with permission from Ref. 26. © The Royal Society of Chemistry.)

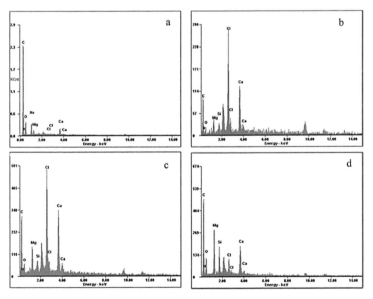

FIGURE 12.17 EDX spectra of PBZ–SiO$_2$ hybrids: (a) PBZ:S0, (b) PBZ:S1, (c) PBZ:S3, and (d) PBZ:S5. (Reproduced with permission from Ref. 26. © The Royal Society of Chemistry.)

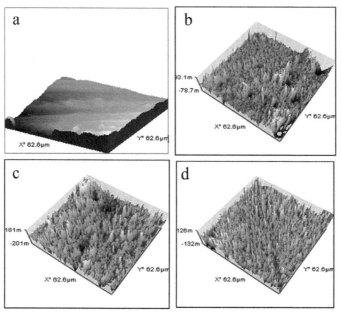

FIGURE 12.18 AFM images of PBZ–SiO$_2$ hybrids: (a) PBZ:S0, (b) PBZ:S1, (c) PBZ:S3, and (d) PBZ:S5. (Reproduced with permission from Ref. 26. © The Royal Society of Chemistry.)

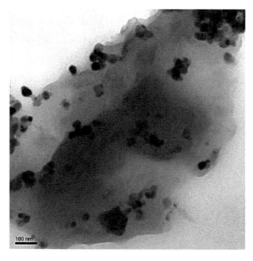

FIGURE 12.19 TEM image of PBZ–SiO$_2$ hybrids of PBZ:S5. (Reproduced by permission from Ref. 26. © The Royal Society of Chemistry.)

range (about 20–30 nm in size) and there is no aggregation of the silica particles, showing uniform distribution.

12.4.3 OTHER NANOCOMPOSITES

12.4.3.1 UPR/ALUMINA NANOCOMPOSITES

Baskaran et al. in 2011 prepared nanocomposites of UPR and nanoalumina by casting technique.[27] Alumina nanoparticles (60–70 nm) were prepared by the sol–gel technique using citric acid and aluminum nitrate. A typical TEM micrograph of synthesized alumina particles is shown in Figure 12.20a. Alumina particles are spherical and most of them were dispersed in the primary particle form with a diameter of about 60–70 nm. Figure 12.20b,c shows TEM micrographs of UPR/alumina nanocomposites with 5 and 7 wt% of nanoalumina, respectively. For 5 wt% alumina nanocomposite, a good dispersion is achieved. Most of the alumina particles are uniformly distributed as nanosized particles in the UPR matrix. The particles appear to be agglomeration free and the individual particles can be identified very clearly.

However, more aggregates are found in the UPR/alumina nanocomposite with 7 wt% of alumina nanoparticles, which suggests a poor dispersion. The poor dispersion may be due to the interaction between particles leading to agglomeration. This is reasonable, considering that at high nanoalumina concentration, the interparticle distance is small, and hence, flocculation of these nanoparticles can occur.

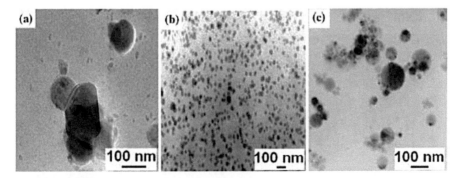

FIGURE 12.20 TEM image of (a) alumina nanoparticles, (b) UPR/5 wt% nanoalumina, and (c) UPR/7 wt% nanoalumina. (Reproduced by permission from Ref. 27. © 2011 Springer.)

The SEM pictures of tensile fractured surfaces of UPR/alumina nanocomposites are given in Figure 12.21. The possible origins of crack initiation in a composite material are air bubble or voids, resin-rich area, foreign matter such as dust particles, particle size, and poor particle matrix adhesion. The fractured surface of the unfilled resin (Fig. 12.21a) shows a brittle failure. At low levels of nanoalumina in UPR, good adhesion between the particle and polymer matrix can be seen from the fact that there is not much particle pull out and subsequent cavity formation (Fig. 12.21b). Another possible mode of failure (Fig. 12.21c) is noted in nanocomposites wherein agglomeration of the nanoalumina particles is seen. Since the nanoalumina particles are randomly oriented, large numbers of them are subjected to tensile stresses acting perpendicular to the plane and crack propagation occurs parallel to the plane. It is clearly seen that in Al_2O_3 nanocomposite with 7% Al_2O_3, the particles in UPR/7 (Fig. 12.21c) have experienced a high level of debonding and particle pull out.

12.4.2.2 UPR/NANO-ZnO NANOCOMPOSITES

Baskaran et al. in 2011 fabricated UPR/nano-ZnO nanocomposites with variable wt% of nano-ZnO by casting process.[28] Homogenous precipitation technique was used for the synthesis of nanozinc oxide. TEM image of nano-ZnO shows that the shape of the particles is spherical with an average diameter of about 40–50 nm. It can be seen from the TEM image of UPR matrix having 3 wt% of nano-ZnO that the ZnO nanoparticles are uniformly dispersed in the UPR matrix. The spherical shape of the nanoparticles is also clearly seen. When the nanoparticle loading is 5 wt% in the UPR matrix the degree of dispersion of the ZnO nanoparticles becomes rather poor and

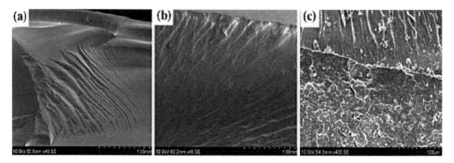

FIGURE 12.21 SEM image of (a) pure polyester, (b) UPR/5 wt% nanoalumina, and (c) UPR/7 wt% nanoalumina. (Reproduced by permission from Ref. 27. © 2011 Springer.)

particle agglomerations are observed. Therefore, for nanocomposites with a high concentration of nanoparticles, aggregation can take place easily. It is also clear that the vast majority of these agglomerates are still in the nanometer size range.

The well-dispersed nanoparticles in UPR matrix toughen the system and there is good interfacial bonding between the ZnO particles and the UPR matrix as seen from the SEM images of UPR and fractured surface of nanocomposites. Though the fracture surface of the UPR nanocomposites containing 5 wt% nano-ZnO is rough, the impact strength decreased. The agglomeration, voids, high stress concentrations, etc., have played a definite role in decreasing the impact strength at higher loading of nano-ZnO (>5 wt%) in UPR matrix. The agglomerated particles serve as weak points and lower the stress required for the composite to fracture.

12.4.2.3 UPR/CaCO$_3$ NANOCOMPOSITES

Baskaran et al. in 2011 prepared nanocomposites composed of UPR and nano-CaCO$_3$. The nanosized calcium carbonate synthesized using in situ deposition technique is used as reinforcement in UPR matrix resin by casting techniques.[29] It can be seen from the TEM micrographs of the synthesized nano-CaCO$_3$ and UPR/nano-CaCO$_3$ binary composites that the size of CaCO$_3$ particles is about 50–60 nm and they are spherical in shape. It is noted that there is good dispersion of nano-CaCO$_3$ in the UPR matrix at 5 wt% of nano-CaCO$_3$. The homogenous dispersion of CaCO$_3$ in the UPR matrix indicates good compatibility between nano-CaCO$_3$ and UPR matrix. Large aggregates are found when the content of nano-CaCO$_3$ is increased to 7 wt%. It is also clear that the vast majority of these agglomerates are still in the nanometer size range. At higher nano-CaCO$_3$ content, the distance between the nanoparticles is smaller, thereby increasing the reuniting chances of the added nanoparticles, leading to decreased dispersion of the CaCO$_3$ particles and hence increased aggregation.

The fracture surface of the virgin polyester specimen is relatively smooth as seen from the SEM images. This indicates that the resistance to crack propagation is less and leads to brittle failure. The crack surface is rough in the UPR/CaCO$_3$ nanocomposites containing less than 5 wt% nano-CaCO$_3$. At lower loading of nano-CaCO$_3$ in the UPR matrix, the presence of uniformly distributed particles obstructs the crack propagation in the matrix. The nano-CaCO$_3$ particles guide the crack to propagate in a torturous path, leading to the high strength of composites. UPR with 5 wt% nano-CaCO$_3$ shows scratches at the fracture surface. The scratches are due to the particle peel-off from the material. During this process, the particle could have scratched the surface and tried to resist the propagating crack. This offers some resistance to the crack propagation and increases the strength of the nanocomposites. The agglomeration, voids, high stress concentrations, etc., have decreased the impact strength of 7 wt% nano-$CaCO_3$-filled nanocomposites.

12.4.2.4 CYANATE ESTER/CNT NANOCOMPOSITES

Samikannu Rakesh et al. in 2015 fabricated polycyanurate/CNT composite material. Three different dicyanate ester monomers 1,3-bis(4-cyanatophenyl)prop-2-en-1-one (propyl-CE), 1,5-bis(4-cyanato phenyl)penta-1,4-dien-3-one (pentyl-CE), and 2,6-bis(4-cyanatobenzylidene)cyclohexanone (cyclohexyl-CE) were prepared.[30] The xerogels of the three monomers were prepared by magnetic stirring and heating a dilute solution of the monomer in DMF taken in a vial and quenching in an icebath when DMF was completely immobilized by the cyanate ester forming a gel. The same procedure was carried out for the preparation of cyanate ester gels in the presence of CNT (CNT-CE gel) using a dispersion of CNT in DMF. Figure 12.22 shows the SEM images of xerogels of propyl-CE and cyclohexyl-CE showing the fibrous morphology. The SEM images of cross-sections of the cured resins using neat monomer pentyl-CE (clump or powder) show smooth fracture surface, without any particular features (Fig. 12.23a,b). Figure 12.23c–f shows different views and magnifications of the cross-sections of the cured resin from the xerogel of pentyl-CE. The resin obtained by curing the xerogels shows network of fibers and thus a porous morphology.

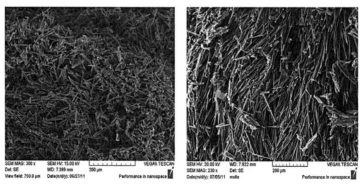

FIGURE 12.22 SEM images of gels of (a) propyl-CE and (b) cyclohexyl-CE. (Reprinted with permission from Ref. 30. © 2015 Elsevier.)

Microstructural Analysis of Polymer Blends, Composites

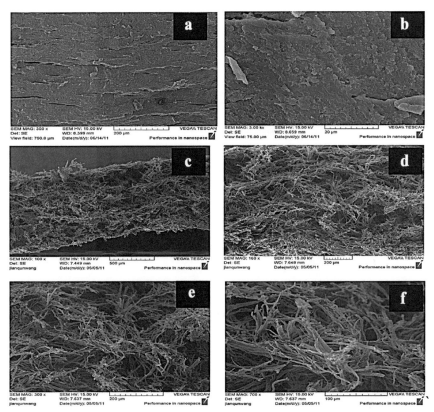

FIGURE 12.23 SEM images of cured cyanate ester resins: (a and b) cured from neat powder of pentyl-CE with two magnifications. (c–f) Cured xerogel of pentyl-CE. Different views and magnifications. (Reprinted with permission from Ref. 30. © 2015 Elsevier.)

Figure 12.24 shows the SEM images of the xerogels of monomers pentyl-CE and cyclohexyl-CE, with CNT incorporated during gelation. The CNT intertwined with the monomer gel fibers are clearly seen in Figure 12.24a (as indicated by the arrows). That these are CNT is confirmed by their width of about 75 nm, while the monomer fibers are a few microns in thickness.

Figure 12.25 shows the SEM images of the cross-sections of polymerized xerogels of pentyl-CE with CNT concentration varying from 1 to 4 wt%. The CNT forms a network with the polymerized gel fibers as seen from the images in Figure 12.25e,f. The thickness of the features denoted by arrows in Figure 12.25e is about 60 nm. The stepwise curing protocol mentioned above was used for the CNT/monomer gels as well. Thus, incorporating the CNT in the cyanate ester resin was accomplished by introducing it in the monomer gelation stage.

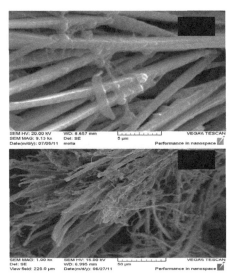

FIGURE 12.24 SEM images of xerogels of monomer/CNT: (a) pentyl-CE/CNT (3 wt%) and (b) cyclohexyl-CE/CNT (3 wt%). Arrows indicate the CNT bridging the fibers. (Reprinted with permission from Ref. 30. © 2015 Elsevier.)

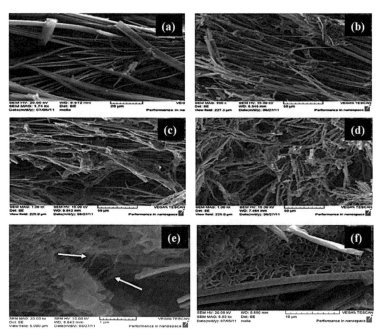

FIGURE 12.25 SEM images of polymerized cyanate ester pentyl-CE/CNT xerogels with (a) 1 wt%, (b) 2 wt%, (c) 3 wt%, and (d) 4 wt% CNT. Higher magnification images with (e) 2 wt% and (f) 4 wt% CNT are also shown. Arrows indicate the CNT. (Reprinted with permission from Ref. 30. © 2015 Elsevier.)

12.4.2.5 POLYTRIAZOLEIMIDE/GO NANOCOMPOSITES

Balasubramanian et al. in 2015 fabricated polytriazoleimide (PTAI)/GO nanocomposites.[31] The SEM images in Figure 12.26 provide morphological information for the PTAI hybrids with different GO contents. Neat PTAI was rather flat and smooth, and some bright dots were obvious at higher magnification. For the PTAI/GO nanocomposites, the fractured surfaces were relatively rougher compared with that of neat PTAI. These samples exhibit platelet morphology with the uniform orientation. Because of the difference between the scattering densities of the GO and the polymer matrix, the GO dispersion can be readily observed in the SEM images. Furthermore, the layered structures are aligned parallel to the surface of the PTAI/GO nanocomposite films. Due to the planar orientation, GO sheets are embedded tightly within the polymer indicating good compatibility and strong adhesion between GO and polymer matrix.

Nonuniform dispersion of GO will significantly reduce its effectiveness in a nanocomposites. Fortunately, nanocomposites produce the desired interface, homogenous dispersion, good interfacial compatibility,

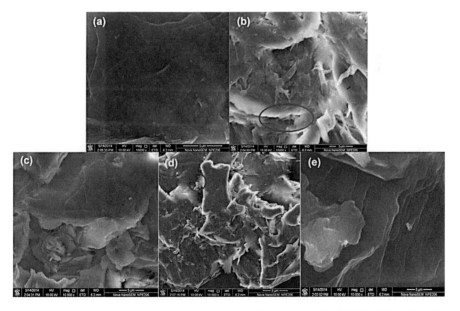

FIGURE 12.26 SEM images of (a) PTAI, (b) PTAI/GO (0.5 wt%), (c) PTAI/GO (1.0 wt%), (d) PTAI/GO (2.0 wt%), and (e) PTAI/GO (3.0 wt%). (Reprinted with permission from Ref. 31. © 2016 Wiley.)

and strong interfacial regions. These characteristics combined with a uniform arrangement of GO result in efficient transfer of stress across the interfaces, resulting in high performance polymers/GO nanocomposites. Figure 12.27 shows TEM images of the GO/PTAI nanocomposites with 0.5 and 1.0 wt% of GO, respectively. These images of the PTAI/GO nanocomposites show that the graphene sheets are well exfoliated and homogenously dispersed throughout the polymer matrix with almost no large agglomerates. This morphology is very consistent with the SEM observation.

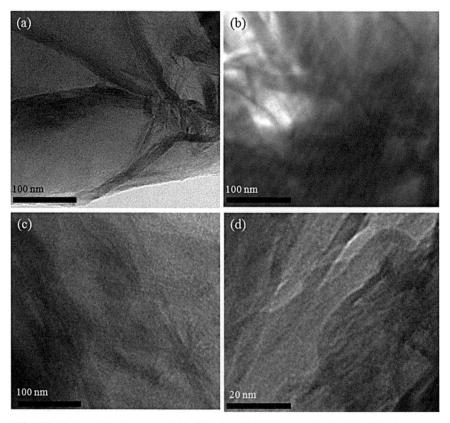

FIGURE 12.27 TEM images of (a) GO, (b) PTAI/GO, and (c, d) PTAI/GO 1 wt% (at different resolutions). (Reprinted with permission from Ref. 31. © 2016 Wiley.)

KEYWORDS

- **microstructural analysis**
- **polymer blends**
- **composites**
- **nanocomposites**
- **microscopy**
- **EDX**

REFERENCES

1. Goldstein, J. *Scanning Electron Microscopy and X-Ray Microanalysis*; Kluwer Academic/Plenum Publishers: Dordrecht, 2003; p 689.
2. Remier, L. *Scanning Electron Microscopy: Physics of the Image Format Ion and Microanalysis*; Springer: Berlin, 1998; p 527.
3. Egerton, R. F. *Physical Principles of Electron Microscopy: An Introduction to TEM, SEM, and AFM*; Springer: Berlin, 2005; p 202.
4. Clarke, A. R. *Microscopy Techniques for Materials Science*; CRC Press (Electronic Resource): Boca Raton, FL, 2002.
5. Reimer, L. *Scanning Electron Microscopy: Physics of Image Formation and Microanalysis*, 2nd ed.; Springer-Verlag: Berlin, 1998.
6. Scanning Electron Microscopy (SEM). http://acept.la.asu.edu/ PiN/rdg/elmier/elimer.shtml.
7. Scanning Electron Microscope Lab at Ohio Wesleyan University: Delaware, OH. http://go.own.edu/~semlab/.
8. Flegler, S. L.; Heckman, Jr., J. W.; Klomparens, K. L. *Scanning and Transmission Electron Microscopy: An Introduction*. W. H. Freeman and Company: New York, 1993.
9. Bhardwaj, R.; Gupta, V. *2nd International Conference on Role of Technology in Nation Building (ICRTNB-2013)*, 2013. ISBN:97881925922-1-3.
10. Mukhopadhyay, S. M. In *Sample Preparation Techniques in Analytical Chemistry*; Mitra, S., Eds.; ISBN 0-471-32845-6, John Wiley & Sons, Inc.: Hoboken, NJ, 2003.
11. Milica, M.; Vlahovic, I.; Predrag Jovanic, B. *Current Microscopy Contributions to Advances in Science and Technology*; Mendez-Vilas, A., Ed.; 2012, pp 1171–1182.
12. Sakthidharan, C. P.; Sundararajan, P. R.; Sarojadevi, M. Odd-Even Effect on the Thermal Properties of Schiff Base Functionalized Cynate Ester/Epoxy Blends. *RSC Adv.* **2015**, *5*, 73363–73372.
13. Hu, J. T.; Gu, A.; Liang, G.; Zhuo, D.; Yuan, L. Preparation and Properties of Maleimide-Functionalized Hyperbranched Polysiloxane and Its Hybrid Based on Cyanate Ester Resin. *J. Appl. Polym. Sci.* **2012**, *126* (1), 205–215.
14. Sakthidharan, C. P.; Sundararajan, P. R.; Sarojadevi, M. Thermal and Mechanical Properties of Azomethine Functionalized Cynate Ester/Epoxy Blends. *RSC Adv.* **2015**, *5*, 19666–19674.

15. Guhanathan, S.; Hariharan, R.; Sarojadevi, M. Studies on Castor Oil-Based Polyurethane/Polyacrylonitrile Interpreting Polymer Network for Toughening of Unsaturated Polyester Resin. *J. Appl. Polym. Sci.* **2004**, *92*, 817–829.
16. Jayakumari, L. S.; Thulasiraman, V.; Sarojadevi, M. Synthesis and Characterization of Bis(4-Cynanato-3,5-Dimethylphenyl) Naphthyl Methane/Epoxy/Glass Fiber Composites. *Polym. Compos.* **2008**, *29*, 709–716.
17. Sampath, P. S.; Murugesan, V.; Sarojadevi, M.; Thanigaiyarasu, G. Mode I and Mode II Delamination Resistance and Mechanical Properties of Woven Glass/Epoxy-PU IPN Composites. *Polym. Compos.* **2008**, *29*, 1227–1234.
18. Selladurai, M.; Sundararajan, P. R.; Sarojadevi, M. Synthesis, Thermal and Mechanical Properties of Modified PMR/Carbon Fiber Composites. *Chem. Eng. J.* **2012**, *203*, 333–347.
19. Thirukumaran, P.; Sathyamoorthi, R.; Shakila Parven, A.; Sarojadevi, M. New Benzoxazines from Renewable Resources for Green Composites Application. *Polym. Compos.* **2016**, *37*, 573–582.
20. Thirukumaran, P.; Shakila Parveen, A.; Kumudha, K.; Sarojadevi, M. Synthesis and Characterization of New Polybenzoxazines from Renewable Resources for Bio-composite Applications. *Polym. Compos.* **2014**. DOI:10.1002/PC.23356.
21. Balaji, S.; Sarojadevi, M. Synthesis and Characterization of Optically Active Polyimidies and their Octa(aminophenyl)silsesquioxane Nanocomposites. *High Perform. Polym.* **2015**, *28*, 547–561.
22. Rakesh, S.; Sakthidharan, C. P.; Selladurai, M.; Sudha, V.; Sundarrajan, P. R.; Sarojadevi, M. Thermal and Mechanical Properties of POSS–Cynate Ester/Epoxy Nanocomposites. *High Perform. Polym.* **2012**, *25*, 87–96.
23. Thirukumaran, P.; Shakila Parveen, A.; Sarojadevi, M. Synthesis of Eugenol-Bases Polybenzoxzine–POSS Nanocomposites for Low Dielectric Applications. *Polym. Compos.* **2015**, *11*, 1973–1982.
24. Thirukumaran, P.; Shakila Parveen, A.; Sarojadevi, M. New Benzoxazines Containing Polyhedral Oligomeric Silesquioxane from Eugenol, Guaiacol and Vanillin. *N. J. Chem.* **2015**, *39*, 1691–1702.
25. Baskaran, R.; Sarojadevi, M.; Vijayakumar, C. T. Mechanical and Thermal Properties of Unsaturated Polyester–Silica Nanocomposites. *Nanosci. Nanotechnol.* **2010**, *4*, 1–5.
26. Shakila Parveen, A.; Thirukumaran, P.; Sarojadevi, M. Fabrication of Highly Durable Hydrophobic PBZ/SiO$_2$ Surfaces. *RSC Adv.* **2015**, *5*, 43601–43610.
27. Baskaran, R.; Sarojadevi, M.; Vijayakumar, C. T. Unsaturated Polyester Nanocomposites Filled with Nano Alumina. *J. Mater. Sci.* **2011**, *46*, 4864–4871.
28. Baskaran, R.; Sarojadevi, M.; Vijayakumar, C. T. Effect of Nano Zinc Oxide on Mechanical and Thermal Properties of Unsaturated Polyester/Nano Zinc Oxide Composites. *J. Polym. Mater.* **2011**, *28*, 217–231.
29. Baskaran, R.; Sarojadevi, M.; Vijayakumar, C. T. Mechanical and Thermal Properties of Unsaturated Polyester/Calcium Carbonate Nanocomposites. *J. Reninforced Plast. Compos.* **2011**, 1–8.
30. Rakesh, S.; Sakthidharan, C. P.; Sarojadevi, M.; Sundarajan, P. R. Monomer Self-Assembly and Organo-Gelation as a Route to Fabricate Cyanate Ester Resins and Their Nanocomposites with Carbon Nanotubes. *Eur. Polym. J.* **2015**, *68*, 161–174.
31. Balasubramanian, R.; Balaji, S.; Sarojadevi, M. Synthesis and Properties of Polytriazoleimide (by Click Reaction), Polyetherimides and Their Nanocomposites with Graphene Oxide. *Polym. Compos.* **2016**. DOI: 10.1002/PC24184.

CHAPTER 13

GRAPHENE-MODIFIED CARBON MICROSURFACES IN VOLTAMMETRIC SENSING APPLICATIONS

GURURAJ KUDUR JAYAPRAKASH[1*],
BANANAKERE NANJEGOWDA CHANDRASHEKAR[2], and
BAHADDURGHATTA ESHWARASWAMY KUMARA SWAMY[3*]

[1]*Departamento de Ingeniería de Proyectos, Centro Universitario de Ciencias Exactas e Ingenierías, Blvd. Marcelino García Barragán 1421, C. P. 44430, Guadalajara Jal., México*

[2]*Department of Materials Science and Engineering, Southern University of Science and Technology, Shenzhen, Guangdong 518055, P. R. China*

[3]*Department of P.G. Studies and Research in Industrial Chemistry, Kuvempu University, Shankaraghatta 577451, Shimoga, Karnataka, India*

*Corresponding authors. E-mail: kumaraswamy21@gmail.com; rajguru97@gmail.com

ABSTRACT

Graphene is an interesting carbon material with unique redox properties. Graphene and its composites are promising candidates for fabricating current and next-generation electrochemical sensors. Since giving electric connections for graphene is difficult, usually, they are used to modify existing electrode materials. Graphene properties such as large active surface area, wide potential window, low detection limit, high stability and biocompatibility make to select it as better electrochemical sensor. This small chapter reviews the applications of graphene in sensor applications from voltammetric methods; especially, we have discussed applications for detection of dopamine, paracetamol, glucose, metal ions, DNA, and protein analysis. Still working scientific communities synthesize graphene with new functional groups to modify graphene texture/microscopic structures. Despite the vast amount of research already conducted on graphene for various sensor applications, the field is still growing and many questions remain unanswered. I have mentioned only few important articles deals with graphene voltammetric sensor applications.

13.1 INTRODUCTION

Carbon materials are always important and curios to scientific community because they are very abundantly available in Earth's surface. Carbon also has unique characteristics (e.g., it can form single, double, and triple bonds) that allow forming long chains of the same type of atoms. Since carbon–carbon bonds are strong and stable, they facilitate the production of an enormous number of molecular forms with special properties.[1] Carbon materials are used in designing electrodes for electrochemical devices, since they provide many advantageous properties to electrodes such as conductivity,[2] high surface area,[3] low background current,[4] chemical resistance,[2] low cost, possibilities of miniaturization, and relatively environmentally friendly characteristics.[5] Therefore, carbon electrodes are used in many important electrochemical applications (fuel cells,[6] sensors,[7] batteries,[8] and solar cells[9]).

Graphene is one of the most important carbon nanomaterials. Graphene is a layer of carbon atoms arranged in a two-dimensional (2D) sp^2-hybridized network of carbon atoms (carbon atoms are arranged in a honeycomb pattern). From its first report, it has gained huge attention from the scientific community.[10,11] It is the world's thinnest, strongest, and stiffest material[12] with attractive properties, such as high surface area (2630 m^2/g),[13] high thermal conductivity (≈5000 W/m),[14] fast charged carrier mobility (≈200,000 $cm^2 V^{-1} s^{-1}$),[14] strong Young's modulus (≈1 TPa),[15] as well as being an excellent conductor of both heat and electricity.[16] From the application point of view, graphene is considered as the most important carbon allotrope.[12] We are using graphene as precursor model for graphite toward the understanding of electrochemical phenomena. Graphite is the main constituent of the most popular carbon electrodes such as carbon paste, glassy carbon, and highly ordered pyrolytic graphite (HOPG). Although the properties of graphene are currently the main attraction for chemical community, its chemistry remains largely unexplored.[17]

13.2 REACTIVITY OF IDEAL GRAPHENE

Recently, we have proved the reactivity of ideal graphene from the analytical Fukui functions.[18] This function is defined as the derivative of the electron density $\rho(r)$ with respect to the total number of electrons of the system N, under a constant external potential, $v(r)$.[19–22] Fukui results of ideal graphene are shown in Figure 13.1.[18] At ideal state, graphene bond angle and bond lengths of all carbon atoms are nearly equal (periodic nature of the system), therefore each carbon atom of graphene shows equal reactivity for electron acceptance or electron removal. Terminal carbon atoms of the graphene are more reactive than the atoms which were in the inner shell[23,24]; however, we are not interested in analysis of terminal carbon atom reactivity because we have terminated graphene atoms artificially from H atoms. In Figure 13.1, we have captured one of three equivalent states (all states differ only by translations along the graphene surface).[18] If ideal graphene is used as a working electrode in voltammetric experiments, redox reactivity of all carbon atoms will be same apart from terminal carbon atoms (graphene microsurface has uniform probability for redox reactions).

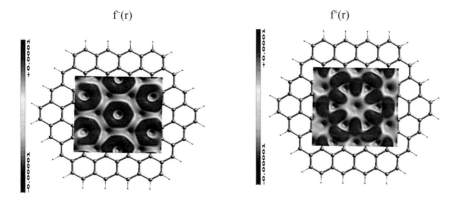

FIGURE 13.1 Fukui results for ideal graphene. Source: Adapted with permission from Ref. [18]. Copyright reserved to the American Chemical Society (2016).

Planar nature (opened π surface) of the ideal graphene makes their reactivity different than the fullerenes and carbon nanotubes.[25] Single-layer graphene reactivity will be different from the multilayer graphene.[26] We have to note that, graphene surface is sensitive to its surrounding environment. Numerous voltammetric experiments already have been done on graphene materials to utilize them in electrochemical sensing.

It is difficult to connect metallic wire to graphene from external power source. Electrochemists solved this problem by modifying well-utilized carbon electrodes (glassy carbon, carbon paste, screen printed, and diamond electrode) from the graphene materials. Once after modification, graphene materials improve the electrode sensing abilities. We have to take note that presence of impurities, defects, and functional groups on the graphene further changes their electron transfer properties. In many cases, presence of defects, dopants, and functional groups improves their electron transfer properties.

13.3 BENEFITS OF USING GRAPHENE AS WORKING ELECTRODE MATERIAL

As I have discussed in the Introduction section, graphene is more suitable candidate as working electrode material in electrochemical analysis. Carbon materials have been utilized in both analytical and industrial electrochemistry, and in fields they showed superior performance than the noble metals.[27] I have mentioned the specific properties of graphene received from microscopic structural alterations and its uses in sensing of particular type of analytes in the following sections. For example, $\pi-\pi$ interactions with phenyl groups help it to sense selectively and sensitively organic moieties with phenyl group. Increased active surface area of graphene-modified electrodes is advantageous for analyzing analytes such as glucose. High adsorptivity and conductivity of graphene composites help them to analyze DNA sequences. Increased charge transfer rates and high active surface area make it as better choice for immunosensor. Graphene also acts as an adsorbent and transducer which help it to sense heavy metal ions easily. According to D. A. C. Browson and C. E. Banks, graphene is more advantageous for electrochemical sensing than the carbon nanotubes.[27] Currently, graphene is a rising star in electrochemical analysis; in next decades, their applications in sensors field will further be improved.

13.3.1 FOR ANALYSIS OF DOPAMINE AND PARACETAMOL

Dopamine (DA) has been of great interest to neuroscientists and chemists, since from its discovery. DA is an important neurotransmitter in the central nervous system of mammals; it plays a vital role in maintaining the functional activities of cardiovascular and hormonal systems. Abnormal quantities of DA are linked to many diseases such as schizophrenia, Parkinson's, gout, scurvy, and cardiovascular diseases.[28] Paracetamol is a well-known active ingredient in several pharmaceutical products, widely used as an analgesic.[29] Paracetamol overdose is a frequent cause of fulminating hepatic failure, hepatic toxicity, and renal failure.[30,31] Thus, developing sensors for DA and paracetamol is helpful in pharmaceutical industry.

Graphene's unique physical and chemical properties (attractive π–π interaction and strong adsorptive capability) give it electrocatalytic properties.[32] Graphene-modified electrodes are widely used to detect organic molecules such as paracetamol, DA, uric acid (UA), ascorbic acid (AA), etc. In organic molecules, particularly DA and paracetamol, detection is more attractive to electrochemists. DA and paracetamol have phenyl π–π interactions between the phenyl structure of these organic compounds and the planar hexagons of the graphene increase the charge transfer at modified surface.[33] There is week π–π interactions between AA and graphene, therefore, oxidation of AA is not feasible. Hence, the simultaneous detection of DA/paracetamol and AA is possible at graphene-modified surface.[33]

Mohan Kumar et al.[34] synthesized graphene oxide and have modified carbon paste electrode with it to sense DA in presence of AA. The prepared sensors were good to analyze DA in pharmaceutical samples. Vidya et al.[35] modified carbon paste electrodes with graphene oxides and sodium dodecyl sulfate for sensing DA in presence of AA and UA. According to them, modified electrode showed good sensitivity and selectivity for DA. Shang et al.[36] has developed a sensor to detect DA, AA, and UA simultaneously from the multilayer graphene nanoflake films. Shang et al.'s[36] electrode have improved sensing abilities mainly due to edge plane sites/defects (generally, alteration of graphene surface by creating microsurface defects will increase its electron transfer properties[18]). Alwarappan et al.[37] chemically synthesized and characterized electrochemical properties of graphene nanosheets for sensing DA in presence of AA and serotonin. They reported that, more sp^2-like planes are present in the graphene nanosheets than the single-wall carbon nanotubes. Therefore, graphene nanosheets are better than the single-wall carbon nanotubes for electrochemical sensing of DA. Kang et al.[32] developed electrochemical sensor from the graphene-modified glassy carbon electrode (GCE) for detecting paracetamol from cyclic voltammetry and square-wave voltammetry. They proved that graphene has electrocatalytic activity toward paracetamol sensing.

13.3.2 FOR ANALYSIS OF GLUCOSE

Diabetes is a common worldwide problem in all age group of humans, as it is creating serious disability, complete treatment for diabetes is difficult, and during the diagnosis, it requires careful monitoring.[38] Development of economical, simple, and rapid methods for monitoring glucose is important in many areas such as pharmaceutical, food, clinical, and biological industry.[39] Graphene is an inexpensive material, as compared to other nanomaterials, with advantageous properties for sensors (high electrical conductivity, good mechanical, and thermal properties). Along with these properties, large surface area to volume ratio and good biocompatibility make it as better material for sensing biomolecules.[40] Worden et al.[41] have developed sensitive and fast-responding glucose biosensor from exfoliated graphite as an alternative to carbon nanotube; their electrodes showed excellent sensitivity, resistance to interference, and long-term stability. Dong et al.[42] synthesized on three-dimensional (3D) graphene composites for enzyme-free glucose detection; they have sensed glucose with detection limit less than 25 nM. Authors proposed that, increased active surface area at composite electrodes is responsible for reliable glucose sensing. Liang et al.[43] studied electron transfer between reduced carboxyl graphene-modified GCE and glucose oxidase. The developed sensors from it have detection limit of 0.02 µM and their method is facile, rapid, and eco-friendly. Alwarappan et al.[44] developed glucose biosensor from enzyme-doped graphene nanosheets. The presence of functional groups on the edge plane of graphene makes glucose oxidase to bind on it easily from covalent amide linkage. Their results show that the 2D graphene holds great promise to be conjugated with a variety of enzymes for biosensing. From the results of all above articles, we may conclude that microscopic changes in graphene surfaces are beneficial for graphene sensing properties.

13.3.3 DNA ANALYSIS

Electrochemical methods offer high sensitivity, high selectivity, rapid and economical way for DNA analysis.[45,46] Addition of modifiers such as chitosan, dyes, and metal oxides to graphene surfaces will be beneficial for increasing sensitivity of graphene-based DNA sensor. In this small section, we mention few notable works done on DNA sensors by altering graphene surfaces. Bo et al.[47] developed electrochemical DNA sensors based on graphene and polyaniline nanowires; charge transfer properties at the developed electrodes were good to give differential pulse voltammetric responses. In addition to this, Bo et al.[47] characterized oxidized graphene and polyaniline nanowires from scanning electron microscopy, cyclic voltammetric, and electrochemical impedance spectroscopy methods. Bo et al.[48] synthesized another DNA sensor from graphene paste electrode modified with Prussian blue and chitosan. Prussian blue helps to improve the graphene paste electrode voltammetric response for sequence analysis. Fan et al.[49] developed TiO_2–graphene nanocomposite electrodes for sensing adenine and guanine. TiO_2–graphene nanocomposite electrodes have showed good sensing properties for these species, which were due to the high adsorptivity and conductivity of TiO_2–graphene nanocomposite. Lim et al.[50] used anodized epitaxial graphene on simultaneous detection of four DNA bases (guanine, adenine, thymine, and cytosine). Zhou et al.[51] have proposed chemically reduced graphene oxide to sense four bases of DNA. Their electrodes are able to simultaneously sense four bases from differential pulse voltammetry. Rasheed et al.[52] developed a DNA sensor to detect BRCA1 gene (which is related to breast cancer). Their sensors could find applications in early cancer diagnosis.

13.3.4 PROTEIN ANALYSIS

Developing sensors to detect protein especially disease-related proteins is very important and essential in pharmaceutical and biochemical studies. Sensing cancer biomarkers will help in future to diagnose cancer in early stages.[53,54] Du et al.[55] proposed new type of electrochemical immunosensor to α-fetoprotein using graphene sheet sensor platform and functionalized carbon nanospheres labeled with horseradish peroxidase-secondary antibodies. The developed methods were reliable with amplified response. One of the main reasons for the amplification of response is graphene sheets (increased active surface area). Xie et al.[56] developed a new p53 immunosensor based on graphene-modified electrode. The developed method showed good performance for detecting protein with high accuracy and reproducibility. They even propose that graphene materials are advantageous over carbon nanotube in purity and in economical ways. Zhu et al.[57] designed method for simultaneous electrochemical determination of alpha-fetoprotein, carcinoembryonic, and streptococcus suis serotype. The designed immunosensor exhibited high selectivity and sensitivity in simultaneous determination of three analytes from differential pulse voltammetry. Zhao et al.[58] developed a new enzyme-free immunosensor for detecting α-fetoprotein. In their immunosensor, they have modified graphene surfaces from N-doping, which increased electron transfer rates; as a result of this, more signals were generated and enzyme-free detection α-fetoprotein was achieved.

13.3.5 FOR ANALYSIS OF HEAVY METAL IONS (CADMIUM AND LEAD)

Monitoring heavy metal ions (Cd^{2+} and Pb^{2+}) is useful for the environmental and food chemists.[45,46] Generally, nanomaterials-based electrode modification is beneficial for electrode modification. Nanomaterials-based electrodes are capable of providing more sensitive electrode interfaces due to increased surface area, charge transfer, and mass transfer process at electrode interfaces. Electrodes which were modified from the graphene nanomaterials have been showed to have superior sensing capabilities when compared to metal nanoparticles.[26]

Graphene simultaneously can act as an adsorbent and transducer, therefore detection of heavy metal ions will be easy at graphene or graphene-modified electrode surfaces.[26,59] Wang et al.[60] developed stannum film/poly(p-aminobenzene sulfonic acid, p-ABSA)/graphene composite-modified GCE for sensing Cd^{2+} ions from square-wave anodic stripping voltammetry. Stripping current signals are greatly increased and distinguished at graphene electrodes; this will be advantageous to detect heavy metal ions at these electrodes.[45] Graphene sheets nature such as nanothickness and size along with increased conductivity makes ions to adsorb on it strongly; this is also advantageous in voltammetric analysis.[45] Li et al.[61,62] proposed a voltammetric sensors from Nafion–graphene composite electrode for the analysis of Pb^{2+} and Cd^{2+} using differential pulse anodic stripping voltammetry. They reported that Nafion–graphene composite electrode showed limit of detection of 0.02 μgL^{-1} for both Cd^{2+} and Pb^{2+}. Brownson et al.[63] used commercially available graphene for sensing Cd^{2+} ions.

13.4 CONCLUDING REMARKS

Graphene is mother of all other carbon allotrope (e.g., carbon nanotubes, fullerenes, graphite, etc.). Graphene properties are very sensitive to its preparation methods[26] and surrounding environment. Microscopic changes in the graphene surface can lead to significant changes in its electron transfer properties.[18] Graphene has showed excellent performance in direct electrochemistry of organic molecules (DA, AA, and UA), pharmaceutical molecules (paracetamol), biomolecules (DNA and protein), and inorganic molecules (cadmium and lead) for their electroanalysis, which makes it as more suitable candidate to select it as a working electrode material for voltammetric sensors. For detection of few molecules, graphene exhibited superior sensing properties than its close relatives (carbon nanotubes). To develop futuristic graphene electrode material, one has to understand its properties from Quantum chemical ways.[18]

Graphene properties such as π–π interactions with phenyl moieties of organic molecules (DA and paracetamol) help it to sense these organic analytes sensitively and selectively. Large surface area to volume ratio makes it as a better choice to sense biomolecules such as glucose and protein. High adsorptivity and conductivity of graphene composites help it to analyze DNA sequences. Increased charge transfer rates and high active surface area make it as better choice for immunosensor. Graphene also acts as an adsorbent and transducer which help it to sense heavy metal ions easily. Apart from these properties, wide potential window, availability, cost, and purity make it a suitable choice as working electrode materials for sensing applications.

In summary, graphene is an excellent electrode material for electroanalysis and electrocatalysis. Despite the wide research on graphene, many questions of graphene fundamental properties and modified graphene electron transfer properties are still unanswered. Solving these questions will further improve their sensor applications in future.

KEYWORDS

- graphene
- sensors
- voltammetry
- surface properties
- electrochemical analysis

REFERENCES

1. Dinadayalane, T. C.; Leszczynski, J. Remarkable Diversity of Carbon–Carbon Bonds: Structures and Properties of Fullerenes, Carbon Nanotubes, and Graphene. *Struct. Chem.* **2010,** *21,* 1155–1169.

2. Švancara, I.; Vytřas, K.; Barek, J.; Zima, J. Carbon Paste Electrodes in Modern Electroanalysis. *Crit. Rev. Anal. Chem.* **2001**, *31*, 311–345.
3. Shankar, S. S.; Swamy, B. E. K.; Pandurangachar, M.; Chandra, U.; Chandrashekar, B. N.; Manjunatha, J. G.; Sherigara, B. S. Electrocatalytic Oxidation of Dopamine on Acrylamide Modified Carbon Paste Electrode: A Voltammetric Study. *Int. J. Electrochem. Sci.* **2010**, *5*, 944–954.
4. Wang, J.; Naser, N.; Angnes, L.; Wu, H.; Chen, L. Metal-dispersed Carbon Paste Electrodes. *Anal. Chim. Acta* **1992**, *64*, 1285–1288.
5. Švancara, I.; Walcarius, A.; Kalcher, K.; Vytřas, K. Carbon Paste Electrodes in the New Millennium. *Open Chem.* **2009**, *7*, 598–656.
6. Takeuchi, R. M.; Santos, A. L.; Padilha, P. M.; Stradiotto, N. R. Copper Determination in Ethanol Fuel by Differential Pulse Anodic Stripping Voltammetry at a Solid Paraffin-based Carbon Paste Electrode Modified with 2-Aminothiazole Organofunctionalized Silica. *Talanta* **2007**, *71*, 771–777.
7. Gilbert, O.; Swamy, B. E. K.; Chandra, U.; Sherigara, B. S. Simultaneous Detection of Dopamine and Ascorbic Acid Using Polyglycine Modified Carbon Paste Electrode: A Cyclic Voltammetric Study. *J. Electroanal. Chem.* **2009**, *636*, 80–85.
8. Švancara, I.; Vytřas, K.; Kalcher, K.; Walcarius, A.; Wang, J. Carbon Paste Electrodes in Facts, Numbers, and Notes: A Review on the Occasion of the 50-Years Jubilee of Carbon Paste in Electrochemistry and Electroanalysis. *Electroanalysis* **2009**, *21*, 7–28.
9. Kureishi, Y.; Tamiaki, H.; Shiraishi, H.; Maruyama, K. Photoinduced Electron Transfer from Synthetic Chlorophyll Analogue to Fullerene C 60 on Carbon Paste Electrode: Preparation of a Novel Solar Cell. *Bioelectrochem. Bioenerg.* **1999**, *48*, 95–100.
10. Novoselov, K. S.; Geim, A. K.; Morozov, S. V.; Jiang, D.; Zhang, Y.; Dubonos, S. V.; Grigorieva, I. V.; Firsov, A. A. Electric Field Effect in Atomically Thin Carbon Films. *Science* **2004**, *306*, 666–669.
11. Wang, H.; Maiyalagan, T.; Wang, X. Review on Recent Progress in Nitrogen-doped Graphene: Synthesis, Characterization, and Its Potential Applications. *ACS Catal.* **2012**, *2*, 781–794.
12. Georgakilas, V.; Otyepka, M.; Bourlinos, A. B.; Chandra, V.; Kim, N.; Kemp, K. C.; Hobza, P.; Zboril, R.; Kim, K. S. Functionalization of Graphene: Covalent and Non-covalent Approaches, Derivatives and Applications. *Chem. Rev.* **2012**, *112*, 6156–6214.
13. Stoller, M. D.; Park, S.; Zhu, Y.; An, J.; Ruoff, R. S. Graphene-based Ultracapacitors. *Nano Lett.* **2008**, *8*, 3498–3502.
14. Balandin, A. A.; Ghosh, S.; Bao, W.; Calizo, I.; Teweldebrhan, D.; Miao, F.; Lau, C. N. Superior Thermal Conductivity of Single-layer Graphene. *Nano Lett.* **2008**, *8*, 902–907.
15. Lee, C.; Wei, X.; Kysar, J. W.; Hone, J. Measurement of the Elastic Properties and Intrinsic Strength of Monolayer Graphene. *Science* **2008**, *321*, 385–388.
16. Szewczyk, R.; Zieliński, C.; Kaliczyńska, M. Progress in Automation, Robotics and Measuring Techniques: Volume 3 Measuring Techniques and Systems. In *Advances in Intelligent Systems and Computing*; Springer International Publishing: Switzerland, 2015.
17. Miao, M.; Shi, H.; Wang, Q.; Liu, Y. The Ti 4 Cluster Activates Water Dissociation on Defective Graphene. *PCCP* **2014**, *16*, 5634–5639.
18. Kudur Jayaprakash, G.; Casillas, N.; Astudillo-Sánchez, P. D.; Flores-Moreno, R. Role of Defects on Regioselectivity of Nano Pristine Graphene. *J. Phys. Chem. A* **2016**, *120*, 9101–9108.

19. Ayers, P. W. The Dependence on and Continuity of the Energy and Other Molecular Properties with Respect to the Number of Electrons. *J. Math. Chem.* **2008**, *43*, 285–303.
20. Fuentealba, P.; Florez, E.; Tiznado, W. Topological Analysis of the Fukui Function. *J. Chem. Theory Comput.* **2010**, *6*, 1470–1478.
21. Flores-Moreno, R.; Melin, J.; Ortiz, J. V.; Merino, G. Efficient Evaluation of Analytic Fukui Functions. *J. Chem. Phys.* **2008**, *129*, 224105/1–6.
22. Guillén-Villar, R. C.; Vargas-Álvarez, Y.; Vargas, R.; Garza, J.; Matus, M. H.; Salas-Reyes, M.; Domínguez, Z. Study of the Oxidation Mechanisms Associated to New Dimeric and Trimeric Esters of Ferulic Acid. *J. Electroanal. Chem.* **2015**, *740*, 95–104.
23. Davies, T. J.; Moore, R. R.; Banks, C. E.; Compton, R. G. The Cyclic Voltammetric Response of Electrochemically Heterogeneous Surfaces. *J. Electroanal. Chem.* **2004**, *574*, 123–152.
24. Davies, T. J.; Hyde, M. E.; Compton, R. G. Nanotrench Arrays Reveal Insight into Graphite Electrochemistry. *Angew. Chem. Int. Ed.* **2005**, *44*, 5121–5126.
25. Rodriguez-Perez, L.; Herranz, M. a. A.; Martin, N. The Chemistry of Pristine Graphene. *Chem. Commun.* **2013**, *49*, 3721–3735.
26. Ambrosi, A.; Chua, C. K.; Bonanni, A.; Pumera, M. Electrochemistry of Graphene and Related Materials. *Chem. Rev.* **2014**, *114*, 7150–7188.
27. Brownson, D. A. C.; Banks, C. E. Graphene Electrochemistry: An Overview of Potential Applications. *Analyst* **2010**, *135*, 2768–2778.
28. Velasco, M.; Luchsinger, A. Dopamine: Pharmacologic and Therapeutic Aspects. *Am. J. Ther.* **1998**, *5*, 37–44.
29. SĂČndulescu, R.; Mirel, S.; Oprean, R. The Development of Spectrophotometric and Electroanalytical Methods for Ascorbic Acid and Acetaminophen and Their Applications in the Analysis of Effervescent Dosage Forms. *J. Pharm. Biomed. Anal.* **2000**, *23*, 77–87.
30. ShangGuan, X.; Zhang, H.; Zheng, J. Electrochemical Behavior and Differential Pulse Voltammetric Determination of Paracetamol at a Carbon Ionic Liquid Electrode. *Anal. Bioanal. Chem.* **2008**, *391*, 1049–1055.
31. Gururaj, K.; Swamy, B. K. Electrochemical Synthesis of Titanium Nano Particles at Carbon Paste Electrodes and Its Applications as an Electrochemical Sensor for the Determination of Acetaminophen in Paracetamol Tablets. *SNL* **2013**, 3.
32. Kang, X.; Wang, J.; Wu, H.; Liu, J.; Aksay, I. A.; Lin, Y. A Graphene-based Electrochemical Sensor for Sensitive Detection of Paracetamol. *Talanta* **2010**, *81*, 754–759.
33. Brownson, D. A. C.; Banks, C. E. *The Handbook of Graphene Electrochemistry*; Springer, 2014.
34. Kumar, M.; Swamy, B. K.; Asif, M. M.; Viswanath, C. Preparation of Alanine and Tyrosine Functionalized Graphene Oxide Nanoflakes and Their Modified Carbon Paste Electrodes for the Determination of Dopamine. *Appl. Surf. Sci.* **2017**, *399*, 411–419.
35. Vidya, H.; Swamy, B. K. Voltammetric Determination of Dopamine in the Presence of Ascorbic Acid and Uric Acid At Sodium Dodecyl Sulphate/Reduced Graphene Oxide Modified Carbon Paste Electrode. *J. Mol. Liq.* **2015**, *211*, 705–711.
36. Shang, N. G.; Papakonstantinou, P.; McMullan, M.; Chu, M.; Stamboulis, A.; Potenza, A.; Dhesi, S. S.; Marchetto, H. Catalyst-free Efficient Growth, Orientation and Biosensing Properties of Multilayer Graphene Nanoflake Films with Sharp Edge Planes. *Adv. Funct. Mater.* **2008**, *18*, 3506–3514.

37. Alwarappan, S.; Erdem, A.; Liu, C.; Li, C.-Z. Probing the Electrochemical Properties of Graphene Nanosheets for Biosensing Applications. *J. Phys. Chem. C* **2009**, *113*, 8853–8857.
38. Siangproh, W.; Dungchai, W.; Rattanarat, P.; Chailapakul, O. Nanoparticle-based Electrochemical Detection in Conventional and Miniaturized Systems and Their Bioanalytical Applications: A Review. *Anal. Chim. Acta* **2011**, *690*, 10–25.
39. Newman, J. D.; Turner, A. P. Home Blood Glucose Biosensors: A Commercial Perspective. *Biosens. Bioelectron.* **2005**, *20*, 2435–2453.
40. Unnikrishnan, B.; Palanisamy, S.; Chen, S.-M. A Simple Electrochemical Approach to Fabricate a Glucose Biosensor Based on Graphene–Glucose Oxidase Biocomposite. *Biosens. Bioelectron.* **2013**, *39*, 70–75.
41. Lu, J.; Drzal, L. T.; Worden, R. M.; Lee, I. Simple Fabrication of a Highly Sensitive Glucose Biosensor Using Enzymes Immobilized in Exfoliated Graphite Nanoplatelets Nafion Membrane. *Chem. Mater.* **2007**, *19*, 6240–6246.
42. Dong, X.-C.; Xu, H.; Wang, X.-W.; Huang, Y.-X.; Chan-Park, M. B.; Zhang, H.; Wang, L.-H.; Huang, W.; Chen, P. 3D Graphene Cobalt Oxide Electrode for High-performance Supercapacitor and Enzymeless Glucose Detection. *ACS Nano* **2012**, *6*, 3206–3213.
43. Liang, B.; Fang, L.; Yang, G.; Hu, Y.; Guo, X.; Ye, X. Direct Electron Transfer Glucose Biosensor Based on Glucose Oxidase Self-assembled on Electrochemically Reduced Carboxyl Graphene. *Biosens. Bioelectron.* **2013**, *43*, 131–136.
44. Alwarappan, S.; Liu, C.; Kumar, A.; Li, C.-Z. Enzyme-doped Graphene Nanosheets for Enhanced Glucose Biosensing. *J. Phys. Chem. C* **2010**, *114*, 12920–12924.
45. Shao, Y.; Wang, J.; Wu, H.; Liu, J.; Aksay, I.; Lin, Y. Graphene Based Electrochemical Sensors and Biosensors: A Review. *Electroanal.* **2010**, *22*, 1027–1036.
46. Kuila, T.; Bose, S.; Khanra, P.; Mishra, A. K.; Kim, N. H.; Lee, J. H. Recent Advances in Graphene-based Biosensors. *Biosens. Bioelectron.* **2011**, *26*, 4637–4648.
47. Bo, Y.; Yang, H.; Hu, Y.; Yao, T.; Huang, S. A Novel Electrochemical DNA Biosensor Based on Graphene and Polyaniline Nanowires. *Electrochim. Acta* **2011**, *56*, 2676–2681.
48. Bo, Y.; Wang, W.; Qi, J.; Huang, S. A DNA Biosensor Based on Graphene Paste Electrode Modified with Prussian Blue and Chitosan. *Analyst* **2011**, *136*, 1946–1951.
49. Fan, Y.; Huang, K.-J.; Niu, D.-J.; Yang, C.-P.; Jing, Q.-S. TiO_2–Graphene Nanocomposite for Electrochemical Sensing of Adenine and Guanine. *Electrochim. Acta* **2011**, *56*, 4685–4690.
50. Lim, C. X.; Hoh, H. Y.; Ang, P. K.; Loh, K. P. Direct Voltammetric Detection of DNA and pH Sensing on Epitaxial Graphene: An Insight into the Role of Oxygenated Defects. *Anal. Chem.* **2010**, *82*, 7387–7393.
51. Zhou, M.; Zhai, Y.; Dong, S. Electrochemical Sensing and Biosensing Platform Based on Chemically Reduced Graphene Oxide. *Anal. Chem.* **2009**, *81*, 5603–5613.
52. Rasheed, P. A.; Sandhyarani, N. Graphene–DNA Electrochemical Sensor for the Sensitive Detection of BRCA1 Gene. *Actuator B: Chem.* **2014**, *204*, 777–782.
53. Kitano, H. Systems Biology: A Brief Overview. *Science* **2002**, *295*, 1662–1664.
54. Srinivas, P. R.; Kramer, B. S.; Srivastava, S. Trends in Biomarker Research for Cancer Detection. *Lancet Oncol.* **2001**, *2*, 698–704.
55. Du, D.; Zou, Z.; Shin, Y.; Wang, J.; Wu, H.; Engelhard, M. H.; Liu, J.; Asay, I. A.; Lin, Y. Sensitive Immunosensor for Cancer Biomarker Based on Dual Signal Amplification Strategy of Graphene Sheets and Multienzyme Functionalized Carbon Nanospheres. *Anal. Chem.* **2010**, *82*, 2989–2995.

56. Xie, Y.; Chen, A.; Du, D.; Lin, Y. Graphene-based Immunosensor for Electrochemical Quantification of Phosphorylated p53 (S15). *Anal. Chim. Acta* **2011**, *699*, 44–48.
57. Zhu, Q.; Chai, Y.; Yuan, R.; Zhuo, Y.; Han, J.; Li, Y.; Liao, N. Amperometric Immunosensor for Simultaneous Detection of Three Analytes in One Interface Using Dual Functionalized Graphene Sheets Integrated with Redox-probes as Tracer Matrixes. *Biosens. Bioelectron.* **2013**, *43*, 440–445.
58. Zhao, L.; Li, S.; He, J.; Tian, G.; Wei, Q.; Li, H. Enzyme-free Electrochemical Immunosensor Configured with Au–Pd Nanocrystals and N-doped Graphene Sheets for Sensitive Detection of AFP. *Biosens. Bioelectron.* **2013**, *49*, 222–225.
59. Aragay, G.; MerkoÃği, A. Nanomaterials Application in Electrochemical Detection of Heavy Metals. *Electrochim. Acta* **2012**, *84*, 49–61.
60. Wang, Z.; Wang, H.; Zhang, Z.; Yang, X.; Liu, G. Sensitive Electrochemical Determination of Trace Cadmium on a Stannum Film/Poly(p-aminobenzene Sulfonic Acid)/Electrochemically Reduced Graphene Composite Modified Electrode. *Electrochim. Acta* **2014**, *120*, 140–14.
61. Li, J.; Guo, S.; Zhai, Y.; Wang, E. High-sensitivity Determination of Lead and Cadmium Based on the Nafion–Graphene Composite Film. *Anal. Chim. Acta* **2009**, *649*, 196–201.
62. Li, J.; Guo, S.; Zhai, Y.; Wang, E. Nafion–Graphene Nanocomposite Film as Enhanced Sensing Platform for Ultrasensitive Determination of Cadmium. *Electrochem. Commun.* **2009**, *11*, 1085–1088.
63. Brownson, D. A.; Banks, C. E. Graphene Electrochemistry: Surfactants Inherent to Graphene Inhibit Metal Analysis. *Electrochem. Commun.* **2011**, *13*, 111–113.

PART II
Life Sciences

CHAPTER 14

FUNCTIONALIZED NANOMATERIALS FOR BIOLOGICAL AND CATALYTIC APPLICATIONS

K. ANAND[1*], R. M. GENGAN[2], and A. A. CHUTURGOON[1]

[1]*Discipline of Medical Biochemistry and Chemical Pathology, School of Laboratory Medicine and Medical Sciences, College of Health Sciences, University of KwaZulu-Natal, Durban, South Africa*

[2]*Faculty of Applied Science, Department of Chemistry, Durban University of Technology, Durban, South Africa*

*Corresponding author. E-mail: organicanand@gmail.com

ABSTRACT

Organic chemistry moves toward a multidisciplinary environment blending biology, physics, and materials sciences. This book chapter will be based on the preparation of organic small molecules by using nanocatalyst and nanodrug formulations. Nature provides abundant resources to prepare nanomaterials with novel structures and properties. Also, synthesis and stabilization of nanoparticles using biocompatible capping agents from natural resources such as microorganisms (e.g., alga and fungi) and plant extract provide great advancement over existing chemical and physical methods since they are renewable, cost effective, and eco-friendly. The synthesis and utility of nitrogen containing heterocyclic, fluorinated bioactive small molecules and biomolecules interaction and functionalized nanomaterials for catalytic and biological applications are discussed.

14.1 INTRODUCTION

Highly skilled organic chemists are venturing deeper into the unknown! When reading this statement, retinal (1), an organic compound, converts visible light into nerve impulses. As this dissertation is physically lifted, sugar molecules in the muscles causes chemical reactions. During this process, gaps between the brain cells are being bridged by neurotransmitter amines such as serotonin (2). This allows nerve impulses to travel around the brain. Although these processes occur in the human body, they are not fully understood. To date, no one, however brilliant, understands the detailed mechanism that occurs in the human mind and body. As a result, there is an opportunity to venture into the unknown.[1]

1
11-*cis*-retinal
absorbs light when we see

2
serotonin
human neurotransmitter

FIGURE 14.1 Representative example of a light-absorbing and human neurotransmitter. (Adapted from Ref. [1])

Organic chemistry research is a dynamic field mainly concerned with developing and applying new methods for making molecules, understanding their reactivity, and enhancing knowledge in other research fields. The application of organic chemistry research and education programs into an integrated transdisciplinary program is important for good research. In 2012, two biologists, Professor Robert J. Lefkowitz and Professor Brian Kobika, were awarded the Nobel Prize for Chemistry. They unveiled the signaling mechanism of G protein-coupled receptors (GPCRs). These proteins belong to one of the largest families of human proteins. They are involved in many physiological activities and therefore, are targets of a number of drugs. Determination of the molecular structures of this class of receptors helps researchers to understand the actual mechanism of different cellular processes. Also lifesaving and more effective drugs can be designed.

Recently, nano- and organic chemists have captured the high-resolution images of a molecule as it breaks and reforms chemical bonds. Using a high-resolution atomic force microscope, these scientists have taken the first atom-by-atom picture, which shows how a molecule's structure changes during a reaction. In the past, scientists have obtained this type of information from spectroscopic analysis. This great achievement was only realized through interdisciplinary collaboration between organic chemists and nanotechnologists.[2]

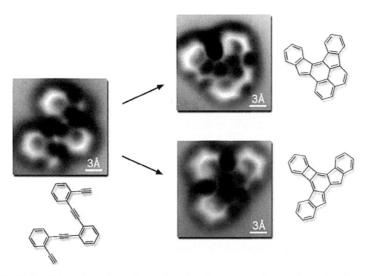

FIGURE 14.2 Direct imaging of covalent bond structure in single-molecule chemical reactions. (Adapted from Ref. [2])

During the 20th century, total synthesis of complex natural product was the main theme of organic chemistry.[3] Fundamental information into reactivity and selectivity principles were obtained by these synthetic projects. Nowadays, synthetic chemists can construct molecules which are complex. However, a change in the perception defining organic synthesis as an art is needed. In 1975, a key issue was addressed by Hendrickson. He defined an "ideal synthesis" as one which[4]:

"…Creates a complex molecule … In a sequence of only construction reactions involving no intermediary refunctionalizations, and leading directly to the target, not only its skeleton but also its correctly placed functionality."

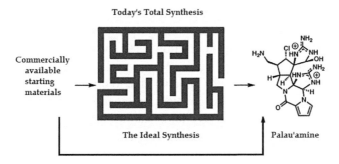

FIGURE 14.3 A schematic diagram showing a pathway of ideal synthesis. (Reproduced with permission from Ref. 72. © 2010 American Chemical Society.)

Aromatic heterocycles are highly important structural units in a vast number of biological active natural compounds, pharmaceuticals, and materials. They are also important intermediates in organic synthesis, often providing access to other highly desirable structures. Thus, there is a compelling need to develop novel and more general methods for the synthesis of heterocycles. Functionalized heterocycles are excellent scaffolds for preparing diversity-oriented compounds for medicinal and pharmaceutical applications.[5] This is due to their ability to mimic the structure of peptides and their protein binding potential. Among the heterocycles, the nitrogen-based ones such as quinoline and carbazole derivatives are one of the most indispensable structural motifs. These are widespread in nature and are key structural component in several families of bioactive compounds.[6] Several alkaloids isolated from natural resources or prepared synthetically are important for medicinal and biomedical use.

Fluorine, a small and highly electronegative atom, has a huge impact in medicinal chemistry. The presence of fluorine in a therapeutic or diagnostic small molecule enhances many pharmacokinetic and physicochemical properties. These include improved metabolic stability and enhanced membrane permeation. Increased binding affinity of fluorinated drug candidates enhances interaction with biomolecules and increases blood–brain barrier (BBB) permeability, which is important in CNS active drugs.[7] Therefore, fluorinated drugs are exploited for therapeutic applications.

Ionic liquids (ILs) are organic nitrogen-containing heterocyclic cations with inorganic anions or organic salts with low melting points. These are being used for many fields of chemistry and industry, due to their potential as "Green" recycle alternatives to the traditional organic solvents. They have a wide liquid range, in some cases, in excess of 400°C. Their very favorable properties, such as high polarity, negligible vapor pressure, high ionic conductivity, and thermal stability, make them effective in catalysis. Recently, hydrated ILs have been identified as an ideal medium for long-term DNA and protein storage. Hence, understanding the binding characteristics and molecular mechanism of interactions of ILs with DNA and protein is of paramount importance.

The power of chemistry lies in being able to create new forms of matter and/or nanostructures. Both the covalent and noncovalent bond are essential in fabricating nanostructures. By understanding the kinetics and thermodynamics of many fundamental chemical processes, nanosized molecules, and nanomaterials can be produced, thereby leading to discoveries of new phenomena. Figure 14.4 shows some synthetic nanosized molecules, size-dependent properties of quantum dots, and organic nanoparticles.

Nanomaterial have gained focus due to their potential applications in drug delivery, sensing, imaging, and chemotherapy; drug delivery includes polymeric nanoparticles, dendrimers, liposomes, and quantum dots.[8]

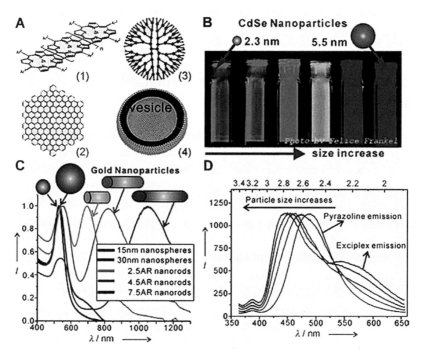

FIGURE 14.4 (A) Examples of nanosized molecules and assemblies: (1) fused porphyrins, (2) molecular graphene, (3) dendrimers, and (4) a self-assembled vesicle, (B) fluorescence emission of CdSe quantum dots with different sizes, (C) absorption spectra of gold nanoparticles with various sizes and shapes, and (D) fluorescence emission spectra of organic nanoparticles with different sizes.[9]

Metal nanoparticles are recently being widely explored. In particular, design of drug functional gold nanomaterial is of interest because of a variety of potential applications ranging from chemistry to biological sciences. Silver, gold, and palladium nanoparticles have unique optical, thermal, hetero- and homogeneous catalytic properties.[10] After the discovery of carbon nanotubes (CNT), it was suggested that carbon is not a unique element being able to form nanotubes; boron nitride (BN) appears as a potential material for this class in view of the structural similarity of graphite and bulk BN.[11] Hexagonal boron nitride (h-BN) is well-known as one important ceramic material with outstanding thermal and electrical properties. Furthermore, it has excellent chemical stability, good resistance to corrosion, low density, and high melting point. Boron nitride nanotubes are a good alternative to carbon nanotubes and other kinds of inorganic materials because of their improved chemical properties. They possess good heterogeneous catalytic activity and theoretically guarantee better stability and compatibility for safe drug delivery and targeting carriers.[12]

14.2 SYNTHESIS AND UTILITY OF NITROGEN HETEROCYCLE SYSTEM

Heterocyclic compounds are cyclic compounds in which one or more of the ring carbons are replaced by another atom. The noncarbon atoms in such rings are referred to as "heteroatoms." Heterocycles form the largest class of organic compounds. The majority of pharmaceutical products that mimic natural products with biological activity are heterocycles. Therefore, researchers are designing and producing better pharmaceuticals, pesticides, insecticides, rodenticides, and weedicides by following natural models. Other important practical applications of these compounds are use in additives and modifiers in a wide variety of industries including cosmetics, reprography, information storage, plastics, solvents, antioxidants, and vulcanization accelerators. Hence, heterocyclic chemistry is an inexhaustible resource of novel compounds. Many combinations of carbon, hydrogen, and heteroatoms can be designed, providing compounds with the most diverse physical, chemical, and biological properties.[13,14] Among the approximately 20 million chemical compounds identified by the end of the second millennium, more than two-thirds are fully or partially aromatic and approximately one-half are heteroaromatic. Heterocyclic compounds, especially nitrogen heterocycles, are the most important class of compounds in the pharmaceutical and agrochemical industries, in which heterocycles comprising around 60% are covered as a drug substances. 5-membered N-heterocycles such as pyrroles, indoles, and carbazoles are important structural motifs and are present in an extensive number of biologically active compounds.[15] The 5-membered N-heterocycles are of exceptional interest in the pharmaceutical industry, as they appear in the core structure of several drugs. Six membered heterocycles such as substituted pyridines exhibit a broad range of biological activity. They are used to modulate hypertension and angina pectoris, act as Ca^{2+} channel blockers, and are antidiabetic, heptaprotective, and show antitumor properties.[16] The fused quinoline moiety is also present in an extensive number of naturally occurring sources and are biologically active.[17] In addition, pyridine derivatives are also used as organic bases and organocatalysts in organic synthesis. Six-membered aromatic rings containing two nitrogen atoms, such as phthalazinones, quinazolinones, pyrimidines, and pyrimidinones possess a broad spectrum of biological activities and are therefore of interest as target compounds in pharmaceutical and medicinal chemistry.[18] Six-membered rings containing three nitrogen atoms, such as 1,3,5-triazines are used as a templates in supramolecular chemistry and dendrimer synthesis, due to their unlike C3 symmetric core structure.[19] Seven- and higher membered nitrogen containing compounds, for example, benzodiazepines, show interesting anticancer properties and inhibit HIV-1 reverse transcriptase.[20]

The following few sections cover the review of the nitrogen heterocyclic compounds as a core area of research. Several quinoline derivatives isolated from natural resources or prepared synthetically are significant with respect to medicinal chemistry and biomedical use. Compounds containing pyridine, quinoline, carbazole, and indole motif are most widely used as antimalarials,[21] antibacterials,[22] antifungals,[23] and anticancer[24] agents. Furthermore, quinoline derivatives are used in the synthesis of fungicides, virucides, biocides, rubber chemicals, and flavoring agents.[25]

Kimpe De et al.[26] reported the high-yielding, asymmetric synthesis of novel 4-formyl-1-(2- and 3-haloalkyl) azetidin-2-ones **3**. It was developed as a valuable starting material for the synthesis of different enantiomerically enriched bicycles azetidin-2-ones **4**. These include preparation of morpholine and 1,4-diazepane annulled β-lactam derivatives.

R : Ph

R$_1$: Secondary amine

n: 1, 2
X: Br, Cl

Li et al.[27] reported the iodine-mediated intramolecular cyclization of enamines **5**. This leads to various 3H-indole derivatives **6** containing multifunctional groups. Transition metal-free reaction conditions are used and yields are good.

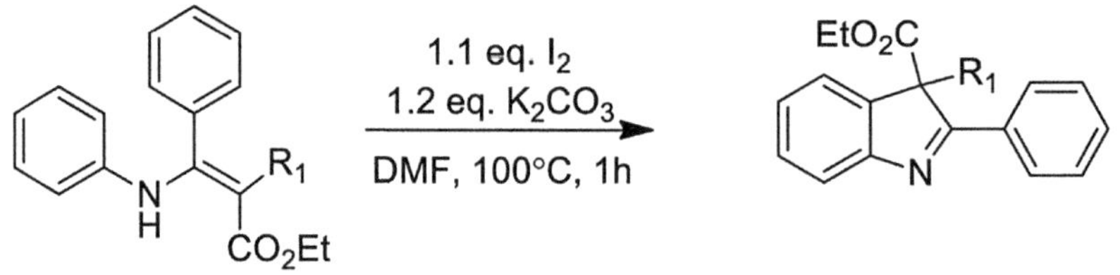

[Scheme showing compound 5 → 6 with conditions: 1.1 eq. I₂, 1.2 eq. K₂CO₃, DMF, 100°C, 1h]

5 → **6**

R₁ : alkyl, Bn. CO₂Et, p-NO₂-C₆H₄

Chauhan et al.[28] reported the synthesis of 2, 3-dihydro/spiroquinazolin-4(1H)-ones **10** via three-component cyclocondensation reaction of isatoic anhydride **7**, amines **8**, and isonicotinaldehyde **9**.

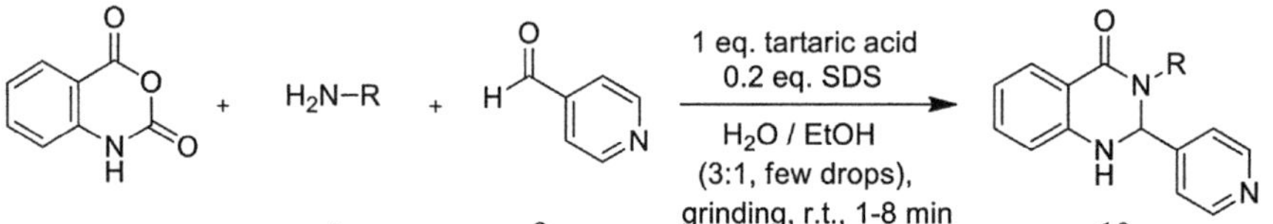

[Scheme: 7 + 8 + 9 → 10, conditions: 1 eq. tartaric acid, 0.2 eq. SDS, H₂O / EtOH (3:1, few drops), grinding, r.t., 1-8 min]

R: Alkyl, Ph. H

Perumal et al.[29] reported an environmentally friendly method for the cyclization of 2-aminochalcones **11** to 2-aryl-2,3-dihydroquinolin-4(1H)-ones **12** on the surface of silica gel impregnated with indium(III) chloride. They used microwave irradiation without a solvent.

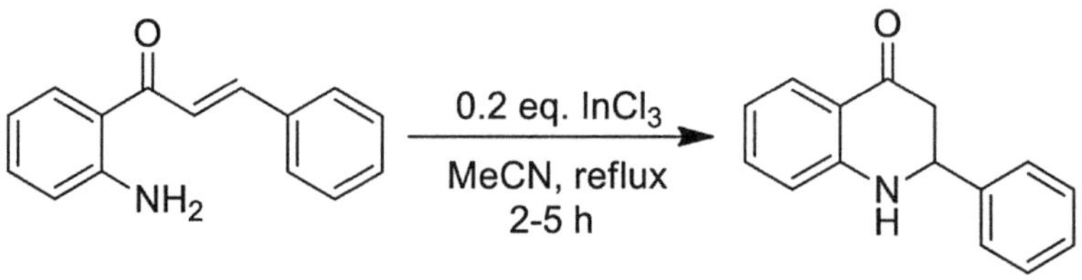

[Scheme: 11 → 12, conditions: 0.2 eq. InCl₃, MeCN, reflux, 2-5 h]

Chuchi et al.[30] reported an efficient procedure for the stereo controlled construction of 2H-thiopyrano[2, 3-b]quinoline **15** starting from simple compounds. The domino Michael/Aldol reactions between 2-mercapto benzaldehydes **13** and 3-phenylprop-2-enal **14**, promoted by chiral diphenylprolinol TMS ether, proceeded with excellent chemo- and enantioselectivity.

A synthetically useful and pharmaceutically valuable 2H-thiopyrano[2,3-b] quinolines in high yields with 90–99% was obtained.

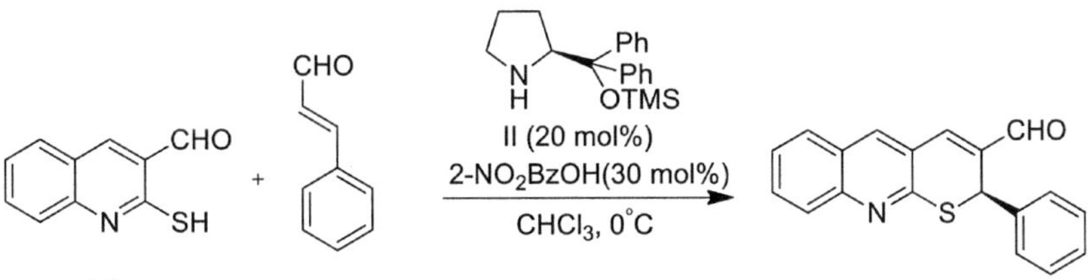

13 + **14** → **15**

Mane et al.[31] reported a one-pot water-mediated synthesis via 2-chloro 3-formyl quinoline **16** and phenyl hydrazine **17** for pyrazolo[3,4-b]quinolines **18**. They used microwave energy irradiation. This route is convenient, eco-friendly, and can be scaled-up easily.

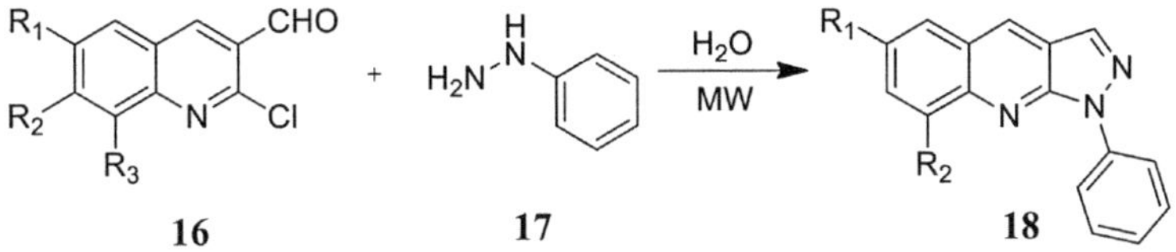

16 + **17** → **18**

R_1 = H, R_2 = H, R_3 = H, Me
R_1 = OMe, R_2 = H, R_3 = H, Me
R_1 = Me, R_2 = H, R_3 = H, Me
R_1 = H, R_2 = Et, R_3 = H, Me

Shika et al.[32] reported the synthesis of diversely substituted chalcones **20** derived from 2-chloro-3-formylquinoline **16**, with appropriately substituted 2-hydroxyl acetophenones **19** via Claisen condensation. All derivatives showed promising potent *Plasmodium falciparum* lactate dehydrogenase activity.

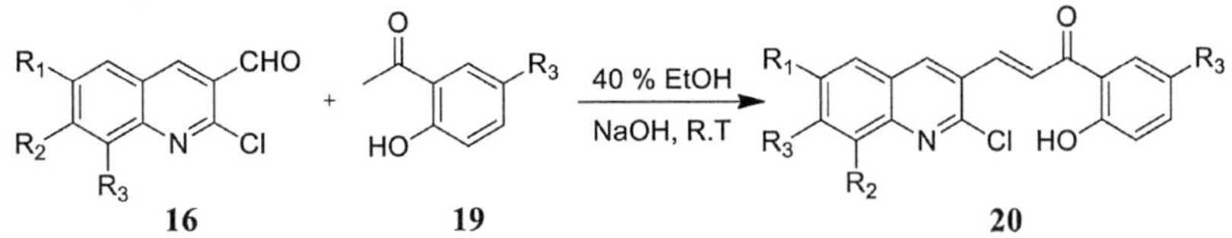

16 + **19** → **20**

R_1 = H, R_2 = H, R_3 = H, Me
R_1 = OMe, R_2 = H, R_3 = H, Me
R_1 = Me, R_2 = H, R_3 = H, Me
R_1 = H, R_2 = Et, R_3 = H, Me

14.2.1 MULTICOMPONENT REACTIONS

Among the known strategies of drug discovery, high throughput screening is one that serve most efficient to the pharmaceutical industry. To guarantee the success on acquiring lead compounds via this kind of screening tactic, sources of large amount of molecular libraries are important preconditions. With the advent of multiresistant strains of bacteria, viruses, and cancer cells, the search for new drugs is paramount.[33] Although the use of monoclonal antibodies as drugs is on the rise, the majority of new drugs are still and are likely to continue to be small molecules. A key step in the drug discovery process is the generation of novel chemical entities that can serve as potential drug candidates.[34–36] Successful drug development relies on high efficiency, low cost, and short cycles of design make test and therefore, requires short and efficient synthetic sequences for lead discovery. Multicomponent reactions (MCRs) are a one-pot process that involves the reaction of at least three components to form a single product that incorporates essentially all the atoms of the starting materials. These reactions are atom economic, step efficient, and have high exploratory power with regard to chemical space and are, therefore, ideally suited for the generation of libraries of compounds.[37]

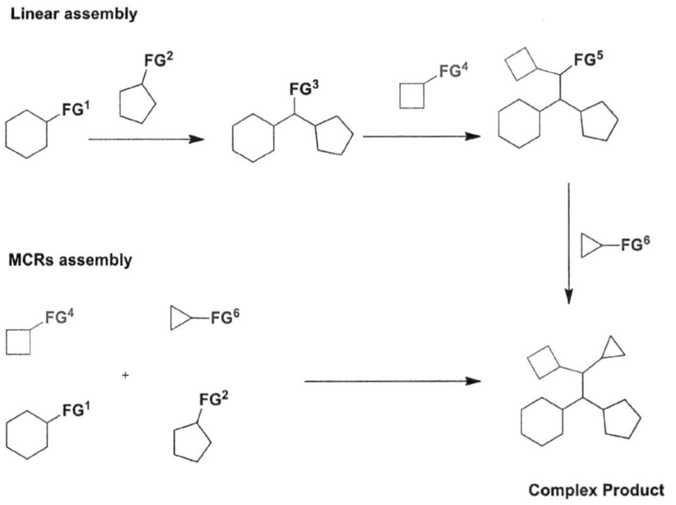

SCHEME 14.1 Traditional linear synthesis (3 steps) versus multicomponent assembly (1 step). (Adapted from Ref. [17].)

Zhen et al.[38] reported a novel functionalized quinolines via Ugi and Pd-catalyzed intramolecular arylation reactions. To prepare this interesting ring system, the authors made use of an Ugi-Heck MCRs aryl–aryl coupling which occurs in the second step.

The second most important isocyanide-based MCR is the Ugi four component reaction. This elegant four-component reaction between cyclcohexyl isocyanide **24**, 3-iodobenzo[b]thiophene-2-carboxylic acid **23**, benzaldehyde **21**, and toluidine **22** afford dipeptide-like structures **25**. Song et al.[39] prepared pseudopeptides bearing a difluoromethyl **30**. A Ugi reaction using aniline **27**, trimethylacetonitrile **29**, 4-fluorobenzaldehyde **26**, and 2, 2-difluoro-2-(phenylthio) acetic acid **28** as one component is used to prepare **31**.

Nonisocyanide-based MCRs usually involve an activated carbonyl species. Biginelli reaction discovered, in 1891, used tolualdehyde **32**, (thio) ureas **33**, and ethyl acetoacetate **34** to synthesize dihydro pyrimidinones **35**.

The products of the Biginelli reaction are widely used in the pharmaceutical industry as calcium channel blockers,[40] antihypertensive agents, and alpha-1-a-antagonists.

X = O, S

14.3 FLUORINE-SUBSTITUTED HETEROCYCLIC NITROGEN SYSTEMS

The incorporation of fluorine atom(s) within organic molecule enhances their biopotency, bioavailability, metabolic stability, and lipophilicity.[41] It has shown to modulate the stereo electronic parameters of organic molecules,[42,43] alters the electronic environment, but also influences the p value of neighboring Bronsted acid/base centers, polarity, and influences lipophilicity. Organofluorine interactions with protein residues are used to enhance protein–ligand binding affinity and selectivity[44] and hence, synthesis of bioactive compounds for the treatment of infectious diseases[45,46] are being undertaken.

R_1 = H, CH_3. Cl, OCH_3 R_2 = H, CH_3
R_3 = R_5 = R_6 = H, F R_4 = H, F, CF_3

Patel et al.[47] reported a one-pot three-component cyclocondensation of betaaryloxyquinoline-3-carbaldehydes **36**, malononitrile **37**, and beta-enaminones **38**. They used a catalytic amount of piperidine. In vitro antimicrobial activity and antituberculosis activity were evaluated against *Mycobacterium tuberculosis* H37Rv.

Microwave irradiation was used for the synthesis of 2-pentafluorophenyl-quinoline derivatives **43** by Zhang and coworkers.[48] This was a one-pot reaction of pentafluorobenzaldehyde **40**, aromatic anilines **41**, and alkynes **42** on the surface of montmorillonitle clay impregnated with a catalytic amount of CuBr under solvent-free conditions. Improved yields and enhanced reaction rates were recorded.

14.4 ORGANIC NITROGEN-CONTAINING HETEROCYLIC CATIONS AND INORGANIC FLUORINE ANIONS (ILs)

In recent years, interest in ILs has grown rapidly. Fluorine-containing heterocycles is a class of organofluoro molten salts. The synthesis and properties of quaternary alkyl substituted ammonium, imidazolium, triazolium, and pyridinium salts with fluorine-containing anions have unique applications in biological and catalysis field.[49]

Prajapati et al.[50] reported an eco-friendly method for the synthesis of phenyl-substituted quinolines **46** via Meyer–Schuster rearrangement of 2-aminoaryl ketones **44** and phenylacetylenes **45** in the presence of a catalytic amount of zinc trifluoromethane sulfonate in the IL 1-butyl-3-methylimidazolium hexafluorophosphate [bmim] PF_6.

Zhu et al.[51] reported the interactions of bovine serum albumin (BSA) with two alkyl imidazolium-based ILs, 1-butyl-3-methylimidazolium tetrafluoroborate ([bmim]BF4) and 1-butyl-3-methylimidazolium hexafluorophosphate [bmim] PF$_6$, in buffer solutions at pH ~7.0. These were analyzed by isothermal titration calorimetry (ITC) and circular dichroism (CD); spectra showed that the two ILs changed the secondary structure of BSA.

14.5 GOLD–SULFUR NANOPARTICLES

The covalent bond between gold and sulfur stabilizes nanostructures and transmits electronic interactions between gold- and sulfur-containing molecules.[52] These interactions are mediated through the sulfhydryl (SH) functional group in thiols (RSH). A wide variety of studies in molecular biology, inorganic chemistry, surface science, and materials science are undertaken. Potential applications include in site specific bioconjugate labeling and sensing,[53] drug delivery and medical therapy,[54] molecular recognition, and molecular electronics.[55] The interplay between experiment and theory has aided in developing the understanding of the gold–sulfur nanointerface.

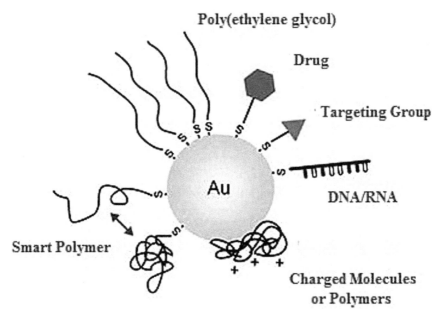

FIGURE 14.5 Gold nanoparticle.

14.5.1 BIOACTIVE THIOLATED MOLECULES-CONJUGATED GOLD NANOPARTICLES

Self-assembly of organothiols (OTs) and thiolated biomolecules is used extensively for gold nanoparticles (AuNPs) surface modification. This improves functionality, biocompatibilities, and target specificities.[56-60]

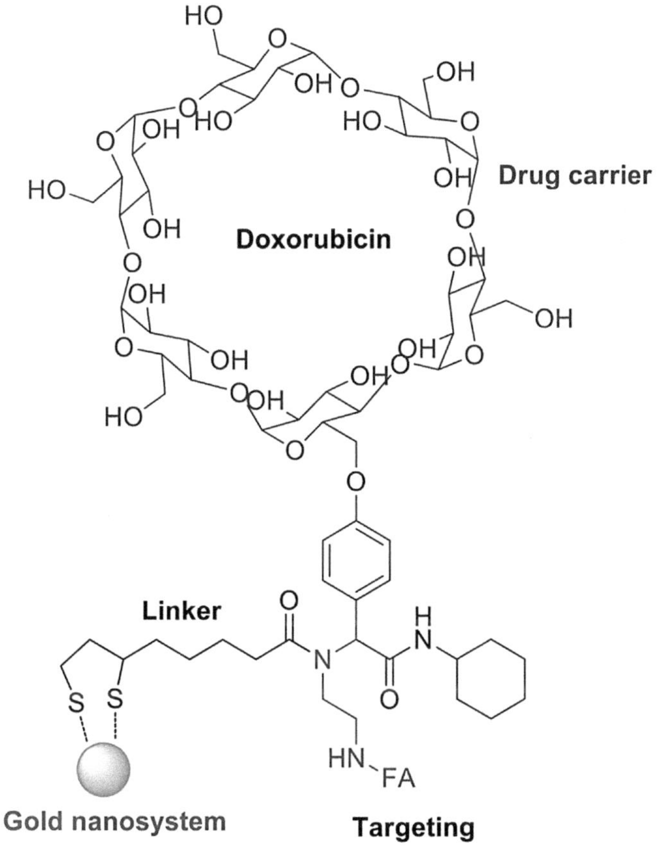

FIGURE 14.6 Bioactive thiolated molecule-conjugated gold nanoparticle.

Bing et al.[61] reported that multifunctionalized nanoparticles are important for nanobiomedical applications. However, synthesis is tedious and time-consuming. Hence, MCRs on nanostructures is a good way to prepare nanomaterials. Gold nano-OT system, illustrated in the scheme, was made. This shows enhanced cancer cell targeting and killing.

Singha et al.[62] reported 11-mercaptoundecanoic acid-modified gold nanoparticles (~7 nm). These were conjugated with chloroquine to assess their potential application in cancer therapeutics. The anticancer activity of chloroquine–gold nanoparticle conjugates (GNP–ChlQ) was demonstrated in MCF-7 breast cancer cells.

14.6 METAL-LOADED BORON NITRIDE

Elemental palladium, nickel, and rhodium have a high affinity for hydrogen and form hydrides in reaction with hydrogen. They are used in hydrogen storage technology and as catalysts for hydrogenation due to their high hydrogen solubility, diffusivity, and corrosion resistance.[63] Boron nitride (BN) is a catalyst support. It is a benign powder possessing a hexagonally shaped crystal structure composed of continuous boron–nitrogen bonds. The numerous lone pairs on the nitrogen atoms can coordinate with Pd metal and suppress catalyst activity for hydrogenation and coupling reactions.[64]

Hironao et al.[64] reported a hydrogenation catalyzed heterogeneous reaction by Pd on boron nitride (Pd/BN) in methanol. Hydrogenation of azides **47** and alkynes **49** in the presence of other reducible functionalities such as benzyl ethers, aryl halides, aryl ketones, and nitro groups were obtained. Also, the semihydrogenation of alkynes could be achieved without the reduction of other coexisting reducible functionalities.

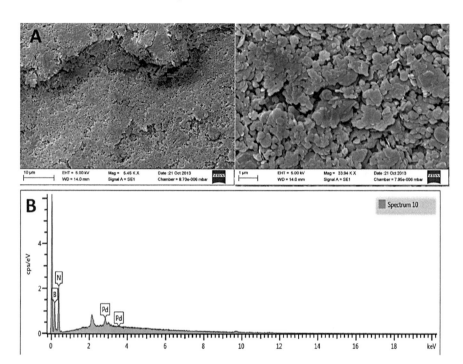

FIGURE 14.7 Palladium-loaded boron nitride.

14.7 BIOMOLECULES-FUNCTIONALIZED NANOPARTICLES

Living organisms have high potential for nanoparticles[65] production. Hence, green synthetic methods using biological extracts have gained importance in nanoparticle synthesis.[66] However, the mechanism of involvement of biomolecules is not well understood. Plants, algae, diatoms, heterotrophic human cell lines, and some biocompatible agents are reported to synthesize greener nanoparticles. Metals such as cobalt, copper, silver, gold, bimetallic alloys, silica, palladium, platinum, iridium, and magnetite are used. However, the use of phototrophic and heterotrophic eukaryotes and biocompatible agents for the synthesis of nanomaterials is yet to be fully explored.[67]

Fu et al.[68] reported the biosynthesis of silver nanoparticles using aqueous aloe leaf extract. Their results suggest that nanoparticles can be used as effective growth inhibitors against the test microorganisms. Greater bactericidal activity was observed for silver nanoparticles. The *Escherichia coli*, a Gram-negative bacterium, was reported as more efficient for gold and silver nanoparticles synthesis than *Staphylococcus aureus*, a Gram-positive bacterium.

Mata et al.[69] reported the bioreduction of Au(III) to Au(0) using biomass of the brown alga *Fucus vesiculosus*. Bioreduction with *F. vesiculosus* could be an alternative and environmentally friendly process. It can be used for recovering gold from dilute hydrometallurgical solutions and leachates of electronic scraps and for the synthesis of gold nanoparticles of different size and shape.

Venkataraman et al.[70] reported the synthesis of biomolecules functionalized gold nanoparticles using edible mushroom *Pleurotus florida* by photoirradiation method. The mixture containing aqueous gold ions and mushroom extract was kept in sunlight.

Sharma et al.[71] reported on metal nanoparticles having catalytic activities. However, they are difficult to recover after downstream processing of chemical reactions. Therefore, plant-based nanoparticles have enormous advantages.

Also, an increased awareness toward green chemistry has necessitated the development of simple, cost-effective, and eco-friendly procedures.

14.8 FUTURE PERSPECTIVES

This book chapter gives the brief introduction on the synthesis and utility of nitrogen containing heterocyclic, fluorinated bioactive small molecules, and IL biomolecules interaction and functionalized nanomaterials for catalytic and biological applications. Organic chemistry moves toward a multidisciplinary environment blending biology, physics, and materials sciences.

KEYWORDS

- graphene
- nanoparticles
- neurotransmitter
- ionic liquids

REFERENCES

1. Clayden, J.; Greeves, N.; Warren, S. Eds. *Organic Chemistry*; Oxford Press: UK, 2010.
2. De Oteyza, D. G.; Gorman, P.; Chen, Y. C.; Wickenburg, S.; Riss, A.; Mowbray, D. J.; Etkin, G.; Pedramrazi, Z.; Tsai, H. Z.; Rubio, A.; Crommie, M. F.; Fischer, F. R. *Science* **2013**, *340*, 1434.
3. (a) Corey, E. J.; Cheng, X. M., Eds. *The Logic of Chemical Synthesis*; John Wiley: New York, 1989. (b) *Classics in Total Synthesis: Targets, Strategies, Methods*; Nicolaou, K. C., Sorensen, E. J., Eds.; VCH: Weinheim: New York, 1996. (c) *Classics in Total Synthesis II: More Targets, Strategies, Methods*; Nicolaou, K. C., Snyder, S. A., Eds.; Wiley-VCH: Weinheim, 2003. (d) *Molecules That Changed the World: A Brief History of the Art and Science of Synthesis and Its Impact on Society*; Nicolaou, K. C., Montagnon, T., Eds.; Wiley: New York, 2008.
4. Hendrickson, J. B. *J. Am. Chem. Soc.* **1975**, *97*, 5784.
5. Dolle, R. E.; Nelson, K. H. Jr. *J. Comp. Chem.* **1999**, *1*, 235.
6. Nordell, P.; Lincoln, P. *J. Am. Chem. Soc.* **2005**, *127*, 9670.
7. Elliot, A. J.; Filler, R.; Kobayashi, Y., Eds. *Biomedicinal Aspects of Fluorine Chemistry*; Elsevier Biomedical Press: Amsterdam, 1982.
8. Tamara, M.; Lorna, R. R.; Vitaly P, Eds. *Advanced Drug Delivery Reviews*. 2013; Vol. *65*(13), pp 1665.
9. Kostarelos, K. *Nanomedicine* **2006**, *1*, 1.
10. Yuki, Y.; Yoshinari, S.; Tsuyoshi, Y.; Saori, N.; Yasunari, M.; Hironao, S. *Chem. Cat. Chem.* **2013**, *5*, 2360.
11. Sousa, A.; Sousa, E. M. B. *J. Non Cryst. Solids* **2006**, *352*, 3451.
12. Xiao, D.; Xi, L.; Yang, W.; Fu, H.; Shuai, Z.; Fang, Y.; Yao, J. *J. Am. Chem. Soc.* **2003**, *125*, 6740.
13. (a) Katritzky, A. R.; Ress, C. W.; Scriven, E. F. V., Eds.*Comprehensive Heterocyclic Chemistry II*; Pergamon: Oxford, U.K, 1996, 1. (b) Katritzky, A. R.; Ramsden, C. A.; Scriven, E. F. V.; Taylor, R. J. K. *Comprehensive Heterocyclic Chemistry III*; Eds.; Pergamon: Oxford, U.K, 2008, 1. (c) Martins, M. A. P.; Cunico, W.; Pereira, C. M. P.; Flores, A. F. C.; Bonacorso, H. G.; Zanatta, N. *Curr. Org. Synth.* **2004**, *1*, 39. (d) Drunhinin, S. V.; Balenkova, E. S.; Nenajdenko, V. G. *Tetrahedron* **2007**, *63*, 7753. (e) Martins, M. A. P.; Frizzo, C. P.; Moreira, D. N.; Zanatta, N.; Bonacorso, H. G. *Chem. Rev.* **2008**, *104*, 2015.
14. Balaban, A. T.; Oniciu, D. C.; Katrizky, A. R. *Chem. Rev.* **2004**, *104*, 2777.
15. (a) Torok, B.; Abid, M.; London, G.; Esquibel, G. K. S. *Angew. Chem. Int. Ed.* **2005**, *44*, 3086. (b) Estevez, V.; Villacampa, M.; Menendez, J. C. *Chem. Soc. Rev.* **2010**, *39*, 4402. (c) Sheikh, K. D.; Banerjee, P. P.; Jagadeesh, S.; Grindrod, S. C.; Zhang, C.; Paige, M.; Brown, M. C. *J. Med. Chem.* **2010**, *53*, 2376.
16. (a) Kumar, R.; Mittal, A.; Ramachandran, U. *Bioorg. Med. Chem. Lett.* **2007**, *17*, 4613. (b) Boecker, R. H.; Guengerich, F. P. *J. Med. Chem.* **1986**, *29*, 1596. (c) Humphries, P. S.; Almaden, J. V.; Barnum, S. T.; Carlson, T. J.; Do QQ, T.; Fraser, J. D.; Hess, M.; Kim, Y. H.; Ogilvie, K. M.; Sun, S. *Bioorg. Med. Chem. Lett.* **2006**, *16*, 6116. (d) Son, J. K.; Zhao, L. X.; Thapa, P.; Karki, R.; Na, Y.; Jahng, Y.; Jeong, T. C.; Jeong, B. S.; Lee, C. S.; Lee, E. S. *Eur. J. Med. Chem.* **2008**, *43*, 675.
17. (a) Solmon, V. R.; Haq, W.; Srivastava, K.; Puri, S. K.; Katti, S. B. *J. Med. Chem.* **2007**, *50*, 394. (b) Chauhan, P. M. S.; Srivastava, S. K. *Curr. Med. Chem.* **2001**, *8*, 1535. (c)

Michael, J. P. *Nat. Pro. Rep.* **2004**, *21*, 650. (d) Michael, J. P. *Nat. Pro. Rep.* **2007**, *24*, 223. (e) Kim, J. I.; Shim, I. S.; Lee, J. K. *J. Am. Chem. Soc.* **2005**, *127*, 1614.
18. (a) Achterrathtuckermann, U.; Weischer, C. H.; Szelenyi, I. *Pharmacology* **1988**, *36*, 265. (b) Druker, B. J.; Tamura, S.; Buchdunger, E.; Ohno, S.; Segal, G. M.; Fanning, S.; Zimmermann, J.; Lydon, N. B. *Nat. Med.* **1996**, *2*, 561. (c) Kappe, C. O. *Eur. J. Med. Chem.* **2000**, *35*, 1043. (d) Lee, Y.; Shacter, E. *J. Biol. Chem.* **1999**, *274*, 19792. (e) McNeely, W.; Wisemann, L. R. *Drugs* **1998**, *56*, 91. (f) Normann, M. H.; Rigdon, G. L.; Navas, F.; Cooper, B. R. *J. Med. Chem.* **1994**, *37*, 2552. (g) Ralevic, V.; Burnstock, S. G. *Pharmacol. Rev.* **1998**, *50*, 413. (h) Studer, A.; Hadida, S.; Ferrito, R.; Kim, S. Y.; Jegger, P.; Wipf, P.; Curran, D. P. *Science* **1997**, *275*, 823. (i) Wolfe, J. F.; Rathman, T. L.; Sleevi, M. C.; Campbell, J. A.; Greenwood, T. D. *J. Med. Chem.* **1990**, *33*, 161. (j) Yamaguchi, M.; Kamei, K.; Koga, T.; Akima, M.; Kuroki, T.; Ohi, N. *J. Med. Chem.* **1993**, *36*, 4052.
19. (a) McCallien, D. W.; Sanders, J. K. M. *J. Am. Chem. Soc.* **1995**, *117*, 6611. (b) Chouai, A.; Simanek, E. E. *J. Org. Chem.* **2008**, *73*, 2357.
20. (a) Wang, J.; Shen, Y.; Hu, W.; Hsieh, M.; Lin, F.; Hsu, M. K.; Hsu, M. H. *J. Med. Chem.* **2006**. 49, 1442. (b) DeCorte, B. L. *J. Med. Chem.* **2005**, *48*, 1689.
21. Roma, G.; Braccia, M. D.; Grossi, G.; Mattioli, F.; Ghia, H. *Eur. J. Med. Chem.* **2000**, *35*, 1021.
22. Benkovic, S. J.; Baker, S. J.; Alley, M. R. K.; Youn, H. W.; Yong, K. Z.; Tsutomu, A.; Weimin, M.; Justin, B.; Ravi, R. P. T.; Mark, W.; Lyn Sue, K.; Ali, T.; Lucy, S. *J. Med. Chem.* **2005**, *48*, 7468.
23. Vargas, M. L. Y.; Castelli, M. V.; Kouznetsov, V. V.; Urbina, G. J. M.; Lopez, S. N.; Sortino, M.; Enriz, R. D.; Ribas, J. C.; Zacchino, S. *Bioorg. Med. Chem.* **2003**, *11*, 1531.
24. Bailly, C. *Biochemistry* **1999**, *38*, 7719.
25. Jones, G.; Kartrizky, A. R.; Rees, C. W.; Scriven, E. F., Eds. *Comprehensive Heterocyclic Chemistry II*; Pergamon: Oxford, **1996**, *5*, 167.
26. Van Brabandt, W.; Vanwalleghem, M.; Dhooghe, M.; De Kimpe, N. *J. Org. Chem.* **2006**, *71*, 7083.
27. He, Z.; Li, H.; Li, Z. *J. Org. Chem.* **2010**, *75*, 4296.
28. Sharma, R.; Pandey, A. K.; Chauhan, P. M. S. *Synlett* **2012**, *23*, 2209.
29. Kumar, K. H.; Muralidharan, D.; Perumal, P. T. *Synthesis* **2004**, 63.
30. Lulu, W.; Youming, W.; Haibin, S.; Liangfu T.; Zhenghong, Z.; Chuchi, T. *Asian J. Chem.* **2013**, *8*, 2204.
31. Mali, J. R.; Pratap, U. R.; Jawale, D. V.; Mane, R. A. *Tetrahedron Lett.* **2010**, *51*, 3980.
32. Shika, S. D.; Ajay, M. G.; Anjali, M. R.; Mukund, S. C.; Chavoun, P. M. S.; Kumkum, S. *Ind. J. Chem.* **2009**, *48*, 1780.
33. Coates, A.; Hu, Y.; Bax, H. R.; Page, C. *Nat. Rev. Drug Discov.* **2002**, *1*, 895.
34. Kirkpatrick, P. *Nat. Rev. Drug Discov.* **2002**, *1*, 97.
35. Szakacs, G.; Paterson, J. K.; Ludwig, J. A.; Booth-Genthe, C.; Gottesman, G. M. *Nat. Rev. Drug Discov.* **2006**, *5*, 219.
36. Keser, G. M.; Makara, G. M. *Drug Discov. Today* **2006**, *11*, 741.
37. Orru, R. V. A.; De Greef, M. *Synthesis* **2003**, *10*, 1471.
38. Zhibo, M.; Zheng, X.; Tuoping, L.; Kui, L.; Zhibin, X.; Jiahua, C.; Zhen, Y. *J. Comb. Chem.* **2006**, *8*, 696.
39. Jingjing, W.; Hui, L.; Song, C. *Beilstein. J. Org. Chem.* **2011**, *7*, 1070.
40. Rovnyak, G. C.; Atwal, K. S.; Hedberg, A.; Kimball, S. D.; Moreland, S.; Gougoutas, J. Z.; O'Reilly, B. C.; Schwartz, J.; Malley, M. F. *J. Med. Chem.* **1992**, *35*, 3254.

41. Saeed, A.; Shaheen, U.; Hameed, A.; Kazmi, F. *J. Fluorine Chem*. **2010**, *131*, 333.
42. Smart, B. E. *J. Fluorine Chem*. **2001**, *109*, 1, 3.
43. Bonacorso, H. G.; Wentz, A. P.; Lourega, R. V. *J. Fluorine Chem*. **2006**, *127*, 8, 1066.
44. Olsen, J. A.; Banner, D. W.; Seiler, P. *Chem Bio Chem*. **2004**, *5*(5), 666.
45. Abdel R, R. M.; Al Footy, K. O.; Aqlan, F. M. *Int. J. ChemTech. Res*. **2011**, *3*(1), 423.
46. Abdel, R.; Makki, R. M.; Bawazir, M. S. I. T. *J. Chem*. **2011**, *8*(1), 405.
47. Harshad, G. K.; Manish, P. P. *E. J. Med. Chem*. **2013**, *63*, 675.
48. Zhang, J. M.; Yang, W.; Song, L. P.; Cai, X.; Zhu, S. Z. *Tet. Lett*. **2004**, *45*, 5771.
49. Dongbin, Z.; Min, W.; Yuan, K.; Enze, M. *Catal. Today* **2002**, *74*, 157.
50. Sarma, R.; Prajapati, D. *Synlett* **2008**, *56*, 3001.
51. Lan, Y. Z.; Guang, Q. L.; Fu, Y. Z. *J. Biophy. Chem*. **2011**, *2*, 146.
52. Dubois, L. H.; Nuzzo, R. G. *Annu. Rev. Phys. Chem*. **1992**, *43*, 437.
53. Ackerson, C. J.; Powell, R. D.; Hainfeld, J. F., Eds. *Cryo-EM Part A: Sample Preparation and Data Collection. Methods in Enzymology*; Elsevier, 2010; pp 481, 2–410.
54. Bowman, M. C. *J. Am. Chem. Soc.* **2008**, *103*, 6896.
55. Demers, L. M. *Science* **2002**, *296*, 1836.
56. Kang, B.; Mackey, M. A.; El Sayed, M. A. *J. Am. Chem. Soc.* **2010**, *132*, 1517.
57. Daniel, M. C.; Astruc, D. *Chem. Rev.* **2004**, *104*, 293.
58. Huang, X.; El Sayed, I. H.; Qian, W.; El Sayed, M. A. *J. Am. Chem. Soc.* **2006**, *128*, 2115.
59. Kim, B.; Han, G.; Toley, B. J.; Kim, C. K.; Rotello, V. M.; Forbes, N. S. *Nat. Nano.* **2010**, *5*, 465.
60. Gong. *J. Chem. Rev.* **2012**, *112*, 2987.
61. Hongyu, Z.; Gaoxing, S.; Peifu, J.; Bing, Y. *Asian. J. Chem*. **2012**, *18*(18), 5501.
62. Joshi, P.; Chakraborti, S.; Ramirez Vick, J. E.; Ansari, Z. A.; Shanker, V.; Pinak, Chakrabarti, P.; Singh, S. P. *Colloids Surf. B Biointerfaces* **2012**, *95*, 195.
63. Zhang, L. P.; Wu, P.; Sullivan, M. B. *J. Phys. Chem*. **2011**, *115*, 4289.
64. Yuki, Y.; Tsuyoshi, Y.; Saori, N.; Yoshinari, S.; Yasunari, M.; Hironao, S. *Adv. Synth. Catal*. **2012**, *354*, 1264.
65. Sastry, M.; Ahmad, A.; Khan, M. I.; Kumar, R. *Curr. Sci.* **2003**, *85*, 170.
66. Noruzi, M.; Zare, D.; Davoodi, D. *Spectrochim. Acta A* **2012**, *94*, 84.
67. Kannan, B. N.; Natarajan, S. *Adv. Colloid. Inter. Sci.* **2011**, *169*, 59.
68. Zhang, Z.; Cheng, X.; Zhang, Y.; Xue, X.; Fu, Y. *Colloids Surf. B Biointerfaces* **2013**, *423*, 63.
69. Mata, Y. N.; Torres, E.; Blazquez, M. L.; Ballester, A.; Gonzalez, F.; Munoz, J. A. *J. Hazard. Mater*. **2009**, *166*, 612.
70. Bhat, R.; Sharanabasava, V. G.; Deshpande, R.; Shetti, U.; Sanjeev, G.; Venkataraman, A. *J. Photochem. Photobiol. Biol. B* **2013**, *125*, 63.
71. Sharma, N. C.; Sahi, S. V.; Nath, S.; Parsons, J. G.; Gardea, T. J. L.; Pal, T. *Environ. Sci. Technol.* **2007**, *41*, 5137.
72. Gaich, T.; Baran, P.S. Aiming for the ideal synthesis. J. Org. Chem., 2010, 75 (14), pp 4657–4673.

CHAPTER 15

REMOVAL OF ORGANIC CANCER CARCINOGENS FROM WASTEWATER USING GREEN SYNTHESIS NANOPARTICLES

K. ANAND[1*], K. G. MOODLEY[2], and A. A. CHUTURGOON[1*]

[1]*Discipline of Medical Biochemistry and Chemical Pathology, School of Laboratory Medicine and Medical Sciences, College of Health Sciences, University of KwaZulu-Natal, Durban, South Africa*

[2]*Faculty of Applied Science, Department of Chemistry, Durban University of Technology, Durban, South Africa*

*Corresponding authors. E-mail: organicanand@gmail.com; chutur@ukzn.ac.za

ABSTRACT

Due to limited water for human consumption in many countries, recycling of used water is very often a necessity. The limitation in the amount of water available is due to many factors such as limited rainfall, insufficient storage dams coupled with very short rainy periods. Recycling or reuse of water is fraught with many obstacles such as costs of collection, storage, and purification. Cleanup of water is not as simple as removal of solids by filtration and addition of disinfectants to destroy germs. The reason for this is that this type of treatment does not deal with all types of pollutants in water. Pollutants enter water from activities, mainly, in the following sectors: industrial, agricultural, and domestic. Water is polluted by insecticides, pesticides, fungicides, heavy metals, dissolved organic compounds, bacteria and viruses, to name just a few.

Many industries use water as the lifeblood of their operations. Textile manufacturers are one class of industries which use large volumes of water.

Some fabric makers use dyes as part of their industrial processing. The effluent water from these dyeing plants may contain only small amounts of dyes but they impart a color to the water which is aesthetically unacceptable. However, the more pressing reason for removing dyes from water is that some of them cause illnesses such as cancer. Dyes can be removed from water by biological, chemical, and biochemical methods. Among chemical methods, the technique using adsorption has been investigated quite extensively. The high cost of production and treatment with activated carbon as adsorbent has led to the use of nanoparticles as adsorbents. A very large number of nanoparticles have been synthesized and used for a variety of purposes. This paper will describe the methods of preparation of nanoparticles (by green chemistry principles which keeps pollution of by-products to a minimum) and their applications to removal of dyes and heavy metals from wastewater. An important aspect of the synthesis of nanoparticles would be the use of a biomass resource, destined for the waste dump, to generate nanoparticles from various metal salts. This will be linked to beneficiation of waste, in general, and the use of green chemistry, in particular, to decrease environmental pollution including pollution of water.

15.1 INTRODUCTION

"The next world war will be fought with water as the point of dispute" or similar statements feature quite regularly during national and international conferences on water and water-related themes. These statements and similar observations in media articles are based on the frightening prospect that demand for water will outstrip supply in many countries. There are several reasons why potable water is very scarce even in countries with adequate rainfall. Increases in urbanization is due to population increase and migrations from rural areas, poor management of water utilities, lack of good control of water usage by domestic and industrial users, and most importantly, pollution of grounds and rivers by effluents from all categories of users. Even a short list of the pollutants which are found in domestic and industrial wastewater will serve to highlight the enormity of the task and attendant costs in wastewater cleanup for recycling purposes: insecticides, pesticides, fungicides, and waste from animals (from agricultural activities); heavy metals such as Pb, Cr, As, and Hg (mostly from mining and metallurgical plants); dyes and similar chemicals (mostly from textile industry) and bacteria/viruses. It should therefore be understood that it would be difficult to find a single technique or a very small combination of techniques to remove all pollutants from raw or wastewater. Furthermore, the eco-friendliness of the selected methods needs to be factored into the considerations of the various methods.

15.1.1 KINDS OF DYES, PROPERTIES, AND ENTRY INTO WATER BODIES

There are two general or layman's classifications of dyes, namely, natural dyes and synthetic dyes. Natural dyes have been used from ancient times. A variety of sources of natural dyes were available. A ready source of natural dyes is the myriad of colored plants and flowers. Synthetic dyes were invented in the late 1856 when W. H. Perkins accidently made the first synthetic dye, mauveine. Dyes are used to increase the aesthetic appeal of articles/items used by humans. Since dyes are generally water soluble, the dyeing process is invariably carried out in water, as the medium. The wastewater containing unused dye (effluent) is disposed into municipal wastewater systems, rivers, and holding ponds of industries and onto undeveloped land. The main reason for this wanton discarding of effluents is that it is the cheaper option. Since the effluents contain relatively low concentrations of dyestuff, it is not cost-effective (for the users) to recover the unused dyes. While dyes are needed to make the dyed articles appealing to human eyes, the same dyes in wastewater are an undesirable nuisance.

15.1.2 DYES CAN ALSO BE CLASSIFIED IN OTHER WAYS

15.1.2.1 ON THE BASIS OF ACID–BASE PROPERTIES

Acidic dyes are also known as anionic dyes. They commonly used for dyeing of silks, wool, and nylon. Basic dyes are cationic dyes which find widespread use in dyeing of acrylic materials and in adding color to paper. However, they need an acid such as acetic acid that needs to be added to the dyeing bath.

15.1.2.2 ON THE BASIS OF TECHNIQUE USED IN THE DYEING PROCESS

Direct dyeing is done in neutral or basic solution. A salt such as sodium chloride or sodium sulfate needs to be added. In this way, metals are added to the dyeing mixture and become part of the effluent at end of the process. Indirect dyeing which requires a mordant (usually a powder of a heavy metal compound).

15.1.2.3 ON THE BASIS OF SPECIFIC PROPERTIES

1. Azo dyes and nitro aromatic dyes compounds,
2. Disperse dyes are dyes which need to be ground with a dispersing agent,
3. Kinds of dye structures, properties, and health issues.

Dyes in the above figure (Fig. 15.1)[1] are known as azo dyes on account of having –N=N-grouping known as an azo group. This particular dye is known as disperse blue 183. Although azo dyes are a very widely used group of dyes, one of their negative attributes is that they are classified as carcinogens.

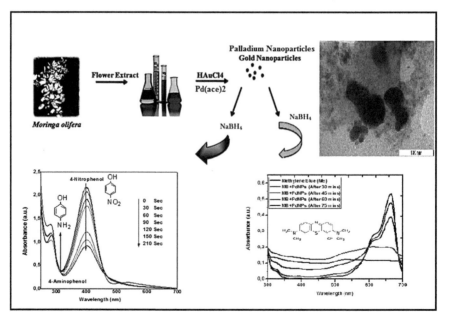

FIGURE 15.1 Green synthesis of metal nanoparticles and their catalytic studies of methylene blue and p-nitrophenol. (Adapted from Ref. [17]).

Dyes give wastewater a color which is unacceptable for domestic use; as the dyes may have impact on health of humans. Two main options used to deal with the undesirable presence of dyes in effluent water are:

- To remove the dye from the water
- Decolorize or degrade the dye

15.1.3 COMMON METHODS FOR REMOVING DYES

Simple materials that have been used to remove dyes from wastewater are, inter alia: rice husks, moringa seed pods, and chicken feathers, whereas the use of synthetic ion exchanges, membranes, and various types of adsorbents need costlier setups. A very brief survey of the literature is given below to indicate the diversity of methods used to remove dyes from water. Activated carbon has been used over a long period and is still being used as an adsorbent because of its low cost and ease of use to remove dyes from wastewater. Recently, Nair and Vinu have reported[3] the use of pyrolysis char as a suitable substitute. As an alternative to removing the dye by adsorption, a degradation method[4] using microbes been reported. Another method of degrading the residual dye has used a strain of bacteria.[5] Of the several applications of nanomaterials in dye removal, the use of cyclodextrin as a carrier of magnetic nanoparticles[6] is a good example of combining properties to solve a problem (Fig. 15.2).

In the study undertaken by the authors of this paper, nanotechnology was developed for the biocatalysis of the degradation of a commonly used dye, namely, methylene blue and p-nitrophenol. Nanocatalysis is a recent important technology leading to the total mineralization of most organic effluents.[7] The utilization of silver, platinum, gold, and palladium metal nanoparticles have numerous applications in heterogeneous and homogeneous catalysis.[8] The development of one-step green synthesis method for monodispersed palladium nanocrystals using cheap and nontoxic chemicals, environmentally benign solvents, and renewability is an important green strategy. Organic reactions in water are environmentally friendly; the Suzuki coupling reaction has recently been used to synthesize compounds for industry. It is used in the synthesis of intermediates for pharmaceuticals or fine chemicals.[9–12]

In quest of innovative eco-friendly healthcare, bionanotechnology is currently an active area of research. It focuses on the rapid biosynthesis of benign nanoparticles using antidiabetic medicinal plants containing important active compounds that serve as a natural reducing agents as well as possessing medicinal properties of health significance.[13] This has led to many researchers developing newer and more efficient green methods for synthesis of biocompatible and inert nanoparticles.[14] Chemical, physical, and microbial methods of synthesizing gold (Au), silver (Ag), palladium (Pd), platinum (Pt), indium oxide (In_2O_3), magnetite (Fe_3O_4), and zinc oxide (ZnO) nanoparticles currently exist.[15] However, these methods are costly and time-consuming, and difficult.[15,16] In light of this, nanomedicine is making use of phytotherapy and nanoforms of noble metals such as gold, silver, and palladium.

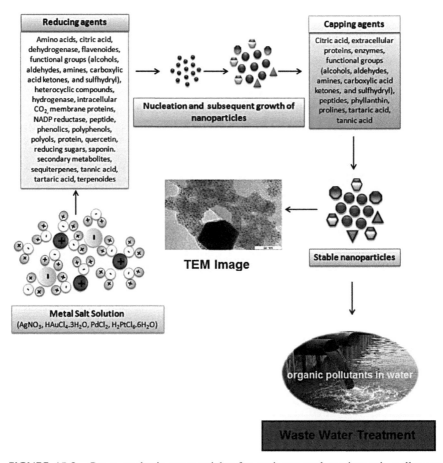

FIGURE 15.2 Green synthesis nanoparticles for environmental carcinogenic pollutants removing treatment.

15.2 SYNTHESIS AND CHARACTERIZATION OF NANO MATERIALS

15.2.1 MATERIALS AND METHODS

Monoclinic crystalline palladium acetate [Pd(OAc)$_2$] and gold (III) chloride trihydrate (HAuCl$_4$•3H$_2$O, ACS reagent) was purchased from Sigma-Aldrich (South Africa). All aqueous solutions were prepared using distilled water. *Moringa oleifera* flower was collected from Phoenix agricultural farm, Durban, South Africa, and identified at the KwaZulu-Natal Herbarium.

15.2.2 PREPARATION AND CHARACTERIZATION OF NANOSTRUCTURED BIOPALLADIUM AND GOLD

Fresh flowers were collected and washed several times with distilled water; 10 g of the flower were finely cut and 100 mL of distilled water was added. The flower with the water was heated to boil for 20 min, allowed to cool, filtered through a Whatman No.1 filter paper, and 1 mL of the supernatant leaf extract was added to 49 mL of 1 mM palladium acetate and gold (III) chloride aqueous solution at 60°C and the solution was stirred. The color of the solution gradually turned from dark brown to light yellow for palladium, yellow to purple color for gold within 6 h, at pH ~7. The palladium nanoparticles (PdNPs) and gold nanoparticles (AuNPs) were characterized using to obtain the particle size and shape, 1 μL of the PdNPs and AuNPs was placed on formvar-coated grids, air dried, and viewed at 100 kV to conduct transmission electron microscopy (TEM) (JEOL 1010 TEM using a Mega view III camera and iTEM software) studies.

15.2.3 CATALYTIC REDUCTION OF METHYLENE BLUE

An aqueous ice-cold solution of NaBH$_4$ (0.5 mL of 0.1 M) was mixed with aqueous stock solution of methylene blue (MB) (2 mL, 10^{-5} M) in a standard quartz cuvette and PdNPs (25 μL) were added. The final volume was adjusted with Millipore water to get the final volume as 3 mL. The mixture was shaken well and the reaction was monitored with the help of a UV–Visible spectrophotometer in the scanning range of 300–700 nm at room temperature.

15.2.4 CATALYTIC REDUCTION OF 4-NITROPHENOL AND 4-NITROANILINE

Ice-cold aqueous solution of $NaBH_4$ (1 mL, 15 mM) was mixed with 4-nitrophenol (1.7 mL, 0.2 mM) in a standard quartz cuvette. The light yellow color of the 4-nitrophenol changed yellowish green due to the formation of 4-nitrophenolate ion. An aliquot of AuNPs prepared at room temperature (0.3 mL) was added to the resulting solution, and the time-dependent absorbance spectra peak were recorded with a time interval of 0.5 min in the scanning range of 200–700 nm at room temperature. The same procedure was followed for the catalytic study using 4-nitroaniline.

15.3 NANOPARTICLE COMPOSITION AND SIZE DISTRIBUTION

The optical signature of PdNPs was determined; the distribution of sizes and shapes were conducted by TEM images (Fig. 15.3). Representative TEM images recorded from the PdNPs colloidal solution showed that most of the particles are spherical or near spherically shaped and exhibited a distribution of sizes in the range 3–5 nm and some very large images of PdNPs with linked monometallic shapes were also observed. The pictogram of PdNPs shown particle size of 5 nm is abundant.

TEM images recorded from the AuNPs colloidal solution (Fig. 15.3A) showed that most of the particles are spherical or near spherically shaped. The particles were polydispersed in the colloidal solution and exhibited a distribution of sizes in the range 3–5 nm and few images of AuNPs with projected shapes of a hexagon and a triangle. Under careful examination, it was evident that a thin layer of materials surrounded some of the particles which suggested the presence of organic-based capping agents inherent in the aqueous extract. Figure 15.3B shows the pictogram of AuNPs; particle size of 3 nm is abundant.

Figure 15.4 shows the scanning electron microscopy (SEM) with energy-dispersive X-ray spectroscopy (EDX) pattern; the crystalline nature of the AuNps was confirmed and the nanosize structures of AuNPs by SEM spectroscopy. The particle dimensions of nanomaterial were observed at 200 and 100 nm range. AuNPs showed spherical morphology of particle with size 100 nm. EDS pattern for *M. oleifera* shows signals for AuNPs at five different places.

Removal of Organic Cancer Carcinogens from Wastewater 349

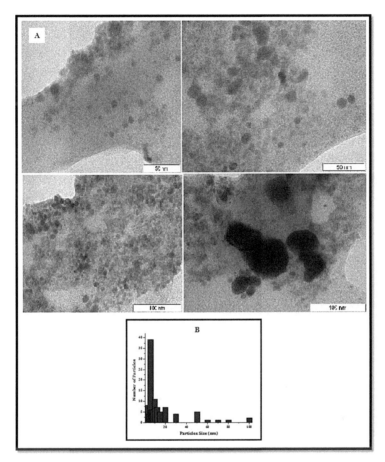

FIGURE 15.3 (A) Representative TEM micrograph of PdNPs biosynthesized by aqueous flower extracts of *Moringa oleifera* and (B) histogram representation of size distribution of PdNPs.1

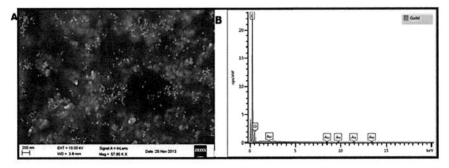

FIGURE 15.4 SEM with EDX AuNPs synthesized using *Moringa oleifera* flower extract.

15.3.1 CATALYTIC DEGRADATION OF METHYLENE BLUE

The catalytic effectiveness of the flower extract-mediated synthesized PdNPs were tested for the reduction of MB using NaBH$_4$ as a reducing agent. The reduction was carried out at room temperature and the catalytic degradation process was monitored by using a UV–Visible spectrophotometer. The characteristic absorption peak occurs at 665 nm for MB in the UV–Visible spectrum (Fig. 15.5). The degradation of dyes is indicated by the decolorization of the solution. MB, initially blue in color in an oxidizing environment, became colorless in the presence of reducing agent (NaBH$_4$) indicating the reduction of MB to Leucomethylene blue (LMB).[17]

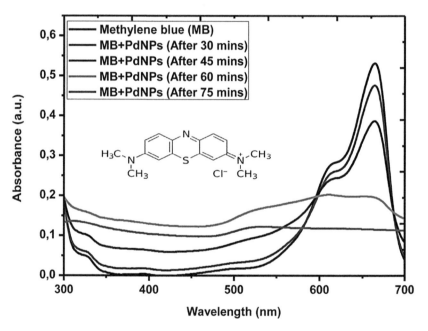

FIGURE 15.5 Catalytic study of methylene blue.

The catalytic reduction of 4-nitrophenol and 4-nitroaniline is perhaps the most often used reaction to test the catalytic activity of AuNPs in aqueous solution. Hence, we selected two chemical reactions that qualify as model reactions, that is, the reduction of 4-nitrophenol and 4-nitroaniline by sodium borohydride.[17, 18] This reaction was easily monitored by UV–Visible spectroscopy (Figs. 15.6 and 15.7), of NaBO$_2$; sodium metaborate has low toxicity when ingested, inhaled, or in contact with the skin.[19]

Removal of Organic Cancer Carcinogens from Wastewater

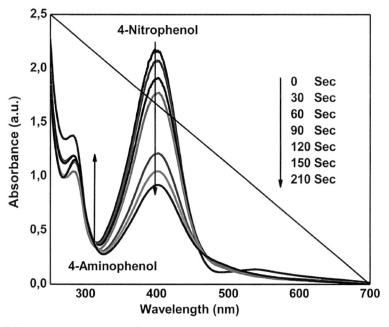

FIGURE 15.6 Catalytic study of p-nitrophenol.

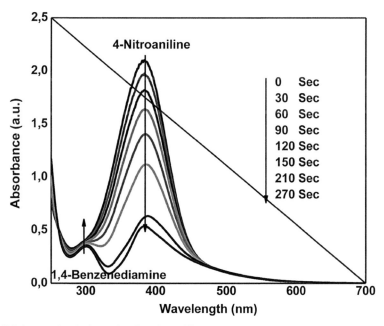

FIGURE 15.7 Catalytic study of p-nitroaniline.

15.4 CONCLUSION

A facile method which is simple, quick, cost-effective, and reproducible was developed to green synthesize PdNPs and AuNPs in vitro using the aqueous flower extract of *Moringa oleifera*. The PdNPs and AuNPs were characterized by TEM. UV–Visible spectroscopy proved to be useful for confirming catalytic reduction reactions. The PdNPs and AuNPs displayed catalytic activity in reducing organic cancer carcinogens MB and aromatic nitro compounds rapidly. The excellent catalytic reduction of azo dyes and nitro aromatics performances make them a promising toxic organic material to remove from synthetic wastewater.

KEYWORDS

- **wastewater**
- **pollutants**
- **dyes**
- **heavy metals**
- **removal**
- **nanoparticles**
- **beneficiation**

REFERENCES

1. Rajabi, A. A.; Yamini, Y., Faragi, M.; Noormohammmadian, F. Modified Magnetite Nanoparticles with Cetyltrimethylammonium Bromide as Superior Adsorbent for Rapid Removal of the Disperse Dyes from Wastewater of Textile Companies. *Nanochem. Res.* **2016,** *1*(1), 49–56.
2. Li, H.; Liu, S.; Zhao, J.; Feng, N. Removal of Reactive Dyes from Wastewater Assisted with Kaolin Clay by Magnesium Hydroxide Coagulation Process. *Colloids Surf. A* **2016,** *494*, 222–227.
3. Nair, V.; Vinu, R. Peroxide-assisted Microwave Activation of Pyrolysis Char for Adsorption of Dyes from Wastewater. *Bioresour. Technol.* **2016,** *216*, 511–519.
4. Chengalroyen, M. D.; Dabbs, E. R. The Microbial Degradation of Azo Dyes: Mini Review. *World J. Microbiol. Biotechnol.* **2013,** *29*, 389–399.
5. Li, G.; Ying, C.; Thavamani, P.; Zuliang, C.; Mallavarapu, M.; Ravendra, N. Pathways of Reductive Degradation of Crystal Violet in Wastewater Using Free-strain *Burkholderia vietnamiensis* C09V. *Environ. Sci. Pollut. Res.* **2014,** *21*, 10339-10348.

6. Arslan, M.; Sayin, S.; Yilmaz, M. Removal of Carcinogenic Azo Dyes from Water by New Cyclodextrin–Immobilized Iron Oxide Magnetic Nanoparticles. *Water Air Soil Pollut*. **2013**, *224*, 1527–1230.
7. Dahl, J. A.; Maddux, B. L. S.; Hutchison, J. E. Toward Greener Nanosynthesis. *Chem. Rev.* **2007**, *107*, 2228–2269.
8. Astruc, D.; Lu, F.; Aranzaes, J. R. Nanoparticles as Recyclable Catalysts: The Frontier Between Homogeneous and Heterogeneous Catalysis. *Chem. Int. Ed.* **2005**, *44*, 7852–7872.
9. Thompson, D. Chemical Hypothesis for Arsenic Methylation in Mammals. *Chem. Biol. Interact.* **1993**, *88*, 89–114.
10. Sastry, M.; Ahmad, A.; Khan, M. I.; Kumar, R.; Niemeyer, C. M.; Mirkin, C. A. Microbial Nanoparticle Production. In *Nanobiotechnology: Concepts, Applications and Perspective*. Weinheim: Wiley-VCH, 2004; p 126.
11. Bhattacharya, D.; Gupta, R. Nanotechnology and Potential of Microorganisms. *Biotechnol. Crit. Rev*. **2005**, *25*, 199–204.
12. Mohanpuria, P.; Rana, N. K.; Yadav, S. K. Biosynthesis of Nanoparticles: Technological Concepts and Future Applications. *J. Nanopart. Res*. **2008**, *10*, 507–517.
13. Asthana, S. Which Doctor for Primary Health Care? Quality of Care and Non-physician Clinicians in India. *Soc. Sci. Med*. **1982**, *102*, 201.
14. Ahmed, D.; Sharma, M.; Mukerjee, A.; Ramteke, P. W.; Kumar, V. Improved Glycemic Control, Pancreas Protective, and Hepatoprotective Effect by Traditional Poly-herbal Formulation "Qurs Tabasheer" in Streptozotocin Induced Diabetic Rats. *BMC Complement. Altern. Med*. **2013**, *13*, 10–15.
15. Iravani, S. Green Synthesis of Metal Nanoparticles Using Plants. *Green Chem*. **2011**, *13*, 2638–2650.
16. Gengan, R. M.;. Anand, K.; Phulukdaree, A.; Chuturgoon, A. A549 Lung Cell Line Activity of Biosynthesized Silver Nanoparticles Using *Albizia adianthifolia* Leaf. *Colloids Surf. B: Biointerfaces* **2013**, *105*, 87–92.
17. Anand, K.; Tiloke, C.; Phulukdaree, A.; Ranjan, B.; Chuturgoon, A.; Gengan, R. M. Biosynthesis of Palladium Nanoparticles by Using *Moringa oleifera* Flower Extract and Their Catalytic and Biological Properties. DOI.org/10.1016/j.jphotobiol.2016.09.039.
18. Anand, K.; Gengan, R. M.; Phulukdaree, A.; Chuturgoon, A. Agroforestry Waste *Moringa oleifera* Petals Mediated Green Synthesis of Gold Nanoparticles and Their Anti-cancer and Catalytic Activity. *J. Ind. Eng. Chem*. **2015**, *21*, 1105–1111.
19. Weir, R. J.; Fisher, R. S. Toxicologic Studies on Borax and Boric Acid. *Toxicol. Appl. Pharmacol*. **1972**, *23*, 351–353.

CHAPTER 16

ASSESSING ADIPOSE TISSUE ENGINEERING IN VITRO AND IN VIVO: A MICROSCOPIC APPROACH

BALU VENUGOPAL[1], FRANCIS B. F.[2], SUSAN MANI[2], HARIKRISHNAN V. S.[3], VARMA H. K.[4], and ANNIE JOHN[2*]

[1]*Tissue Culture Lab, Biomedical Technology Wing, Sree Chitra Tirunal Institute for Medical Sciences and Technology, Trivandrum 695012, Kerala, India*

[2]*Transmission Electron Microscopy Lab, Biomedical Technology Wing, Sree Chitra Tirunal Institute for Medical Sciences and Technology, Trivandrum 695012, Kerala, India*

[3]*Division of Laboratory Animal Science, Biomedical Technology Wing, Sree Chitra Tirunal Institute for Medical Sciences and Technology, Trivandrum 695012, Kerala, India*

[4]*Bio-ceramic Laboratory, Biomedical Technology Wing, Sree Chitra Tirunal Institute for Medical Sciences and Technology, Trivandrum 695012, Kerala, India*

*Corresponding author. E-mail: karippacheril@gmail.com

ABSTRACT

Microscope is one of the most fundamental and versatile tools of the biologist used in its various adaptations, ranging from the simple light microscope to the advanced electron microscope. Microscopy enables to elicit details of anatomical structures at cellular and subcellular levels to understand the cell/tissue architecture better and hence, an appropriate evaluation tool for tissue-engineered living constructs. Tissue engineering (TE) is defined logically as an approach to develop "biological substitutes to restore, maintain, or improve functions" of a normal or pathological tissue or organ. The triad in tissue engineering includes—cells, a structural support or a scaffold and the micro environmental cues provided for the fabricated/developed living construct. Herein, an attempt is made to demonstrate adipogenesis (the genesis of fat tissue) using adipose-derived stem cells (ASCs) on two distinct scaffolds—bioceramic and collagen, with microscopy as the premier evaluation tool. This work aims to reconstruct soft tissue voids and deformities caused by tumors, congenital defects, or trauma using ASCs seeded on ceramic (hard) and collagen (soft) scaffolds, and inducing them toward the adipogenic lineage with special induction cocktails to ultimately achieve adipogenic differentiation. For this, various microscopic techniques were successfully relied upon to achieve the abovementioned objectives in this work.

As a first step, ASCs and their transition to adipose cells was demonstrated using phase contrast microscopy. Simultaneously, the noncytotoxic nature of the scaffold materials were also demonstrated microscopically using confocal microscopy. The morphological change of ASCs from fibroblastic to spherical during the adipogenic differentiation was further depicted using fluorescent microscopic techniques. Again, accumulation of small spherical lipid globules within the cytoplasm was visualized using Nile red staining and imaged using confocal microscopy. Furthermore, subcellular morphology was analyzed using transmission electron microscopy which showed empty globules within the differentiated cells. Differential staining techniques employed indicated the peripheral positioned nucleus, characteristic of adipocytes. Apart from the in vitro demonstration of the hard and soft scaffolds, in vivo efficacy of the construct was also evaluated histologically in rat models as demonstrated by light microscopy. Scanning electron micrographs of the retrieved (implanted) construct similarly depicted spherical cells on the scaffold. Microscopy being the sole unique technique, the experimental pipeline of a tissue-engineered construct, that is, the process from precursor cell morphology to tracking cell differentiation and analyzing the in vivo effectiveness of the tissue-engineered construct was demonstrated and evaluated. Thus, using various adaptations of microscopy, the assessment lifecycle of a tissue-engineered construct from in vitro to in vivo was undertaken successfully. Microscopy is a state of graphic art and will remain always as the oldest and the most relevant tool in biological science.

16.1 INTRODUCTION

The new era of modern medicine witnessed remarkable progress in the field of tissue engineering (TE) and regenerative medicine. Even though the historical background falls back to decades, it could be identified as an emerging field aiming to develop and demonstrate prototypes of various tissues and organs. TE is defined logically as an approach to develop "biological substitutes to restore, maintain, or improve functions" of a normal or pathological tissue or organ[18] by making use of autologous or allogenic cell sources, engineered materials as suitable scaffolds,[13] and identifying the biochemical and physicochemical factors that could help the construct in mimicking the biological system[11]—altogether, this artificially created supporting system is intended to mimic the respective organ and perform its function.

Adipose tissue engineering (ATE) is one among the significant fields in TE with immense applications, as the current therapeutic strategies do not meet patient requirements with unpredictable end results.[8,14] ATE is an attempt to remodel the tissue voids formed due to any type of deformities caused by congenital defects, trauma, or tumor resection.[2] This aims at filling the tissue voids using an in vitro fabricated construct which itself could induce adipogenesis—the differentiation of stem cells to adipocytes.[23] Adipose-derived mesenchymal stem cells (ASCs), is one among the easily available, and proven stem cell type for TE applications. ASCs are easily accessible and readily expandable, obtained from adipose tissue by liposuction, a comparatively simple method to bone marrow aspiration. Adipose stem cells are isolated from white fat by collagenase digestion method.[27] There are two types of fats in our body. White fat is widely seen in body and is the main site for fat metabolism and storage, whereas the other type, brown fat is mainly associated with heat management and is mainly seen in new born and disappears with time.

This chapter is focused in demonstrating the events toward developing a tissue-engineered construct utilizing diverse visualization techniques. The major challenge while using microscopy for demonstrating tissue-engineered constructs is its material property hindering imaging techniques. For this reason, two inherently distinct scaffolds supporting adipose tissue regeneration were fabricated—a soft collagen sponge and a hard ceramic scaffold, both having discrete material properties. An adipose tissue-engineered construct was fabricated from these scaffolds using ASCs as the stem cell source and was demonstrated in vitro and in vivo using microscopic techniques.

Thereby, an ATE construct is evinced by means of visual demonstration techniques using microscopy as the primary tool to evaluate a TE construct. Irrespective of the developments in science and technology, microscopy is still the fundamental evaluation tool in cell biology since it relies on the concept "Seeing truly believes." Most of the in vitro and in vivo demonstration techniques employ microscopy along with other quantitative modes of evaluation which involves biochemical, genomic, and proteomic methods. Most of the experimental observations and conclusions still rely on phase-contrast, immunofluorescence, and electron microscopic techniques as pillars on which other data are built upon.[3] Imaging techniques as a tool has far-reaching benefits and applications while engineering biological tissues.

This chapter explains the extensive use of a variety of microscopic techniques to demonstrate the whole experimental pipeline in the development of an ATE construct. To brief, the chapter has been divided in to three phases:

1. In vitro *adipogenesis* including adipose stem cell isolation, scaffold surface properties, and cell-scaffold interactions to demonstrate the noncytotoxic nature of the scaffold.
2. In vivo *adipogenesis* demonstrated the interaction of the ATE construct with the in vivo microenvironment. Histological evaluation of soft and hard tissue samples using paraffin and plastic-embedded sections, respectively, are detailed, which can be applied in practice depending on the texture and properties of the construct.
3. *Fluorescent in situ hybridization* (*FISH*) for tracing the implanted cells in vivo to identify their postimplantation fate validating the end results.
4. *Histomorphometry* for comparing adipogenesis induced and supported by both the scaffolds.

Even though, a lot of other additional data can fit and support the question in focus, none other than microscopic evaluation can serve exclusively as an independent data set to demonstrate and evaluate the entire experimental pipeline. Thus in this chapter, we have relied only on microscopic-based analysis to demonstrate the entire pipeline in developing tissue-engineered construct with different scaffold properties.

16.2 IN VITRO ADIPOGENESIS

16.2.1 ISOLATION AND EXPANSION OF ASCs

The cell source used for the study was ASCs isolated from the subcutaneous fat pad of male Sprague Dawley rats. ASCs, a well-established stem cell source[9] was isolated from rat subcutaneous fat pad based on their plastic adherence.[10]

To brief the isolation protocol, the fat tissue from the subcutaneous fat pad was minced well and digested using collagenase type I. The digested solution was filtered and the filtrate was centrifuged, plated into tissue culture treated flasks with Dulbecco's modified eagle medium supplemented with 10% fetal bovine serum and 1% antibiotics. This medium will be referred to as growth medium here after. ASCs from the isolated cell population adhered to plastic surfaces, whereas nonadherent heterogeneous population of cells was washed off on further medium changes. ASCs exhibited characteristic spindle shape morphology when cultured in plastic dishes as evident from phase-contrast images (Fig. 16.1). The adherent stem cell population when passaged gives a homogenous population with extensive proliferation potential in the subsequent passages. The cells maintained their morphological features for at least 6–7 passages and later appeared to have more elongated and stretched morphology which can be implied as cellular aging.[26]

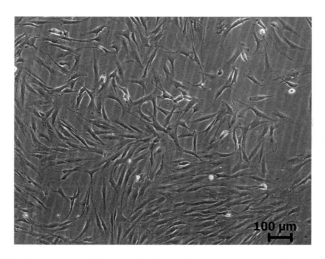

FIGURE 16.1 Phase-contrast image of adipose-derived mesenchymal stem cells (passage 4) exhibiting spindle morphology.

16.2.2 DETERMINING THE MULTILINEAGE POTENTIAL OF MESENCHYMAL STEM CELLS

To determine their multilineage potential or stem cell properties, ASCs were induced to multiple lineages—adipo-, osteo-, and chondro lineages.

For adipogenic induction, the growth media was supplemented with isobutylmethylxanthine, dexamethasone, indomethacin, and insulin, whereas for osteogenic induction the growth medium was supplemented with β-glycerophosphate, dexamethasone, and L-ascorbic acid. chondrogenic induction was achieved by supplementing the growth medium with TGF-β1, ITS premix, and dexamethasone. Apart from their plastic adherence property, the cell population isolated also showed the capacity to differentiate into multiple lineages, such as adipocytes, osteoblasts, and chondroblasts. The adipocyte induction was confirmed using triglyceride specific staining dye Nile red. This dye was taken up by the lipid-filled globules in the cytoplasm. Oil inclusions in the cytoplasm fluoresce red in adipogenic induced ASCs when visualized in confocal laser scanning microscope (CLSM) (Fig. 16.2A). Abundant mineralized matrix deposition of calcium and phosphorous along with formation of a collagen-rich matrix was the indication toward osteogenesis which was evident from von Kossa staining (Fig. 16.2B), alizarin red (Fig. 16.2C), and Masson's trichrome (Fig. 16.2D) staining, respectively, and was imaged using bright-field microscope. For chondrogenic induction, Alcian blue stained acid mucosubstances and acetic mucins located on the cartilage permitted the examination of cartilage formation (Fig. 16.2E). Safranin O stained the proteoglycan deposition (Fig. 16.2F) on chondrogenic-induced mesenchymal stem cells demonstrating a shift to chondrogenic lineage by accumulation of the relevant extracellular matrix components.

ASCs were confirmed to have multilineage potential by their differentiation to adipo-, osteo-, and chondro lineages.[28] Adipose differentiation was evident by lipid droplet accumulation in the cells while matrix mineralization by cell secretions guided osteo- and chondrodifferentiation. Calcium and phosphorous deposition as well as collagen secretion were evident in osteodifferentiation,[20] whereas glycosamino glycan and proteoglycan secretions were observed in chondrogenic differentiation.[7]

16.2.3 SURFACE TOPOLOGY OF SCAFFOLDS

Two characteristically different scaffolds were chosen as cell carriers. One is a soft scaffold, collagen sponge (Nitta Gelatin—gifted by Prof. Yasuhiko Tabata, Institute for Frontier Medical Sciences, Kyoto University, Japan) and the other is a hard scaffold—a ceramic composite (developed by Bio ceramic laboratory BMT Wing SCTIMST). Scaffolds of choice (ceramic and collagen) were characterized for surface topology using scanning electron microscopy (SEM). Both the scaffolds were gold-coated and examined using

Assessing Adipose Tissue Engineering In Vitro and In Vivo

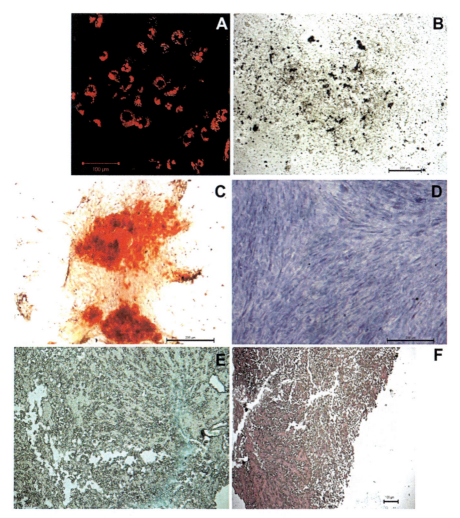

FIGURE 16.2 In vitro adipogenesis: (A) Nile red-stained lipid globules within the cell cytoplasm indicating adipogenesis, (B) von Kossa stain showing phosphorous deposition by cells, (C) alizarin red staining indicating calcium deposition by cells, (D) Masson's trichrome staining indicating collagen secretion by osteogenic induced cells, (E) Alcian blue stains acid mucosubstances and acetic mucins, and (F) safranin orange stains mucopolysaccrides indicating chondrogenesis.

SEM. Scanning electron micrographs showed rough and irregular surface topography for the ceramic composite, exhibiting its porous nature—both micro- and macropores (Fig. 16.3A), while the SEM of the collagen sponge depicted an interwoven matted fibrous structure (Fig. 16.3B).

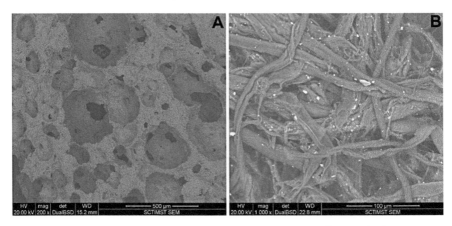

FIGURE 16.3 Scanning electron micrographs: (A) ceramic scaffold indicating a porous structure with micro- and macropores and (B) collagen sponge showing a fibrous matted structure.

16.2.4 CELL-MATERIAL INTERACTION

Cell–material interaction is a critical parameter in the tissue-engineered triad that determines the success of a functional tissue construct.[21] Within the limits of microscopy, cell-material interaction was elucidated using acridine orange/ethidium bromide staining of cells on scaffolds. ASCs were seeded at a density of 10,000 cells per scaffold and cultured in growth medium for 48 h. The cells were stained with acridine orange at a concentration of 10 μg/mL (Ex 502 nm/Em 526 nm). The viability and distribution of ASCs seeded on both the scaffolds were observed under CLSM after 48 h in culture. Both in case of ceramic (Fig. 16.4A) and collagen (Fig. 16.4B) scaffolds, almost all cells took up acridine orange and fluoresced green establishing their viable nature in classic fibroblastic morphology. Dead cells took up ethidium bromide (Ex 518 nm/Em 605 nm) when used at a concentration of 15 μg/mL. The excitation was carried out with Argon 2 laser. Microscopic evaluation of the cell-seeded construct showed excellent cell proliferation around the scaffolds. The initial adhesion of ASCs on both the scaffolds was visualized using fluorescence microscopy imaging.

Acridine orange is a nucleic acid-binding cationic dye. It is cell permeable and intercalates to the double-stranded DNA fluorescing green; Ethidium bromide also intercalates to the double-stranded DNA which fluoresces with a red-orange color. Since ethidium bromide is excluded from viable cell, live cells will not take up ethidium bromide and so the dead cells fluoresce red. Uniformly distributed viable cell population (green color) on the scaffold surface visually demonstrated the cell friendly nature of the scaffold. In ceramic scaffolds, the cells rim around the macropores neither shows pore occlusion nor bridging, and infiltrate into the pores. The cells spread over and profusely distribute within the scaffold perimeter, ensuring that the scaffold does not have a negative or cytotoxic effect on cells. Biochemical assays can evaluate cell proliferation and viability. Most of them rely on spectrophotometry which takes into account a lot of parameters[17] proportionally increasing the chance of errors. On the contrary, microscopic techniques help in visualization of viable cells on scaffold, enabling to derivate more observations on cell health, morphology, cell area, spreading, etc., even when they are still intact on the scaffold. Colocalization studies can also be carried out to visualize stress fiber formation, nuclear dimensions, and evaluating cellular health giving more inputs on cell behavior at the scaffold interface. Imaging techniques are well versed in determining the compatibility of a cell to survive, adhere, and proliferate on a scaffold surface.

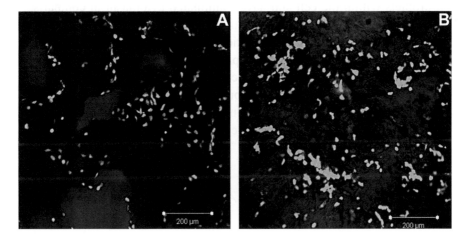

FIGURE 16.4 Acridine orange emitting green fluorescence showing viable cells and ethidium bromide emits red fluorescence indicating dead cells: (A) ceramic and (B) collagen.

16.2.5 ADIPOGENIC INDUCTION

ASCs were induced to adipogenic lineage by incubating with adipogenic induction cocktail. Induction cocktail contains growth medium with added supplements having a final working volume concentration of 0.5 mM isobutylmethylxanthine, 1 µM dexamethasone, 50 µM indomethacin, and 5 µg/mL insulin.[27] The cells were treated for 48 h in induction cocktail and then maintained in growth medium with a final concentration of 10 µg/mL insulin. The cells were maintained at 37°C, 5% CO_2, and 95% relative humidity for 21 days in a CO_2 incubator. ASCs induced to adipogenic lineage showed a clear change in morphology, where the spindle-shaped cells start accumulating triglycerides and fat-soluble substances in small cytoplasmic inclusions called lipid droplets, thus acquiring a spherical morphology[22] in contrast to their characteristic spindle shape. Mesenchymal stem cells gradually gets transformed in its shape, function, and properties with time influenced by the milieu around them.[5]

16.2.5.1 PHASE-CONTRAST IMAGING

The phase-contrast micrographs showed adipogenic induced as well as uninduced cells in the same population. Most of the cells have accumulated oil droplets within the cell and have illuminated cytoplasm in comparison with normal MSCs when imaged with phase-contrast microscope. Almost all the cells have lost their characteristic mesenchymal stem cell morphology and has spread out from a classical spindle shape to an elongated spherical structure (Fig. 16.5A, B). The lipid accumulation process is observed from 4 to 5 days postinduction as minute globular structures within the cell giving it a spherical shape. Post 21 days of induction, cells depicted evident globular structures with a typical peripheral nucleus indicating characteristic matured adipocyte morphology.[19]

16.2.5.2 DEMONSTRATION OF ADIPOGENESIS USING FLUORESCENT DYES

In vitro demonstration of adipogenesis is also well explained with lipid soluble dyes and simple phase-contrast images. Nile red stained differentiated cell population-depicted globular structures within cell membrane clearly indicating lipid accumulation.[27] Earlier studies have correlated the peripheral nucleus with the increased accumulation of Nile red as a characteristic indicator of adipogenesis.[12]

Here, adipogenic induction was demonstrated using Nile red staining, a lipid specific stain with an excitation at 485 nm and emission at 525 nm. The induced ASCs were cultured for 21 days; fixed with 4% paraformaldehyde before staining with Nile red.[24] Nile red was used at a concentration of 0.5 μg/mL in glycerol-water mixture and counter-stained with DAPI. The stained cells were observed using CLSM (Carl Zeiss LSM 510 Meta). Excitation lasers used are 405 and 514 nm. Staining with Nile red showed intense red globular structures inside the cell (Fig. 5C, D). The confocal images clearly demonstrated the transition of ASCs from spindle to spherical morphology with visible globular structures within the cell with peripherally shifted nucleus (Fig. 5E).

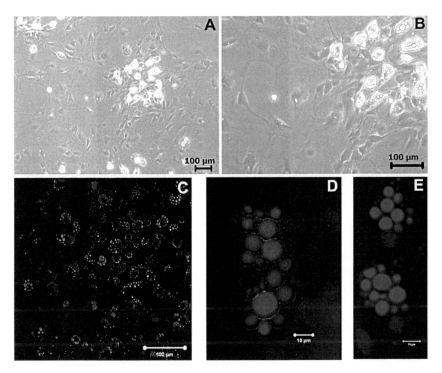

FIGURE 16.5 Demonstrating in vitro adipogenesis. A and B: phase-contrast image indicating adipogenesis showing globular structures within cell cytoplasm; (A) 10× and (B) 20×. C–E: fluorescent Nile red staining for induction of adipogenesis; (C) 10× image, (D) 63× image of single cell showing lipid globules, and (E) induced cells showing peripheral nucleus characteristic of adipocytes (63×).

16.2.5.3 TRANSMISSION ELECTRON MICROSCOPY

Further, in vitro demonstration with transmission electron microscopy also supported adipogenesis incidence. ASCs were induced to adipogenic lineage in culture inserts, fixed in 3% gluteraldehyde, and stained with 1% osmium tetroxide. This was then dehydrated with Acetone, infiltrated and embedded in Polybed 812. Subsequently ultrathin sections (50–70 nm) were cut using a diamond knife (Diatome, Switzerland) in the ultramicrotome and collected onto copper grids of 300 mesh size, stained with uranyl acetate-lead citrate. The grids were then dried and viewed under the transmission electron microscope (TEM) (Hitachi-7650, Japan) at an accelerating voltage of 80 kV. Adipogenesis in vitro demonstrated using TEM depicted globular structures within cells on par with observations from phase-contrast and fluorescent imaging techniques (Fig. 16.6A, B). The ultrastructure of induced cells showed an array of lipid globules within the cytoplasm. TEM micrographs indicated the presence of cells with evacuated spherical structure within.[19] The empty structures indicate the loss of globular fat during the tissue-processing steps for electron microscopy that include loss of lipid layers. Similar investigations into the role of collagen in tissue fibrosis related to adipose have noted similar cellular micromorphology in native rat adipose tissue.[16] Similarity of induced tissue to native adipose tissue at the cellular level indicates optimal maturation and subcellular organization. Analogous cellular anatomy demonstrates ability of construct to assimilate and adapt to the surrounding tissue milieu at a system level indicating successful adaptation on site.

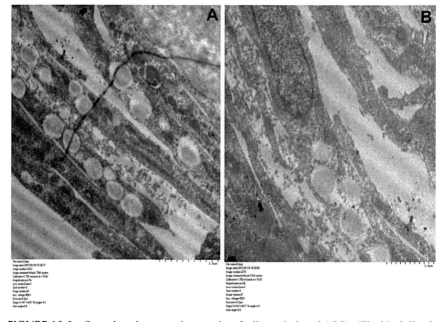

FIGURE 16.6 Scanning electron micrographs of adipose induced ASCs. (The black line in Fig. 16.5A is an artifact while processing).

16.3 IN VIVO ADIPOGENEISIS

The in vitro evaluated construct has to be assessed for its efficiency in vivo. In vivo adipogenesis of the adipose induced constructs (both collagen and ceramic) was evaluated histologically 21 days postimplantation in rat muscle model. Subsequently survival and postimplantation fate of both the constructs were assessed. Most importantly to infer whether the implanted construct has performed its intended function of adipose tissue regeneration was further evaluated from the retrieved tissues.

In total, 1×10^4 cells were seeded on both the scaffolds and then was induced to adipogenic lineage for 4 days in vitro. After 4 days, these cells were implanted in the dorsal muscle of rats. The animal experiments were monitored by Institutional Animal Ethics committee (IAEC), following the guidelines of Committee for the Purpose of Control and Supervision of Experiments on Animals (CPCSEA). The in vivo implantation was carried out on female Sprague Dawley rats with sample size, $n = 6$ per scaffold, age 4 months, and weighed approximately 200 g. Rats were anesthetized with ketamine@70 mg/kg body weight and xylazine@5 mg/kg body weight as intramuscular injection. Approximately 1-cm pocket was made on the dorsal muscle region on both side of the vertebral column and the cell-seeded scaffolds (both collagen and ceramic) were placed in the right dorsum and secured in situ with sterile nonabsorbable braided silk sutures in simple interrupted pattern. The left dorsum had the scaffold alone control. Twenty-one days postimplantation, the implant with adjacent muscle tissue was retrieved and analyzed histologically.

In vivo conditions are well known for extraneous factors[25] and foils in vitro predicted properties of material-cell construct. In vitro observations might give insights to the in vivo behavior; however, it solely depends on the implant performance in vivo under the influence of numerous secreted and recruited factors which determines the survival of the construct. Histological evaluation often is pivotal in understanding the fate of implanted cells in the dynamic in vivo system. Diverse staining methods were exploited to understand in detail the influence of the micro environment on the implanted construct, functional efficiency of the construct, and the implant-tissue integration in living models.

16.3.1 IN VIVO HISTOLOGY

For detailed histologic analysis, the retrieved constructs has to be processed in different ways due to their difference in material properties. Collagen sponge as the scaffold material is amenable to paraffin embedding and post-processing, whereas brittle ceramic constructs are excluded. So for ceramic sections, poly(methyl methacrylate) (PMMA)-based processing was done and evaluated. Optimized techniques are available for preservation of tissue architecture, prevention of artifact formation, and facilitating in situ evaluation of retrieved tissue constructs within the restrictions imposed by varied scaffold structures.

16.3.2 PARAFFIN PROCESSING AND H AND E STAINING OF CELL-COLLAGEN CONSTRUCT

The retrieved implants were fixed in 10% neutral buffered formalin, washed in water, dehydrated in graded ascending series of acetone (Merck India) infiltrated in xylene (Merck, India) in Leica TP-1020 Histokinette (Germany), and embedded in paraffin wax (SLEE MP3/P1 Paraffin wax embedder). Thin sections (5 µm) were cut using the microtome (Leica RM2255, Germany) and kept at 37°C overnight. The sections were deparaffinized with xylene; rehydrated in descending grade of ethanol (Merck, India), washed in tap water; stained with Harris's Haematoxylin (Sigma-Aldrich); differentiated in 1% acid alcohol, and blued with 0.2% ammonia water. It was then rinsed with tap water, counterstained with 1% eosin; dehydrated in ethanol, cleared in xylene, and mounted in DPX to be viewed under light microscope (Leica DM 6000 M, DFC300 FX camera). "Haematoxylin and eosin," the classical staining system was used to demonstrate cellular morphology at a basic level. Nuclear and cytoplasmic morphology is best understood using this classical system. Empty spaces indicate vacuoles preoccupied by fatty accumulation and lost during tissue processing (alcohol processing). As described in contemporary references, the underlying morphological similarity may be crucial in preventing abnormal function, which in the in vivo system, will relate to implant attrition and absorption.[1]

16.3.3 PICROSIRIUS AND MASSON'S TRICHROME STAINING ON PARAFFIN-PROCESSED CELL-COLLAGEN CONSTRUCT

To detail the observations extracted from microscopic evaluations of haematoxylin stained sections, paraffin-processed collagen constructs were stained with different histologic stains—Picrosirius and Masson's trichrome. Picrosirius stains collagen fibers red and was used to evaluate the implanted area. Picrosirius stained sections of collagen construct clearly showed adipogenesis, wherein vacuoles were visualized within the red-stained scaffold structures. Masson's trichrome is a differential stain, which simultaneously stains muscles, collagen, and cells. Trichrome stain depicts differentiated fat-like cells within the collagen network along with adjacent muscle tissues. Muscle is stained red, while collagen fibers are stained blue. Implanted or differentiated cells again showed up as empty vacuolated area due to loss of cytoplasmic inclusions during processing. Another interesting observation evident from the haematoxylin staining of the collagen construct is its ability to support angiogenesis. Haematoxylin- and eosin-stained sections showed abundant angiogenic sites within the adipose network area. Sprouting capillaries containing RBCs attributed enhanced vasculature support that helps in long-term survival of the construct.

16.3.4 PMMA PROCESSING AND H AND E STAINING OF CELL-CERAMIC CONSTRUCT

The ceramic composites being hard solid structures, it has to be processed through PMMA-processing method. The cell–ceramic construct was dehydrated with graded concentration of isopropanol and then embedded in PMMA. The plastic sections of less than 200 µm were sliced with a linear precision saw (Isomet™ 5000 Precision Saw, Buehler, Germany), thereafter polished to fine structures (Ecomet R 3000, Buehler, Germany), and were stained with haematoxylin and eosin. Later viewed and imaged under the automated Leica digital microscope (Leica DM 6000 M, Germany) with DFC 300 FX camera.

Paraffin processing of retrieved tissue construct (cell–collagen) and its demonstration with H and E staining is widely described but H and E staining of a PMMA-processed cell–ceramic tissue-engineered construct maintaining the integrity and architecture as such for adipogenesis is described in this chapter. This study could retrieve excellent data from PMMA-processed histological sections which showed both the ceramic construct and adipose cell formation within the implant perimeter and the vicinity in situ. Embedding of the scaffold *en masse* with surrounding tissue offers a keen understanding of cell-material interaction at the tissue level. The sections of 180 μm thickness were polished prior to the staining process. Rough sections of PMMA-processed samples were compared with the paraffin-processed ones. Guided by this, the fine polishing step in the PMMA sections was carried out where the tissue embedded in the PMMA is finely polished to reduce its thickness to around 70–90 μm. This thickness range of the PMMA (cell-ceramic) sections and its hydrophobic nature has been a stumbling block compared to the 4–5 μm sections of the paraffin-embedded (cell–collagen) sections. Processing, polishing, and staining have not resulted in the distortion, alteration of histological features, or hindrance in the retrieval of histological data. Resin sections primarily help in the preservation of cellular architecture by preventing tissue shrinkage and distortion.

16.3.5 HISTOLOGICAL ANALYSIS

Microscopic analysis showed a network of spherical or more hexagonal-shaped cells. The cell-collagen construct stained with H and E showed well-defined adipocytes morphology (Fig. 16.7A). The presence of collagen construct was further confirmed using Picrosirius staining which stained the collagen fibers red (Fig. 16.7B) with similar results observed in Masson's trichrome staining (Fig. 16.7C). Masson's trichrome stained collagen network blue and muscle tissue pale pink. A network of adipocytes was found enveloped within the collagen scaffold. As the implantation was done within the dorsal muscle, the muscular architecture was also evidently seen in the histological analysis and is an ectopic site for adipogenesis.

In the cell ceramic construct, a similar observation was visible with a cluster of fat cells (Fig. 16.7D) in the vicinity of the scaffold (Fig. 16.7E). The PMMA-embedding technique showed comparable results as in paraffin-embedding technique. The muscle architecture embedding the fat cells were also clearly visible within the cell–ceramic construct (Fig. 16.7F). The H and E staining showed typical adipocytes network in the cell–ceramic construct as in the cell–collagen construct. The staining techniques employed also showed angiogenesis by staining the sprouting blood vessels in both PMMA-embedded ceramic construct (Fig. 16.7G) and paraffin-embedded collagen construct (Fig. 16.7H). Rat skin was used as a control to compare collagen architecture (Fig. 16.7I).

Assessing Adipose Tissue Engineering In Vitro and In Vivo 371

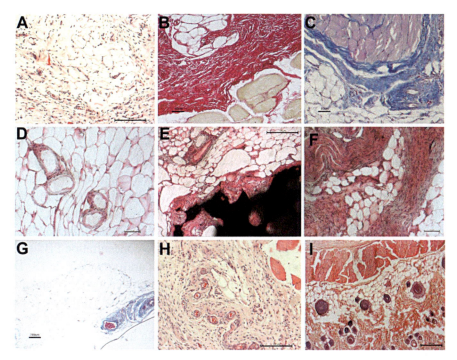

FIGURE 16.7 Bright field images of postimplantation histology for in vivo adipogenic induction in rat models. A–C: adipogenesis on collagen scaffold post paraffin processing (A) H and E showing a network of *de novo* adipose tissue, (B) Picrosirius stained collagen scaffold with unstained adipose cells and muscle tissue in the vicinity perimeter, and (C) Masson's trichrome stains collagen scaffold blue and muscle tissue pink with unstained adipose cells in the vicinity. D–F: Adipogenic induction on ceramic scaffold post-PMMA processing (D) H and E showing adipocyte network morphology, (E) H and E showing ceramic scaffold along with induced adipose cell population, and (F) H and E of induced cell with surrounding muscle tissue. (G) Paraffin embedded section after ceramic removal stained with Masson's trichrome shows capillary formation, (H) paraffin embedded section of collagen sponge indicating neoangiogenesis, and (I) rat skin histology showing native adipocyte network morphology.

16.4 FLUORESCENT IN SITU HYBRIDIZATION

The efficiency of transplantation was visualized using the FISH technique, where implanted donor cells (male) were tracked in the recipient (female) rats 21 days postimplantation. The male rat cells were labeled with Y chromosome probes and were stained using FISH technique. This ensured that transplantation was efficient and the transplanted cells were viable till explantation, evidently supporting adipogenesis. FISH technique was efficiently used to track the implanted cells in vivo and determine their postimplantion fate. This technique displayed the stability and adaptability of the construct in a living system supporting or performing its intended function—regeneration.

The Y chromosome positive cells (male donor ASCs) in the female recipients were labeled with rat Y/12 chromosome paint (Star-FISH, Cat no: CA-1631-BF, Cambio, Cambridge, UK), as per manufacturer's instructions. For the hard ceramic constructs, the implant was carefully excised and the surrounding tissue was paraffin blocked as described above. The paraffin-embedded ceramic and collagen-implanted tissues were deparaffinized with xylene (Merck, India) incubated in 1 M sodium thiocyanate for 10 min at 80°C, washed in *Phosphate-buffered saline (*PBS), permeabilized with 0.4% pepsin (Himedia, India) in 0.1 M HCl (Merck, India) for 10 min at 37°C. The reaction was quenched in 0.2% glycine in double-concentration PBS and the sections were rinsed in PBS, postfixed in 4% paraformaldehyde (S. D Fine Chem. Ltd., India) in PBS, dehydrated through graded alcohols, and air dried. The sections were then incubated with FITC-labeled Y chromosome paint (Star-FISH, Cambio, Cambridge, UK) at 60°C for 10 min, and later the temperature was set to 37°C overnight. Following hybridization, a series of washes were performed at 37°C with 50% deionized formamide (Sisco research Labs, India) in 2× sodium chloride–sodium citrate (SSC) for 15 min; 2× SSC for 15 min, and 4× SSC containing 0.1 mL 10% Tween 20 for 10 min. The slides were washed in PBS and the sections were mounted using DAPI (4',6-diamidino-2-phenylindole). All slides were analyzed with CLSM (Carl Zeiss 510 meta). The Y-FISH probe was validated using rat testes.[6,27]

FISH was done to ensure the presence and survival of implanted cells 21 days postimplantation. FISH analysis was evaluated by the presence of Y chromosome positive cells on both the ceramic construct (Fig. 16.8A) and collagen construct (Fig. 16.8B) in the implanted area. The Y chromosome positive cells will be from the male donor cells which were implanted in female recipients; hence, demonstrating the fate of donor cells in the recipient microenvironment. The Y chromosomes were fluorescently labeled with Y chromosome probes and were visualized with the CLSM. The nucleus was stained with DAPI that ensured the positive labeling of Y chromosome localized within the nucleus. Consecutive sections when scanned with confocal microscope with high laser power that showed collagen autofluorescence. This helped in visualizing the collagen network architecture with empty ovular or spherical voids which can be attributed as the presence of spherical cells on the implanted collagen construct (Fig. 16.8C).

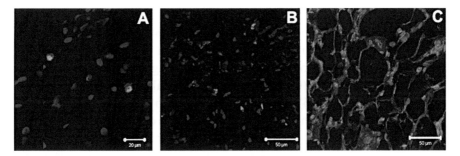

FIGURE 16.8 A and B: fluorescent in situ hybridization showing the presence of male donor cells in female recipients. FITC labeled Y chromosome tag colocalized within DAPI stained nucleus (A) paraffin embedded section after ceramic removal and (B) paraffin embedded section of collagen sponge. (C) captured auto fluorescence of collagen sponge embedding vacuolated spaces indicating adipocyte morphology. Blue color indicates DAPI staining of nucleus.

16.5 HISTOMORPHOMETRY

Histomorphometry was done using the image J—image processing and analysis software (NIH) and was used to calculate the area of the de novo-generated fat cells in both soft collagen sponge and the hard ceramic scaffold. The histology images were used for histomorphometry analysis and only the two-dimensional view of the cells were analyzed considering it as a cross section of the cell network. Data were then compared to the native fat cells in the subcutaneous skin of rat models. Statistical analysis was done using graph pad software and the data was expressed in terms of mean and standard deviation, the results seemed to be statistically significant (p value < 0.05).

Histomorphometry analysis of adipogenesis induced both the constructs showed comparable size of cell network and cross-sectional area per induced cell. Even though the values of the adipogenic-induced mesenchymal stem cells on collagen and ceramic scaffolds when comparing with native fat pad of rat skin showed significant differences, most of the values were overlapping and was within limit. The analysis was done by determining the cross-sectional area of 100 cells, each from collagen-seeded cells, ceramic-seeded cells, and from the native fat pad (Fig. 16.9). Cross-sectional area of cells induced by collagen construct was 1655 ± 505 µm², ceramic was 1722.7 ± 392.8 µm,² while the cross-sectional area of adipose cells in native rat skin was 1426 ± 329 µm². Calculating and comparing the area of 100 individual fat cells from the de novo-generated fat cell networks on both ceramic and collagen scaffolds does not show much significant difference and the p value was 0.480, which was not statistically significant at a confidence interval of 95%. But when comparing this with the native fat cell network on rat skin, the cross-sectional area of fat cells generated using the in vitro-synthesized scaffolds was far larger (Ceramic vs rat skin, p value—0.0001, collagen vs rat skin, p value—0.0085). This can be attributed to the native stiffness of the adipose tissue which cannot be exactly replicated in case of in vitro-fabricated scaffolds. The mechanical forces that the cells feel from its surroundings play a major role in their fate determination and characteristics.[4] But still, most of the individual values of both the scaffolds where overlapping with the cross-sectional area of rat skin.

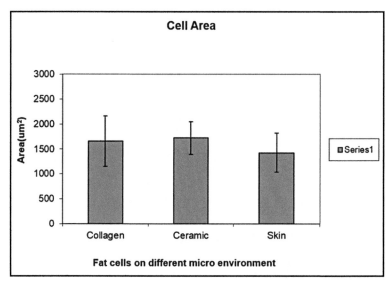

FIGURE 16.9 Graph showing area of *de novo* regenerated adipose cells in different micro environment. $N = 100$. **Significant p value ≤ 0.05 (0.0085). *Very significant p value ≤ 0.005 (0.001).

16.6 SUMMARY

The broad field of TE has gone beyond limits and dreams with its potential applications and far-reaching benefits in the field of human health care. Stem cells together with the scaffold form a tissue construct, which when implanted in living systems, integrate with the tissue architecture supporting regeneration and with restoration of the physiological functionality of the entire organ).[15] This pipeline of development and evaluation of a tissue construct (in vitro and in vivo) is demonstrated in this study using microscopy, the most relevant tool in cell biology.

Herein, adipose tissue-engineered constructs were used to demonstrate the above said concept, with ASCs as the cell source and using two distinct scaffolds in terms of texture, mechanical integrity, and properties—bioactive ceramic and the collagen sponge. Two different scaffolds were introduced to imply the fact that the change in texture and other material properties barely affects the mode of demonstration by visual means. Primary focus was given to the morphologic and histologic parameters of the cells journey in its process of differentiation from ASC to differentiated adipocyte on different scaffolds. Furthermore, going behind the concept of "seeing is truly believing," this chapter also aims to suggest microscopic techniques as an effective tool for a first-level evaluation of TE constructs. As a first step in evaluating the efficiency of a tissue-engineered construct, diverse microscopic techniques have been employed such as bright field, phase-contrast, fluorescence, and electron microscopy. These techniques helped us in detailing out the transformation of ASCs to a differentiated cell by observing the adipose-related morphological transitions by using specific dyes and suitable microscopic techniques. After elucidating cell–material interactions visually, elaborate histological analysis has been done to demonstrate the in vivo set of experiments and the efficiency of transplantation (postimplantation fate of donor cells) is determined using FISH. Roles of hard and soft scaffolds in inducing adipogenesis in vivo have been evaluated using histomorphometry analysis. From the scaffold point of view, the histological and histomorphometry data suggest that stiffness or texture of the scaffolds does not make statistically significant difference in morphology or in supporting fat cell formation. The histology data suggest that both the ceramic and the collagen scaffolds well support adipogenesis in vivo. So, ultimately utilizing diverse techniques of microscopy, we were able to describe and detail the developmental pipeline of a tissue-engineered construct. These constructs were, thus, evaluated in vitro and in vivo along with demonstrating its functional relevance and survival efficiency. Two different adipose constructs were evaluated in vivo and in vitro providing substantial data including exhaustive histological analysis including FISH data. Thereby demonstrating, an adipose construct which was developed in vitro and demonstrated in vivo for its efficiency in supporting adipogenesis. Thus a suitable work flow for the evaluation of constructs in vitro and in vivo using microscopic techniques was validated.

ACKNOWLEDGMENT

The authors acknowledge the Director SCTIMST, Head BMT, for facilities to carry out the work; Dr. Anil Kumar T. V. and Mr. Thulasidharan for CLSM and histology; Dr. Suresh Babu S. S. for Ceramic Synthesis; Mr. M. A. Sanoj for SEM; Mrs. Beena G. M. for embedding and sectioning of PMMA samples; and Mr. Joseph S. for technical advice.

KEYWORDS

- phase contrast microscopy
- light microscopy
- confocal microscopy
- scanning electron microscopy
- transmission electron microscopy
- adipogenesis
- tissue engineering
- fluorescent in situ hybridization

REFERENCES

1. Chen, H. C.; Farese, R. V. Jr. Determination of Adipocyte Size by Computer Image Analysis. *J. Lipid Res.* **2002,** *43*(6), 986–989.
2. Choi, Y. S.; Park, S. N.; Suh, H. Adipose Tissue Engineering Using Mesenchymal Stem Cells Attached to Injectable PLGA Spheres. *Biomaterials* **2005,** *26*(29), 5855–5863.
3. Cooper, G. M. *The Cell: A Molecular Approach.* (Tools of Cell Biology); 2nd Ed.; Sinauer Associates: Sunderland (MA), 2000.
4. Darling, E. M.; Topel, M.; Zauscher, S.; Vail, T. P.; Guilak, F. Viscoelastic Properties of Human Mesenchymally-derived Stem Cells and Primary Osteoblasts, Chondrocytes, and Adipocytes. *J. Biomech.* **2008,** *41*(2), 454–464.
5. Djouad, F.; Delorme, B.; Maurice, M.; Bony, C.; Apparailly, F.; Louis-Plence, P.; Canovas, F.; Charbord, P.; Noël, D.; Jorgensen, C. Microenvironmental Changes During Differentiation of Mesenchymal Stem Cells Towards Chondrocytes. *Arthritis Res. Ther.* 2007, *9*(2), R33.
6. Evans, C. H.; Liu, F. J.; Glatt, V.; Hoyland, J. A.; Kirker-Head, C.; Walsh, A.; Betz, O.; Wells, J. W.; Betz, V.; Porter, R. M.; Saad, F. A.; Gerstenfeld, L. C.; Einhorn, T. A.; Harris, M. B.; Vrahas, M. S. Use of Genetically Modified Muscle and Fat Grafts to Repair Defects in Bone and Cartilage. *Eur. Cell Mater.* **2009,** *18*, 96–111.

7. Fernandez, F. B.; Shenoy, S.; Suresh Babu, S.; Varma, H. K.; John A. Short-term Studies Using Ceramic Scaffolds in Lapine Model for Osteochondral Defect Amelioration. *Biomed. Mater.* **2012,** *7*(3), 035005.
8. Flynn, L.; Prestwich, G. D.; Semple, J. L.; Woodhouse, K. A. Adipose Tissue Engineering with Naturally Derived Scaffolds and Adipose-derived Stem Cells. *Biomaterials* **2007,** *28*(26), 3834–3842.
9. Flynn, L.; Woodhouse, K. A. Adipose Tissue Engineering with Cells in Engineered Matrices. *Organogenesis* 2008, *4*(4), 228–235.
10. Gimble, J. M.; Katz, A. J.; Bunnell, B. A. Adipose-derived Stem Cells for Regenerative Medicine. *Circ. Res.* **2007,** *100*(9), 1249–1260.
11. Gomillion, C. T.; Burg, K. J. Stem Cells and Adipose Tissue Engineering. *Biomaterials* **2006,** *27*(36), 6052–6063.
12. Gonzales, A. M.; Orlando, R. A. Role of Adipocyte-derived Lipoprotein Lipase in Adipocyte Hypertrophy. *Nutr. Metab. (Lond)* **2007,** *4*, 22.
13. Griffith, L. G.; Naughton, G. Tissue Engineering—Current Challenges and Expanding Opportunities. *Science* **2002,** *295*(5557), 1009–1014.
14. Hong, L.; Peptan, I. A.; Colpan, A.; Daw, J. L. Adipose Tissue Engineering by Human Adipose-derived Stromal Cells. *Cells Tissues Organs* **2006,** *183*(3), 133–140.
15. Horch, R. E.; Kneser, U.; Polykandriotis, E.; Schmidt, V. J.; Sun, J.; Arkudas, A. Tissue Engineering and Regenerative Medicine—Where Do We Stand? *J. Cell Mol. Med.* **2012,** *16*(6), 1157–1165.
16. Khan, T.; Muise, E. S.; Iyengar, P.; Wang, Z. V.; Chandalia, M.; Abate, N.; Zhang, B. B.; Bonaldo, P.; Chua, S.; Scherer, P. E. Metabolic Dysregulation and Adipose Tissue Fibrosis: Role of Collagen VI. *Mol. Cell Biol.* **2009,** *29*(6), 1575–1591.
17. Kruger, N. J. Errors and Artifacts in Coupled Spectrophotometric Assays of Enzyme Activity. *Phytochemistry* **1995,** *38*(5), 1065–1071.
18. Langer, R.; Vacanti, J. P. Tissue Engineering. *Science* **1993,** *260*, 920.
19. Mirancea, N.; Mirancea, D. The Ultra Structure of the White Subcutaneous Adipose Tissue. *Proc. Rom. Acad. Series B* **2011,** *3*, 206–211.
20. Mohan, B. G.; Shenoy, S. J.; Babu, S. S.; Varma, H. K.; John, A. Strontium Calcium Phosphate for the Repair of Leporine (*Oryctolagus cuniculus*) Ulna Segmental Defect. *J. Biomed. Mater. Res. A* **2013,** *101*(1), 261–271.
21. Murphy, C. M.; O'Brien, F. J.; Little, D. G.; Schindeler, A. Cell-scaffold Interactions in the Bone Tissue Engineering Triad. *Eur. Cell Mater.* **2013,** *26*, 120–132.
22. Niemela, S.; Miettinen, S.; Sarkanen, J. R.; Ashammakhi, N (Eds.). Adipose Tissue and Adipocyte Differentiation: Molecular and Cellular Aspects and Tissue Engineering Applications (Chapter 4). In *Topics in Tissue* Engineering; Ashammakhi, N.; Reis, R.; Chiellini, F. (Eds.); 2008; Vol. 4, pp 1–26.
23. Patrick, C. W. Jr. Tissue Engineering Strategies for Adipose Tissue Repair. *Anat. Rec.* **2001,** *263*(4), 361–366.
24. Pittenger, M. F.; Mackay, A. M.; Beck, S. C.; Jaiswal, R. K.; Douglas, R.; Mosca, J. D.; Moorman, M. A.; Simonetti, D. W.; Craig, S.; Marshak, D. R. Multilineage Potential of Adult Human Mesenchymal Stem Cells. *Science* **1999,** *284*(5411), 143–147.
25. Rameshwar, P. Microenvironment at Tissue Injury, a Key Focus for Efficient Stem Cell Therapy: A Discussion of Mesenchymal Stem Cells. *World J. Stem Cells* **2009,** *1*(1), 3–7.
26. Sethe, S.; Scutt, A.; Stolzing, A. Aging of Mesenchymal Stem Cells. *Ageing Res. Rev.* **2006,** *5*(1), 91–116.

27. Venugopal, B.; Fernandez, F. B.; Babu, S. S.; Harikrishnan, V. S.; Varma, H. K.; John, A. Adipogenesis on Biphasic Calcium Phosphate Using Rat Adipose-derived Mesenchymal Stem Cells: In Vitro and In Vivo. *J. Biomed. Mater. Res. A.* **2012,** *100*(6), 1427–1437.
28. Zuk, P. A.; Zhu, M.; Ashjian, P.; De Ugarte, D. A.; Huang, J. I.; Mizuno, H.; Hedrick, M. H. Human Adipose Tissue Is a Source of Multipotent Stem Cells. *Mol. Biol. Cell* **2002,** *13*(12), 4279–4295.

CHAPTER 17

THE CONTRIBUTION OF LIGHT MICROSCOPY TO STUDY MALE REPROTOXICITY OF CADMIUM

MARIA DE LOURDES PEREIRA[1*], JOSÉ COELHO[1],
RENATA TAVARES[2], HENRIQUE M. A. C. FONSECA[3],
VIRGÍLIA SILVA[4], PAULA P. GONÇALVES[4], and
FERNANDO GARCIA E COSTA[5]

[1]*Department of Biology and CICECO—Aveiro Institute of Materials, University of Aveiro, Aveiro 3810-193, Portugal*

[2]*Biology of Reproduction and Stem Cell Research Group, Center for Neuroscience and Cell Biology, Department of Life Sciences, University of Coimbra, 3004-517 Coimbra, Portugal*

[3]*Department of Biology and GeoBioTec, University of Aveiro, 3810-193 Aveiro, Portugal*

[4]*Department of Biology and CESAM, University of Aveiro, 3810-193 Aveiro, Portugal*

[5]*Department of Morphology and Function, CIISA—Interdisciplinary Centre of Research in Animal Health, Faculty of Veterinary Medicine, University of Lisbon, Lisbon, Portugal*

*Corresponding author. E-mail: mlourdespereira@ua.pt

ABSTRACT

Cadmium is one of the most toxic heavy metals widespread in industrial and environmental settings and a well-recognized endocrine disruptor agent. Cadmium compounds induce the generation of reactive oxygen species, which accounts for degenerative changes in several tissues and organs, including testis. The goal of this chapter is to describe the main effects of cadmium on male reproductive functions with focus on morphological changes in testis and spermatozoa. Several recent studies conducted on laboratory animals are listed to confirm the deleterious effects of cadmium on male reproductive function.

17.1 INTRODUCTION

Cadmium (Cd), a well-documented environmental toxicant, is classified as a prostate and testicular carcinogen by the International Agency for the Research on Cancer.[1] It also acts as an endocrine disruptor[2] and reproductive toxicant.[3-5] As a result, legislative guidelines and procedures to reduce industrial discharges of cadmium and environmental contamination have been implemented, as well as the dissemination of information on the health effects of cadmium exposure among the population groups most at risk and in the general population.[6-10]

Humans are exposed to cadmium compounds in different scenarios, for example, in the workplace that includes agricultural and industrial activities and especially those related to the handling of cadmium compounds (phosphate fertilizers, color pigments, plastic stabilizers, etc.). However, nutritional sources (food and water) and lifestyle (cigarette smoking) are the most common causes of exposure to cadmium through gastrointestinal, pulmonary, and dermal absorption pathways. Cadmium accumulates mostly in liver, kidney, and bone and also in embryos and reproductive tissues, thereby inducing several pathological conditions.[11-16]

Cadmium binds to sulfur ligands in folded proteins, such as metallothioneins.[17] The formation of cadmium–metallothionein complexes in the liver accounts for systemic redistribution and secretion of cadmium in urine. Thus, the exposure to cadmium induces the synthesis of metallothioneins, which plays a protective role against its toxicity but does not reduce the vulnerability of seminal tract to cadmium.[18] Intrauterine exposure to cadmium might compromise male fertility in adulthood.[19] Moreover, testicular damage is a fast-onset response to acute cadmium intoxication.[20] Hence, there are accentuated advantages of evaluating reproductive endpoints in preference to assess the nephrotoxicant action of cadmium.[21]

At cellular level, the relevant outcomes of cadmium-induced testicular injury include the DNA fragmentation, disorganization of the multicellular layers in the blood–testis barrier, induction of expression of androgen-regulated genes and of ZIP8, exacerbated oxidative stress, and replacement of physiological ionic species in enzymes and regulatory proteins.[11] The loss of germ cells and the disruption of the blood–testis barrier, which appears to be triggered by activation of SAPK/JNK and p38 MAPK signaling pathways,[22] are often observed on cross sections of testis exposed to cadmium and stained with hematoxylin–eosin (HE) and lead citrate.

The deleterious actions of cadmium on steroidogenesis by Leydig cells and spermiation by Sertoli cells[23] are facilitated by enhanced permeation through actin-based atypical adherents junctions, essential for preleptotene spermatocytes transit,[24,25] and through the iron and zinc transporter 8.[26] Similarly to the overexpression of metallothioneins in cadmium accumulating organs, it also promotes elevation of the metallothionein-like protein tesmin, albeit with abnormal promotion of spermatogenesis.[27]

Blockage of sulfhydryl groups in proteins, depletion of glutathione and antioxidants enzymes are among common features of enhanced oxidative stress in the testis that causes infertility. Special enrichment in long-chain polyunsaturated fatty acids of testicular cell types occurs in pubertal mammalians, which is essential for testis maturation and male fertility.[28] Polyunsaturated lipids are particularly susceptible to oxidative stress. Accordingly, morphological alterations of sperm and testis of acutely exposed rats to high concentrations of cadmium chloride are well correlated with steroidogenic and oxidative impairment-related parameters.[29] Morphological examination of spermatozoa and testicular sections under the light microscope revealed decreased sperm counts, increased sperm abnormalities, severe necrosis, partial loss of spermatogenic cells, and defoliation of spermatocytes into the lumen of degenerated seminiferous tubules. The plasma testosterone levels and the activities of the testicular steroidogenic enzymes (Δ5–3β-hydroxysteroid dehydrogenase and 17β-hydroxysteroid dehydrogenase) were decreased significantly by cadmium. Cadmium also compromised the enzymatic (glucose-6-phosphate dehydrogenase, superoxide dismutase, catalase, and glutathione S-transferase, reductase, and peroxidases) and nonenzymatic (the ratio of reduced to oxidized glutathione and total thiols) antioxidant activities[30]. Conversely, the intracellular levels of reactive oxygen species and indicators of lipid peroxidation, protein carboxylation, and DNA fragmentation were elevated in testicular tissue exposed to cadmium[31]. Most important, testicular pathophysiological modifications are partially prevented or reversed by administration of antioxidant compounds.[29,32-39]

Histological examination of the male reproductive tract could be a valuable tool to evaluate the role of cadmium as an endocrine disruptor, since spermatogenesis is not only regulated locally but also by a negative feedback mechanism through the hypothalamic–pituitary–gonadal axis. Cadmium decreases the secretion of the follicle stimulating hormone (FSH) by anterior pituitary, alters blood levels of the Leydig cell stimulating luteinizing hormone (LH) and inhibits the synthesis of testosterone.[2]

The male reproductive system of both humans and other animals is quite sensitive to cadmium insult.[4,40-44] Interdonato and coworkers[45,46] reported delayed onset of puberty and impaired testicular growth in adolescents with increased urinary levels of cadmium from a residential area surrounding industrial plants. In nonoccupationally men exposed to hazardous compounds (e.g., heavy metals and pesticides), the average concentration of cadmium in semen was 4.91 ± 2.12 µg/dL and exhibited poor semen quality with damaged DNA.[47,48] According to the 2011–2012 examination survey of the U.S. National Health and Nutrition Examination Survey, blood cadmium levels are positively correlated with serum testosterone in men.[49] Impaired development of the neurobehavioral system of the progeny of paternal exposure to cadmium[50] also reinforces the relevance of monitoring male reproductive health in human populations. Evidence of the decline of semen quality (low sperm counts, abnormalities in the morphology and motility parameters, and DNA fragmentation), due to the severe effects of environmental toxicants including cadmium on testicular function, is well documented.[51-56] Recently, male reproductive toxicity of low concentrations of cadmium chloride has been demonstrated in vitro by using isolated Sertoli cells from prepubertal pig testes.[57]

In this chapter, we summarize the recent contributions of light microscopy to evaluate cadmium toxicity in male reproductive system and illustrate the effects of cadmium in testis and spermatozoa morphology of mice.

17.2 ANIMAL EXPERIMENTAL STUDIES

In vivo studies with laboratory animals are a valuable contribution to understand the effects of xenobiotics. Table 17.1 outlines some of the more recent in vivo assays carried out on laboratory animals to demonstrate the toxicity of cadmium compounds in the male reproductive system. Although not intended to be an exhaustive list in this field, some important contributions are included, mostly those where light and electron microscopy techniques are particularly relevant to the study of the lesions induced by various compounds of cadmium.

As summarized in Table 17.1, the effects of cadmium compounds—for example, $CdCl_2$, $Cd(CH_3CO_2)_2$, and $CdSO_4$—were studied on different organs and tissues of the male reproductive system of some rodents. In most of these studies, emphasis is given to testis and epididymis morphology,

TABLE 17.1 Some Representative Studies Using Animal Models to Evaluate the Effects of Some Cadmium Compounds on Male Reproductive Function.

Animal model	Exposure route and dosing	Tissue/organ/system	Results	References
Mouse	Subcutaneous injection of $CdCl_2$ (1.50, 1.65, and 1.75 mg/L) and sacrificed after 35 days.	Testis and epididymis	Testis atrophy, hemorrhage in intertubular tissue, decrease of sperm quality.	[58]
Rat	Sertoli cells incubated with $CdCl_2$ (3 μM) for 0, 3, 6, and 9 h.	Testis	Injury of the blood–testis barrier.	[11]
Mouse	Single subcutaneous injection of $CdCl_2$ (1, 2, and 3 mg/kg bw) and sacrificed after 24 h (short term) and after 35 days (long term).	Testis and sperm	Cd levels in testis: short-term effects (increased fraction of spermatozoa with abnormal morphology, premature acrosome reaction, and reduced motility), late-term effects (drastic reduction of sperm cell count, sperm motility, and increase of DNA fragmentation). Lowest dose after 35 days presented microsatellites in the testis.	[59,60]
Mouse	Intraperitoneal injection of $CdCl_2$ (0.25 and 0.5 mg/kg bw) weekly for 6 weeks.	Testis and epididymis	Testicular injury, reduction of sperm count and their motility, and abnormalities in the histology of testis. Mild histopathological changes in testis (congested blood vessels and enlarged amount of interstitial connective tissue at both concentrations).	[61]
Rat	Intraperitoneal injection of $CdCl_2$ (3 mg/kg bw) over 48 h.	Testis	Irreversible disruption of the blood–testis barrier, germ cell loss, and downregulation of the expression of drug transporters at the blood–testis barrier.	[43]
Rat	$CdCl_2$ (15, 20, and 25 mg/L) in drinking water for 6 weeks.	Testis	Deleterious effect on testicular function and biometric parameters.	[62]
Rat	Daily oral dose of $CdCl_2$ (3 mg/kg bw, 1/30 LD50) for 90 consecutive days.	Testis	Severe histopathological degenerative changes on spermatogenic cells, pyknotic nuclei, and congestion of blood vessels.	[63]

TABLE 17.1 (Continued)

Animal model	Exposure route and dosing	Tissue/organ/system	Results	References
Mouse	Single intraperitoneal injection of CdCl$_2$ (4.5 mg/kg) during two different periods of day (ZT6 and ZT18) for 1 week.	Testis and epididymis	Daily fluctuation of Cd effects: at ZT6 sperm motility and number of sperm head were reduced, although no significant changes were observed at ZT18.	[64]
Rat	CdCl$_2$ (30 mg/L) in drinking water for 90 days.	Testis and epididymis	Increased weight of testis and epididymis, reduction of epididymal epithelium, vascular constriction, moderate to severe testicular degeneration and lumen distortion, and/or contraction.	[65]
Rat	Intraperitoneal injection of CdCl$_2$ (1 mg/kg bw) for 16 days with an interval of 48 h between subsequent treatments.	Semen	Significant reduction in sperm count, motility, and vitality.	[66]
Mouse	Intraperitoneal injection of CdCl$_2$ (0.35 mg/kg bw) for 15 days.	Testis, testicular macrophages, and sperm	Accumulation of Cd within testis, increase of inflammatory damage with loss of immune privilege, and alterations on sperm functions.	[67]
Mouse	Intraperitoneal injection of CdCl$_2$ (0.25, 0.5, 1, and 2 mg/kg bw) every 2 days for 28 days.	Testis	Severe histopathologic changes in testis with severe necrosis, Cd exposure can induce MSI in mice.	[68]
Rat	Pregnant rats during GD1 and GD20 were given CdCl$_2$ (0.01 and 0.1 ppm) in drinking water.	Testis and pituitary from fetuses at GD20	Reduced expression of testicular steroidogenic acute-regulatory protein (STAR), mRNA.	[69]
Rat	Single intraperitoneal injection of CdCl$_2$ (2 mg/kg bw).	Testis	Degeneration of seminiferous tubules, lacking of most germ cells, arrest in spermatogenesis.	[56]
Rat	CdCl$_2$ (15 ppm) over 10 weeks or 20 weeks.	Prostate and testis	Inhibition of matrix metalloproteinases (MMP) 2 and 9.	[70]

TABLE 17.1 (Continued)

Animal model	Exposure route and dosing	Tissue/organ/system	Results	References
Mouse	Single intraperitoneal injection of $CdCl_2$ (15 and 25 µM/kg bw) for 48 h.	Testis	Concentration-dependent apoptosis and necrosis in testis, histological and cellular changes, depletion of spermatids and spermatozoa in seminiferous tubules, and severe disorganization.	[71]
Mouse	$CdCl_2$ (0.5 mg/kg bw) for 24 h and 35 days.	Testis and epididymis	No substantial changes of sperm density, acrosome integrity and motility, significant decline of progressive spermatozoa after 35 days, reduction of seminiferous tubules diameter after 24 h, and immature spermatids in the lumen of epididymis after 35 days.	[72]
Rat	Intraperitoneal injection of $CdSO_4$ (3 mg/kg bw) and euthanized after 12, 24, and 48 h.	Testis	Reduction of matrix metalloproteinases (MMP) and respiratory protein complex, reduction of mitochondrial phospholipid content, and increase of phospholipase A2 (PLA2) expression.	[73]
Rat	$CdCl_2$ (22.15 mg/kg bw) gavaged every 2 days for 9 weeks.	Semen and testis of the newborns	Sperm quality affected significantly. Cd levels in testis of the newborns, paternal Cd exposure could exert significant effects on the neurobehavioral system of their offspring.	[50]
Rat	$Cd(CH_3CO_2)_2$ (10 mg/kg bw) was given in drinking water ad libitum during pregnancy and lactation.	Semen of adult progeny	Adverse effects on sperm quality of adult rats.	[19]

where among others, light and electron microscopy techniques were used by different authors to fully characterize the degenerative responses within testis and epididymis. Although different study designs were used, including doses/concentrations, route, and periods of exposure, they generally reported similar sets of deleterious responses. Several studies reported on morphometry of the testis where the evaluation of the diameter of seminiferous tubules was particular relevant to the toxicological evaluation.[74] In addition to the reduction in diameter of seminiferous tubules, also necrosis, testicular atrophy, damage to the testis including hemorrhagic lesions within the intertubular tissue, and the decrease of hormonal levels (testosterone, LH, and FSH) were some of the reported physiological changes induced by cadmium compounds.[56,63,68] Furthermore, the impact of $CdCl_2$ on the spermatogenesis in general and on the disruption of the blood–testis barrier[43] in particular is present in various studies. Disruption of the blood–testis barrier, namely, inter-Sertoli tight junctions, was one of the main features, with subsequent increased permeability through acceleration of endocytosis of occludin and N-cadherin, two blood–testis barrier components.[11] Conspicuous changes in sperm quality were also documented.[50]

Recent studies described some morphological degenerative alterations on testis and epididymis induced by low doses of cadmium (0.5 mg/kg/bw) aimed to evaluate the possibility of recovery.[72] These authors showed that after a spermatogenic cycle without any treatment with cadmium injury retrievals were often observed, which emphasizes the need for additional cellular and molecular studies for a comprehensive understanding of the mechanisms involved in those events.

In short, although the data based on animal studies clearly illustrate the toxic effects of cadmium compounds on male reproductive function, much still eludes our understanding, namely, the mechanisms involved and what triggers observed recoveries.

17.3 CASE STUDY: EFFECTS OF CADMIUM ON MICROSCOPIC FEATURES OF TESTIS AND SPERMATOZOA

The morphology of seminiferous tubules and sperm quality parameters have been studied for their relevance in assessing the impact of cadmium compounds on fertility.

As reported in Table 17.1, previous studies showed that $CdCl_2$ induces testicular and epididymal damage in mice. $CdCl_2$ causes, depending on dosage, degenerative lesions in the seminiferous tubules impairing spermiogenesis and decreasing motility and vitality of spermatozoa. According to previously described methodology,[58] studies were conducted to characterize the effects of $CdCl_2$ on the seminiferous epithelium and seminal parameters of mice. The methods used include morphological analysis of spermatozoa using histological techniques and computer-assisted morphometric analysis of the seminiferous tubules. Particular emphasis was given to the use of light microscopy as an important tool for the characterization of testis lesions and morphology of spermatozoa.

In brief, male ICR-CD1 mice (7 weeks old; 30 ± 37 g) were purchased from Charles Rives Laboratories, Spain. Animal experiments were conducted in accordance with institutional guidelines for ethics in animal experimentation. Mice received a single subcutaneous injection of 0.5 mL $CdCl_2$ (Sigma-Aldrich, St Louis, MO) at doses of 1.50, 1.65, and 1.75 mg/L. The same volume of the vehicle (0.9% NaCl) was administrated to the control group. After 35 days, animals were sacrificed for testis and epididymis sampling. Then, small pieces of testicular tissue were immersed in Bouin's solution for 24 h, dehydrated, and processed for light microscopy studies using standard protocols. HE staining was used. Histological sections were observed using the light microscope Nikon AFX-DX, and photographs were taken with coupled FD-35DX digital system. For all groups, the diameter of seminiferous tubules was measured as described by Guevara and coauthors (2003).[74] For this purpose, 10 photographs were taken from different sections of testis of each treatment, a total of 50 randomly selected seminiferous tubes were then analyzed per concentration of $CdCl_2$. Cauda epididymis from all specimens were dissected out and placed in a Petri dish containing 1 mL of Tyrode's modified medium for the release of spermatozoa. For each smear, 10 µL aliquot of sperm was taken. Fixation was performed with 95% ethanol. The smears were then stained with HE for morphology evaluation of spermatozoa under light microscopy using magnifications of 400× and 1000×. A total of 200 spermatozoa were analyzed for each animal, and classified as: normal; abnormal (head, middle piece, and tail); multiple anomalies; disconnected head and tail.

As shown in Figure 17.1, morphological changes were observed on seminiferous tubules of mice exposed to cadmium with hemorrhagic injuries in the intertubular tissue of testis on cadmium-exposed animals (Fig. 17.1e). The effect of cadmium on seminiferous tubules was also evaluated by comparing their maximum average diameter. $CdCl_2$ exposure induced a decrease in the diameter of the seminiferous tubules of more than 10% compared to control mice (Fig. 17.1f).

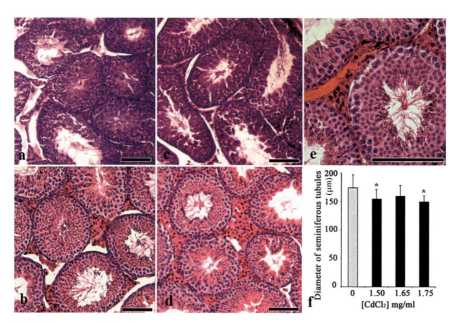

FIGURE 17.1 Histological images of mice seminiferous tubules. (a) Control, (b) 1.5 mg/L CdCl$_2$, (c) 1.65 mg/L CdCl$_2$, and (d and e) 1.75 mg/L CdCl$_2$. (e) Evidence of hemorrhagic lesions within the intertubular tissue. HE staining; bars 100 μm. (f) Effect of different doses of CdCl$_2$ on diameter of the seminiferous tubules. Values are mean + standard deviation (n = 6 for each group). Significantly different from control group ($P \leq 0.001$).

Figure 17.2 shows the morphology of normal (Fig. 17.2a) and abnormal (Fig. 17.2 b–n) mice spermatozoa. In control mice, only 18% ± 9% of spermatozoa released from cauda epididymis exhibited abnormal morphology. This percentage increased up to 42% ± 10% in sperm sample collected from animals exposed to CdCl$_2$.

Cadmium chloride reduced, in a concentration-dependent manner, the number of spermatozoa with normal morphology: control, 82% ± 9%; 1.5 mg/L, 75% ± 11%; 1.65 mg/L, 66% ± 4%, $P < 0.001$; 1.75 mg/L, 58% ± 10%, $P < 0.001$. The abnormalities of mice spermatozoa include the irregular shape of the head and middle piece (Fig. 17.2b–d) and detached head (Fig. 17.2e). Eosin staining also evidenced defects at tail level, such as bended and coiled pieces (Fig. 17.2f–m) and detached tails (Fig. 17.2n).

In conclusion, the administration of a single dose of CdCl$_2$ (1.5, 1.65, and 1.75 mg/L CdCl$_2$) significantly affected the morphology of testis and the microscopic features of spermatozoa. Thus, microscopy techniques, such as histological methods, prove to be useful tools for the evaluation of cadmium effects on morphological parameters of the seminiferous tubules. In addition, methods for assessing sperm morphology, including those of HE staining are simple and accurate in analyzing the shape of the head, the midpiece and the tail—an important parameter to evaluate the quality of a sample of semen. For example, eosin staining allows to easily identify the cytoplasmic droplets and other abnormalities within the tail. On the whole, they represent a relevant contribution to knowledge of the deleterious effects of this reprotoxicant in experimental rodents.

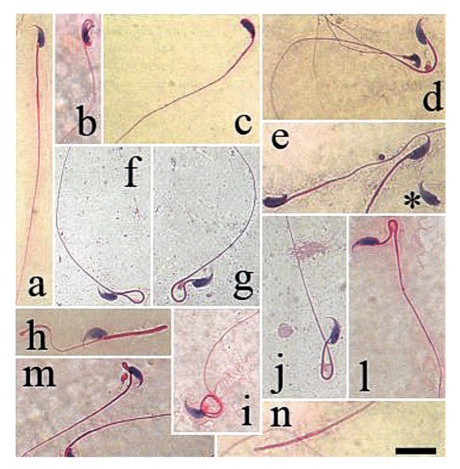

FIGURE 17.2 Representative examples of morphology of spermatozoa: (a) normal spermatozoon. Types of anomalies on spermatozoa from mice exposed to $CdCl_2$: (b–e) anomalies on head and middle piece, "*" indicates detached head; (f–m) coiled/folded/bent tail with droplet; (n) detached tail; HE staining; scale bar = 10 μm.

17.4 CONCLUSIONS AND FUTURE PERSPECTIVES

Cadmium is an environmental and occupational toxic metal with harmful consequences on male reproductive health. An updated overview of toxic effects of cadmium in mice testis and spermatozoa morphology is highlighted in this chapter, with new data that corroborate previous studies showing that this reprotoxicant compromises mammalian fertility potential.

Notwithstanding the existence of laws and regulations to reduce public exposure to cadmium, its impact on men infertility remains a real concern in the human reproductive health. Thereby early monitoring of male fertility—especially aimed at young people—through preventive medicine is necessary to protect human reproductive health. In addition, the harmful effects of cadmium compounds also recommend prenatal safety precautionary measures, another important goal to avoid future risks on male fertility.

ACKNOWLEDGMENT

This research was developed within the scope of the project CICECO-Aveiro Institute of Materials, POCI-01-0145-FEDER-007679 (Foundation for Science and Technology—FCT Ref. UID/CTM/50011/2013), financed by national funds through the FCT/MEC and when appropriate cofinanced by FEDER under the PT2020 Partnership Agreement; and FCT Ref. UID/GEO/04035/2013.

KEYWORDS

- **blood–testis barrier**
- **Sertoli cells**
- **cadmium toxicity**
- **xenobiotics**
- **reprotoxicity**

REFERENCES

1. IARC. Cadmium and Cadmium Compounds. *IARC Monogr. Eval. Carcinog. Risks Hum.* **2012,** *100,* 121–145.

2. Lafuente, A. The Hypothalamic–Pituitary–Gonadal Axis is Target of Cadmium Toxicity. An Update of Recent Studies and Potential Therapeutic Approaches. *Food Chem. Toxicol.* **2013**, *59*, 395–404.
3. Henson, M. C.; Chedrese, P. J. Endocrine Disruption by Cadmium, a Common Environmental Toxicant with Paradoxical Effects on Reproduction. *Exp. Biol. Med.* **2004**, *229*, 383–392.
4. Siu, E. R.; Mruk, D. D.; Porto, C. S.; Cheng, C. Y. Cadmium Induced Testicular Injury. *Toxicol. Appl. Pharmacol.* **2009**, *238*, 240–249.
5. Cheng, C. Y.; Wong, E. W. P.; Lie, P. P. Y.; Li, M. W. M.; Su, L.; Siu, E. R.; Yan, H. H. N.; Mannu, J.; Mathur, P. P.; Bonanomi, M.; Silvestrini, B.; Mruk, D. D. Environmental Toxicants and Male Reproductive Function. *Spermatogenesis* **2011**, *1*, 2–13.
6. World Health Organization. Inorganic Pollutants. In *Air Quality Guidelines*. WHO Regional Publications, European Series, No. 91. 2nd Ed.; World Health Organization: Copenhagen, Denmark, 2000; pp 136–138.
7. Nawrot, T. S.; Staessen, J. A.; Roels, H. A.; Munters, E.; Cuypers, A.; Richart, T.; Ruttens, A.; Smeets, K.; Clijsters, H.; Vangronsveld, J. Cadmium Exposure in the Population: From Health Risks to Strategies of Prevention. *Biometals* **2010**, *23*, 769–782.
8. CDC. 2011. *Fourth National Report on Human Exposure to Environmental Chemicals: Updated Tables*. http://www.cdc.gov/exposurereport/pdf/Updated_Tables.pdf (accessed Feb 27, 2007).
9. Agency for Toxic Substances and Disease Registry (ATSDR). Toxicological Profile for Cadmium, **2012**, 1–430.
10. IRIS. *Cadmium. Integrated Risk Information System*. Environmental Protection Agency. Washington, DC: U.S., 2012. http://www.epa.gov/iris/subst/index.html (accessed Feb 27, 2007).
11. Siu, E. R.; Wong, E. W.; Mruk, D. D.; Sze, K. L.; Porto, C. S.; Cheng, C. Y. An Occludin–Focal Adhesion Kinase Protein Complex at the Blood–Testis Barrier: A Study Using the Cadmium Model. *Endocrinology* **2009**, *150*, 3336–3344.
12. Gallagher, C. M.; Chen, J. J.; Kovach, J. S. Environmental Cadmium and Breast Cancer Risk. *Aging* **2010**, *2*, 804–814.
13. Morais, S., Garcia e Costa, F., Pereira M. L. Heavy Metals and Human Health. In *Environmental Health—Emerging Issues and Practice*; Oosthuizen, J., Ed.; InTech: Croatia, 2012; pp 227–246.
14. Adams, S. V.; Quraishi, S. M.; Shafer, M. M.; Passarelli, M. N.; Freney, E. P.; Chlebowski, R. T.; Luo, J.; Meliker, J. R.; Mu, L.; Neuhouser, M. L.; Newcomb, P. A. Dietary Cadmium Exposure and Risk of Breast, Endometrial, and Ovarian Cancer in the Women's Health Initiative. *Environ. Health Perspect.* **2014**, *122*, 594–600.
15. Ali, I.; Engstrom, A.; Vahter, M.; Skerfving, S.; Lundh, T.; Lidfeldt, J.; Samsioe, G.; Halldin, K.; Åkesson, A. Associations Between Cadmium Exposure and Circulating Levels of Sex Hormones in Postmenopausal Women. *Environ. Res.* **2014**, *134*, 265–269.
16. Belani, M.; Purohit, N.; Pillai, P.; Gupta, S.; Gupta, S. Modulation of Steroidogenic Pathway in Rat Granulosa Cells with Subclinical Cd Exposure and Insulin Resistance: An Impact on Female Fertility. *BioMed. Res. Int.* **2014**, *2014*, 460251.
17. Pulido, P.; Kägi, J. H.; Vallee, B. L. Isolation and some Properties of Human Metallothionein. *Biochemistry* **1966**, *5*, 1768–1777.
18. Klaassen, C. D.; Liu, J.; Diwan, B. A. Metallothionein Protection of Cadmium Toxicity. *Toxicol. Appl. Pharmacol.* **2009**, *238*, 215–220.

19. Banzato, T. P.; Godinho, A. F.; Zacarin, E. C. S.; Perobelli, J. E.; Fernandez, C. B.; Favareto, A. P.; Kempinas, W. G. Sperm Quality in Adult Male Rats Exposed to Cadmium in Utero and Lactation. *J. Toxicol. Environ. Health A* **2012**, *75*, 1047–1058.
20. Santos, F. W.; Oro, T.; Zeni, G.; Rocha, J. B.; do Nascimento, P. C.; Nogueira, C. W. Cadmium Induced Testicular Damage and Its Response to Administration of Succimer and Diphenyl Diselenide in Mice. *Toxicol. Lett.* **2004**, *152*, 255–263.
21. Wang, H. J.; Liu, Z. P.; Jia, X. D.; Chen, H.; Tan, Y. J. Endocrine Disruption of Cadmium in Rats Using the OECD Enhanced TG 407 Test System. *Biomed. Environ. Sci.* **2014**, *27*, 950–959.
22. Yu, X.; Hong, S.; Faustman, E. M. Cadmium-induced Activation of Stress Signaling Pathways, Disruption of Ubiquitin-dependent Protein Degradation and Apoptosis in Primary Rat Sertoli Cell–Gonocyte Cocultures. *Toxicol. Sci.* **2008**, *104*, 385–396.
23. Hew, K. W.; Ericson, W. A.; Welsh, M. J. A Single Low Cadmium Dose Causes Failure of Spermiation in the Rat. *Toxicol. Appl. Pharmacol.* **1993**, *121*, 15–21.
24. Hew, K. W.; Heath, G. L.; Jiwa, A. H.; Welsh, M. J. Cadmium In Vivo Causes Disruption of Tight Junction-associated Microfilaments in Rat Sertoli Cells. *Biol. Reprod.* **1993**, *49*, 840–849.
25. Xiao, X.; Mruk, D. D.; Tang, E. I.; Wong, C. K.; Lee, W. M.; John, C. M.; Turek, P. J.; Silvestrini, B.; Cheng, C. Y. Environmental Toxicants Perturb Human Sertoli Cell Adhesive Function via Changes in F-actin Organization Mediated by Actin Regulatory Proteins. *Hum. Reprod.* **2014**, *29*, 1279–1291.
26. Wang, C. Y.; Jenkitkasemwong, S.; Duarte, S.; Sparkman, B. K.; Shawki, A.; Mackenzie, B.; Knutson, M. D. ZIP8 Is an Iron and Zinc Transporter Whose Cell-surface Expression Is Up-regulated by Cellular Iron Loading. *J. Biol. Chem.* **2012**, *287*, 34032–34043.
27. Matsuura, T.; Kawasaki, Y.; Miwa, K.; Sutou, S.; Ohinata, Y.; Yoshida, F.; Mitsui, Y. Germ Cell-specific Nucleocytoplasmic Shuttling Protein, Tesmin, Responsive to Heavy Metal Stress in Mouse Testes. *J. Inorg. Biochem.* **2002**, *88*, 183–191.
28. Koeberle, A.; Shindou, H.; Harayama, T.; Yuki, K.; Shimizu, T. Polyunsaturated Fatty Acids Are Incorporated into Maturating Male Mouse Germ Cells by Lysophosphatidic Acid Acyltransferase 3. *FASEB J.* **2012**, *26*, 169–180.
29. Manna, P.; Sinha, M.; Sil, P. C. Cadmium Induced Testicular Pathophysiology: Prophylactic Role of Taurine. *Reprod. Toxicol.* **2008**, *26*, 282–291.
30. Sharma, B.; Singh, S.; Siddiqi, N. J. Biomedical Implications of Heavy Metals Induced Imbalances in Redox Systems. Biomed. *Res. Int.* **2014**, *2014*, 640754.
31. Turner T. T.; Lysiak J. J. Oxidative Stress: A Common Factor in Testicular Dysfunction. *J. Androl.* **2008**, *29*, 488–498.
32. Agarwal, A.; Ikemoto, I.; Loughlin, K. R. Prevention of Testicular Damage by Free-radical Scavengers. *Urology* **1997**, *50*, 759–763.
33. Arguelles, N.; Alvarez-Gonzalez, I.; Chamorro, G.; Madrigal-Bujaidar, E. Protective Effect of Grapefruit Juice on the Teratogenic and Genotoxic Damage Induced by Cadmium in Mice. *J. Med. Food* **2012**, *15*, 887–893.
34. Erboga, M.; Kanter, M.; Aktas, C.; Donmez, Y. B.; Erboga, Z. F.; Aktas, E.; Gurel, A. Anti-apoptotic and Anti-oxidant Effects of Caffeic Acid Phenethyl Ester on Cadmium-induced Testicular Toxicity in Rats. *Biol. Trace Elem. Res.* **2015**, 1–9. DOI:10.1007/s12011-015-0509-y.

35. Jahan, S.; Zahra, A.; Irum, U.; Iftikhar, N.; Ullah, H. Protective Effects of Different Antioxidants Against Cadmium Induced Oxidative Damage in Rat Testis and Prostate Tissues. *Syst. Biol. Reprod. Med.* **2014,** *60*, 199–205.
36. Minutoli, L.; Micali, A.; Pisani, A.; Puzzolo, D.; Bitto, A.; Rinaldi, M.; Pizzino, G.; Irrera, N.; Galfo, F.; Arena, S.; Pallio, G.; Mecchio, A.: Germanà, A.; Bruschetta, D.; Laurà, R.; Magno, C.; Marini, H.; Squadrito, F.; Altavilla, D. Flavocoxid Protects Against Cadmium-induced Disruption of the Blood–Testis Barrier and Improves Testicular Damage and Germ Cell Impairment in Mice. *Toxicol. Sci.* **2015,** *148*, 311–329.
37. Ognjanović, B. I.; Marković, S. D.; Ethordević, N. Z.; Trbojević, I. S.; Stajn, A. S.; Saicić, Z. S. Cadmium-induced Lipid Peroxidation and Changes in Antioxidant Defense System in the Rat Testes: Protective Role of Coenzyme Q(10) and Vitamin E. *Reprod. Toxicol.* **2010,** *29*, 191–197.
38. Oguzturk, H.; Ciftci, O.; Aydin, M.; Timurkaan, N.; Beytur, A.; Yilmaz, F. Ameliorative Effects of Curcumin Against Acute Cadmium Toxicity on Male Reproductive System in Rats. *Andrologia* **2012,** *44*, 243–249.
39. Yiin, S. J.; Chern, C. L.; Sheu, J. Y.; Lin, T. H. Cadmium Induced Lipid Peroxidation in Rat Testes and Protection by Selenium. *Biometals* **1999,** *12*, 353–359.
40. Thompson, J.; Bannigan, J. Cadmium: Toxic Effects on the Reproductive System and the Embryo. *Reprod. Toxicol.* **2008,** *25*, 304–315.
41. Fouad, A. A.; Qureshi, H. A.; Al-Sultan, A. I.; Yacoubi, M. T.; Ali, A. A. Protective Effect of Hemin Against Cadmium-induced Testicular Damage in Rats. *Toxicology* **2009,** *257*, 153–160.
42. Fowler, B. A. Monitoring of Human Populations for Early Markers of Cadmium Toxicity: A Review. *Toxicol. Appl. Pharmacol.* **2009,** *38*, 294–300.
43. Su, L.; Mruk, D. D.; Cheng, C. Y. Regulation of Drug Transporters in the Testis by Environmental Toxicant Cadmium, Steroids and Cytokines. *Spermatogenesis* **2012,** *2*, 285–293.
44. Marettová, E.; Maretta, M.; Legáth, J. Toxic Effects of Cadmium on Testis of Birds and Mammals: A Review. *Anim. Reprod. Sci.* **2015,** *155*, 1–10.
45. Interdonato, M.; Bitto, A.; Pizzino, G.; Irrera, N.; Pallio, G.; Mecchio, A.; Cuspilici, A.; Minutoli, L.; Altavilla, D.; Squadrito, F. Levels of Heavy Metals in Adolescents Living in the Industrialised Area of Milazzo-Valle del Mela (Northern Sicily). *J. Environ. Public Health* **2014,** *2014*, 326845.
46. Interdonato, M.; Pizzino, G.; Bitto, A.; Galfo, F.; Irrera, N.; Mecchio, A.; Pallio, G.; Ramistella, V.; De Luca, F.; Santamaria, A.; Minutoli, L.; Marini, H.; Squadrito, F.; Altavilla, D. Cadmium Delays Puberty Onset and Testis Growth in Adolescents. *Clin. Endocrinol.* **2015,** *83*, 357–362.
47. Pant, N.; Pant, A. B.; Chaturvedi, P. K.; Shukla, M.; Mathur, N.; Gupta, Y. K.; Saxena, D. K. Semen Quality of Environmentally Exposed Human Population: The Toxicological Consequence. *Environ. Sci. Pollut. Res. Int.* **2013,** *20*, 8274–8281.
48. Pant, N.; Kumar, G.; Upadhyay, A. D.; Patel, D. K.; Gupta, Y. K.; Chaturvedi, P. K. Reproductive Toxicity of Lead, Cadmium, and Phthalate Exposure in Men. *Environ. Sci. Pollut. Res.* **2014,** *21*, 11066–11074.
49. Lewis, R. C.; Meeker, J. D. Biomarkers of Exposure to Molybdenum and Other Metals in Relation to Testosterone Among Men from the United States National Health and Nutrition Examination Survey 2011–2012. *Fertil. Steril.* **2015,** *103*, 172–178.

50. Zhao, X.; Cheng, Z.; Zhu, Y. I.; Li, S.; Zhang, L.; Luo, Y. Effects of Paternal Cadmium Exposure on the Sperm Quality of Male Rats and the Neurobehavioral System of Their Offspring. *Exp. Ther. Med.* **2015**, *10*, 2356–2360.
51. Balabanič, D.; Rupnik, M.; Klemenčič, A. K. Negative Impact of Endocrine-disrupting Compounds on Human Reproductive Health. *Reprod. Fertil. Dev.* **2011**, *23*, 403–416.
52. Wong, E. W.; Cheng, C. Y. Impacts of Environmental Toxicants on Male Reproductive Dysfunction. *Trends Pharmacol. Sci.* **2011**, *32*, 290–299.
53. Marzec-Wróblewska, U.; Kamiński, P.; Lakota, P. Influence of Chemical Elements on Mammalian Spermatozoa. *Folia Biol.* **2012**, *58*, 7–15.
54. Abarikwu, S. O. Causes and Risk Factors for Male-factor Infertility in Nigeria: A Review. *Afr. J. Reprod. Health* **2013**, *17*, 150–166.
55. Mango, F. P.; Nantia, E. A.; Mathur, P. P. Effect of Environmental Contaminants on Mammalian Testis. *Curr. Mol. Pharmacol.* **2014**, *7*, 119–135.
56. Kheradmand, A.; Alirezaei, M.; Dezfoulian, O. Biochemical and Histopathological Evaluations of Ghrelin Effects Following Cadmium Toxicity in the Rat Testis. *Andrologia* **2015**, *47*, 634–643.
57. Luca, G.; Lilli, C.; Bellucci, C.; Mancuso, F.; Calvitti, M.; Arato, I.; Falabella, G.; Giovagnoli, S.; Aglietti, M. C.; Lumare, A.; Muzi, G.; Calafiore, R.; Bodo, M. Toxicity of Cadmium on Sertoli Cell Functional Competence: An In Vitro Study. *J. Biol. Regul. Homeost. Agents* **2013**, *27*, 805–816.
58. Coelho, J. Regeneration of Spermatogenesis and Sperm Quality in Mice Exposed to $CdCl_2$. Master's Degree in Ecotoxicology and Toxicology, University of Aveiro, 2008.
59. Oliveira, H.; Spanò, M.; Santos, C.; Pereira, Mde L. Adverse Effects of Cadmium Exposure on Mouse Sperm. *Reprod. Toxicol.* **2009**, *28*, 550–555.
60. Oliveira, H.; Lopes, T.; Almeida, T.; Pereira, Mde L.; Santos, C. Cadmium-induced Genetic Instability in Mice Testis. *Hum. Exp. Toxicol.* **2012**, *31*, 1228–1236.
61. Elbetieha, A.; Tbeileh, N.; Darmani, H. Assessment of Six Weeks Exposure to Cadmium Chloride on Fertility of Adult Male Mice. *J. Appl. Biol. Sci.* **2011**, *5(3)*, 13–19.
62. Ekhoye, E. I.; Nwangwa, E. K.; Aloamaka, C. P. Changes in some Testicular Biometric Parameters and Testicular Function in Cadmium Chloride Administered Wistar Rats. *Br. J. Med. Med. Res.* **2013**, *3*(4), 2031–2041.
63. El-Refaiy, A. I.; Eissa, F. I. Histopathology and Cytotoxicity as Biomarkers in Treated Rats with Cadmium and some Therapeutic Agents. *Saudi J. Biol. Sci.* **2013**, *20*, 265–280.
64. Ohtani, K.; Yanagiba, Y.; Ashimori, A.; Takeuchi, A.; Takada, N.; Togawa, M.; Hasegawa, T.; Ikeda, M.; Miura, N. Influence of Injection Timing on Severity of Cadmium-induced Testicular Toxicity in Mice. *J. Toxicol. Sci.* **2013**, *38*, 145–150.
65. Adamkovicova, M.; Toman, R.; Cabaj, M.; Massanyi, P.; Martiniakova, M.; Omelka, R.; Krajcovicova, V.; Duranova, H. Effects of Subchronic Exposure to Cadmium and Diazinon on Testis and Epididymis in Rats. *Sci. World J.* **2014**, *2014*, 1–9.
66. Asadi, M. H.; Zafari, F.; Sarveazad, A.; Abbasi, M.; Safa, M.; Koruji, M.; Yari, A.; Alizadeh Miran, R. Saffron Improves Epididymal Sperm Parameters in Rats Exposed to Cadmium. *Nephrourol. Mon.* **2013**, *6*, e12125.
67. Chakraborty, S.; Gang, S.; Sengupta, M. Functional Status of Testicular Macrophages in an Immunopriviledged Niche in Cadmium Intoxicated Murine Testes. *Am. J. Reprod. Immunol.* **2014**, *72*, 14–21.

68. Du, X.; Lan, T.; Yuan, B.; Chen, J.; Hu, J.; Ren, W.; Chen, Z. Cadmium-induced Microsatellite Instability in the Kidneys and Leukocytes of C57BL/6J Mice. *Environ. Toxicol.* **2015**, *30*, 683–692.
69. Kariyazono, Y.; Taura, J.; Hattori, Y.; Ishii, Y.; Narimatsu, S.; Fujimura, M.; Takeda, T.; Yamada, H. Effect of In Utero Exposure to Endocrine Disruptors on Fetal Steroidogenesis Governed by the Pituitary–Gonad Axis: A Study in Rats Using Different Ways of Administration. *J. Toxicol. Sci.* **2015**, *40*, 909–916.
70. Lacorte, L. M.; Rinaldi, J. C.; Justulin, L. A. Jr.; Delella, F. K.; Moroz, A.; Felisbino, S. L. Cadmium Exposure Inhibits MMP2 and MMP9 Activities in the Prostate and Testis. Biochem. *Biophys. Res. Commun.* **2015**, *457*, 538–541.
71. Niknafs, B.; Salehnia, M.; Kamkar, M. Induction and Determination of Apoptotic and Necrotic Cell Death by Cadmium Chloride in Testis Tissue of Mouse. J. *Reprod. Infertil.* **2015**, *16*, 24–29.
72. Pereira, M. L.; Tavares, R.; Garcia e Costa, F. Light Microscopy of Male Reproductive Organs After CdCl$_2$ Exposure: Focus on Morphometric Analysis of Testis and Sperm Parameters. *IJSER* **2015**, *6*, 39–41.
73. Sivaprakasam, C.; Nachiappan, V. Modulatory Effect of Cadmium on the Expression of Phospholipase A2 and Proinflammatory Genes in Rat Testis. *Environ. Toxicol.* [Online early access]. DOI: 10.1002/tox.22124.
74. Guevara, M. A.; Oliveira, H.; Pereira, M. L.; Morgado, F. A. Segmentation and Morphometry of Histological Sections Using Deformable Models: A New Tool for Evaluating Testicular Histopathology. In *Progress in Pattern Recognition, Speech and Image Analysis*; Sanfeliu, A., Ruiz-Shulcloper, J., Eds.; Springer: Berlin/Heidelberg; Lecture Notes on Computer Science; 2003, 2905, pp 282–290.

CHAPTER 18

MICROSCOPY ASSESSMENT OF EMERGING CONTAMINANTS' EFFECTS ON AQUATIC SPECIES

ÂNGELA BARRETO[1], ANA VIOLETA GIRÃO[2],
MARIA DE LOURDES PEREIRA[3*], TITO TRINDADE[2],
AMADEU MORTÁGUA VELHO DA MAIA SOARES[1], and
MIGUEL OLIVEIRA[1]

[1]*Department of Biology and CESAM, University of Aveiro, Campus Santiago, 3810-193 Aveiro, Portugal*

[2]*Department of Chemistry and CICECO—Aveiro Institute of Materials, University of Aveiro, Campus Santiago, 3810-193 Aveiro, Portugal*

[3]*Department of Biology and CICECO—Aveiro Institute of Materials, University of Aveiro, Campus Santiago, 3810-193 Aveiro, Portugal*

*Corresponding author. E-mail: mlourdespereira@ua.pt

ABSTRACT

The possible adverse effects of emerging contaminants (EC) on the environment and human health are currently one of the highest concerns in the scientific community and legislators. EC include a diverse array of compounds that have recently revealed to occur extensively in the environment and may induce a variety of potential toxic effects. This chapter describes the relevance of microscopy in the ecotoxicology field, in particular, in the study of the effects of some EC on aquatic organisms. Some representative examples are given to illustrate the application of microscopy in the detection of EC-induced lesions. Overall, microscopy coupled to other techniques is a valuable tool for ecotoxicology.

18.1 EMERGING CONTAMINANTS

18.1.1 INTRODUCTION

Anthropogenic activities involve the increasing use of a wide variety of compounds for a large variety of purposes (e.g., production and conservation of food, industrial manufacturing procedures, as well as for human and animal healthcare) that may ultimately be released into the terrestrial and aquatic environments.[1,2] Historically, the industrial activities have been associated with the use of a large variety and quantity of chemicals and generation of chemical residues released to the environment. Until the middle of 20th century, it was expected that released gas would be dispersed in atmosphere and the liquids would be diluted in large water masses and transported away from their production sites. However, environmental disasters associated with contamination (mercury and cadmium), with serious consequences to humans (Minamata and Itai Itai diseases, respectively), emphasized their effects on biota. Similarly, the book *Silent Spring*[3] attracted the attention to the consequences of the accumulation of pesticides in the food chain and their consequences to the wildlife. Since the 1960s, an entire new research discipline, ecotoxicology, has developed and tools to detect and predict the hazards and risks of environmental contaminants have been established.[4]

Ecotoxicology can be defined as the study of the effects of contaminants (chemical, physical, or biological) on biota, at different levels, from organism to ecosystem level. It is an interdisciplinary field concerned with mechanisms involved in transport, distribution, transformation, interactions, fate, and effects of contaminants to different levels of biological organization of an ecosystem.[5]

Over the last decades, the term "emerging contaminants" (EC) has been increasingly used, incorporating not only new synthetic or natural compounds or microorganisms but also those recently found in the environment due to new detection methods[6] and compounds only recently been categorized as contaminants.[1] EC, continuously entering the aquatic environment due to their continuous and widespread use and limited elimination or degradation in wastewater treatments,[7] encompass a varied range of man-made compounds, as well as metabolites and transformation products (e.g., pesticides, cosmetics, personal and domestic care products, pharmaceuticals, engineered nanoparticles (NP), and microplastics). Several reviews have been recently published concerning the occurrence of EC in the environment.[1,2,8–20] EC may induce adverse effects to nontarget organisms[13] leading to an increasing concern in the scientific community to study the environmental fate of EC and their potential toxic effects.[1,2,11,21]

The diversity of compounds currently considered EC led to their classification in different categories based on their characteristics (e.g., purpose and nature). Some common categories of EC are present in Table 18.1.

TABLE 18.1 Examples of Classes of Emerging Contaminants (EC).

Classes	Examples	Definition
Pesticides	Dimethoate and deltamethrin	Agents that deter, incapacitate, kill, or otherwise discourage pests.
Pharmaceuticals	Diclofenac, gemfibrozil, and fluoxetine	Substances used to prevent, diagnose, and treat diseases.
Microplastics	Polyethylene and polystyrene particles	Polymeric particles with sizes smaller than 5 mm.
Nanoparticles	Gold, silver, and zinc oxide	Particles typically smaller than 100 nm.

18.1.2 PHARMACEUTICALS

The occurrence of pharmaceuticals in the environment, defined as chemicals for diagnosis, treatment, alteration, or avoidance of disease, health situation or structure/function of an organism,[22] is currently an issue of great concern. These compounds may be categorized according to the mechanisms of action or therapeutic purposes. Overall, pharmaceuticals may be classified as analgesics, antibiotics, antidiabetics, antihypertensives, contraceptives, lipid regulators, antidepressants, and cytotoxic drugs.[23]

Pharmaceuticals, continuously released into the environment through wastewater treatment plants (WWTP) products such as sewage sludge and wastewaters,[24] runoff from agricultural fields, and animal wastes,[25] as well as inappropriate disposal of expired medication and aquaculture activities,[26] were considered to interact with specific pathways and processes in target organisms. It is estimated that with the progress in medical technology and aging demographics an increased consumption of pharmaceuticals will occur[27] and ultimately their levels in the environment can also increase. As reported above, pharmaceuticals are often detected within WWTP effluents and receiving waters at low concentrations (ng.L^{-1} to µg.L^{-1}),[28,29] although in an Indian wastewater treatment effluent, values as high as 100 µg.L^{-1} have been described.[30]

The concerns of the scientific community on the presence of pharmaceuticals in the environment is highlighted by the fact that the biological systems upon which pharmaceuticals act are not exclusive to the target organisms (e.g., humans) and that these substances may also display unusual dose–effect responses, with effects detected at very low concentrations disappearing at higher concentrations.[31–33] Currently, approximately 5000 active substances are available, with a large number entering the market before the environmental risk assessment guidelines could be established.[34]

The first evidence of the occurrence of pharmaceuticals in the environment can be pointed to the 1960's when hormones were found in WWTP effluents after conventional treatment processes.[35] However, the development of sensitive analytical techniques that permitted detection of pharmaceuticals in low concentrations,[36] and the reports on relevant alterations within aquatic organisms (e.g., behavior and reproduction) emphasized the need for ecotoxicity studies. Despite the ubiquitous use and release of pharmaceuticals in the environment, only a small ratio of prescribed pharmaceuticals has been studied for their environmental impacts.

18.1.3 NANOPARTICLES

In the past decades, there has been great progress in the production and application of NP. NP are fine divided materials whose average sizes are typically less than 100 nm, thus displaying sizes comparable to large biosystems such as enzymes, receptors, and antibodies. Owing to their small sizes and large surface area per volume, NP display a number of new properties as compared to conventional materials. Hence, several physical, chemical, and biological properties of NP can be explored with advantage in diverse areas of human activity including advanced materials, optoelectronics, biomedicine, pharmaceuticals, cosmetics, energy, environmental detection, and monitoring.[37]

Despite the numerous advantages of these new nanotechnologies, the study of the environmental impact of NP is increasingly needed due to their adventitious release to ecosystems. In fact, NP are considered EC[21] due to their increased use, size dependent characteristics, and scarce knowledge on their impact into the environment.[38] NP may reach the environment mainly by direct application to an environmental compartment (either deliberately or through unintentional product degradation), through WWTP effluent and sludge.[39–41] However, so far, it is problematic to estimate the concentrations of NP that will be released at any given time due to the limited data on the current and expected NP prevalence in commercial products.[41–43] Moreover, the alterations of NP in the environment, such as dissolution, agglomeration, and sedimentation may significantly affect the pathway and extent of environmental release.[41] Predicted environmental concentrations of NP are present in Table 18.2.

TABLE 18.2 Predicted Environmental Concentrations (PEC) in Aquatic Media of some Nanoparticles (NP).

Nanoparticles	Predicted environmental concentrations	References
Silver	0.0164–17 $\mu g.L^{-1}$	[44, 45]
Zinc oxide	0.22–10 $\mu g.L^{-1}$	[44]
Titanium dioxide	1–100 $\mu g.L^{-1}$	[44, 46, 47]
Carbon-based	3.69–32.66 $\mu g.L^{-1}$	[40, 44]

In addition to the limited information about the environmental levels of NP and their behavior in the environment, the knowledge about the biological effects of NP exposure is still scarce. The relevance of microscopy in the study of the effects of EC will be presented in the following sections by describing some of the principles, assessed endpoints, and some examples of the detected effects.

18.2 BIOLOGICAL IMPLICATIONS

18.2.1 INTRODUCTION

The success of aquatic organisms in challenging environments requires responsive adjustments at different biological levels, from molecular, cellular, tissue, and individual level. Thus, some responses at lower levels of biological organization may be used in ecotoxicology, as an environmental surveillance tool, helping to identify the beginning of potential pernicious impacts and allow corrective measures before effects are detected at population, community, and ecosystem levels.[48,49] Taking into account that in aquatic systems, a complex mixture of contaminants may be found and that the quantification of contaminants in the water/sediment compartments per se provide scarce information on the threat to biota (e.g., bioavailability and interactions), the analysis of the stressor-induced biological alterations become a valuable tool.

Fish have been widely used in the assessment of the effects of contaminants in the aquatic environment[49,50] due to its high responsiveness to low concentrations of contaminants, importance as a nutritional source for humans, and thus, as an important source of potential transfer to them.[51]

Overall, some of the identified risks of EC in the aquatic environment may include development of microbial resistance (antibiotics), effects on reproduction success,[52,53] feminization or masculinization of aquatic organisms,[54,55] dispersion vehicle of microorganisms and other contaminants, false food satiation, and decrease of predatory behavior (e.g., microplastics).[56-59] Other identified effects include the capacity to interfere with a wide range of physiological functions in aquatic organisms, namely, in biotransformation,[60,61] reactive oxygen species production and regulation,[62] hormone synthesis,[63] and death.[64]

The organ-specific responses to exposure to contaminants are related with their anatomic position, determining its exposure route and distribution of contaminants, as well as tissue regeneration rate and defensive ability. In this perspective, effects are mostly assessed in the liver (due to its significance on the metabolism and storage of "xenobiotics"[65-67]; gills (prime target of contaminants due to the wide surface area in contact with the external medium and reduced distance between internal and external media and relevant route for uptake, bioconcentration, and excretion of "toxicants"[49]; kidney (with a vital role in the maintenance of organisms' internal environment, being the key to the extracellular fluid volume, composition and acid–base balance regulation, presenting an additional function associated to "hematopoiesis"[68]; and blood (vehicle between gills and other target organs).[49,69,70]

Some reviews have been published with information about the toxicity of diverse EC to nontarget organisms.[18,71,72] One of the problems related with the presence of EC in the environment is the unpredictability of their effects. Even pharmaceuticals may cause effects on nontarget animals that cannot be predicted on the basis of their therapeutic mode of action[73] and, even in low concentrations and in short-term exposures, may have profound impacts on biological systems.

Microscopy allows researchers to evaluate a diversity of parameters to evaluate the toxicity of EC. In the case of NP, electron microscopy allows to establish the importance of size and shape in the biological effects of NP; the alterations of NP in different environmental media (e.g., seawater); and the incorporation of NP by cells and tissues and their effects. Different assays, that make use of microscopy, have been validated to test the effects of "classic" contaminants on biota and are currently also used for EC. For example, genotoxicity, may be assessed by quantifying DNA strand breaks using fluorescence microscopy (Comet Assay) and cytogenetic damage based on the quantification of micronuclei and other nuclear abnormalities in the cells, using light microscopy. Endpoints such as alterations in cell size, structure and cell viability, their number (e.g., sexual and blood) and mobility, as well as modifications in the structure of tissues may be assessed by light microscopy. In this perspective, microscopy is a valuable tool in the field of ecotoxicology although the information obtained by microscopy will depend on the methodology employed.

18.2.2 THE ROLE OF MICROSCOPY IN ECOTOXICOLOGICAL STUDIES

Microscopy is one of the most important imaging techniques used in life and materials sciences. It became an essential tool particularly in modern biology and enabled the birth of histology. Moreover, the main value of the microscope relies on the transformation of objects outside the range of the normal eye resolution such as cells or tissues of plants and animals into much larger images. Although the early history of the microscope invention is still surrounded by some uncertainty, it was always the excitement for observing smaller objects and creating better magnified images that became the endeavor of every microscopy developer or contributor.[74] And so, it continues leading to the production of several types of microscopes presenting countless technological advances, used for many different purposes and providing fundamental scientific information. It should be noted that decisive contributions to the development of super-resolution fluorescence microscopy has resulted in the award of the Nobel Prize in chemistry to Betzig, Hell, and Moerner in 2014.[75]

Microscopy plays a main role in the assessment of refined histological and ultrastructural features in organisms exposed to contaminants as well as providing additional parameters in toxicity evaluation on many bioindicators such as fish, earthworms, diplopods, and bacteria, among others.

Histopathological changes may be considered as a result of a large number of physiological processes, allowing the signaling damage in cells, tissues, or organs[62] and providing information related with the fitness of organisms. Histological analysis of gills and liver are broadly used to provide relevant insights about the aquatic environment and are among the most reliable indicators of fish health impairment induced by anthropogenic activities.[66] The selection of the proper tissue for analysis depends on the exposure form (waterborne versus food), tissue function, and properties of the contaminant to be analyzed. In fish, the skin and gills are at the interface between the internal and external media, hence becoming target tissues for waterborne EC-induced damage. Among these tissues, gills are the most studied due to the thin lamellae of the respiratory surface, its function in gas exchanges, and proven higher sensitivity. Nonetheless, histological alterations in skin have also been reported in fish collected from contaminated sites (e.g., partial loss of epidermis, total missing epidermis, separation of dermis from epidermis, separation of muscle from dermis, melanin pigment, and vacuole in muscle and dermis have been reported).[76] In gills, the thickening and lifting of secondary lamellar epithelium owing to hypertrophy may be a first sign that these structures have been exposed to hazardous conditions. Commonly detected effects include missing secondary gill lamellae, hemorrhage, necrosis, hyperplasia and hypertrophy, and gill clubbing. The liver plays a determinant role in the biotransformation of foreign substances, thereby is frequently a target of EC-induced alterations. These alterations may include deposition of body fat, hypertrophy and hyperplasia of hepatocytes, blood vessel disruption and necrosis, nuclear alterations, pyknosis, vacuolation, and fatty degeneration. The renal fish system has an extensive portal circulation structure ensuring that the kidney tubules are exposed to a higher fraction of cardiac circulation. Vacuolation, necrosis, fibrosis, pyknosis, and tubular degeneration have been described in the kidney of fish exposed to contaminated environments. These alterations may have serious consequences for ions regulation, excretion of products, and stress responses. Moreover alterations of gonads development (e.g., delayed maturation, high levels of atresia, or intersexuality) are a major effect detected in complex environments and currently associated with several pharmaceuticals.

The evaluation of effects on DNA is one important endpoint commonly used to assess the effects of contaminants. At a molecular level, DNA damage may be assessed in different cells by detecting DNA strand breaks, for instance, using fluorescent microscopy through the Comet Assay.[77] The incorporation of DNA lesion-specific repair endonucleases allows specific detection of oxidized bases and thus, identifying oxidative DNA damage as a harmful process underlying the genomic integrity loss.[78] The quantification of micronuclei and other nuclear abnormalities, particularly in erythrocytes,[49,67,70,79,80] using light microscopy, has demonstrated a high sensitivity to the exposure to environmental contaminants, providing information on the potential to induce cytogenetic damage.[67] Thus, the study of the interaction contaminants/cells takes advantage of the methodological developments of microscopy techniques (e.g., fluorescent and confocal techniques) and electron microscopy (scanning and transmission). Figure 18.1 shows some of the applications of microscopy.

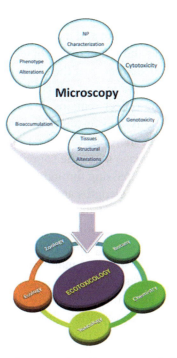

FIGURE 18.1 Scheme showing examples of some research fields related to microscopy and its applications. The role of microscopy on areas such as ecotoxicology, chemistry, and zoology, from NP characterization to their bioaccumulation and possible effects at different levels of biological organization (molecular, cellular, and tissue/organs) is highlighted.

18.2.2.1 LIGHT AND ELECTRON MICROSCOPY

There are important and common concepts inherent to any imaging systems such as image formation, magnification, resolution, depth of field/focus, and lens aberration. An image can basically be formed in three different ways. The simplest one to consider is the projection of an image like the shadow of a person/object and another is that formed by a conventional lens system, which is often called as the optical image. All parts of a projection and optical images are formed simultaneously. In the third case, each point of the picture is presented serially, a scanning image just like that in a television. Magnification of an object consists in the enlargement of its image only in appearance. The ratio between the size of the enlarged image and that of the object is also often denominated as magnification. Resolution or resolving power is defined as the closest spacing of two points which can clearly be seen as two separate entities under the microscope with the highest accuracy. Therefore, the resolution limit is often what defines the convenience of a certain instrument and its application for a certain analysis. Depth of field is known as the range of the object positions for which our eyes can detect unchanged sharpness of the corresponding image and depth of focus is also a range of positions at which the image is viewed without being out of focus and the object remains at a fixed position. Lens aberration leads to loss of quality and resolution of the final image due to its distortion at every point in the object. Aberrations are often divided into chromatic ones which take place when the source of illumination comprises a wavelength range and when passing through the lens is deviated according to the different composing wavelengths and achromatic (or monochromatic) aberrations caused by the different path lengths of different rays from an object point to that of the image including spherical aberration, astigmatism, pincushion, and barrel distortions.[81] Consequently, one of the main concerns for any microscopist and/or instrument manufacturer is the minimization of lenses aberrations in order to improve the resolution of the final image.

As previously stated, light microscopes are one of the most used tools in life sciences. In terms of configuration, the simplest microscope does not provide a large amount of magnification and it is based in single or groups of convex lenses like a magnifying glass, a loupe or an eye piece for more complex systems.[82] The second configuration is the most common one, the compound microscope, as shown in Figure 18.2.

A transmitted light source placed at the bottom of the microscope base is used to illuminate the object of interest. It first makes way through the condenser lens that focus that light and turns it into a more dense and intense one before reaching the specimen. The condenser may include other features such as a diaphragm and/or filters in order to adjust the quality and intensity of the light. The object is fixed onto a platform called stage, usually a mechanical one to enable movement of the specimen under the objective turret or revolving nosepiece. The latter component holds several lenses, which are closest to the object, the objective lens. The light is then collected by the objective lens forming a focused real image. On the side of the microscope arm are the knobs that perform the coarse and fine focus adjustment of the final image by moving the stage up or down. Finally, the focused real image formed after the objective lens is then magnified by another set of lenses, ocular lenses, or eyepiece, giving a final enlarged image of the object. The final magnification at which an image of the specimen is formed results from the product of the powers of the ocular and the objective lenses.[83]

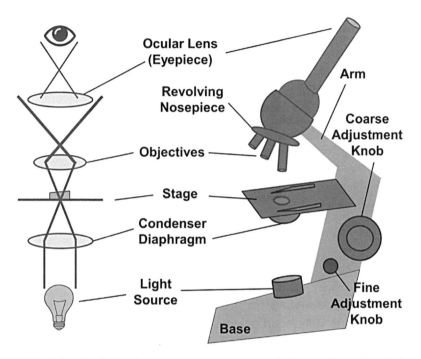

FIGURE 18.2 The light microscope: its components and diagram illustrating the image generation of an object observed under the microscope.

In light microscopy, the contrast of the final image can be controlled according to the chosen illumination technique. The most basic and common microscope is the bright-field one often termed as the light microscope. The light source enters directly in the optical system through the condenser illuminating the object, which will appear darker than its background due to absorbance of light by the specimen. In dark field imaging, a sort of stopper placed below the condenser lens is used preventing the light to enter directly the objective lens: only the sample scattered light enters the objective and is seen as an image. Therefore, the observed specimen is brighter than its background.[82] Figure 18.3 schematically illustrates the difference between bright- and dark-field imaging.

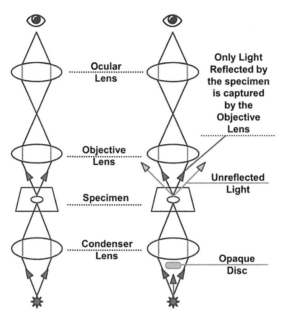

FIGURE 18.3 Diagram showing the difference between bright- and dark-field imaging.

Cross-polarized light illumination is another technique where the contrast is due to rotation of polarized light through the sample, or phase-contrast illumination where interference of different path lengths through the sample defines the contrast of the final image. Finally, the light source can be simply daylight directed by a mirror or, more commonly, a halogen lamp, lasers, or LEDs.[82]

Light microscopy has several limitations and the most important one is the loss of resolution at very high magnifications due to the diffraction limit imposed when using transmitted visible light. In practice, a conventional optical microscope offers a maximum of resolution of around 200 nm. Although there are several other techniques, which have been designed in order to overcome this limitation, surpassing the diffraction limit is still a large restrain in light microscopy.[84] Light microscopy, despite the lower resolution provides an ability to image time-evolving cellular processes.[49] Inevitably, different types of microscopy applying different physical principles in the image generation have been developed. Electron microscopy has become very important since it is widely used in materials research but also in life sciences. The first major difference between light and electron microscopy is the illumination source, which in the latter case is a highly energetic electron beam. The concept of "seeing" using electrons may be a little awkward to understand, perhaps it is easier if we consider light as radiation with wavelength of 400–700 nm and electrons with wavelengths between 0.001 and 0.01 nm. Although electrons are strongly scattered by gases much more than light, they ultimately provide a highly energetic source of illumination consequently enabling a theoretical resolution of around 0.02 nm (for 100 kV electrons). Nevertheless, lens aberration particularly the monochromatic ones are once again the drawback and such subatomic resolution is not, so far, achievable.[81]

Scanning electron microscope (SEM) is a type of electron microscope in which the specimen is scanned with the highly energetic and focused electron beam in a raster scan pattern and an image is created from specific electrons resulting from the interaction between the electron beam and the atoms of the sample. The transmission electron microscope (TEM) analyzes very thin specimens by directing the electron beam through the sample and generating bright/dark field images as well as electron diffraction patterns. The multiple and different signals generated from the interaction between the specimen and the electron beam, denominated as the interaction volume, is one main advantage of electron microscopy.[81,85] A schematic presentation of the light, TEM, and SEM microscopes is shown in Figure 18.4.

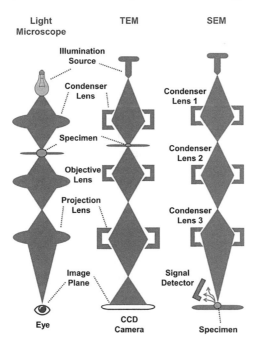

FIGURE 18.4 Schematic presentation of the light, TEM, and SEM microscopes.

In SEM, the interaction volume is much larger than that of TEM and some of the signals are the secondary and backscattered electrons used for imaging purposes or X-rays for elemental analysis. The latter is also generated and detected in TEM for qualitative/quantitative composition calculations. In TEM, different techniques based on unscattered and elastically/inelastically scattered electrons can be explored because in this case, the interaction volume is much smaller. Table 18.3 compares some distinctive features for TEM and SEM.[81]

TABLE 18.3 Characteristics of Scanning Electron Microscope (SEM) and Transmission Electron Microscope (TEM).

SEM	TEM
Signal	
Scattered electrons (secondary/backscattered)	Transmitted electrons
Resolution	
Lower resolution of 0.4 nm	Higher resolution of 0.5 Å
Magnification	
2 million	50 million
Depth of field	
Yes	No
Information	
Topography, morphology, composition, and crystallographic information	Morphology, composition, crystallographic information, dislocations, lattice parameters, single crystal orientation, etc.
Dimensional image	
2D/3D	2D
Image color	
Grey scale	Grey scale
Composition: qualitative analysis	
Available	Available and accurate
Composition: quantitative analysis	
Depends on sample preparation	Available and relatively accurate
Specimen preparation	
Relatively simple: thick/thin samples	Difficult: nearly transparent samples
Accelerating voltages	
1–50 kV	100 kV–300 kV
Cost	
Expensive	Very expensive

The evolving application of both SEM and TEM onto biological studies is becoming wider and very useful, for example, in the evaluation of cell and tissue changes under toxicological conditions. SEM enables a tridimensional image of the cell surface and TEM provides a bidimensional image of the cell relevant for any modification that may take place in cell organization. Although light and electron microscopes share some features, the differences are quite evident as summarized in Table 18.4.[81]

TABLE 18.4 Summary of the Main Differences Between Light and Electron Microscopy.

Light microscopy	Electron microscopy
Illumination source	
Light rays	Electron beam
Resolution	
Below 0.25 up to 0.30 µm	Down to 0.5 Å
Magnification	
500×–1500×	50,000,000×
Depth of field	
Small	Large
Lenses material/vacuum	
Optical glass/no need for vacuum	Electromagnets/high vacuum is needed
Dimensional image	
2D/3D	2D/3D
Image color	
RGB	Grey scale
Specimen preparation	
Simple	Difficult
Applications	
Live/dead specimens	Dead and dried specimens
Cost	
Cheap	Very expensive
Maintenance	
Negligible	Costly
Advantages	
Simple use	Higher resolution to the submicron

Although light and electron microscopy may be used for different purposes, they become essential complementary tools for the assessment of EC effects on biota.

18.2.2.2 PHARMACEUTICALS

As previously mentioned, pharmaceuticals have been detected in the environment in levels of concern. This section will focus on some of pharmaceuticals, such as 17α-ethinylestradiol, 1β-estradiol, and carbamazepine, proving examples of assessed endpoints. Among the diverse pharmaceuticals currently studied in ecotoxicology, particular attention has been given to estrogens due to their ability to impact fish populations at environmentally relevant levels.[86] The effects of these compounds may be assessed by different methodologies at different biological levels. However, microscopy has been a valuable tool in the study of the mechanisms and impacts of these compounds. The combination of light and electron microscopy provides insights on the severity of the impact, reversibility, and mechanism.

The synthetic hormone 17α-ethinylestradiol, which is a derivative of the natural hormone estradiol, is an orally bioactive estrogen, and is one of the most commonly used medications for humans as well as livestock and aquaculture activity.[87] This compound has been detected in effluents in levels as high as 0.047 µg.L^{-1} and in surface water at 0.0027 µg.L^{-1} and has demonstrated the ability to induce a variety of effects such as increase of plasma vitellogenin in fish, intersex, reduced gamete quality, complete feminization of male fish, reduced fertility and fecundity, and also behavioral changes.[29,87,88] Light and electron microscopy allowed the study of the structural alterations induced by this pharmaceutical in gonads, liver, gills, and kidney. Thus, light microscopy allowed the detection of alterations of gonadal architecture and secondary sex characteristics.[89–91] In addition, electron microscopy permitted the detection of effects on spermatogonia, the hepatic levels of glycogen and lipids as well as in the number of rough endoplasmic reticulum and macrophages; in the kidney, degeneration and tubular dilation and glomerulonephritis has been observed.[91]

17β-estradiol is a natural estrogen detected in surface waters at concentrations up to 0.027 µg.L^{-1}.[92] It has been reported to induce alterations in the morphology of seminiferous tubules in fish, hyperplastic and hypertrophied Sertoli cells, and loss of germinal cells. The presence of degenerate spermatozoa and occasional germ cell syncytia were also reported. Electron microscopy revealed enlarged, distended Sertoli cells containing phagocytized, degenerate germ cells, remnants of spermatids, lipids, and cellular debris after exposure to 17β-estradiol.[93] Genotoxic effects of this compound have also been reported after the analysis of fish erythrocytes by the Comet Assay and cytogenetic damage assessed by nuclear abnormalities.[94,95]

Carbamazepine, within the antiepileptic category, is the most detected drug in the environment. It has been found in several environments,[24] such as in surface water at concentrations up to 0.640 µg.L^{-1}.[96] Carbamazepine induces several alterations in fish as detected by microscopy, including changes in kidney tubule structure, alterations in nuclear size or density and the presence of hyaline or hypertrophic/hyperplasic of renal tissue, enlarged mitochondria, increased amount of macrophages, cellular debris within intercellular spaces, and secondary lysosomes in basal portions of cells. In gills, epithelial lifting and hyperplasia has been observed.[52,97] In gonads, atreitic oocytes and apoptotic ovarian follicles have been reported.[52] In blood, alterations in cell numbers have been observed (e.g., increased erythrocyte, monocytes, and neutrophil and decreased lymphocyte).[98] Electron microscopy allowed detections of increased amount of macrophages and membrane material in the liver cytoplasm.[97]

Overall, the magnitude of detected effects is related with the tested species, sex, the mechanism of action of the pharmaceutical, the duration of exposure and concentration, and presence of other contaminants. Nonetheless, the relevance effects detected by microscopy are undeniable.

18.2.2.3 NANOPARTICLES

Engineered NP has been used in different areas of human activities due to their unique properties that make them useful for a wide range of applications. The last decades assisted to a great use of NP, which increased the risk of significant amounts of these materials to reach the environment highlighting the need to investigate not only the behavior of NP in complex environments but also the potential effects to human and environmental health. NP can disturb the normal physiology and alter the structure of tissues and organs of humans and animals.[99–105] The NP effects (e.g., cytotoxic) vary with the core element, size, shape, surface charge, chemistry, composition, and subsequent NP stability. In addition to particle-related factors, experimental conditions, route and dose of administration, extent of tissue distribution, and biological structures or organisms studied are important considerations to take into account in nanotoxicity studies.[103,106] Due to the diverse factors that may influence NP properties, the exact mechanism involved in the NP toxicity is yet unknown,[99,100] and no safety and regulatory guidelines concerning the production and application of NP exist.

Microscopy has been widely used in the study of the properties of NP (e.g., size, shape, and surface coating). Electron microscopy (scanning or transmission) has been used as an irreplaceable tool to characterize the primary particle size.[102,107–113] Figure 18.5 shows TEM images of a typical sample of colloidal gold NP that gives information on the morphology, size, and crystallinity of the particles.

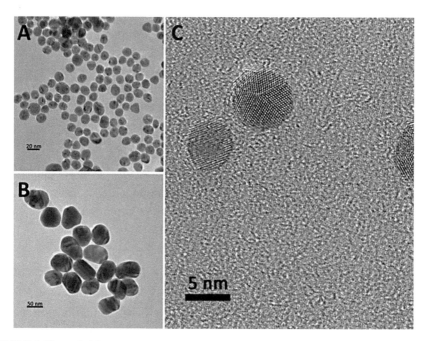

FIGURE 18.5 Gold NP: (A) 15 nm (bright field TEM), (B) 60 nm (bright field TEM), and (C) 5 nm (bright field high-resolution TEM).

NP may interact with environmental/biological systems and may suffer several processes such as agglomeration, sedimentation, and precipitation, which can alter their original characteristics and consequently their toxicity.[38] Several studies have used electron microscopy to study the behavior of diverse NP in different high ionic strength media, such as artificial seawater and biological buffers, representative of the environmental/biological media.[38,102,114,115]

Microscopy has also been used as a tool for investigating NP accumulation in cells. Several studies in different organisms have shown that SEM and TEM are useful to detect the presence of different types of NP in diverse organs/cells.[109,110,115–126] Light microscopy has been also used to identify histological alterations caused by NPs.[127,128] Figure 18.6 shows the histological alterations caused by gold NP in combination with a pharmaceutical, gemfibrozil, on gills and liver of fish *Sparus aurata*.

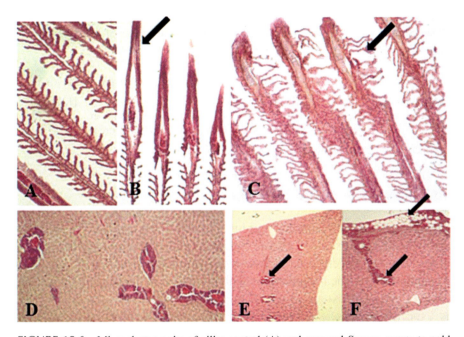

FIGURE 18.6 Microphotographs of gills: control (A) and exposed *Sparus aurata* to gold NP and pharmaceutical gemfibrozil (B and C) displaying several lesions (arrows): fusion of adjacent secondary lamellae that resulted in the obliteration of the interlamellar space between secondary lamellae, hypertrophy of lamellar epithelium, necrosis, and edema. Histological pictures of liver and pancreas from control (D) and exposed fish (E and F) showing several alterations (arrows): disintegration of parenchyma, fatty degeneration, numerous and large lipid vacuoles in liver; disorganization of the normal structure of pancreas, degeneration of pancreatic cells, decrease of acini walls, and increase in the acini lumen; H and E 200×.

Table 18.5 depicts some examples of endpoints assessed in nanotoxicology studies with aquatic organisms using of microscopy.

TABLE 18.5 Examples of Microscopic Applications in the Field of Nanotoxicology.

Organisms/cells	Microscopy	Assessed endpoints	References
Clams—digestive gland and gills *Ruditapes philippinarum, Scrobicularia plana*	TEM	Bioaccumulation of gold NP.	[110, 115, 121]
Fish—liver			
Barbonymus gonionotus —gills	Light microscopy	Accumulation of silver NP.	[122]
Danio rerio	Light microscopy	Histological alterations after the exposure to cooper, silver, and titanium dioxide NP.	[128]
Oncorhynchus mykiss	TEM and light microscopy	Bioaccumulation of silver NP aggregates.	[117, 129]
Sparus aurata		Histological changes on liver and gills after the exposure to gold NP.	[127]

18.3 FINAL CONSIDERATIONS

The relevance of conventional and highly specialized microscopy methods not only to industries but also to medicine and ecotoxicology is undeniable. They are largely used to assess the subcellular location of EC such as nanomaterials, their interactions with biological molecules, and potential toxic effects. The extensive use of microscopy in several areas of biology is expected to increase directly with the escalation of human activities introducing nanoentities with size- and shape-dependent properties into our environment with still uncertain impact.

This chapter emphasized the role of light and electron microscopy techniques in the field of ecotoxicology applied to aquatic species. The relevance of such application of advanced techniques has been clearly demonstrated and their use and development should be encouraged in future studies.

ACKNOWLEDGMENT

This work was supported through the COMPETE—Operational Competitiveness Programme and national funds through FCT—Foundation for Science and Technology, under the projects: "NANOAu—Effects of Gold Nanoparticles to Aquatic Organisms" (FCT PTDC/MAR-EST/3399/2012)

(FCOMP-01-0124-FEDER-029435), "UID/AMB/50017/2013," "Pest-C/CTM/LA0011/2013"; CICECO-Aveiro Institute of Materials "POCI-01-0145-FEDER-007679," "UID/CTM/50011/2013" financed by national funds through the FCT/MEC and cofinanced by FEDER under the PT2020 Partnership Agreement. A. Barreto has a doctoral fellowship from FCT (SFRH/BD/97624/2013), A. V. Girão has a post-doc fellowship from FCT (SFRH/BPD/66407/2009) and M. Oliveira has a post-doc fellowship from FCT (SFRH/BPD/85107/2012) supported by the European Social Fund and national funds from the "Ministério da Educação e Ciência (POPH—QREN—Tipologia 4.1)" of Portugal.

KEYWORDS

- **nanoparticles**
- **ecotoxicology**
- **pharmaceuticals**
- **aquatic species**
- **microscopy**

REFERENCES

1. Lapworth, D. J.; Baran, N.; Stuart, M. E.; Ward, R. S. Emerging Organic Contaminants in Groundwater: A Review of Sources, Fate and Occurrence. *Environ. Pollut.* **2012,** *163,* 287–303.
2. Gavrilescu, M.; Demnerová, K.; Aamand, J.; Agathos, S.; Fava, F. Emerging Pollutants in the Environment: Present and Future Challenges in Biomonitoring, Ecological Risks and Bioremediation. *N. Biotechnol.* **2015,** *32*(1), 147–156.
3. Carson, R. *Silent Spring*. Penguin: Middlesex, UK, 1962.
4. Werner, I.; Hitzfeld, B. 50 Years of Ecotoxicology Since Silent Spring: A Review. *GAIA—Ecol. Perspect. Sci. Soc.* **2012,** *21*(3), 217–224.
5. Relyea, R.; Hoverman, J. Assessing the Ecology in Ecotoxicology: A Review and Synthesis in Freshwater Systems. *Ecol. Lett.* **2006,** *9*(10), 1157–1171.
6. Richardson, S. D.; Ternes, T. A. Water Analysis: Emerging Contaminants and Current Issues. *Anal. Chem.* **2014,** *86*(6), 2813–2848.
7. Petrović, M.; Gonzalez, S.; Barceló, D. Analysis and Removal of Emerging Contaminants in Wastewater and Drinking Water. *TrAC Trends Anal. Chem.* **2003,** *22*(10), 685–696.
8. Jurado, A.; Vazquez-Sune, E.; Carrera, J.; Lopez de Alda, M.; Pujades, E.; Barcelo, D. Emerging Organic Contaminants in Groundwater in Spain: A Review of Sources, Recent Occurrence and Fate in a European Context. *Sci. Total Environ.* **2012,** *440,* 82–94.

9. Ramakrishnan, A.; Blaney, L.; Kao, J.; Tyagi, RD.; Zhang, T. C.; Surampalli, R. Y. Emerging Contaminants in Landfill Leachate and Their Sustainable Management. *Environ. Earth Sci.* **2015**, *73*(3), 1357–1368.
10. Houtman, C. J. Emerging Contaminants in Surface Waters and Their Relevance for the Production of Drinking Water in Europe. *J. Integr. Environ. Sci.* **2010**, *7*(4), 271–295.
11. Thomaidis, N. S.; Asimakopoulos, A. G.; Bletsou, A. A. Emerging Contaminants: A Tutorial Mini-review. *Global Nest J.* **2012**, *14*(1), 72–79.
12. Meffe, R.; de Bustamante, I. Emerging Organic Contaminants in Surface Water and Groundwater: A First Overview of the Situation in Italy. *Sci. Total Environ.* **2014**, *481*, 280–295.
13. Geissen, V.; Mol, H.; Klumpp, E.; Umlauf, G.; Nadal, M.; van der Ploeg, M. et al. Emerging Pollutants in the Environment: A Challenge for Water Resource Management. *Int. Soil Water Conser. Res.* **2015**, *3*(1), 57–65.
14. Zhang, C.; Li, Y.; Wang, C.; Niu, L. H.; Cai, W. Occurrence of Endocrine Disrupting Compounds in Aqueous Environment and Their Bacterial Degradation: A Review. *Crit. Rev. Environ. Sci. Technol.* **2016**, *46*(1), 1–59.
15. Zhang, S.; Zhang, Q.; Darisaw, S.; Ehie, O.; Wang, G. Simultaneous Quantification of Polycyclic Aromatic Hydrocarbons (PAHs), Polychlorinated Biphenyls (PCBs), and Pharmaceuticals and Personal Care Products (PPCPs) in Mississippi River Water, in New Orleans, Louisiana, USA. *Chemosphere* **2007**, *66*(6), 1057–1069.
16. Kasprzyk-Hordern, B.; Dinsdale, R. M.; Guwy, A. J. The Occurrence of Pharmaceuticals, Personal Care Products, Endocrine Disruptors and Illicit Drugs in Surface Water in South Wales, UK. *Water Res.* **2008**, *42*(13), 3498–3518.
17. Murray, K. E.; Thomas, S. M.; Bodour, A. A. Prioritizing Research for Trace Pollutants and Emerging Contaminants in the Freshwater Environment. *Environ. Pollut.* **2010**, *158*(12), 3462–3471.
18. Stuart, M.; Lapworth, D.; Crane, E.; Hart, A. Review of Risk from Potential Emerging Contaminants in UK Groundwater. *Sci. Total Environ.* **2012**, *416*, 1–21.
19. Petrie, B.; Barden, R.; Kasprzyk-Hordern, B. A Review on Emerging Contaminants in Wastewaters and the Environment: Current Knowledge, Understudied Areas and Recommendations for Future Monitoring. *Water Res.* **2015**, *72*, 3–27.
20. Terzić, S.; Senta, I.; Ahel, M.; Gros, M.; Petrović, M.; Barcelo, D. et al. Occurrence and Fate of Emerging Wastewater Contaminants in Western Balkan Region. Sci. Total Environ. 2008, 399(1–3), 66–77.
21. Sauve, S.; Desrosiers, M. A Review of What Is an Emerging Contaminant. *Chem. Cent. J.* **2014**, *8*(1), 15.
22. Jones, O. A. H.; Voulvoulis, N.; Lester, J. N. Human Pharmaceuticals in the Aquatic Environment: A Review. *Environ. Technol.* **2001**, *22*(12), 1383–1394.
23. Huerta, B.; Rodríguez-Mozaz, S.; Barceló, D. Pharmaceuticals in Biota in the Aquatic Environment: Analytical Methods and Environmental Implications. *Anal. Bioanal. Chem.* **2012**, *404*(9), 2611–2624.
24. Oliveira, M.; Cardoso, D.; Soares, A. M. V. M.; Loureiro, S. Effects of Short-term Exposure to Fluoxetine and Carbamazepine to the Collembolan *Folsomia candida*. *Chemosphere* **2015**, *120*, 86–91.
25. Pedersen, J. A.; Soliman, M.; Suffet, I. H. Human Pharmaceuticals, Hormones, and Personal Care Product Ingredients in Runoff from Agricultural Fields Irrigated with Treated Wastewater. *J. Agric. Food Chem.* **2005**, *53*(5), 1625–1632.

26. Jiang, J.-J.; Lee, C.-L.; Fang, M.-D. Emerging Organic Contaminants in Coastal Waters: Anthropogenic Impact, Environmental Release and Ecological Risk. *Marine Poll. Bull.* **2014**, *85*(2), 391–399.
27. Corcoran, J.; Winter, M. J.; Tyler, C. R. Pharmaceuticals in the Aquatic Environment: A Critical Review of the Evidence for Health Effects in Fish. *Crit. Rev. Toxicol.* **2010**, *40*(4), 287–304.
28. Fent, K.; Weston, A. A.; Caminada, D. Ecotoxicology of Human Pharmaceuticals. *Aquat. Toxicol.* **2006**, *76*(2), 122–159.
29. Ternes, T. A.; Stumpf, M.; Mueller, J.; Haberer, K.; Wilken, R. D.; Servos, M. Behavior and Occurrence of Estrogens in Municipal Sewage Treatment Plants—I. Investigations in Germany, Canada and Brazil. *Sci. Total Environ.* **1999**, *225*(1–2), 81–90.
30. Larsson, D. G. J.; de Pedro, C.; Paxeus, N. Effluent from Drug Manufactures Contains Extremely High Levels of Pharmaceuticals. *J. Hazard. Mater.* **2007**, *148*(3), 751–755.
31. Rivetti, C.; Campos, B.; Barata, C. Low Environmental Levels of Neuro-active Pharmaceuticals Alter Phototactic Behaviour and Reproduction in *Daphnia magna*. *Aquat. Toxicol.* **2016**, *170*, 289–296.
32. De Lange, H. J.; Noordoven, W.; Murk, A. J.; Lürling, M.; Peeters, E. T. H. M. Behavioural Responses of *Gammarus pulex* (Crustacea, Amphipoda) to Low Concentrations of Pharmaceuticals. *Aquat. Toxicol.* **2006**, *78*(3), 209–216.
33. Guler, Y.; Ford, A. T. Anti-depressants Make Amphipods See the Light. *Aquat. Toxicol.* **2010**, *99*(3), 397–404.
34. Küster, A.; Adler, N. Pharmaceuticals in the Environment: Scientific Evidence of Risks and Its Regulation. *Philos. Trans. Royal Soc. Lond. B Biol. Sci.* **2014**, *369*, 1–8.
35. Stumm-Zollinger, E.; Fair, G. M. Biodegradation of Steroid Hormones. *J. Water Pollut. Control Fed.* **1965**, *37*(11), 1506–1510.
36. Ternes, T. A. Occurrence of Drugs in German Sewage Treatment Plants and Rivers. *Water Res.* **1998**, *32*(11), 3245–3260.
37. Liu, W.-T. Nanoparticles and Their Biological and Environmental Applications. *J. Biosci. Bioeng.* **2006**, *102*(1), 1–7.
38. Barreto, Â.; Luis, L. G.; Girão, A. V.; Trindade, T.; Soares, A. M. V. M.; Oliveira, M. Behavior of Colloidal Gold Nanoparticles in Different Ionic Strength Media. *J. Nanopart. Res.* **2015**, *17*(12), 1–13.
39. Gottschalk, F.; Sonderer, T.; Scholz, R. W.; Nowack, B. Possibilities and Limitations of Modeling Environmental Exposure to Engineered Nanomaterials by Probabilistic Material Flow Analysis. *Environ. Toxicol. Chem.* **2010**, *29*(5), 1036–1048.
40. Mueller, N C.; Nowack, B. Exposure Modeling of Engineered Nanoparticles in the Environment. *Environ. Sci. Technol.* **2008**, *42*(12), 4447–4453.
41. Maurer-Jones, M. A.; Gunsolus, I. L.; Murphy, C. J.; Haynes, C. L. Toxicity of Engineered Nanoparticles in the Environment. *Anal. Chem.* **2013**, *85*(6), 3036–3049.
42. Batley, G. E.; Kirby, J. K.; McLaughlin, M. J. Fate and Risks of Nanomaterials in Aquatic and Terrestrial Environments. *Acc. Chem. Res.* **2013**, *46*(3), 854–862.
43. Klaine, S. J.; Alvarez, P. J. J.; Batley, G. E.; Fernandes, T. F.; Handy, R. D.; Lyon, D. Y. et al. Nanomaterials in the Environment: Behavior, Fate, Bioavailability, and Effects. *Environ. Toxicol. Chem.* **2008**, *27*(9), 1825–1851.
44. Gottschalk, F.; Sonderer, T.; Scholz, R. W.; Nowack, B. Modeled Environmental Concentrations of Engineered Nanomaterials (TiO_2, ZnO, Ag, CNT, Fullerenes) for Different Regions. *Environ. Sci. Technol.* **2009**, *43*(24), 9216–9222.

45. Blaser, S A.; Scheringer, M.; MacLeod, M.; Hungerbühler, K. Estimation of Cumulative Aquatic Exposure and Risk due to Silver: Contribution of Nano-functionalized Plastics and Textiles. *Sci. Total Environ.* **2008**, *390*(2–3), 396–409.
46. Westerhoff, P.; Song, G.; Hristovski, K.; Kiser, M. A. Occurrence and Removal of Titanium at Full Scale Wastewater Treatment Plants: Implications for TiO_2 Nanomaterials. *J. Environ. Monit.* **2011**, *13*(5), 1195–1203.
47. Praetorius, A.; Scheringer, M; Hungerbühler, K. Development of Environmental Fate Models for Engineered Nanoparticles—A Case Study of TiO_2 Nanoparticles in the Rhine River. Environ. *Sci. Technol.* **2012**, *46*(12), 6705–6713.
48. Monserrat, J. M.; Geracitano, L. A.; Bianchini, A. Current and Future Perspectives Using Biomarkers to Assess Pollution in Aquatic Ecosystems. *Comments Toxicol.* **2003**, *9*(5), 255–269.
49. Oliveira, M.; Maria, V. L.; Ahmad, I.; Serafim, A.; Bebianno, M. J.; Pacheco, M. et al. Contamination Assessment of a Coastal Lagoon (Ria de Aveiro, Portugal) Using Defence and Damage Biochemical Indicators in Gill of *Liza aurata*—An Integrated Biomarker Approach. *Environ. Pollut.* **2009**, *157*(3), 959–967.
50. Zhou, Q.; Zhang, J.; Fu, J.; Shi, J.; Jiang, G. Biomonitoring: An Appealing Tool for Assessment of Metal Pollution in the Aquatic Ecosystem. *Anal. Chim. Acta* **2008**, *606*(2), 135–150.
51. Alsabti, K.; Metcalfe, C. D. Fish Micronuclei for Assessing Genotoxicity in Water. *Mut. Res. Genet. Toxicol. Environ. Mutagen.* **1995**, *343*(2–3), 121–135.
52. Galus, M.; Kirischian, N.; Higgins, S.; Purdy, J.; Chow, J.; Rangaranjan, S. et al. Chronic, Low Concentration Exposure to Pharmaceuticals Impacts Multiple Organ Systems in Zebrafish. *Aquat. Toxicol.* **2013**, *132–133*, 200–211.
53. Galus, M.; Rangarajan, S.; Lai, A.; Shaya, L.; Balshine, S.; Wilson, J. Y. Effects of Chronic, Parental Pharmaceutical Exposure on Zebrafish (*Danio rerio*) Offspring. *Aquat. Toxicol.* **2014**, *151*, 124–134.
54. Agunbiade, F. O.; Moodley, B. Occurrence and Distribution Pattern of Acidic Pharmaceuticals in Surface Water, Wastewater, and Sediment of the Msunduzi River, Kwazulu-Natal, South Africa. *Environ. Toxicol. Chem.* **2016**, *35*(1), 36–46.
55. Nash, J. P.; Kime, D. E.; Van der Ven, L. T. M.; Wester, P. W.; Brion, F.; Maack, G. et al. Long-term Exposure to Environmental Concentrations of the Pharmaceutical Ethynylestradiol Causes Reproductive Failure in Fish. *Environ. Health Perspect.* **2004**, *112*(17), 1725–1733.
56. Luís, L. G.; Ferreira, P.; Fonte, E.; Oliveira, M.; Guilhermino, L. Does the Presence of Microplastics Influence the Acute Toxicity of Chromium (VI) to Early Juveniles of the Common Goby (*Pomatoschistus microps*)? A Study with Juveniles from Two Wild Estuarine Populations. *Aquat. Toxicol.* **2015**, *164*:163–74.
57. Nassef, M.; Matsumoto, S.; Seki, M; Khalil, F.; Kang, I. J.; Shimasaki, Y. et al. Acute Effects of Triclosan, Diclofenac and Carbamazepine on Feeding Performance of Japanese Medaka Fish (*Oryzias latipes*). *Chemosphere* **2010**, *80*(9), 1095–1100.
58. Oliveira, M.; Ribeiro, A.; Hylland, K.; Guilhermino, L. Single and Combined Effects of Microplastics and Pyrene on Juveniles (0+ group) of the Common Goby *Pomatoschistus microps* (Teleostei, Gobiidae). *Ecol. Indic.* **2013**, *34*, 641–647.
59. Rochman, C. M.; Hoh, E.; Kurobe, T.; Teh, S. J. Ingested Plastic Transfers Hazardous Chemicals to Fish and Induces Hepatic Stress. *Sci. Rep.* **2013**, *3*, 3263.

60. Laville, N.; Aït-Aïssa, S.; Gomez, E.; Casellas, C.; Porcher, J. M. Effects of Human Pharmaceuticals on Cytotoxicity, EROD Activity and ROS Production in Fish Hepatocytes. *Toxicology* **2004**, *196*(1–2), 41–55.
61. Ribalta, C.; Solé, M. In Vitro Interaction of Emerging Contaminants with the Cytochrome P450 System of Mediterranean Deep-sea Fish. *Environ. Sci. Technol.* **2014**, *48*(20), 12327–12335.
62. Saddick, S.; Afifi, M.; Abu Zinada, O. A. Effect of Zinc Nanoparticles on Oxidative Stress-related Genes and Antioxidant Enzymes Activity in the Brain of *Oreochromis niloticus* and *Tilapia zillii*. *Saudi J. Biol. Sci.* **2017**, *24*(7), 1672–1678.
63. Mimeault, C.; Woodhouse, A. J.; Miao, X. S.; Metcalfe, C. D.; Moon, T. W.; Trudeau, V. L. The Human Lipid Regulator, Gemfibrozil Bioconcentrates and Reduces Testosterone in the Goldfish, *Carassius auratus*. *Aquat. Toxicol.* **2005**, *73*(1), 44–54.
64. Stepanova, S.; Praskova, E.; Chromcova, L.; Plhalova, L.; Prokes, M.; Blahova, J. et al. The Effects of Diclofenac on Early Life Stages of Common Carp (*Cyprinus carpio*). *Environ. Toxicol. Pharmacol.* **2013**, *35*(3), 454–460.
65. Oliveira, M.; Maria, V. L.; Ahmad, I.; Pacheco, M.; Santos, M. A. Seasonal *Liza aurata* Tissue-specific DNA Integrity in a Multi-contaminated Coastal Lagoon (Ria de Aveiro, Portugal). *Marine Poll. Bull.* **2010**, *60*(10), 1755–1761.
66. Oliveira, M.; Pacheco, M.; Santos, M. A. Cytochrome P4501A, Genotoxic and Stress Responses in Golden Grey Mullet (*Liza aurata*) Following Short-term Exposure to Phenanthrene. *Chemosphere* **2007**, *66*(7), 1284–1291.
67. Oliveira, M.; Serafim, A.; Bebianno, M. J.; Pacheco, M.; Santos, M. A. European Eel (*Anguilla anguilla* L.) Metallothionein, Endocrine, Metabolic and Genotoxic Responses to Copper Exposure. *Ecotoxicol. Environ. Saf.* **2008**, *70*(1), 20–26.
68. Oliveira, M.; Ahmad, I.; Maria, V.; Pacheco, M.; Santos, M. Monitoring Pollution of Coastal Lagoon Using *Liza aurata* Kidney Oxidative Stress and Genetic Endpoints: An Integrated Biomarker Approach. *Ecotoxicology* **2010**, *19*(4), 643–653.
69. Oliveira, M.; Ahmad, I.; Maria, V. L.; Pacheco, M.; Santos, M. Antioxidant Responses Versus DNA Damage and Lipid Peroxidation in Golden Grey Mullet Liver: A Field Study at Ria de Aveiro (Portugal). *Arch. Environ. Contam. Toxicol.* **2010**, 59(3), 454-463.
70. Oliveira, M.; Maria, V. L.; Ahmad, I.; Teles, M.; Serafim, A.; Bebianno, M. J. et al. Golden Grey Mullet and Sea Bass Oxidative DNA Damage and Clastogenic/Aneugenic Responses in a Contaminated Coastal Lagoon. *Ecotoxicol. Environ. Saf.* **2010**, *73*(8), 1907–1913.
71. Pal, A.; Gin, K. Y.-H.; Lin, A. Y.-C.; Reinhard, M. Impacts of Emerging Organic Contaminants on Freshwater Resources: Review of Recent Occurrences, Sources, Fate and Effects. *Sci. Total Environ.* **2010**, *408*(24), 6062–6069.
72. Backhaus, T. Medicines, Shaken and Stirred: A Critical Review on the Ecotoxicology of Pharmaceutical Mixtures. *Philos. Trans. Royal Soc. Lond. B Biol. Sci.* **2014**, *369*, 1-11.
73. Fabbri, E. Pharmaceuticals in the Environment: Expected and Unexpected Effects on Aquatic Fauna. *Ann. N. Y. Acad. Sci.* **2015**, *1340*(1), 20–28.
74. Croft, W. J. *Under the Microscope: A Brief History of Microscopy;* Weiss R. J., Ed.; Series in Popular Sciences; World Scientific Publishing Company: Singapore, 2006.
75. Sahl, S. J.; Moerner, W. E. Super-resolution Fluorescence Imaging with Single Molecules. *Curr. Opin. Struct. Biol.* **2013**, *23*(5), 778–787.
76. Rakhi, S. F.; Reza, A. H. M. M.; Hossen, M. S.; Hossain, Z. Alterations in Histopathological Features and Brain Acetylcholinesterase Activity in Stinging Catfish

Heteropneustes fossilis Exposed to Polluted River Water. *Int. Aquat. Res.* **2013,** *5*(1), 1–18.
77. Jha, A. N. Ecotoxicological Applications and Significance of the Comet Assay. *Mutagenesis* **2008,** *23*(3), 207–221.
78. de Lapuente, J.; Lourenço, J.; Mendo, S. A.; Borràs, M.; Martins, M. G.; Costa, P. M. et al. The Comet Assay and Its Applications in the Field of Ecotoxicology: A Mature Tool That Continues to Expand Its Perspectives. *Front. Genet.* **2015,** *6*, 180.
79. Pacheco, M.; Santos, M. A. Induction of Micronuclei and Nuclear Abnormalities in the Erythrocytes of *Anguilla anguilla* L Exposed Either to Cyclophosphamide or to Bleached Kraft Pulp Mill Effluent. *Fresen. Environ. Bull.* **1996,** *5*(11–12), 746–751.
80. Oliveira, M.; Ahmad, I.; Maria, V. L.; Ferreira, C. S. S.; Serafim, A.; Bebianno, M. J. et al. Evaluation of Oxidative DNA Lesions in Plasma and Nuclear Abnormalities in Erythrocytes Of Wild Fish (*Liza aurata*) as an Integrated Approach to Genotoxicity Assessment. *Mut. Res./Genet. Toxicol. Environ. Mutagen.* **2010,** *703*(2), 83–89.
81. Goodhew, P. J.; Humphreys, F. J.; Beanland, R. *Electron Microscopy and Analysis,* 3rd ed.; Taylor and Francis/CRC Press: New York (London), 2001.
82. Haynes, R. *Optical Microscopy of Materials*; 1984, Springer US: New York.
83. Török, P.; Kao, F.-J. *Optical Imaging and Microscopy Techniques and Advanced Systems*. Springer: Berlin, New York, 2007. http://site.ebrary.com/id/10189332 (acessed Feb 26, 2017).
84. van Putten, E. G.; Akbulut, D.; Bertolotti, J.; Vos, W. L; Lagendijk, A.; Mosk, A. P. Scattering Lens Resolves Sub-100 nm Structures with Visible Light. *Phys. Rev. Lett.* **2011,** *106*(19), 1–4.
85. Williams, D. B.; Carter, C. B. The Transmission Electron Microscope. In *Transmission Electron Microscopy: A Textbook for Materials Science*. Springer: Boston, MA, US, 1996; pp 3–17.
86. Lange, A.; Paull, G. C.; Coe, T. S.; Katsu, Y; Urushitani, H.; Iguchi, T. et al. Sexual Reprogramming and Estrogenic Sensitization in Wild Fish Exposed to Ethinylestradiol. *Environ. Sci. Technol.* **2009,** *43*(4), 1219–1225.
87. Aris, A. Z.; Shamsuddin, A. S.; Praveena, S. M. Occurrence of 17α-ethynylestradiol (EE2) in the Environment and Effect on Exposed Biota: A Review. *Environ. Int.* **2014,** *69*, 104–119.
88. Kolpin, D. W.; Furlong, E. T.; Meyer, M. T.; Thurman, E. M.; Zaugg, S. D.; Barber, L. B. et al. Pharmaceuticals, Hormones, and Other Organic Wastewater Contaminants in U.S. Streams, 1999–2000: A National Reconnaissance. *Environ. Sci. Technol.* **2002,** *36*(6), 1202–1211.
89. Länge, R.; Hutchinson, T. H.; Croudace, C. P.; Siegmund, F.; Schweinfurth, H.; Hampe, P. et al. Effects of the Synthetic Estrogen 17α-ethinylestradiol on the Life-cycle of the Fathead Minnow (*Pimephales promelas*). *Environ. Toxicol. Chem.* **2001,** *20*(6), 1216–1227.
90. Balch, G. C.; Mackenzie, C. A.; Metcalfe, C. D. Alterations to Gonadal Development and Reproductive Success in Japanese Medaka (*Oryzias latipes*) Exposed to 17α-ethinylestradiol. *Environ. Toxicol. Chem.* **2004,** *23*(3), 782–791.
91. Pawlowski S, van Aerle R, Tyler CR, Braunbeck T. Effects of 17α-ethinylestradiol in a Fathead Minnow (*Pimephales promelas*) Gonadal Recrudescence Assay. *Ecotoxicol. Environ. Saf.* **2004,** *57*(3), 330–345.
92. Tabata, A.; Kashiwada, S.; Ohnishi, Y.; Ishikawa, H.; Miyamoto, N; Itoh, M. et al. Estrogenic Influences of Estradiol-17 Beta, p-nonylphenol and Bis-phenol-A on Japanese

Medaka (*Oryzias latipes*) at Detected Environmental Concentrations. *Water Sci. Technol.* **2001**, *43*(2), 109–116.

93. Miles-Richardson, S. R.; Kramer V. J.; Fitzgerald, S. D.; Render, J. A.; Yamini, B.; Barbee, S. J. et al. Effects of Waterborne Exposure of 17 β-estradiol on Secondary Sex Characteristics and Gonads Of Fathead Minnows (*Pimephales promelas*). *Aquat. Toxicol.* **1999**, *47*(2), 129–145.

94. Teles, M.; Pacheco, M.; Santos, M. A. Biotransformation, Stress and Genotoxic Effects of 17β-estradiol in Juvenile Sea Bass (*Dicentrarchus labrax* L.). *Environ. Int.* **2006**, *32*(4), 470–477.

95. Sponchiado, G.; Fortunato de Lucena Reynaldo, E. M.; de Andrade, A. C. B.; de Vasconcelos, E. C.; Adam, M. L.; Ribas de Oliveira, C. M. Genotoxic Effects in Erythrocytes of *Oreochromis niloticus* Exposed to Nanograms-per-Liter Concentration of 17 Beta-estradiol (E-2): An Assessment Using Micronucleus Test and Comet Assay. *Water Air Soil Pollut.* **2011**, *218*(1–4), 353–360.

96. Sacher, F.; Ehmann, M.; Gabriel, S.; Graf, C.; Brauch, H.-J. Pharmaceutical Residues in the River Rhine—Results of a One-decade Monitoring Programme. *J. Environ. Monit.* **2008**, *10*(5), 664–670.

97. Triebskorn, R.; Casper, H.; Scheil, V.; Schwaiger, J. Ultrastructural Effects of Pharmaceuticals (Carbamazepine, Clofibric Acid, Metoprolol, Diclofenac) in Rainbow Trout (*Oncorhynchus mykiss*) and Common Carp (*Cyprinus carpio*). *Anal. Bioanal. Chem.* **2007**, *387*(4), 1405–1416.

98. Li, Z.-H.; Zlabek, V.; Velisek, J; Grabic, R.; Machova, J; Kolarova, J. et al. Acute Toxicity of Carbamazepine to Juvenile Rainbow Trout (*Oncorhynchus mykiss*): Effects on Antioxidant Responses, Hematological Parameters and Hepatic EROD. *Ecotoxicol. Environ. Saf.* **2011**, *74*(3), 319–327.

99. Yildirimer, L.; Thanh, N. T. K.; Loizidou, M.; Seifalian, A. M. Toxicology and Clinical Potential of Nanoparticles. *Nano Today* **2011**, *6*(6), 585–607.

100. Elsaesser, A.; Howard, C. V. Toxicology of Nanoparticles. *Adv. Drug Deliv. Rev.* **2012**, *64*(2), 129–137.

101. Canesi, L.; Ciacci, C.; Fabbri, R.; Marcomini, A.; Pojana, G.; Gallo, G. Bivalve Molluscs as a Unique Target Group for Nanoparticle Toxicity. *Marine Environ. Res.* **2012**, *76*, 16–21.

102. Farkas, J.; Christian, P.; Urrea, J. A. G.; Roos, N.; Hassellöv, M.; Tollefsen, K. E. et al. Effects of Silver and Gold Nanoparticles on Rainbow Trout (*Oncorhynchus mykiss*) Hepatocytes. *Aquat. Toxicol.* **2010**, *96*(1), 44–52.

103. Lewinski, N.; Colvin, V.; Drezek, R. Cytotoxicity of Nanoparticles. *Small* **2008**, *4*(1), 26–49.

104. Lapresta-Fernández, A.; Fernández, A.; Blasco, J. Nanoecotoxicity Effects of Engineered Silver and Gold Nanoparticles in Aquatic Organisms. *TrAC Trends Anal. Chem.* **2012**, *32*, 40–59.

105. Alkilany, A. M.; Murphy, C. J. Toxicity and Cellular Uptake of Gold Nanoparticles: What We have Learned so Far? *J. Nanopart. Res.* **2010**, *12*(7), 2313–2333.

106. Bahadar, H.; Maqbool, F.; Niaz, K.; Abdollahi, M. Toxicity of Nanoparticles and an Overview of Current Experimental Models. *Iran. Biomed. J.* **2016**, *20*(1), 1–11.

107. Sonavane, G.; Tomoda, K.; Makino, K. Biodistribution of Colloidal Gold Nanoparticles After Intravenous Administration: Effect of Particle Size. *Colloids Surf. B Biointerfaces* **2008**, *66*(2), 274–280.

108. Balasubramanian, S. K.; Jittiwat, J.; Manikandan, J.; Ong, C.-N.; Yu, L. E.; Ong, W.-Y. Biodistribution of Gold Nanoparticles and Gene Expression Changes in the Liver and Spleen After Intravenous Administration in Rats. *Biomaterials* **2010**, *31*(8), 2034–2042.
109. Geffroy, B.; Ladhar, C.; Cambier, S.; Treguer-Delapierre, M.; Brèthes, D.; Bourdineaud, J.-P. Impact of Dietary Gold Nanoparticles in Zebrafish at Very Low Contamination Pressure: The Role of Size, Concentration and Exposure Time. *Nanotoxicology* **2012**, *6*(2), 144–160.
110. Joubert, Y.; Pan, J.-F.; Buffet, P.-E.; Pilet, P.; Gilliland, D; Valsami-Jones, E. et al. Subcellular Localization of Gold Nanoparticles in the Estuarine Bivalve *Scrobicularia plana* After Exposure Through the Water. *Gold Bull.* **2013**, *46*(1), 47–56.
111. Tedesco, S.; Doyle, H.; Blasco, J.; Redmond, G.; Sheehan, D. Oxidative Stress and Toxicity of Gold Nanoparticles in *Mytilus edulis*. *Aquat. Toxicol.* **2010**, *100*(2), 178–186.
112. Ates, M.; Demir, V.; Adiguzel, R.; Arslan, Z. Bioaccumulation, Subacute Toxicity, and Tissue Distribution of Engineered Titanium Dioxide Nanoparticles in Goldfish (*Carassius auratus*). *J. Nanomater.* **2013**, *2013*, 6.
113. Kain, J.; Karlsson, H. L.; Möller, L. DNA Damage Induced by Micro- and Nanoparticles—Interaction with FPG Influences the Detection of DNA Oxidation in the Comet Assay. *Mutagenesis* **2012**, *27*(4), 491-500.
114. Mahl, D.; Greulich, C.; Meyer-Zaika, W.; Koller, M.; Epple, M. Gold Nanoparticles: Dispersibility in Biological Media and Cell-biological Effect. *J. Mater. Chem.* **2010**, *20*(29), 6176–6181.
115. García-Negrete, C. A.; Blasco, J.; Volland, M.; Rojas, T. C.; Hampel, M.; Lapresta-Fernández, A. et al. Behaviour of Au-citrate Nanoparticles in Seawater and Accumulation in Bivalves at Environmentally Relevant Concentrations. *Environ. Pollut.* **2013**, *174*, 134–141.
116. Brown, A. P.; Brydson, R. M. D.; Hondow, N. S. Measuring In Vitro Cellular Uptake of Nanoparticles by Transmission Electron Microscopy. *J. Phys. Conf. Ser.* **2014**, *522*(1), 012058.
117. Nghiem, T. H. L.; Thi Tuyen, N.; Emmanuel, F.; Thanh Phuong, N.; Thi My Nhung, H.; Thi Quy, N. et al. Capping and In Vivo Toxicity Studies of Gold Nanoparticles. *Adv. Nat. Sci. Nanosci. Nanotechnol.* **2012**, *3*(1), 015002.
118. Hull, M. S.; Chaurand, P.; Rose, J.; Auffan, M.; Bottero, J.-Y.; Jones, J. C. et al. Filter-feeding Bivalves Store and Biodeposit Colloidally Stable Gold Nanoparticles. *Environ. Sci. Technol.* **2011**, *45*(15), 6592–6599.
119. Uboldi, C.; Bonacchi, D.; Lorenzi, G.; Hermanns, M. I.; Pohl, C.; Baldi, G. et al. Gold Nanoparticles Induce Cytotoxicity in the Alveolar Type-II Cell Lines A549 and NCIH441. *Part. Fibre Toxicol.* **2009**, *6*(1), 1–12.
120. Kwon, D.; Nho, H. W.; Yoon, T. H. Transmission Electron Microscopy and Scanning Transmission X-ray Microscopy Studies on the Bioaccumulation and Tissue Level Absorption of TiO_2 Nanoparticles in *Daphnia magna*. *J. Nanosci. Nanotechnol.* **2015**, *15*(6), 4229–4238.
121. Garcia-Negrete, C. A.; Jimenez de Haro, M. C.; Blasco, J.; Soto, M.; Fernandez, A. STEM-in-SEM High Resolution Imaging of Gold Nanoparticles and Bivalve Tissues in Bioaccumulation Experiments. *Analyst* **2015**, *140*(9), 3082–3089.
122. Yoo-Iam, M.; Chaichana, R.; Satapanajaru, T. Toxicity, Bioaccumulation and Biomagnification of Silver Nanoparticles in Green Algae (*Chlorella sp.*), Water Flea (*Moina macrocopa*), Blood Worm (*Chironomus spp.*) and Silver Barb (*Barbonymus gonionotus*). *Chem. Spec. Bioavailab.* **2014**, *26*(4), 257–265.

123. Gliga, A. R.; Skoglund, S.; Wallinder I. O.; Fadeel, B.; Karlsson, H. L. Size-dependent Cytotoxicity of Silver Nanoparticles in Human Lung Cells: The Role of Cellular Uptake, Agglomeration and Ag Release. *Part. Fibre Toxicol.* **2014,** *11*(1), 1–17.

124. Mu, Q.; Hondow, N. S.; Krzemiński, Ł.; Brown, A. P.; Jeuken, L. J.; Routledge, M. N. Mechanism of Cellular Uptake of Genotoxic Silica Nanoparticles. *Part. Fibre Toxicol.* **2012,** *9*(1), 1–11.

125. Laban, G.; Nies, L. F.; Turco, R. F.; Bickham, J. W.; Sepúlveda, M. S. The Effects of Silver Nanoparticles on Fathead Minnow (*Pimephales promelas*) Embryos. *Ecotoxicology* **2009,** *19*(1), 185–195.

126. AshaRani, P. V.; Low Kah Mun, G.; Hande, M. P.; Valiyaveettil, S. Cytotoxicity and Genotoxicity of Silver Nanoparticles in Human Cells. *ACS Nano* **2009,** *3*(2), 279–90.

127. Barreto, A.; Pereira, M. L.; Luis, L. G.; Trindade, T.; Soares, A. M. V. M.; Oliveira, M. Histological Effects of Gold Nanoparticles and Gemfibrozil on Gilthead Seabream *Sparus aurata*. WCM, India, 2015.

128. Griffitt, R. J.; Hyndman, K.; Denslow, N. D.; Barber, D. S. Comparison of Molecular and Histological Changes in Zebrafish Gills Exposed to Metallic Nanoparticles. *Toxicol. Sci.* **2009,** *107*(2), 404–415.

129. Scown, T. M.; Santos, E. M.; Johnston, B. D.; Gaiser, B.; Baalousha, M.; Mitov, S. et al. Effects of Aqueous Exposure to Silver Nanoparticles of Different Sizes in Rainbow Trout. *Toxicol. Sci.* **2010,** *115*(2), 521–534.

CHAPTER 19

CHROMIUM: THE INTRIGUING ELEMENT. THE BIOLOGICAL ROLE Cr(III)-TRIS-PICOLINATE: IS IT SAFE OR NOT?

TERESA MARGARIDA DOS SANTOS[1,2*], MANUEL FERREIRA[2,3], and MARIA DE LOURDES PEREIRA[2,4]

[1]*Department of Chemistry, University of Aveiro, Campus de Santiago, 3810-193 Aveiro, Portugal*

[2]*CICECO—Aveiro Institute of Materials, University of Aveiro, 3810-193 Aveiro, Portugal*

[3]*Baixo Vouga Hospital Centre, 3810-193 Aveiro, Portugal*

[4]*Department of Biology, University of Aveiro, Campus de Santiago, 3810-193 Aveiro, Portugal*

*Corresponding author. E-mail: teresa@ua.pt

ABSTRACT

The biological behavior of chromium has been the goal of a large body of intense research. Chromium has been designated as an essential trace element for more than 50 years with a crucial role on carbohydrate and lipid metabolism in both humans and animals. This chapter outlines from an historical perspective the evolution of the essential role of chromium, toward the current paradigm as toxic or pharmacologically active element. The essentiality versus the pharmacological relevance of chromium and its compounds is largely debated, and the irrefutable evidence of carcinogenic concerns of Cr(III)-tris-picolinate (CrPic) intake offered. Different scenarios for research in this area, including in vitro and in vivo studies, associated to epidemiological data, are here outlined for the understanding of the biological properties of chromium. This chapter is a contribution for the in vivo data knowledge on CrPic and presents a case study using mice as a model. Examples based on our laboratory experiments had been designed in order to characterize CrPic-damaging effects within distant organs, such as thymus and epididymis, through histopathological techniques.

Although recent new insights have been achieved, some concerns about using dietary supplements based on CrPic are underlined in this chapter. Due to so many viewpoints about the role of chromium(III)-tris-picolinate on human health, an ongoing debate and a significant global polemic remains to be a priority area of investigation.

New concepts for the role of chromium are emerging as acting as a second messenger in the treatment of diabetes. Finally, future research on Cr(III) nutritional supplementation is absolutely required to better clearly define the role of this intriguing element.

19.1 CHROMIUM: ESSENTIAL OR PHARMACOLOGICAL RELEVANT?

Chromium is an intriguing element which all along decades has generated an intense and hot target for research. Its biological behavior, which comprises a multitude of different and parallel facets, has since originated hundreds of scientific publications and keeps on being the goal of many more.

Chromium is an everywhere element found in varying concentrations in air, water, soil, and essentially in all biological tissues. It is known for almost two centuries and at first, it was thought to be only harmful and carcinogenic to living organisms. Still, it has been proposed to be an essential trace element six decades ago and its essentiality, for humans and animals, required for regular carbohydrate and lipid metabolism, was subsequently demonstrated and has since been strongly investigated.[1-4]

Chromium is undoubtingly a peculiar element due to its two distinct faces, that is, in one hand it features essentiality, when as Cr(III), and on the other, as a Cr(VI) chemical species occurs mostly from anthropogenic sources and it is considered a human carcinogen.[5-9]

Chromium has been established as an essential element for over 30 years, due to its "crucial role" on lipid and carbohydrate metabolisms. As a consequence, the biological functions of chromium had been narrowly associated with that of insulin because sufficient dietary chromium was thought to lead to a decrease necessity for insulin and to an enhanced blood lipid profile.

Chromium essentiality story began in 1955,[1] when Mertz and Schwarz showed a relation between this element and glucose after feeding rats a *Torula* yeast-based diet which resulted in the rats apparently developing impaired glucose tolerance in response to a load of intravenous glucose.[1,10] This work became an impressive source for investigation as easily demonstrated by any literature search of that time and in forward years.[11] The next step of the chromium chronicle was the assessment of the intracellular presence of a biologically active form of Cr in vivo, which was believed to be an organic chromium(III) complex that was postulated to possess adequate properties in order to interfere in carbohydrates and lipid metabolisms and to be effective against dysfunctional diseases relative to their regulation. This complex has been named glucose tolerance factor (GFT).[10,12] A short time later, Mertz and collaborators (1977) claimed to have isolated "GTF" from brewer's yeast.[13] This work was followed by others who obtained similar results, and therefore the identification and synthesis of GTF had been among the highest priorities of chromium research at the time.[14–18]

Several different compositions and structural features were then proposed for GTF. It should be a Cr(III) complex coordinated to nicotinate and to the amino acids glycine, cysteine, and glutamic acid, or an alternative similar Cr(III) complex with nicotinate and glutathione, a tripeptide composed precisely by the amino acids glycine, cysteine, and glutamic acid. This GTF model composition gave rise to the synthesis and characterization of many analogous Cr(III)-complexes. Some of them had been well characterized with nicotinate coordinated to Cr(III) in diverse manners, that is, bridging through the carboxylate[19] or being carboxylate or pyridine monodentate.[20–26] Nevertheless, the amount and the variety of the prepared compounds did not bring any success to obtain a "synthetic" GTF as well as any clear advantage for the understanding of GTF mechanism of action. But 20 years later, GTF has been shown to be just an artifact of its isolation method.[4] To cut the long story short, which surrounds GTF, among the multitude of produced studies the works from Simonoff[16] and also from Hwang[17] showed that the Brewer's yeast fractions that stimulate glucose oxidation by rat adipocytes in the presence of insulin were distinct from the chromium-containing fractions. Probably due to the lack of suitable reproducibility for total Cr analysis in foods, tissues, and body fluids, in result from the experimental methodologies available and used at the time, chromium total concentrations, in the range of parts per billion, were very vulnerable to contamination, matrix effects and to nonspecific binding to pipettes, graphite tubes, just to nominate some of the possible causes which turned difficult to determinate accurate data.[27]

Besides all the previously contradictory results, the controversy statements for the essentiality of chromium, and certainly the looking for a consensual knowledge, an almost infinite number of scientific publications, major reviews, book chapters, and entire books, authored by internationally reputed researchers has been produced.[11,28–34]

Since the alleged GTF as a vector for the effectiveness of chromium essential biological role, the hypothesis that another bioactive form of Cr should exist was largely investigated. Probably the first most viable candidate has been an oligopeptide, that is, a low-molecular weight chromium-binding substance (LMWCr), designated by chromodulin. The prime function of LMWCr might be the stimulation of the insulin receptor tyrosine kinase in response to insulin as it showed ability, in a Cr dose-dependent mode, to potentiate insulin by stimulation of the glucose metabolism in isolated rat adipocytes.[35–40] Functional biomimetic compounds for LMWCr have also been designed and synthesized and some of them have shown potential as antidiabetic therapeutic drugs.[41–44] However, posterior studies of long-term supplementation with this type of compounds did not aimed to significantly affect glucose metabolism in rats.[45–47]

Based on all the knowledge reported above, which is, at least apparently, well supported and documented, nutritional chromium supplements, and, namely, Cr(III)-tris-picolinate, [Cr(Pic)$_3$], has attained an astonishing importance in what the chromium issue is concerned. But relatively recent research and reanalysis of the chromium status broke down that aura and launched again the controversy older than 20 years. Is it any truly clear evidence for the essentially and for a specific role for chromium?[48,49] Are chromium(III) species safe or toxic? Is there a need for chromium nutritional supplementation? Although latest studies have shown that chromium (in its trivalent state) is no more an essential trace element, it can generate beneficial effects at pharmacologically relevant doses on insulin sensitivity and cholesterol levels. Nowadays believes lay on the statement that this element should only currently be considered as potentially pharmacologically active. There are studies indicating clearly that the addition of (supra) nutritional amounts of chromium to a diet can only be considered as having some pharmacological effects,[4,30–33] although beneficial health effects have to follow specific intake recommendations. Effectively, for the general public, available data do not warrant routine use of chromium supplements, whose risk–benefit meaning has not yet been demonstrated.[48–52] Nevertheless, some studies had enlighten, at least partially, a positive role for chromium in lipid metabolism through the gene evaluation involved in fat biosynthesis and lipid metabolism in different tissue types, in domestic animals feed on different chromium levels.[53–57]

At the molecular level, the role of Cr(III) is still an area of active discussion. Despite the movement of Cr(III) in the body, particularly in response to changes in insulin concentrations, there is a very recent suggestion that Cr(III) could act as a second messenger, augmenting insulin signaling by insulin sensitivity intensification.[58,59]

19.2 BIOLOGICAL PROPERTIES OF CHROMIUM

The effects of chromium compounds in vivo depend predominantly on their Cr oxidation state and chemical speciation. While Cr(III) complexes are considered relatively nontoxic because of their poor bioavailability, the oxidized form, Cr(VI), is a proven human carcinogen and a significant environmental pollutant.[6–9,60,61] Cr(VI) is toxic and ultimately deleterious in the nucleus. The carcinogenicity and mutagenicity of various chromium compounds have been found to be noticeably dependent on the oxidation state of the metal, showing an evident relationship between the metabolism of chromate, Cr(VI), and its interaction with nucleic acids.[62–65]

Cr(VI) species easily cross in the cellular membrane by misleading the specific sulfate or phosphate channels due to their structural and charge features analogy as these anions are all tetrahedral and highly negatively charged. CrO_4^{2-}, which is the predominate chromium chemical species inside the cells at physiological pH, is a very strong oxidant being promptly reduced by intracellular components, mainly cysteine, glutathione, and ascorbate, originating different chromium reduced species.[66–73] Also, in some cases, oxygen or hydrogen peroxide species have an important role in these chemical processes.[74–77] The interaction of Cr(VI) with sugar cellular components or models had been investigated as well due to the obvious nucleotides composition and because it is known that the degradative oxidation of aldoses by action of chromate can stabilize Cr(V) intermediates.[78–80]

Even though Cr(V) complexes are generally considered unstable, chromium intermediates had been proved to be relevant chromium-active species. Cr(V) complexes are able to undergo a variety of reactions, which are habitually rapid as compared with the subsequent reduction of Cr(V) by organic substrates or can disproportionate to Cr(VI) and to Cr(III).[81] A great number of Cr(V) complexes (intermediates) have been demonstrated to be obtained after Cr(VI) intracellular reduction and some of them were isolated and characterized.[82–93]

In the intracellular reductive way of Cr(VI) to Cr(III), which compounds traditionally have been looked as stable and inert, the new formed intracellular Cr(III) chemical species undoubtedly interact with DNA by the formation of single-strand breaks, chromium-DNA adducts, by means of the DNA phosphodiester backbone, chromium-mediated amino acids, peptide and protein DNA cross-links, interaction with abasic sites, by free radical generation, or by covalence, provoking different modes of DNA damage.[66,70,94–100] All these types of DNA interactions or damage effects have been detected and reported both in vitro[101–109] and in vivo studies.[110–113]

A plethora of different chromium-containing species in initial, intermediate, or final oxidation states, that is, Cr(III), Cr(V), and/or Cr(III), have been synthesized and characterized. Physiologically significant thiol compounds have been used to mediate reactions between chromate and DNA devising to modulate and investigate the intracellular Cr(VI) redox reactions and to get a better understanding of the ways chromate can damage DNA.[114–124] Some of these model compounds had also been studied in what concerns their solution behavior.[125–127]

Chromium(VI), as chromate,[128–131] and Cr(V), in the chemical form of a physiologically stable Cr(V) compound, $[Cr^V\text{-}BT]^{2-}$ [BT = bis(hydroxyethyl) aminotris(hydroxymethyl)methane][130,132–134] had been also the aim of in vivo toxicity studies using animal models. Severe deleterious effects in several organs, such as reproductive,[128,132–136] spleen,[129] liver,[110,130,131] kidney,[110,132] and lung[110,137–140], were observed corroborating their toxic effects.

Conclusions from the previous related observations are in parallel with studies concerning human occupational exposure to chromium(VI). This human exposure occurs from natural or industrial sources. In several industries, particularly in chromate related industrial processes, workers are exposed to high concentrations of Cr(VI). Although the available data are somewhat disperse,[140–147] the carcinogenic potential of oral exposure to Cr(VI) in humans is supported by a number of very recent studies.[147–156] Injurious alterations in several organs, mainly those in lung cancer and in the gastrointestinal and reproductive tracts, are rather appropriated and point toward an high cancer risk assessment of the occupational exposed industrial workers.

It is crucial to look at chromium(III) compounds as they have to be stressed here in what concerns the generic biological properties of the element chromium in oxidation state +3. Despite their "famous" nontoxicity and poor bioavailability many researchers had been soon alerted for their possible offensive effects, especially in vivo. The explanation of the activity differences of Cr(VI) compared to Cr(III) in cellular and subcellular systems have long been done by the "uptake-reduction" model for chromate carcinogenicity recommended by Wetterhahn, in 1989.[157] Further publications illustrated the same model in more detailed schemes aiming to explain comprehensively the major pathways involved in the formation of genetic lesions by chromium as well as the interrelationships between metabolism and genotoxicity.[5–9,28,56,60,62–66,158–161]

The damages induced by Cr(VI) can conduct to dysfunctional DNA replication and transcription, as well as dysregulation of DNA repair mechanisms, inflammatory responses, and to the failure of regulatory gene networks which are the responsible key for balancing cell survival or death. All may play important roles in Cr(VI) carcinogenesis.[162] Consequently only Cr(VI) compounds were burdened for all the carcinogenic risks and for some time they seemed to be the blamed ones. A remarkable amount of research effort has been devoted to clarify the human health effects associated to Cr(VI) exposure and to the involved molecular mechanisms.[157,160,163,164] Yet, important matters remain not understood, mainly the contributions of nuclear DNA damages and mitochondrial dysfunctions provoked by Cr toxicity.

However, results about problems caused by Cr(III), itself, began to be published.[51,109,165–170] Dubious data probably have been triggered and augmented by the enormous amounts of nutritional chromium supplements that nowadays are utilized all over the world.[34,52,54,168–174] The hypothesis of genotoxicity and carcinogenicity of Cr(VI) and the controversial activity of some "antidiabetic" Cr(III) supplements raised safety concerns and risks of long-term Cr(III) nutritional supplementation. Significant intracellular local concentrations of strong oxidants, such as H_2O_2, are formed during cell signaling, including insulin signaling, which may be responsible for Cr(III) oxidation to Cr(V) and/or to Cr(VI) chemical species.[74–77,80,83,111,174,175]

As a final point, the inside cellular localization of the products resulting from the carcinogenic capacities of Cr(VI) species and/or from Cr(III) compounds had allowed unquestionable proofs of the ultimate malignant effects of Cr(III). Applications, in vitro and in vivo, of novel and updated techniques and methods, namely, X-ray microprobes,[176–179] atomic force microscopy, electrochemistry, DNA electrophoresis,[180] fluorescent probes,[181,182] vibrational mapping and imaging of tissues and cells,[183] and high-resolution EPR,[184] are being successful to enlighten over indications and suggestions sometimes difficult to demonstrate. In what concerns chromium utilization in nutrition issues, as an example, the involved major drawbacks may have now the possibility/opportunity to be better clarified.

19.3 CHROMIUM AND NUTRITIONAL SUPPLEMENTS

Chromium, or Cr-containing compounds, despite of the nowadays tough controversy encompassing Cr(III) nutritional supplements, over the last decades became very popular for weight loss, muscle growth, and antidiabetic purposes then having an elevated degree of consummation worldwide.[48,185–190] The chemical compositions of these type of supplements include a variety of Cr(III) chemical forms, which comprise several Cr(III) coordination compounds, that is, Cr-tris-picolinate (CrPic), Cr-histidinate (CrHis), Cr-nicotinate (CrNic), and a Cr(III) complex of D-phenylalanine [Cr(D-phen-ala)$_3$]. All of these chromium(III) nutritional supplements have been used, patented, manufactured, and promoted in huge quantities wideworld reaching.

Among those available chemical formulations of Cr(III), Cr(III)-tris-picolinate, CrPic/[Cr(pic)$_3$], is probably the most "famous" and one of the most utilized Cr-nutritional supplements, acting as a therapeutic nutritional supplement, globally used not only in human diet but also for cattle and swine.[2,3,4,22,27,191–195] Advertisements for Cr(III)-tris-picolinate "beneficial" effects can be found everywhere in publications related with "health," "healthy good practices," "dietary supplements," "obesity," "diabetes," and other associated problems. CrPic is available in different forms, such as tablets, sports drinks, and chewing gums.

The relevance of CrPic among other chromium nutritional supplements is highlighted in Figure 19.1, which is a mere graphical representation with two entries collected from a literature database search,[196] using the key words "Cr-nutritional" and "Cr-picolinate." Yet, serious concerns about the safety of CrPic had begun to appear in scientific publications since decades,[197–201] although initially a lack of toxicity has been reported.[202] Well-documented results pointed out to deleterious side effects on human health issues, resulting from its indiscriminate utilization as a nutritionally supplement. "The Cr(III)-tris-picolinate affair" is actually an area of active debate, which is far from a simple resolution.[4,28,30–34,50–52,107,162,166,169,172,173–179,181,182,191–195,200,203–209]

Essential nutrients are chemical substances necessary for life maintenance, which can be identified when low intakes resulted in organism failure in order to grow or to reproduce or caused a pathological modification. If these elements are needed in very low quantities, they are called "tracers."[2] Essential nutrients have to be found in foods as the body cannot synthesize them. Chromium (Cr) had long been placed into the essential trace elements list,[2,4,11,52] as it was believed to perform benefic bioactivity by acceptance that a poor diet in Cr would increase the risk for chronic disable diseases foremost to a premature dead. This element or some of its compounds were supposed to be beneficial to subjects with varying degrees of glucose intolerance, ranging from hypoglycemia to insulin-dependent diabetics, thus participating in carbohydrate and lipid metabolisms.[5,8,185-187,190-192,208,210-212]

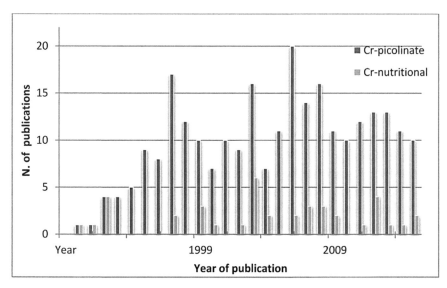

FIGURE 19.1 Number of publications per year versus year of publication of "Cr-nutritional" and "Cr-picolinate" (orange and blue bars, respectively). Source: Adapted with permission from Scopus Data Base, Ref. [196] (accessed Jan 23, 2016).

After the celebration of its 55 anniversary as an essential element,[5,32,33] a sort of veil-covered chromium essentiality issue. Very recent publications have put serious interrogations over its actual effects, preferring to call it a pharmacologically relevant element,[32,33] instead of doubting if it is beneficial or toxic both in animals and humans.[31,32,52-55,57] Lastly, as nonconcordant results about the effectiveness of trivalent chromium compounds were being published,[33-37,191,192,195,204,206,208,213] the polemic has been definitely set up, and in consequence, the dietary guidelines for the European Union (EU) community, provided by the European Food Safety Authority, recently concluded that the requirements for chromium could not be established and chromium has been taken out of the list of the essential trace elements.[214]

Supplementation with Cr has been recommended to result in advantageous responses in mammals with demonstrated glucose intolerance or insulin insensitivity, including Type II diabetes, cardiovascular disease, and related conditions. Nevertheless, studies in humans tend to be negative or at best ambiguous.[191,192,206] There are reports saying that Cr(III)-tris(picolinate), CrPic, the "star" of the chromium(III) nutritional supplements, apparently implies a relatively high Cr intake (>200 µg/d) in order to get a beneficial effect on the glucose metabolism,[51,185] but possible adverse effects on liver and kidney were originated when it was consumed in amounts >1000 µg/d for several months, especially in individuals with preexisting renal and liver diseases.[215–217] Other investigations were ambiguous,[193,194,200,203,204,210–213] and similar effects in animals have not been found.[195,218] There are available evidences in vivo suggesting that genotoxic effects are very unlikely to occur in humans or animals exposed to nutritional, or to moderate recommended supplemental levels of Cr(III). However, an excessive intake of Cr(III) supplements does not appear to be defensible.[205] Thus, the reduction in the risk of chronic disease should be included in formulations of the dietary reference intakes.[51,192–194,204,205,208]

In order to investigate the true influence of Cr(III) supplements in the reduction of body weight, a few years ago, a meta-analysis was undertaken, and the obtained results shown that only a relative small reduction compared with a placebo situation was observed. Although the data were statistically significant, they were not clinically meaningful, then they were not convincing.[190] Another recent meta-analysis that includes reports of clinical studies since 2007 and comprises improved methodologies has found no noteworthy effectiveness of Cr supplementation.[206,213] Consequently, at this point the American Diabetes Association recommendation is that "There is insufficient evidence to support the routine use of micronutrients, such as chromium, to improve glycemic control in people with diabetes."[207] Nonetheless, several clinical studies has reported Cr supplementation efficacy in novel therapeutic targets and in the increase of the understanding of how to optimize chromium use in insulin resistance and Type II diabetes treatment.[52,186,191,192,208]

At the moment, the situation is that evidences about the effectiveness and safety of nutritional chromium supplements are limited and scientifically controversial. As clastogenic and mutagenic features resulting from their widespread use have been reported the possible risks for long-term effects of added chromium have to be seriously taken in account.[32,159,161,162,169,179,181,185,192,219]

The adversarial effect provoked specifically by Cr(III)-tris(picolinate) is "the" main important issue thus it cannot be omitted from this chapter. About 20 years ago, research conducted by Stearns and coworkers, point for chromosome damage due to CrPic effects.[197] This work was followed by others both from the same authors[198,220,221] and from other research groups looking for in vitro and in vivo alterations provoked by this nutritional supplementation.[161,172,173,199,215–217,222–226] Due to a worldwide intense research effort, a multitude of different results/conclusions had been produced. Unfortunately, the enormous number of publications on this, or related subjects, just expanded the controversial debate and conflicted information about the efficacy, the safety and/or the risks consequential of following a CrPic nutritional supplementation.

In one hand, the chemical properties of CrPic have been deeply studied,[200,201,227] and on the other, the adverse effects in vitro[41,200] and/or in vivo[201] have clearly emerged with evidences at the molecular level,[178] and finally some usefulness for taking CrPic has also been shown, mainly in animals, although additional studies are necessary to investigate the effects of long-term CrPic supplementation.[194,203] The significance of these studies to long-term CrPic diabetic human exposure is questionable, given the long latency time of Cr-induced cancers and because they did not use animals under oxidative stress, thus combined inorganic, biochemical, and nutritional studies of chromium(III) are almost compulsory.[204]

As a final remark, the availability of a nutritional literacy review is urgently recommended, perhaps sponsored by national and international authorities, because it is a public health apprehension, in order to support clarified knowledge about taking or not chromium nutritional supplementations, particularly in what CrPic does concern. And, of course, it has to be underscored that the true role of chromium in biological systems is not yet definitively established. Very recently, it has been highlighted that the role of a second messenger probably is the pharmacological mode of action of chromium(III).[58,59] The changing views on the biological role of Cr(III), combined with its rather weak and variable antidiabetic activity, as well as safety concerns over the likely formation of carcinogenic Cr(VI) as a Cr(III) metabolite, meaning the promotion of Cr(III) compounds as nutritional supplements or antidiabetics may need to be reassessed.

Our previous studies described the deleterious effects of Cr(III)-trispicolinate on mice spermatogenesis.[224] Along the next entry of this chapter, a case study is presented showing some in vivo effects of this compound in epididymis and thymus, studied by light microscopy.

19.4 CASE STUDY: Cr(III)-PICOLINATE ON EPIDIDYMIS AND THYMUS—AN HISTOLOGY STUDY

The assessment of data from both in vitro and in vivo assays to relevant epidemiological studies headed to significant advances into the mechanisms of action of chromium compounds. In addition, the key objective of several in vivo studies on toxicological research remains noteworthy to identify possible damage within several tissues and organs. The contribution of these toxicological studies added to other relevant approaches is then helpful to get some insights into the role of these compounds. Given to the widespread use of nutritional chromium(III) supplements, mainly Cr(III)-tris-picolinate, where scientific proof for its effectiveness and safety is severely restricted and debatable, as discussed above in this chapter, numerous authors reported on the side effects of CrPic based on animal studies.[217,224,228–239] Table 19.1 evidences some selected contributions from the enormous extent of published studies which investigate the CrPic supplement effects.

As shown in Table 19.1, it is important to underline that a wide range of study designs and methodologies have been utilized to characterize the effects of CrPic on different animal species (i.e., laboratory, farm, and fish animals), under dissimilar physiological situations.

For example, several studies conducted using healthy animals aiming to detect or to evaluate possible toxic effects of CrPic on relevant organs have not found similar effects.[224,232,235,238,239] The growth performance of some fishery farm species were similarly conducted on healthy specimens supplemented with CrPic on their basal diet twice a day for 58 days.[230] However, no positive properties on improving growth were noted by these authors. Nevertheless, as listed in Table 19.1, other trials were carried out not only on obese and diabetic rats[217,229,234] but also using hypercholesterolemic animal models,[236] aiming to evaluate the effectiveness of CrPic. In the last group of papers, some authors demonstrated that CrPic had no beneficial effects,[217] although other researchers recommended their use.[234,236] For example, Huang and coauthors[234] described some improvements on the morphology of pancreas, and recovery of the function of β-cells on Type II diabetes mellitus rats dosed with 50 and 100 μg/kg CrPic for 15 weeks.

Another example of absence of adverse effects on renal function were reported by Mahamood and coworkers[216] on chronic studies using unilaterally nephrectomyzed rats supplemented with CrPic where benefits for the remaining kidney were denoted. Long-term dietary CrPic supplementation of obese rats decreased oxidative stress and inflammation.[217]

For instance, Ferreira and coauthors[224,232,238] investigated by histopathology the effects of different doses of chromium(III)-tris-picolinate supplementation on mice in order to look for possible toxicological effects. Then, several organs have been selected, including those with metabolic and excretory functions (liver and kidney), immune (spleen), and reproductive functions (testis and epididymis). These authors have used light microscopy

TABLE 19.1 Some of Illustrative Animal Studies on Nutritional Supplements Based on [Cr(pic)₃].

Animal model	Exposure route and dosing	Tissue/organ/system	Results	References
Rats	Obese Zucker rats 5 and 10 mg/kg Cr(pic)₃ for 20 weeks in diet.	Kidney, pancreas, and blood.	Long-term CrPic consumption does not have beneficial effects on glycemic control or indices of growth of OZR, have mild-moderate beneficial effects on several markers of oxidative stress and inflammation.	217
Pigs	(1000 or 2000 μg/kg)	Growth performance, respiratory rate.	No effect on growth performance.	228
Rats	Zucker obese and Zucker diabetic fatty rats; 3 μg Cr/kg/bm	Several organs and tissues.	Any beneficial effects of Cr supplementation.	229
Rainbow trouts	1.6 mg/kg on basal diet twice a day for 58 days.	Serum	No favorable effects on improving growth performance and feed conversion ratio on *Oncorhynchus mykiss*.	230
Mice	Each male, previously fed a diet with 200 mg/kg/day [Cr(pic)₃] was mated with not treated females with sacrifice on gestation day 17.	Litters	Paternal dietary exposure does not account for prenatal mortality, fetal weight, or gross or skeletal morphology.	231
Mice	25 mg/kg and 50 mg/kg/bw/daily orally for 2 weeks.	Liver, spleen, kidney, epididymis, and testis.	Degenerative changes on seminiferous tubules and atrophy on mice exposed to highest dose; multiorgan adverse effects.	224, 232
Sheeps	0.250, 0.375, and 0.500 mg of CrPic/animal/day for 84 days; bw 22.89 ± 2.23 kg.	Blood	No significant effects on humoral immunity, although adverse effect on cellular immunity ($p < 0.05$) with increase of CrPic supplementation.	233
Rats	Type II diabetes mellitus (T2DM) rats supplemented with CrPic 50 μg/kg and 100 μg/kg for 15 weeks.	Pancreas	Some improvements on the morphology of pancreas, no inflammatory cell infiltration, and recovery of the function of β-cells.	234

TABLE 19.1 *(Continued)*

Animal model	Exposure route and dosing	Tissue/organ/system	Results	References
Turkeys	250, 500, and 750 g CrPic/kg in basal diet	Semen	Significant improvement in semen quality.	235
Rats	CrPic (200 μg /day) by gavage for 10 weeks to hypercholesterolemic rats	blood	Improvement of the lipid profile and recovery of platelet hyper-aggregability to control levels.	236
Nile tilapia	200, 400, 600, 800, 1000, and 1200 μg CrPic kg^{-1} on diet for 12 weeks.	blood and dorsal muscles.	CrPic at 400 μg kg^{-1} diet is the most appropriate level for *Oreochromis niloticus* L.	237
Mice	25 mg/kg and 50 mg/kg/bw/daily orally for 2 weeks	Liver, kidney, and spleen.	Multiorgan damage.	238
Lambs	0.250, 0.375, and 0.500 mg CrPic/animal/day for 84 days.	Blood and serum, liver, kidney, heart, lung, and testis	Morphological changes in the liver, kidney, and testis.	239

bm, body mass; bw, body weight; DM, diabetes mellitus; CrPic, chromium(III)-tris-picolinate; ip, intraperitoneal; sc, subcutaneous; CRP, serum C-reactive protein; OZR, obese Zucker rats; ZDF, Zucker diabetic fatty rats.

to analyze the morphology of those organs and reported on some degenerative changes, although neither macroscopic changes nor differences on the behavior pattern had been noticed among different groups of animals. In fact, those light microscopy studies demonstrated multiorgan damage in two groups of mice exposed to 25 and 50 mg/kg doses of CrPic, for a period of 14 days.[224,238] Spermatogenesis was strongly affected, and the most conspicuous adverse effects on seminiferous tubules included vacuolation, release of immature germ cells into the lumen, and atrophic seminiferous tubules.[224] Further microscopy studies also revealed toxic effects on liver, kidney, and spleen.[238] Hepatic tissue displayed some hemorrhagic focus through the tissue being more severe on mice exposed to the higher dose of CrPic. The rupture of Bowman capsule and hemorrhage was demonstrated on animals exposed to 25 and 50 mg/kg of CrPic, respectively. Finally spleen of mice exposed to both doses of CrPic has shown disorganization of pulps and cell depletion. As a result of the microscopic changes reported herein, impairment of the multiorgan functionality was recognized.

Under the light of those reports, new studies have been undertaken on male adult mice aiming to describe possible toxic effects on epididymis and thymus using a similar study design.[232] All procedures were done in accordance with standard ethical guides for the care and use of laboratory animals. Concisely, CrPic synthesized according literature[240–242] was orally given to Swiss albino mice (7 weeks old) provided from Harlan Laboratories (Spain) in their drinking water for a fortnight. NaCl (0.9%)-dosed animals were used as controls ($n = 5$ per time point). After sacrifice samples of organs (epididymis and thymus) were fixed in Bouin's solution for 24 h and dehydrated in ascending concentrations of ethanol. Impregnation and inclusion of samples in paraffin (melting bridge 42–44°C and 52–58°C, respectively) was then performed. Slices (5–7 μm thick) of each organ was sectioned using a rotary microtome (Leitz Model 1512), and stained with hematoxylin and eosin (H&E). A light microscope (Olympus BX41TF) coupled to photographic system was used.

Microscopic alterations on epididymis compared with controls showing a normal morphology are displayed in Figure 19.2. At a dose of 25 mg/kg of CrPic, no significant injuries were noted within tubules (data not shown) whose features are similar to healthy controls. However at the dose of 50 mg/kg of CrPic, spermatozoa condensation in some tubules but the absence in others was noted. These microscopic changes observed on epididymis are in agreement with the lesions previously described for the testis under a similar design protocol. Although sperm quality was not evaluated, authors suggest that both adverse effects on spermatogenesis will result on semen quality decay. However, other authors reported on an improvement of sperm parameters on turkeys fed with this chromium supplement although higher doses (250, 500, and 750 g CrPic/kg) had been given in the basal diet.[235] The antioxidative action of chromium was appointed by these authors for the improvement in semen physical properties.

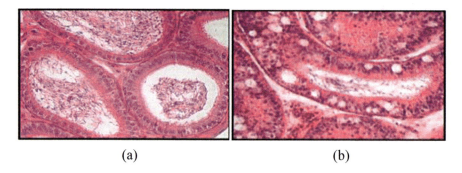

FIGURE 19.2 Histological pictures of mice epididymis. (a) control and (b) representative H- and E-stained histopathological picture of the epididymis from 50 mg/kg [Cr(pic)$_3$]; original magnification 400×.

In Figure 19.3, the microscopic anatomy of thymus in both control and CrPic-exposed mice are displayed.

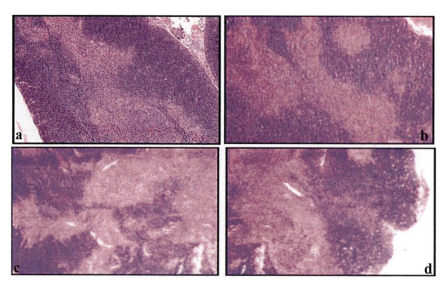

FIGURE 19.3 Histological sections of thymus: (a) control, (b) 25 mg/kg [Cr(Pic)$_3$], and (c–d) 50 mg/kg [Cr(Pic)$_3$]; (H and E); original magnification 400×.

Histological sections of the control group showed normal thymus histology, with Hassall corpuscles, and distinct lobes surrounded by connective tissue capsule (Fig. 19.3a). Both the cortical and medullary areas were distinct. Nevertheless cell depletion and disorganization of thymus structure were noted in animals exposed to oral supplementation with CrPic in a concentration-dependent manner (Fig. 19.3b–d). The analysis of Table 19.1 reflects different scenarios for research on the effectiveness of CrPic supplementation on healthy and pathological laboratory and farm species. Histopathological and biochemistry approaches were the main techniques used to evaluate the possible toxic effects of CrPic on distant organs and tissues. Although the effectiveness of CrPic supplements varied significantly among studies, in the majority of the examples mentioned above, it is clear that CrPic have deleterious effects on different tissues, organs, and systems in a dose, and time duration-dependent manner. These approaches based on microscopy techniques represent an important role for the characterization of morphological lesions within several organs.

19.5 CONCLUSIONS AND SUGGESTIONS FOR FURTHER RESEARCH

As can be seen along this chapter, despite the efforts exerted over the past years in studies with chromium(III)-tris-picolinate, CrPic, there is still a discrepancy of results relating to its effects, which make difficult to establish their safety.

Significant progress on the knowledge of CrPic has been made through in vitro or in vivo toxicological and epidemiological studies where several methodological approaches (morphological, biochemical, and molecular) were utilized. Although new insights have been achieved, concerns about using dietary supplements based on CrPic were underlined.

Some of the results focused in this chapter deal with the adverse effects of CrPic on male mice reproductive health. For this reason, future areas on reproductive health research toward critical stages of exposure of these type of supplements during pregnancy or postnatal period, need more attention to be taken.

Given the analysis of recent reports on the nutritional chromium(III) supplement CrPic and looking through all the research contributions obtained to date, considerable dispute in terms of its use and safety are evident. Even though indications of both beneficial and toxic effects were shown, the problem is not yet well understood and conclusions cannot be postulated. Due to so many perspectives about the role of CrPic on human health, an ongoing discussion and a remarkable global controversy remain to be a major area of investigation. Nowadays, the polemic is even more and more pertinent because the essentiality of chromium has been effaced, at least by EU regulations. After the discussion about the essentiality versus pharmacological relevance of chromium and its compounds, and the irrefutable proof of carcinogenic consequences of CrPic intake a new paradigm has emerged from the new role for chromium effects acting as a second messenger in the treatment of diabetes, that is, in relation to insulin. A personalized dose control of Cr supplementation is absolutely necessary. Further conclusions are strongly dependent on reliable human tests, which surely will take a long time.

ACKNOWLEDGMENTS

This chapter was developed within the scope of the project CICECO—Aveiro Institute of Materials, POCI-01-0145-FEDER-007679 (Foundation for Science and Technology—FCT Ref. UID/CTM/50011/2013), financed by national funds through the FCT/MEC and when appropriate cofinanced by FEDER under the PT2020 Partnership Agreement.

KEYWORDS

- Cr(III)-tris-picolinate
- glucose tolerance factor
- bioavailability
- Cr-carcinogenicity
- Cr(III)-nutritional supplement

REFERENCES

1. Mertz, M.; Schwartz, K. Impaired Intravenous Glucose Tolerance as an Early Sign of Dietary Necrotic Liver Degeneration. *Arch. Biochem. Biophys.* **1955**, *58*, 504–506.
2. Anderson, R. A. Chromium. In *Trace Elements in Human and Animal Nutrition*, 4th Ed.; Academic Press: New York, 1977; pp 258–270.

3. Jeejeebhoy, K. N.; Chu, R. C.; Marliss, E. B.; Greenberg, G. R.; Bruce-Robertson, A. Chromium Deficiency, Glucose Intolerance, and Neuropathy Reversed by Chromium Supplementation, in a Patient Receiving Long-term Total Parenteral Nutrition. *Am. J. Clin. Nutr*. **1977**, *30*, 531–538.
4. Vincent, J. B. Chromium: Celebrating 50 Years as an Essential Element? *Dalton T.* **2010**, *39*, 3787–3794.
5. Costa, M.; Klein, C. B. Toxicity and Carcinogenicity of Chromium Compounds in Humans. *Crit. Rev. Toxicol*. **2006**, *36*,155–163.
6. Agency for Toxic Substances and Disease Registry-ATSDR; 2000 Toxicological Profile for Chromium. Atlanta: US Department of Health and Human Services.
7. Dayan, A. D.; Paine, A. J. Mechanisms of Chromium Toxicity, Carcinogenicity and Allergenicity: Review of the Literature from 1985 to 2000. *Hum. Exp. Toxicol*. **2001**, *20*, 439–451.
8. Ding, M.; Shi, X. Molecular Mechanisms of Cr(VI)-induced Carcinogenesis. *Mol. Cell. Biochem*. **2002**, *234*, 293–300.
9. International Agency for Research on Cancer (2003) Overall Evaluations of Carcinogenicity to Humans. http://www.iarc.fr (accessed Jan 23, 2016).
10. Schwarz, K.; Mertz, W. A Glucose Tolerance Factor and Its Differentiation from Factor 3. *Arch. Biochem. Biophys*. **1957**, *72*, 515–518.
11. Vincent, J. B.; Stallings, D. Introduction: A History of Chromium Studies (1955–1995), Chapter 1. In *The Nutritional Biochemistry of Chromium(III)*; Vincent, J., Ed.; Elsevier: Amsterdam, 2007.
12. Schwarz, K.; Mertz, W. Chromium(III) and the Glucose Tolerance Factor. *Arch. Biochem. Biophys*. **1959**, *85*, 292–295.
13. Toepfer, E. W.; Mertz, W.; Polansky, M. M.; Roginsky, E. E.; Wolf, W. R. Preparation of Chromium-containing Material of Glucose Tolerance Factor Activity from Brewer's Yeast Extracts and by Synthesis. *J. Agric. Food Chem*. **1977**, *25*, 162–162.
14. Vincent, J. B. The Story of Tolerance Glucose Factor, Chapter 3. In: *The Bioinorganic Chemistry of Chromium(III)*; Vincent, J., Ed.; John Wiley and Sons: Sussex, UK, 2012.
15. Ducros, V. Chromium Metabolism. A Literature Review. *Biol. Trace Elem. Res*. **1992**, *32*, 65–77.
16. Simonoff, M.; Shapcott, D.; Alameddine, S.; Sutter-Dub, M. T.; Simonoff, G. The Isolation of Glucose Tolerance Factors from Brewer's Yeast and Their Relation to Chromium. *Biol. Trace Elem. Res*. **1992**, *32*, 25–38.
17. Hwang, D. L.; Lev-Ran, A.; Papoian, T.; Beech, W. K. Insulin-like Activity of Chromium-binding Fractions from Brewer's Yeast. *J. Inorg. Biochem*. **1987**, *30*, 3, 219–222.
18. Barrett, J.; O´Brien, P.; Pedrosa de Jesus, J. Chromium(III) and the Glucose Tolerance Factor. *Polyhedron* **1985**, *4*, 1–14.
19. Haylock, S. J.; Buckley, P. D.; Blackwell, L. F. Separation of Biologically Active Chromium-containing Complexes from Yeast Extracts and Other Sources of Glucose Tolerance Factor (GTF) Activity. *J. Inorg. Biochem*. **1983**, *18*, 195–211.
20. Gonzalez-Vergara, E.; Hegenauer, J.; Saltman, P.; Sabat, M.; Ibers, J. A. Synthesis and Structure of a Trinuclear Chromium(III)–Nicotinic Acid Complex. *Inorg. Chim. Acta* **1982**, *66*, 115–118.
21. Chang, J. C.; Gerdom, L. E.; Baenziger, N. C.; Goff, H. M. Synthesis and Molecular Structure Determination of Carboxyl Bound Nicotinic Acid (Niacin) Complexes of Chromium(III). *Inorg. Chem*. **1983**, *22*, 1739–1744.

22. Barrett, J.; Kormoh, M. K.; O'Brien, P. Preparation and Characterization of some Cr(III) Chloro-complexes with Nicotinic Acid Esters. *Inorg. Chim. Acta* **1985**, *107*, 269–274.
23. Broderick, W. E.; Legg, J. I. Synthesis of the First Stable Nitrogen Coordinated Nicotinic Acid Chromium(III) Complexes: cis- and trans-H[Cr(mal)$_2$(nic-N)$_2$]. *Inorg. Chem.* **1985**, *24*, 3724–3725.
24. Shara, M.; Kincaid, A. E.; Limpach, A. E., Sandstrom, R.; Barrett, L.; Norton, N.; Bramble, J. D.; Yasmin, T.; Tran, J.; Chatterjee, A.; Bagchi, M.; Bagchi, D. Long Term Safety Evaluation of a Novel Oxygen Coordinated Niacin-bound Chromium(III) Complex. *J. Inorg. Biochem.* **2007**, *101*, 1059–1069.
25. Sreejayan, N.; Marone, P. A.; Lau, F. C.; Yasmin, T.; Bagchi, M.; Bagchi, D. Safety and Toxicological Evaluation of a Novel Chromium(III) Dinicocysteinate Complex. *Toxicol. Mech. Meth.* **2010**, *20*, 321–333.
26. Sushil, K. J.; Croad, J. L.; Velusamy, T.; Rains, J. L.; Bul, R. Chromium Dinicocysteinate Supplementation Can Lower Blood Glucose, CRP, MCP-1, ICAM-1, Creatinine, Apparently Mediated by Elevated Blood Vitamin C and Adiponectin and Inhibition of NFκB, Akt, and Glut-2 in Livers of Zucker Diabetic Fatty Rats. *Mol. Nutr. Food Res.* **2010**, *54*, 1371–1380.
27. Mertz, W. Chromium History and Nutritional Importance. *Biol. Trace Elem. Res.* **1992**, *32*, 3–8.
28. Lay, P. A.; Levina, A. Chromium: Biological Relevance. In *Encyclopedia of Inorganic and Bioinorganic Chemistry*; 2012. DOI: 10.1002/9781119951438.eibc0040.pub2.
29. Lay, P. A.; Levina, A. Chromium. In: *Comprehensive Coordination Chemistry II. From Biology to Nanotechnology*; Mc-Cleverty, J. A., Meyer, T. J., Eds.; Elsevier: Amsterdam, 2004; Vol. 4, pp 313–413.
30. Love-Rutledge, S. T. Disproving a Fifty-five Year Old Myth: Chromium the Essential Element, Chapter 1. Ph.D. Dissertation, Department of Chemistry, Graduate School, The University of Alabama, USA, 2014.
31. Di Bona, K. R.; Love, S.; Rhodes, N. R.; McAdory, D.; Sinha, S. H.; Kern, N.; Kent, N. J.; Strickland, J.; Wilson, A.; Beaird, J.; Ramage, J.; Rasco, J. F.; Vincent, J. B. Chromium is Not an Essential Trace Element for Mammals: Effects of a "Low-chromium Diet." *J. Biol. Inorg. Chem.* **2011**, *16*, 381–390.
32. Vincent, J. B. Chromium: Is It Essential, Pharmacologically Relevant, or Toxic? *Met. Ions Life Sci.* **2013**, *13*, 171–198.
33. Vincent, J. B. Is Chromium Pharmacologically Relevant? *J. Trace Elem. Med. Bio.* **2011**, *28*, 397–405.
34. Vincent, J. B. Toxicology of Chromium(III), Chapter 9. In *The Bioinorganic Chemistry of Chromium*. Vincent, J. B., Ed.; John Wiley and Sons: Sussex, UK, 2012.
35. Davis, C. M.; Vincent, J. B. The Biologically Active Form of Chromium Contains a Multi Nuclear Chromium(III) Assembly. *Arch. Biochem. Biophys.* **1997**, *339*, 335–343.
36. Davis, C. M.; Royer, A. C.; Vincent, J. B. Synthetic Multinuclear Chromium Activates Insulin Receptor Kinase: Functional Model Activity Assembly for Low-molecular-weight Chromium-binding Substance. *Inorg. Chem.* **1997**, *36*, 5316–5320.
37. Vincent, J. B. Mechanisms of Chromium Action: Low-molecular-weight Chromium-binding Substance. *J. Am. Coll. Nutr.* **1999**, *18*, 6–12.
38. Jacquamet, L.; Sun, Y.; Hatfield, J.; Gu, W.; Cramer, S. P.; Crowder, M. W.; Lorigan, G. A.; Vincent, J. B.; Latour, J.-M. Characterization of Chromodulin by X-ray Absorption and Electron Paramagnetic Resonance Spectroscopies and Magnetic Susceptibility Measurements. *J. Am. Chem. Soc.* **2003**, *125*, 774–780.

39. Viera, M.; Davis-McGibony, C. M. Isolation and Characterization of Low-molecular-weight Chromium-binding Substance (LMWCr) from Chicken Liver. *Protein J.* **2008**, *27*, 371–375.
40. Speetjens, J. K.; Paran, A.; Crowder, M. W.; Vincent, J. B.; Woski, S. A. Low-Molecular-weight Chromium-binding Substance and Biomimetic $[Cr_3O(O_2CCH_2CH_3)_6(H_2O)_3]^+$ Do not Cleave DNA Under Physiologically-relevant Condition. *Polyhedron* **1999**, *18*, 2617–2624.
41. Vincent, J. B. The Biochemistry of Chromium. *J. Nutr.* **2000**, *130*, 715–718.
42. Clodfelder, B. J.; Emamaullee, J.; Hepburn, D. D.; Chakov, N. E.; Nettles, H. S.; Vincent, J. B. The Trail of Chromium(III) In Vivo from the Blood to the Urine: The Roles of Transferrin and Chromodulin. *J. Biol. Inorg. Chem.* **2001**, *6*, 608–17.
43. Sun, Y.; Clodfelder, B. J.; Shute, A.; Irvin, T.; Vincent, J. B. Oral Administration of the Biomimetic $[Cr_3O(O_2CCH_2CH_3)_6(H_2O)_3]^+$ Decreases Plasma Insulin, Cholesterol, and Triglycerides in Healthy and Type 2 Diabetic Rats but Not Type I Diabetic Rats. *J. Biol. Inorg. Chem.* **2002**, *7*, 852–862.
44. Clodfelder, B. J.; Gullick, B. M.; Lukaski, H. C.; Neggers, Y.; Vincent, J. B. Oral Administration of the Biomimetic $[Cr_3O(O_2CCH_2CH_3)_6(H_2O)_3]^+$ Increases Insulin Sensitivity and Improves Blood Plasma Variables in Healthy and Type 2 Diabetic Rats. *J. Biol. Inorg. Chem.* **2005**, *10*, 119–130.
45. Levina, A.; Lay, P. A. Metal-based Anti-diabetic Drugs: Advances and Challenges. Dalton T. 2011, 40, 11675–11686.
46. Herring, B. J.; Logsdon, A. L.; Lockard, J. E; Miller, B. M.; Kim, H.; Calderon, E. A.; Vincent, J. B.; Bailey, M. M. Long-term Exposure to $[Cr_3O(O_2CCH_2CH_3)_6(H_2O)_3]^+$ in Wistar Rats Fed Normal or High-fat Diets Does not Alter Glucose Metabolism. *Biol. Trace Elem. Res.* **2013**, *151*, 406–414.
47. Levina, A.; Harris, H.; Lay, P. X-ray Absorption and EPR Spectroscopic Studies of the Biotransformations of Chromium(VI) in Mammalian Cells. Is Chromodulin an Artifact of Isolation Methods? *J. Am. Chem. Soc.* **2007**, *129*, 1065–1075.
48. Porter, D. J. Chromium: Friend or Foe? *Arch. Fam. Med.* **1999**, *8*, 386–390.
49. Stearns, D. M. Is Chromium a Trace Essential Metal? *Biofactors* **2000**, *11*, 149–162.
50. Nielsen, F. H. The Clinical and Nutritional Importance of Chromium—Still Debated After 50 Years of Research, Chapter 13. In *The Nutritional Biochemistry of Chromium(III)*; Vincent, J., Ed.; Elsevier: Amsterdam, 2007; pp 265–275.
51. Nielsen, F. H. Should Bioactive Trace Elements not Recognized as Essential, but with Beneficial Health Effects, Have Intake Recommendations. *J. Trace Elem. Med. Biol.* **2014**, *28*, 406–408 (ISTERH-X Conference, Tokyo: Trace Element Research in Health and Disease).
52. Stearns, D. M. Multiple Hypotheses for Chromium(III) Biochemistry: Why the Essentiality of Chromium(III) Is Still Questioned, Chapter 3. In *The Nutritional Biochemistry of Chromium(III)*; Vincent, J., Ed.; Elsevier: Amsterdam, 2007; pp 57–70.
53. Sadeghi, M.; Najaf Panah, M. J.; Bakhtiarizadeh, M. R.; Emami, A. Transcription Analysis of Genes Involved in Lipid Metabolism Reveals the Role of Chromium in Reducing Body Fat in Animal Models. *J. Trace Elem. Med. Biol.* **2015**, *32*, 45–51.
54. Maples, N. L.; Bain, L. Trivalent Chromium Alters Gene Expression in the Mummichog (*Fundulus heteroclitus*). *Environ. Toxicol. Chem.* **2004**, *23*, 626–63.
55. Sadeghi, M.; Najaf Panah, M. J. Analysis of Immune-relevant Genes Expressed in Spleen of Capra Hircus Kids Fed with Trivalent Chromium. *Biol. Trace Elem. Res.* **2013**, *156*, 124–129.

56. Roling, J. A.; Baldwin, W. S. Alterations in Hepatic Gene Expression by Trivalent Chromium in *Fundulus heteroclitus*. *Mar. Environ. Res.* **2006**, *62*, S122–S127.
57. Najafpanah, M. J.; Sadeghi, M.; Zali, A.; Moradi-Shahrebabak, H.; Mousapour, H. Chromium Downregulates the Expression of Acetyl CoA Carboxylase 1 Gene in Lipogenic Tissues of Domestic Goats: A Potential Strategy for Meat Quality Improvement. *Gene* **2014**, *543*, 253–258.
58. Vincent, J. B., Bennett, R. Potential and Purported Roles for Chromium in Insulin Signaling: The Search for the Holy Grail, Chapter 7. In *The Nutritional Biochemistry of Chromium(III)*; Vincent, J., Ed.; Elsevier: Amsterdam, 2007; pp 57–71.
59. Vincent, J. B. Is the Pharmacological Mode of Action of Chromium(III) as a Second Messenger? *Biol. Trace Elem. Res.* **2015**, *166*, 7–12.
60. O´Brien, P.; Kortenkamp, A. The Chemistry Underlying Chromate Toxicity. *Transit. Metal Chem.* **1995**, *20*, 636–642.
61. Codd, R.; Dillon, C.; Levina, A.; Lay, P. Studies on the Genotoxicity of Chromium: From the Test Tube to the Cell. *Coord. Chem. Rev.* **2001**, *216–217*, 537–582.
62. Jennette, K. W. Chromate Metabolism in Liver Microsomes. *Biol. Trace Elem. Res.* **1979**, *1*, 55–62.
63. Dayan, A. D.; Paine, A. J. Mechanisms of Chromium Toxicity, Carcinogenicity and Allergenicity: Review of the Literature from 1985 to 2000. *Hum. Exp. Toxicol.* **2001**, *20*, 439–451.
64. Levina, A.; Lay, P. A. Mechanistic Studies of Relevance to the Biological Activities of Chromium. *Coord. Chem. Rev.* **2005**, *249*, 281–298.
65. Bagchi, D.; Stohs, S. J.; Downs, B. W.; Bagchi, M.; Preuss, H. G. Cytotoxicity and Oxidative Mechanisms of Different Forms of Chromium. *Toxicology* **2002**, *180*, 5–22.
66. Connett, P. H.; Wetterhahn, K. E. Metabolism of the Carcinogen Chromate by Cellular Constituents. *Struct. Bond.* **1983**, *54*, 93–124.
67. Brauer, S. L.; Wetterhahn, K. E. Chromium(VI) Forms a Thiolate Complex with Glutathione. *J. Am. Chem. Soc.* **1991**, *113*, 3001–3007.
68. Moghaddas, S.; Gelerinter, E.; Bose, R. N. Mechanisms of Formation and Decomposition of Hypervalent Chromium Metabolites in the Glutathione–Chromium(VI) Reaction. *J. Inorg. Biochem.* **1995**, *57*, 135–146.
69. Levina, A.; Lay, P. A. Solution Structures of Chromium(VI) Complexes with Glutathione and Model Thiols. *Inorg. Chem.* **2004**, *43*, 324–335.
70. Wiegand, H. J.; Ottenwalder, H.; Bolt, H. M. The Reduction of Chromium(VI) to Chromium(III) by Glutathione: An Intracellular Redox Pathway in the Metabolism of the Carcinogen Chromate. *Toxicology* **1984**, *33*, 341–348.
71. Quievryn, G.; Goulart, M.; Messer, J.; Zhitkovich, A. Reduction of Cr (VI) by Cysteine: Significance in Human Lymphocytes and Formation of DNA Damage in Reactions with Variable Reduction Rates. *Mol. Cell. Biochem.* **2001**, *222*, 107–118.
72. Lay, P. A.; Levina, A. Kinetics and Mechanism of Chromium(VI) Reduction to Chromium(III) by L-Cysteine in Neutral Aqueous Solutions. *Inorg. Chem.* **1996**, *35*, 7709–7717.
73. Zhang, L.; Lay, P. A. EPR Spectroscopic Studies of the Reactions of Cr(VI) with L-Ascorbic Acid, L-Dehydroascorbic Acid, and 5,6-*O*-Isopropylidene-L-ascorbic Acid in Water. Implications for Chromium(VI) Genotoxicity. *J. Am. Chem. Soc.* **1996**, *118*, 12624–12637.
74. Sugden, K. D.; Burris, R. B.; Rogers, S. J. An Oxygen Dependence in Chromium Mutagenesis. *Mutat. Res.* **1990**, *244*, 239–244.

75. Lay, P. A.; Levina, A. Activation of Molecular Oxygen During the Reactions of Chromium (VI/V/IV) with Biological Reductants: Implications for Chromium-induced Genotoxicities. *J. Am. Chem. Soc.* **1998**, *120*, 6704–6714.
76. Kortenkamp, A.; Casadevall, M.; Faux, S.; Jenner, A.; Shayer, R.; Woodbridge, N.; O'Brien, P. A Role for Molecular Oxygen in the Formation of DNA Damage During the Reduction of the Carcinogen Chromium(VI) by Glutathione. *Arch. Biochem. Biophys.* **1996**, *329*, 199–207.
77. Aiyar, J.; Berkovits, H. J.; Floyd, R. A.; Wetterhahn, K. E. Reaction of Chromium(VI) with Hydrogen Peroxide in the Presence of Glutathione: Reactive Intermediates and Resulting DNA Damage. *Chem. Res. Toxicol.* **1990**, *3*, 595–603.
78. Signorella, S., Garcia, S., Rizzotto, M., Levina, A., Lay, P., Sala, L. The EPR Pattern of Cr-V Complexes of D-Ribose Derivatives. *Polyhedron* **2005**, *24*, 1079–1085.
79. Signorella, S.; Santoro, M. I.; Sala, L. Oxidation of D-Gluconic Acid by Chromium(VI) in Perchloric Acid. *Can. J. Chem.* **1994**, *72*, 398–402.
80. Kent, D. Sugden, K. D.; Wetterhahn, K. E. Direct and Hydrogen Peroxide-induced Chromium(V) Oxidation of Deoxyribose in Single-stranded and Double-stranded Calf Thymus DNA. *Chem. Res. Toxicol.* **1997**, *10*, 1397–1406.
81. Levina, A.; Lay, P. A.; Dixon, N. E. Disproportionation of a Model Chromium(V) Complex Causes Extensive Chromium(III)-DNA Binding In Vitro. *Chem. Res. Toxicol.* **2001**, *14*, 946–950.
82. Jennette, K. E. W. Microsomal Reduction of the Carcinogen Chromate Produces Chromium(V). *J. Am. Chem. Soc.* **1982**, *104*, 874–875.
83. Farrell, R. P.; Lay, P. A. New Insights into the Structures and Reactions of Chromium(V) Complexes: Implications for Chromium(VI) and Chromium(V) Oxidations of Organic Substrates and the Mechanisms of Chromium-induced Cancers. Comment. *Inorg. Chem.* **1992**, *13*, 133–175.
84. Farrell, R.; Judd, R.; Lay, P. A. Ligand Exchange and Reduction Reactions of Oxochromate(V) Complexes: Characterization of the Common Chromium(V) Intermediates in the Reductions of Chromium(VI) and of Trans-bis(2-ethyl-2-hydroxybutanoato(2-)oxochromate(V) by Oxalic Acid. *Inorg. Chem.* **1989**, *28*, 3401–3403.
85. Rossi, S.; Wetterhahn, K. E. Chromium(V) is Produced Upon Reduction of Chromate by Mitochondrial Electron Transport Chain Complexes. *Carcinogenesis* **1989**, *10*, 913–920.
86. O'Brien, P.; Barrett, J.; Swanson, F. Chromium(V) can be Generated in the Reduction of Chromium(VI) by Glutathione. *Inorg. Chim. Acta* **1985**, *108*, L19–L20.
87. Stearns, D. M.; Wetterhahn, K. E. Reaction of Chromium(VI) with Ascorbate Produces Chromium(V), Chromium(IV), and Carbon-based Radicals. *Chem. Res. Toxicol.* **1994**, *7*, 219–230.
88. Goodgame, D. M. L.; Joy, A. M. Relatively Long-lived Chromium(V) Species Are Produced by the Action of Glutathione on Carcinogenic Chromium(VI). *J. Inorg. Biochem.* **1986**, *26*, 219–224.
89. Rossi, S. C.; Gorman, N.; Wetterhahn, K. E. Mitochondrial Reduction of the Carcinogen Chromate: Formation of Chromium(V). *Chem. Res. Toxicol.* **1988**, *1*, 101–107.
90. Codd, R.; Lay, P. Chromium(V)–Sialic (Neuraminic) Acid Species are Formed from Mixtures of Chromium(VI) and Saliva. *J. Am. Chem. Soc.* **2001**, *123*, 11799–11800.
91. Codd, R.; Lay, P. A. Competition Between 1,2-Diol and 2-Hydroxy Acid Coordination in Cr(V)-Quinic Acid Complexes: Implications for Stabilization of Cr(V) Intermediates of Relevance to Cr(VI)-induced Carcinogenesis. *J. Am. Chem. Soc.* **1999**, *121*, 7864–7876.

92. Goodgame, D. M. L.; Joy, A. M. Formation of Chromium(V) During the Slow Reduction of Carcinogenic Chromium(VI) by Milk and some of Its Constituents. *Inorg. Chim. Acta* **1987**, *135*, L5–L7.
93. Branca, M.; Micera, G.; Dessí, A. Reduction of Chromium (VI) by D-Galacturonic Acid and Formation of Stable Chromium (V) Intermediates. *Inorg. Chim. Acta* **1988**, *153*, 61–65.
94. Borges, K. M.; Wetterhahn, K. E. Chromium Cross-links Glutathione and Cysteine to DNA. *Carcinogenesis* **1989**, *10*, 2165–2168.
95. Zhitkovich, A. Importance of Chromium-DNA Adducts in Mutagenicity and Toxicity of Chromium(VI). *Chem. Res. Toxicol.* **2005**, *18*, 3–11.
96. Casadevall, M.; da Cruz Fresco, P.; Kortenkamp, A. Chromium(VI)-mediated DNA Damage: Oxidative Pathways Resulting in the Formation of DNA Breaks and Abasic Sites. *Chem. Biol. Interact.* **1999**, *123*, 117–132.
97. da Cruz Fresco, P.; Kortenkamp, A. The Formation of DNA Cleaving Species During the Reduction of Chromate by Ascorbate. *Carcinogenesis* **1994**, *15*, 1773–1778.
98. Stearns, D. M.; Courtney, K. D.; Giangrande, P. H.; Phieffer, L. S.; Wetterhahn, K. E. Chromium(VI) Reduction by Ascorbate: Role of Reactive Intermediates in DNA Damage In Vitro. *Environ. Health Perspect.* **1994**, *102*(Suppl. 3), 21–25.
99. Kortenkamp, A.; O'Brien, P. The generation of DNA Single-strand Breaks During the Reduction of Chromate by Ascorbic Acid and/or Glutathione In Vitro. *Environ. Health Perspect.* **1994**, *102*(Suppl. 3), 237–241.
100. Luo, H.; Lu, Y.; Mao, Y.; Shi, X.; Dalal, N. S. Role of Chromium (IV) in the Chromium (VI)-related Free Radical Formation, dG hydroxylation, and DNA Damage. *J. Inorg. Biochem.* **1996**, *64*, 25–35.
101. Stearns, D. M.; Kennedy, L. J.; Courtney, K. D.; Giangrande, P. H.; Phieffer, L. S.; Wetterhahn, K. E. Reduction of Chromium(VI) by Ascorbate Leads to Chromium-DNA Binding and DNA Strand Breaks In Vitro. *Biochemistry* **1995**, *34*, 910–909.
102. Zhitkovich, A.; Voitkun, V.; Costa, M. Glutathione and Free Amino Acids form Stable Complexes with DNA Following Exposure of Intact Mammalian Cells to Chromate. *Carcinogenesis* **1995**, *16*, 904–913.
103. Salnikow, K.; Zhitkovich, A.; Costa, M. Analysis of the Binding Sites of Chromium to DNA and Protein *In Vitro* and in Intact Cells. *Carcinogenesis* **1992**, *13*, 2341–2346.
104. Pattison, D.; Levina, A.; Lay, P.; Davies, M.; Dixon, N. Chromium(VI) Reduction by Catechol(amine)s Results in DNA Cleavage In Vitro: Relevance to Chromium Genotoxicity. *Chem. Res. Toxicol.* **2001**, *14*, 500–510.
105. Whiting, R. F.; Stich, H. F.; Koropatnick, D. J. DNA Damage and DNA Repair in Cultured Human Cells Exposed to Chromate. *Chem. Biol. Interact.* **1979**, *26*, 26–280.
106. Branca, M.; Dessi, A.; Kozlowski, H.; Micera, G.; Serra, M. V. In Vitro Interaction of Mutagenic Chromium(VI) with Red Blood Cells. *FEBS Lett.* **1989**, *257*, 52–54.
107. Macfie, A.; Hagan, E.; Zhitkovich, A. Mechanism of DNA–Protein Cross-linking by Chromium. *Chem. Res. Toxicol.* **2010**, *23*, 341–347.
108. Voitkun, V.; Zhitkovitch, A.; Costa, M. Cr(III) Mediated Cross-linkings of Glutathione or Amino Acids to the DNA Phosphate Backbone Are Mutagenic in Human Cells. *Nucleic Acids Res.* **1998**, *26*, 2024–2030.
109. Santos, T. M.; Pedrosa de Jesus, J. D.; Beyersmann, D. In *On the Mechanisms of Chromium Toxicity: Interaction of Cr(III) Complexes with Isolated Calf Thymus Nuclei.* Book of Abstracts Poster PB-15, 12th Portuguese Chemistry Society Conference, Lisbon, Portugal, Jan 29–Feb 1, 1992.

110. Tsapakos, M. J.; Hampton, T. H.; Wetterhahn, K. E. Chromium(VI)-induced DNA Lesions and Chromium Distribution in Rat Kidney, Liver, and Lung. *Cancer Res.* **1983**, *43*, 5662–5667.
111. Liu, K. J.; Shi, X. In Vivo Reduction of Chromium (VI) and Its Related Free Radical Generation. *Mol. Cell. Biochem.* **2001**, *222*, 41–47.
112. Hamilton, J. W.; Wetterhahn, K. E. Chromium (VI)-induced DNA Damage in Chick Embryo Liver and Blood Cells In Vivo. *Carcinogenesis* **1986**, *7*, 2085–2088.
113. Yuann, J. M.; Liu, K. J.; Hamilton, J. W.; Wetterhahn, K. E. In Vivo Effects of Ascorbates and Glutathione on the Uptake of Chromium, Formation of Cr(V), Cr–DNA Binding and 8-hydroxy-2'-deoxyguanosine in Liver and Kidney of Osteogenic Disorder Shionogi Rats Following Treatment with Cr(VI). *Carcinogenesis* **1999**, *20*, 1267–1275.
114. O'Brien, P.; Kortenkamp, A. Chemical Models Important in Understanding the Ways in Which Chromate Can Damage DNA. Environ. *Health Perspect.* **1994**, *102*(Suppl 3), 3–10.
115. Santos, T. M. 1995. Chromium Coordination Chemistry Related with Cr(III) and Cr(VI) Compounds Biological Effects. Ph.D. Dissertation, Chapter 4, Department of Chemistry, University of Aveiro, Aveiro, Portugal.
116. Levina, A.; Zhang, L.; Lay, P. A. Structure and Reactivity of a Chromium(V) Glutathione Complex. *Inorg. Chem.* **2003**, *42*, 767–784.
117. Dmitriy-Krepkiy, D.; Antholine, W. E.; Myers, C.; Petering, D. H. Model reactions of Cr (VI) with DNA Mediated by Thiol Species. *Mol. Cell. Biochem.* **2001**, *222*, 213–219.
118. Headlam, H.; Weeks, C.; Turner, P.; Hambley, T.; Lay, P. Dinuclear Chromium(V) Amino Acid Complexes from the Reduction of Chromium(VI) in the Presence of Amino Acid Ligands: XAFS Characterization of a Chromium(V) Amino Acid Complex. *Inorg. Chem.* **2001**, *40*, 5097–5105.
119. Levina, A.; Zhang, L.; Lay, P. Formation and Reactivity of Chromium(V)–Thiolato Complexes. A Model for the Intracellular Reactions of Carcinogenic Chromium(VI) with Biological Thiols. *J. Am. Chem. Soc.* **2010**, *132*, 8720–8731.
120. Barnard, P.; Levina, A.; Lay, P. Chromium(V) Peptide Complexes: Synthesis and Spectroscopic Characterization. *Inorg. Chem.* **2005**, *44*, 1044–1053.
121. Bartholomaus, R.; Harms, K.; Levina, A.; Lay, P. Synthesis and Characterization of a Chromium(V) cis-1,2-Cyclohexanediolato Complex: A Model of Reactive Intermediates in Chromium-induced Cancers. *Inorg. Chem.* **2012**, *51*, 11238–11240.
122. Bartholomaus, R.; Irwin, J.; Shi, L.; Smith, S. M.; Levina, A., Lay, P. Isolation, Characterization, and Nuclease Activity of Biologically Relevant Chromium(V) Complexes with Monosaccharides and Model Diols. Likely Intermediates in Chromium-induced Cancers. *Inorg. Chem.* **2013**, *52*, 4282–4292.
123. Codd, R.; Lay, P. A. Oxochromium(V) Species Formed with 2,3-dehydro-2-deoxy-N-acetylneuraminic or N-acetylneuraminic (Sialic) Acids: An In Vitro Model System of Oxochromium(V) Species Potentially Stabilized in the Respiratory Tract upon Inhalation of Carcinogenic Chromium(VI) Compounds. *Chem. Res. Toxicol.* **2003**, *16*, 881–892.
124. Branca, M.; Dessi, A.; Kozlwski, H.; Micera, G.; Swiatk, J. Reduction of Chromate Ions by Glutathione Tripeptide in the Presence of Sugar Ligands. *J. Inorg. Biochem.* **1990**, *39*, 217–220.
125. Santos, T. S.; Pedrosa de Jesus, J. D.; O´Brien, P. Solution Studies of some Chromium(III) Complexes with Cr–S Bonds. Part 2: Kinetic and Equilibrium Studies of Cysteinato and Penicillaminatochromate(III) Complexes. *Polyhedron* **1992**, *11*, 1687–1695.

126. O´Brien, P.; Pedrosa de Jesus, J.; Santos, T. M. A Kinetic and Equilibrium Study of the Reaction of K- and Na-biscysteinato(N,O,S)-Cr(III) in Moderately Acidic Solutions. *Inorg. Chim. Acta* **1987**, *131*, 5–7.
127. Levina, A.; Zhang, L.; Lay, P. Structure and Reactivity of a Chromium(V) Glutathione Complex. *Inorg. Chem.* **2003**, *42*, 767–784.
128. Oliveira, H.; Spanò, M.; Guevara, M. A.; Santos, T. M.; Santos, C.; Pereira, M. L. Evaluation of In Vivo Reproductive Toxicity of Potassium Chromate in Male Mice. *Exp. Toxicol. Pathol.* **2010**, *62*, 391–404.
129. das Neves, R. P.; Santos, T. M.; Pereira, M. L.; Pedrosa de Jesus, J. Cr(VI) Induced Alterations in Mouse Spleen Cells. A Short-term Assay. *Cytobios* **2001**, *106*, S1, 27–34.
130. das Neves, R. P.; Santos, T. M.; Pereira, M. L.; Pedrosa de Jesus, J. Comparative Histological Studies on Liver of Mice Exposed to Cr(VI) and Cr(V) Compounds. *Hum. Exp. Toxicol.* **2002**, *21*, 365–369.
131. Sánchez-Martín, F. J.; Fan, Y.; Carreira, V.; Ovesen, J. L.; Vonhandorf, A.; Xia, Y.; Puga, A. Long-term Coexposure to Hexavalent Chromium and B[a]P Causes Tissue-specific Differential Biological Effects in Liver and Gastrointestinal Tract of Mice. *Toxicol. Sci.* **2015**, *146*, 52–64.
132. Oliveira, H.; Santos, T. M.; Ramalho-Santos, J.; Pereira, M. L. Histopathological Effects of Hexavalent Chromium in Mouse Kidney. *Bull. Environ. Contam. Tox.* **2006**, *76*, 977–983.
133. Pereira M. L.; das Neves, R. P.; Santos, T. M.; Pedrosa de Jesus, J. Cr(V) Involvement in the Toxicity Pathway of Testis Damage. *Asian J. Androl.* **2002**, *4*, 153–155.
134. Pereira, M. L.; Santos, T. M.; Garcia e Costa, F.; Jesus, J. P. Functional Changes of Mice Sertolli Cells Induced by Cr(V). *Cell Biol. Toxicol.* **2004**, *40*, 285–291.
135. Pereira M. L.; das Neves, R. P.; Oliveira, H.; Santos, T. M.; Pedrosa de Jesus, J. Effect of Cr(V) on Reproductive Organ Morphology and Sperm Parameters: An Experimental Study in Mice. *Environ. Health* **2005**, *4*, 9. DOI:10.1186/1476-069X-4-9.
136. Li, H.; Chen, Q.; Li, S.; Yao, W.; Li, L.; Shi, X.; Wang, L.; Castranova, V.; Vallyathan, V.; Ernst, E.; Chen, C. Effect of Cr(VI) Exposure on Sperm Quality: Human and Animal Studies. *Ann. Occup. Hyg.* **2001**, *45*, 505–511.
137. Li, P.; Li, Y.; Zhang, J.; Yu, S. F.; Tong, W.; Hu, X.; Jia, G. Biomarkers for Lung Epithelium Injury in Occupational Hexavalent Chromium-exposed Workers. *J. Occup. Environ. Med.* **2015**, *57*, e45–e50.
138. Park, R. M.; Bena, J. F.; Stayner, L. T.; Smith, R. J.; Gibbs, H. J.; Lees, P. S. Hexavalent Chromium and Lung Cancer in the Chromate Industry: A Quantitative Risk Assessment. *Risk Anal.* **2004**, *24*, 1099–1108.
139. van Wijngaarden, E.; Mundt, K. A.; Luippold, R. S. Evaluation of the Exposure–Response Relationship of Lung Cancer Mortality and Occupational Exposure to Hexavalent Chromium Based on Published Epidemiological Data. *Nonlinearity Biol. Toxicol. Med.* **2004**, *2*, 27–34.
140. Geller, R., Chromium. In *Clinical Environmental Health and Toxic Exposures*, 2nd Ed; Sullivan, J. B. Jr., Krieger, G. R., Eds.; Lippincott, Williams, and Wilkins: Philadelphia, PA, 2001.
141. Langard, S. One Hundred Years of Chromium and Cancer: A Review of Epidemiological Evidence and Selected Case Reports. *Am. J. Ind. Med.* **1990**, *17*, 189−215.
142. Liu, K.; Husler, J.; Ye, J.; Leonard, S. S.; Cutler, D.; Chen, F.; Wang, S.; Zhang, Z.; Ding, M.; Wang, L.; Shi, X. On the Mechanism of Cr(VI)-induced Carcinogenesis: Dose Dependence of Uptake and Cellular Responses. *Mol. Cell Biochem.* **2001**, *222*, 221–229.

143. Medeiros, M. G.; Rodrigues, A. S.; Batoréu, M. C.; Laires, A.; Rueff, J.; Zhitkovich, A. Elevated Levels of DNA–Protein Crosslinks and Micronuclei in Peripheral Lymphocytes of Tannery Workers Exposed to Trivalent Chromium. *Mutagenesis* **2003**, *18*, 19–24.
144. De Mattia, G., Bravi, M. C.; Laurenti, O.; De Luca, O.; Palmeri, A.; Sabatucci, A.; Mendico, G.; Ghiselli, A. Impairment of Cell and Plasma Redox State in Subjects Professionally Exposed to Chromium. *Am. J. Ind. Med.* **2004**, *46*, 120–125.
145. Costa, M.; Klein, C. B. Toxicity and Carcinogenicity of Chromium Compounds in Humans. *Crit. Rev. Toxicol.* **2006**, *36*, 155–163.
146. OSHA. OSHA Issues Final Standard on Hexavalent Chromium. National News Release: 06-342-NAT (accessed Feb 27, 2006).
147. Zhitkovich, A. Chromium in Drinking Water: Sources, Metabolism and Cancer Risks. *Chem. Res. Toxicol.* **2011**, *24*, 1617–1629.
148. Thompson, C. M.; Proctor, D. M.; Suh, M.; Haws, L. C.; Kirman, C. R.; Harris, M. A. Assessment of the Mode of Action Underlying Development of Rodent Small Intestinal Tumors Following Oral Exposure to Hexavalent Chromium and Relevance to Humans. *Crit. Rev. Toxicol.* **2013**, *43*, 244–274.
149. Langard, S.; Costa, M., Chromium, Chapter 33. In *Handbook on the Toxicology of Metals*; Nordberg, G. F., Fowler, B. A., Nordberg, M., Eds.; Academic Press: *Massachusetts*, 2014; pp 717–742.
150. Proctor, D. M.; Suh. M.; Campleman, S. J.; Thompson, C. M. Assessment of the Mode of Action for Hexavalent Chromium-induced Lung Cancer Following Inhalation Exposures. *Toxicology* **2014**, *325*, 160–179.
151. Berardi, R.; Pellei, C.; Valeri, G.; Pistelli, M.; Onofri, A.; Morgese, F.; Caramanti, M.; Mirza, R. M.; Santoni, M. M.; De Lisa, M.; Savini, A.; Ballatore, Z.; Giuseppetti, G. M.; Cascinu, S. Chromium Exposure and Germinal Embryonal Carcinoma: First Two Cases and Review of the Literature. *J. Toxicol. Environ. Heal. A* **2015**, *1–2*, 1–6.
152. Welling, R.; Beaumont, J. J.; Petersen, S. J.; Alexeeff, G. V.; Steinmaus, C. Chromium VI and Stomach Cancer: A Meta-analysis of the Current Epidemiological Evidence. *Occup. Environ. Med.* **2015**, *72*, 151–159.
153. Thompson, C. M.; Seiter, J.; Chappell, M. A.; Tappero, R. V.; Proctor, D. M.; Suh, M.; Wolf, J. C.; Haws, L. C.; Vitale, R.; Mittal, L.; Kirman, C. R.; Hays, S. M.; Harris, M. A. Synchrotron-based Imaging of Chromium and γ-H2AX Immunostaining in the Duodenum Following Repeated Exposure to Cr(VI) in Drinking Water. *Toxicol. Sci.* **2015**, *143*, 16–25.
154. Sun, H.; Brocato, J.; Costa, M. Oral Chromium Exposure and Toxicity. *Curr. Environ. Health Rep.* **2015**, *2*, 295–303.
155. Proctor, D. M.; Suh, M.; Mittal, L.; Hirsch, S.; Valdes-Salgado, R.; Bartlett, C.; van Landingham, C.; Rohr, A.; Crump, K. Inhalation Cancer Risk Assessment of Hexavalent Chromium-based on Updated Mortality for Painesville Chromate Production Workers. *J. Expo. Sci. Env. Epid.* **2016**, *26*, 224–231.
156. Lei, T.; He, Q.-Y.; Cai, Z.; Zhou, Y.; Wang, Y.-L.; Si, L.-S.; Cai, Z.; Chiu, J.-F. Proteomic Analysis of Chromium Cytotoxicity in Cultured Rat Lung Epithelial Cells. *Proteomics* **2008**, *8*, 2420–2429.
157. Wetterhahn, K. E.; Hamilton, J. W. Molecular Basis of Hexavalent Chromium Carcinogenicity: Effect on Gene Expression. *Sci. Total Environ.* **1989**, *86*, 113–129.
158. Mattagajasingh, S. N.; Misra; B. R.; Misra, H. P. Carcinogenic Chromium(VI)-induced Protein Oxidation and Lipid Peroxidation: Implications in DNA–Protein Crosslinking. *J. Appl. Toxicol.* **2008**, *28*, 978–977.

159. Salnikow, K.; Zhitkovich, A. Genetic and Epigenetic Mechanisms in Metal Carcinogenesis and Cocarcinogenesis: Nickel, Arsenic, and Chromium. *Chem. Res. Toxicol.* **2008**, *21,* 28–44.
160. Jennette, K. W. The Role of Metals in Carcinogenesis: Biochemistry and Metabolism. *Environ. Health Perspect.* **1981**, *40*, 233–252.
161. De Flora, S.; Wetterhahn, K. E. Mechanisms of Chromium Metabolism and Genotoxicity. *Life Chem. Rep.* **1989**, *7*, 169–244.
162. Nickens, K. P.; Steven, R. Patierno, S. R.; Ceryak, S. Chromium Genotoxicity: a Double-edged Sword. *Chem. Biol. Interact.* **2010**, *188*, 276–288.
163. De Flora, S. Threshold Mechanisms and Site Specificity in Chromium(VI) Carcinogenesis. *Carcinogenesis* **2000**, *21*, 533–541.
164. O'Brien, T. J.; Ceryak, S.; Patierno, S. R. Complexities of Chromium Carcinogenesis: Role of Cellular Response, Repair and Recovery Mechanisms. *Mutat. Res.* **2003**, *533*, 3–36.
165. Morse, J. L.; Luczak, M. W.; Zhitkovich, A. Chromium(VI) Causes Interstrand DNA Cross-linking in Vitro but Shows No Hypersensitivity in Cross-link Repair-deficient Human Cells. *Chem. Res. Toxicol.* **2013**, *26*, 1591−1598.
166. Blasiak, J.; Kowalik, J. A Comparison of the In Vitro Genotoxicity of Tri- and Hexavalent Chromium. *Mutat. Res.* **2000**, *469*, 135–145.
167. Beyersmann, D., Köster, A. On the Role of Trivalent Cr in Cr Genotoxicity. *Toxicol. Environ. Chem.* **1987**, *14*, 11–22.
168. Dillon, C. T.; Lay, P. A.; Bonin, A. M.; Cholewa, M.; Legge, G. J. F. Permeability, Cytotoxicity, and Genotoxicity of Cr(III) Complexes and Some Cr(V) Analogues in V79 Chinese Hamster Lung Cells. *Chem. Res. Toxicol.* **2000**, *13*, 742–748.
169. Arakawa, H.; Tang, M. S. Recognition and Incision of Cr(III) Ligand-conjugated DNA Adducts by the Nucleotide Excision Repair Proteins UvrABC: Importance of the Cr(III)–Purine Moiety in the Enzymatic Reaction. *Chem. Res. Toxicol.* **2008**, *21*, 1284–1289.
170. Snow, E. T. A Possible Role for Chromium(III) in Genotoxicity. *Environ. Health Perspect.* **1991**, *92*, 75–81.
171. Anderson, R. A. Nutritional Role of Chromium. *Sci. Total Environ.* **1981**, *17*, 13–29.
172. Mulyani, I.; Levina, A.; Lay, P. A. Biomimetic Oxidation of Chromium(III): Does the Antidiabetic Activity of Chromium(III) Involve Carcinogenic Chromium(VI)? *Angew. Chem. Int. Ed.* **2004**, *43*, 4504–4507.
173. Levina, A., Mulyani, I., Lay, P. Redox Chemistry and Biological Activities of Chromium(III) Complexes. In: *The Nutritional Biochemistry of Chromium(III)*; Vincent, J. B., Ed.; Elsevier: Amsterdam, Netherlands, 2007; pp 225–256.
174. Vincent, J. B. The Potential Value and Toxicity of Chromium Picolinate as a Nutritional Supplement, Weight Loss Agent and Muscle Development Agent. *Sports Med.* **2003**, *33*(3), 213–230.
175. Fang, Z.; Zhao, M.; Zhen, H.; Chen, L.; Ping, S.; Huang, Z. Genotoxicity of Tri- and Hexavalent Chromium Compounds In Vivo and Their Modes of Action on DNA Damage In Vitro. *PLoS One* **2014**, *9,* e103194 (Open Access).
176. Dillon, C. T.; Lay, P. A.; Cholewa, M.; Legge, G. J. F.; Bonin, A. M.; Collins, T. J.; Kostka, K. L.; Shea-McCarthy, G. Microprobe X-ray Absorption Spectroscopic Determination of the Oxidation State of Intracellular Chromium Following Exposure of V79 Chinese Hamster Lung Cells to Genotoxic Chromium Complexes. *Chem. Res. Toxicol.* **1997**, *10,* 533–535.

177. Dillon, C.; Lay, P.; Kennedy, B.; Stampfl, A.; Cai, Z.; Ilinski, P.; Rodrigues, W.; Legnini, D.; Lai, B.; Maser, J. (2002). Hard X-ray Microprobe Studies of Chromium(VI)-treated V79 Chinese Hamster Lung Cells: Intracellular Mapping of the Biotransformation Products of a Chromium Carcinogen. *J. Biolog. Inorg. Chem.* **2002**, *7*, 640–645.
178. Nguyen, A.; Mulyani, I.; Levina, A.; Lay, P. Reactivity of Chromium(III) Nutritional Supplements in Biological Media: An X-ray Absorption Spectroscopic Study. *Inorg. Chem.* **2008**, *47*, 4299–4309.
179. Wu, L. E.; Levina, A.; Harris, H. H.; Cai, Z.; Lai, B.; Vogt, S.; James, D. E.; Lay, P. A. Carcinogenic Chromium(VI) Compounds Formed by Intracellular Oxidation of Chromium(III) Dietary Supplements by Adipocytes. *Angew. Chem. Int. Ed.* **2016**, *55*, 1742–1745.
180. Yang, P.-H.; Gao, H.-Y.; Cai, J.; Chiu, J.-F.; Sun, H.; He, Q.-Y. The Stepwise Process of Chromium-induced DNA Breakage: Characterization by Electrochemistry, Atomic Force Microscopy, and DNA Electrophoresis. *Chem. Res. Toxicol.* **2005**, *18*, 1563–1566.
181. DeLoughery, Z.; Luczak, M. W.; Zhitkovich, A. Monitoring Cr Intermediates and Reactive Oxygen Species with Fluorescent Probes During Chromate Reduction. *Chem. Res. Toxicol.* **2014**, *27,* 843–851.
182. Shoulkamy, M.; Nakano, T.; Ohshima, M.; Hirayama, R.; Uzawa, A.; Furusawa, Y.; Ide, H. Detection of DNA–Protein Crosslinks (DPCs) by Novel Direct Fluorescence Labeling Methods: Distinct Stabilities of Aldehyde and Radiation-induced DPCs. *Nucleic Acids Res*. **2012**, *40*, e143.
183. Carter, E.; Tam, K.; Armstrong, R.; Lay, P. A. Vibrational Spectroscopic Mapping and Imaging of Tissues and Cells. *Biophys. Rev.* **2009**, *1*, 95–103.
184. Levina, A., Codd, R., Lay, P. Chromium in Cancer and Dietary Supplements. In *Biological Magnetic Resonance: High Resolution EPR Applications to Metalloenzymes and Metals in Medicine*; Hanson, G. R., Berliner, L. J., Eds.; Springer: NY, USA, 2009; pp 551–579.
185. Cefalu, W. T. Clinical Effect of Chromium Supplements on Human Health. In *The Nutritional Biochemistry of Chromium(III)*; Vincent, J., Ed.; Elsevier: Amsterdam, 2007; pp 163–183.
186. Brownley, K. A.; Boettiger, C. A.; Young, L.; Cefalu, W. T. Dietary Chromium Supplementation for Targeted Treatment of Diabetes Patients with Comorbid Depression and Binge Eating. *Med. Hypotiheses* **2015**, *85*, 45–48.
187. Lukaski, H. C.; Siders, W. A.; Penland, J. G. Chromium Picolinate Supplementation in Women: Effects on Body Weight, Composition, and Iron Status. *Nutrition* **2007**, *23*, 187–195.
188. Pittler, M. H.; Ernst, E. Dietary Supplements for Body-weight Reduction: A Systematic Review. *Am. J. Clin. Nutr.* **2004**, *79*, 529–536.
189. Mulyani, I.; Levina, A.; Lay, P. Chromium(III) Complexes Used as Nutritional Supplements: Structures and Reactivities. *J. Inorg. Biochem.* **2003**, *961*, 196.
190. Cefalu, W. T.; Hu, F. B. Role of Chromium in Human Health and in Diabetes. *Diabets Care* **2004**, *27*, 2741–2751.
191. Hua, Y.; Clark, S.; Ren, J.; Sreejaya, N. Molecular Mechanisms of Chromium in Alleviating Insulin Resistance. *J. Nutr. Biochem.* **2012**, *23*, 313–319.
192. Golubnitschaja, O.; Yeghiazaryan, K. Opinion Controversy to Chromium Picolinate Therapy's Safety and Efficacy: Ignoring "Anecdotes" of Case Reports or Recognising Individual Risks and New Guidelines Urgency to Introduce Innovation by Predictive Diagnostics? *EPMA J.* **2012**, *3*, 10.

193. Levina, A., Codd, R., Dillon, C., Lay, P. Chromium in Biology: Toxicology and Nutritional Aspects. In: *Progress in Inorganic Chemistry*; Karlin, K. D., Ed.; John Wiley and Sons: Hoboken, 2003; Vol. 51, pp 145–250.
194. Lewicki, S.; Zdanowski, R.; Krzyzowska; Lewicka, A.; Debski, B.; Niemcewicz, M.; Goniewicz, M. The Role of Chromium(III) in the Organism and Its Possible Use in Diabetes and Obesity Treatment. *Ann. Agric. Environ. Med.* **2014**, *21*, 331–335.
195. Zhang, H.; Dong, B.; Zhang, M.; Yang, J. Effect of Chromium Picolinate Supplementation on Growth Performance and Meat Characteristics of Swine. *Biol. Trace Elem. Res.* **2011**, *141*, 159–169.
196. www.scopus.com (Scopus Data Base; accessed Jan 23, 2016).
197. Stearns, D. M.; Wise, J. P.; Patierno, S. R.; Wetterhahn, K. E. Chromium(III) Picolinate Produces Chromosome Damage in Chinese Hamster Ovary Cells. *FASEB J.* **1995**, *9*, 1643–1649.
198. Stearns, D. M.; Silveira, S. M.; Wolf, K. K.; Luke, A. M. Chromium(III) Tris(picolinate) is Mutagenic at the Hypoxanthine (Guanine) Phosphoribosyltransferase Locus in Chinese Hamster Ovary Cell. *Mutat. Res.* **2002**, *513*, 135–142.
199. Speetjens, J. K.; Collins, R. A.; Vincent, J. B.; Woski, S. A. The Nutritional Supplement Chromium(III) Tris(picolinate) Cleaves DNA. *Chem. Res. Toxicol.* **1999**, *12*, 483–487.
200. Trumbo, P. R.; Ellwood, K. C. Chromium Picolinate Intake and Risk of Type 2 Diabetes: An Evidence-based Review by the United States Food and Drug Administration. *Nutr. Rev.* **2006**, *64*, 357–363.
201. Andersson, M. A.; Grawe, K. V. P.; Karlsson, O. M.; Abramsson-Zetterberg, L. A. G.; Hellman, B. E. Evaluation of the Potential Genotoxicity of Chromium Picolinate in Mammalian Cells In Vivo and In Vitro. *Food Chem. Toxicol.* **2007**, *45*, 1097–1106.
202. Anderson, R. A.; Bryden, N. A.; Polansky, M. M. Lack of Toxicity of Chromium Chloride and Chromium Picolinate in Rats. *J. Am. Coll. Nutr.* **1997**, *16*, 273–279.
203. Paiva, A. N.; de Lima, J. G.; Medeiros, A. C. Q.; Figueiredo, H. A. O.; Andrade, R. L.; Ururahye, M. A. G.; Rezendef, A. A.; Brandão-Neto, J.; Almeida, M. G. Beneficial Effects of Oral Chromium Picolinate Supplementation on Glycemic Control in Patients with Type 2 Diabetes: A Randomized Clinical Study. *J. Trace Elem. Med. Biol.* **2015**, *32*, 66–72.
204. Vincent, J. B.; Love, S. T. The Need for Combined Inorganic, Biochemical, and Nutritional Studies of Chromium(III). *Chem. Biodiv.* **2012**, *9*, 1923–1941.
205. Eastmond, D. A.; Macgregor, J. T.; Slesinski, R. S. Trivalent Chromium: Assessing the Genotoxic Risk of an Essential Trace Element and Widely Used Human and Animal Nutritional Supplement. *Crit. Ver. Toxicol.* **2008**, *38*, 173–190.
206. Bailey, C. H. Improved Meta-analytic Methods Show no Effect of Chromium Supplements on Fasting Glucose. *Biol. Trace Elem. Res.* **2014**, *157*, 1–8.
207. American Diabetes Association. Clinical Practice Recommendations. *Diabetes Care* **2014**, *37*(Suppl. 1), S155.
208. Yeghiazaryan, K.; Peeva, V.; Shenoy, A.; Schild, H. H.; Golubnitschaja, O. Chromium–Picolinate Therapy in Diabetes Care: Molecular and Subcellular Profiling Revealed a Necessity for Individual Outcome Prediction, Personalised Treatment Algorithms and New Guidelines. *Infect. Disord. Drug Targets* **2011**, *11*, 188–195.
209. Dai, H.; Liu, J.; Malkas, L. H.; Catalano, J.; Alagharu, S; Hickey, R. J. Chromium Reduces the *In Vitro* Activity and Fidelity of DNA Replication Mediated by the Human Cell DNA Synthesome. *Toxicol. Appl. Pharmacol.* **2009**, *236*, 154–165.

210. Amato, P.; Morales, A. J.; Yen, S. S. C. Effects of Chromium Picolinate Supplementation on Insulin Sensitivity, Serum Lipids, and Body Composition in Healthy, Nonobese, Older Men and Women. *J. Gerontol. Med. Sci.* **2000**, *55*(5), M260–M263.
211. Balk, E. M.; Tatsioni, A.; Lichtenstein, A. H.; Lau, J.; Pittas, A. G. Effect of Chromium Supplementation on Glucose Metabolism and Lipids: A Systematic Review of Randomized Controlled Trials. *Diabetes Care* **2007**, *30*, 2154–2163.
212. Gunton, J. E.; Cheung, N. W.; Hitchman, R.; Hams, G.; O´Sullivan, C.; Foster-Powell, K. McElduff, A. Chromium Supplementation does Not Improve Glucose Tolerance, Insulin Sensitivity, or Lipid Profile: A Randomized, Placebo-controlled, Double-blind Trial of Supplementation in Subjects with Impaired Glucose Tolerance. *Diabets Care* **2005**, *28*, 712–713.
213. Kleefstra, N.; Houweling, S. T.; Bakker, S. J. L.; Verhoeven, S.; Gans, R. O. B.; Meyboom-De Jong, B.; Bilo, H. J. G. Chromium Treatment Has no Effect in Patients with Type 2 Diabetes in a Western Population: A Randomized, Double-blind, Placebo-controlled Trial. *Diabetes Care* **2007**, *30*, 1092–1096.
214. European Food Safety Authority (EFSA) Panel on Dietetic Products, Nutrition and Allergies. Scientific Opinion on Dietary Reference Values for Chromium. *EFSA J.* **2014**, *12*, 10.
215. Wasser, W. G.; Feldman, N. S.; D'Agati, V. D. Chronic Renal Failure After Ingestion of Over-the-Counter Chromium Picolinate. *Ann. Intern. Med.* **1997**, *26*, 410–411.
216. Mahmood, S.; Mozaffaria, T.; Patela, C.; Ballasa, C.; Schaffer, S. W. Effects of Chronic Chromium Picolinate Treatment in Unnephrectomized Rat. *Metab. Clin. Experim.* **2005**, *54*, 1243–1249.
217. Mozaffari, M.; Sayed, R.; Liu, J.; Wimborne, H.; El-Remessy, A.; El-Marakby, A. Effects of Chromium Picolinate on Glycemic Control and Kidney of the Obese Zucker Rat. *Nutr. Metab.* **2009**, *6*, 51.
218. Sales, J.; Jancík, F. Effects of Dietary Chromium Supplementation on Performance, Carcass Characteristics, and Meat Quality of Growing-finishing Swine: A Meta-analysis. *J. Anim. Sci.* **2011**, *89*, 4054–4067.
219. Wise, S. S.; Wise, J. P. Sr. Chromium and Genomic Stability. *Mutat. Res.* **2012**, *733*, 78–82.
220. Manygoats, K. R.; Yazzie, M.; Stearns, D. M. Ultrastructural Damage in Chromium Picolinate-treated Cells: A TEM Study. *J. Biol. Inorg. Chem.* **2002**, *7*, 791–798.
221. Coryell, V. H.; Stearns, D. M. Molecular Analysis of HPRT Mutations Induced by Chromium Picolinate in CHO AA8 Cells. *Mutat. Res.* **2006**, *610*, 114–123.
222. Stallings, D. M.; Hepburn, D. D. D.; Hammah, M.; Vincent, J. B.; O´Donnell, J. Nutritional Supplement Picolinate Generates Chromosomal Aberrations and Impedes Progeny Development in *Drosophila melanogaster*. *Mut. Res.* **2006**, *610*, 101–113.
223. Refaie, F. M.; Esmat, A. Y.; Mohamed, A. F.; Nour, W. H. A. Effect of Chromium Supplementation on the Diabets Induced-oxidative Stress in Liver and Brain of Adult Rats. *Biometals* **2009**, *22*, 1075–1087.
224. Ferreira, M.; Santos, T. M.; Pereira, M. L. Light Microscopy Studies on Mice Testis After the Nutritional Supplement Chromium(III)-tris(picolinate). *Microsc. Microanal.* 2013, *19*(S4), 47–48.
225. Shinde, U. A.; Goyal, R. K. Effect of Chromium Picolinate on Histopathological Alterations in STZ and Neonatal STZ Diabetic Rats. *J. Cell Mol. Med.* **2003**, *7*, 322–329.

226. Sachin, W.; Weskamp, C.; Marple, J.; Spry, L. Acute Tubular Necrosis Associated with Chromium Picolinate-containing Dietary Supplement. *Ann. Pharmacother.* **2006**, *40*, 563–566.
227. Levina, A.; Lay, P. A. Chemical Properties and Toxicity of Chromium(III) Nutritional Supplements. *Chem. Res. Toxicol.* **2008**, *21*, 563–571.
228. Kim, B. G.; Lindemann, M. D.; Cromwell, G. L. The Effects of Dietary Chromium(III) Picolinate on Growth Performance, Blood Measurements, and Respiratory Rate in Pigs Kept in High and Low Ambient Temperature. *J. Anim. Sci.* **2009**, *87*, 1695–1704.
229. Rhodes, N. R.; McAdory, D. E.; Love, S.; Di Bona, K. R.; Chen, Y.; Ansorge, K.; Hira, J.; Kern, N.; Kent, J.; Lara, P.; Rasco, J. F.; Vincent, J. B. Urinary Chromium Loss Associated with Diabetes Is Offset by Increases in Absorption. *J. Inorg. Biochem.* **2010**, *104*, 790–797.
230. Selcuk, Z.; Tiril, S. U.; Alagil, F.; Belen, V.; Salman, M.; Cenesiz, S.; Muglali, O. H.; Yagci, F. B. Effects of Dietary L-carnitine and Chromium Picolinate Supplementations on Performance and some Serum Parameters in Rainbow Trout (*Oncorhynchus mykiss*). *Aquacult. Int.* **2010**, *18*, 213–221.
231. McAdory, A.; Rhodes, N. R.; Briggins, F.; Bailey, M. M.; Di Bona, K. R.; Goodwin, C.; Vincent, J. B.; Rasco, J. F. Potential of Chromium(III) Picolinate for Reproductive or Developmental Toxicity Following Exposure of Male CD-1 Mice Prior to Mating. *Biol. Trace Element Res.* **2011**, *143*, 1666–1672.
232. Ferreira, M. Nutritional Supplements: The Controversy of Chromium(III) Tris-picolinate. Master Thesis in Molecular and Cell Biology, University of Aveiro, Aveiro, Portugal, 2012.
233. Dallago, B. S. L.; McManus, C. M.; Caldeira, D. F.; Campeche, A.; Burtet, R. T.; Paim, T. P.; Gomes, E. F.; Branquinho, R. P.; Braz, S. V.; Louvandini, H. Humoral and Cellular Immunity in Chromium Picolinate-supplemented Lambs. *Biol. Trace Elem. Res.* **2013**, *154*, 196–201.
234. Huang, S.; Peng, W.; Jiang, X.; Shao, K.; Xia, L.; Tang, Y.; Qiu, J. The Effect of Chromium Picolinate Supplementation on the Pancreas and Macroangiopathy in Type II Diabetes Mellitus Rats. *J. Diabetes Res.* **2014**, *2014*, 717219. DOI: 10.1155/2014/717219.
235. Biswas, A.; Divya, S.; Mandal, A. B.; Majumdar, S.; Singh, R. Effects of Dietary Supplementation of Organic Chromium (Picolinate) on Physical and Biochemical Characteristics of Semen and Carcass Traits of Male Turkeys. *Animal Reprod. Sci.* **2014**, *151*, 237–243.
236. Seif, A. Chromium Picolinate Inhibits Cholesterol-induced Stimulation of Platelet Aggregation in Hypercholesterolemic Rats. *Ir. J. Med. Sci.* **2015**, *184*, 291–296.
237. Mehrim, A. Physiological, Biochemical and Histometric Responses of Nile Tilapia (*Oreochromis niloticus* L.) by Dietary Organic Chromium (Chromium Picolinate) Supplementation. *J. Adv. Res.* **2014**, *5*, 303–310.
238. Pereira, M. L.; Ferreira, M.; Santos, T. M. Cr(III)-tris(picolinate) Induced Multi-organ Toxicity on Mice: Microscopy Studies. *Int. J. Sci. Eng. Res.* **2015**, *6*, 42–46.
239. Dallago, B. S. L.; Braz, S. V.; Marçola, T. G.; McManus, C.; Caldeira, D. F.; Campeche, A.; Gomes, E. F.; Paim, T. P.; Borges, B. O.; Louvandini, H. Blood Parameters and Toxicity of Chromium Picolinate Oral Supplementation in Lambs. *Biol. Trace Elem. Res.* **2015**, *168*, 91–102.
240. Steams, D. M.; Armsbong, W. H. Mononuclear and Binuclear Chromium(III) Picolinate Complexes. *Inorg. Chem* **1992**, *31*, 5178–5184.

241. Chakov, N. E.; Collins, R. A.; Vincent, J. B. A Reinvestigation of the Electronic Spectra of Chromium(III) Picolinate Complexes and High Yield Synthesis and Characterization of $Cr_2(\mu-OH)_2(pic)_4$ $5H_2O$ (Hpic = picolinic acid). *Polyhedron* **1999,** *18*, 2891–2897.
242. Chaudhary, S., Pinkston, J., Rabile, M. M., Van Horn, J. D. Unusual Reactivity in a Commercial Chromium Supplement Compared to Baseline DNA Cleavage with Synthetic Chromium Complexes. *J. Inorg. Biochem.* **2005,** *99*, 787–794.

CHAPTER 20

TREATMENT OF PHARMACEUTICAL WASTEWATER: A CASE STUDY ON DEGRADATION IN ELECTROCHEMICAL OXIDATION

SAPTARSHI GUPTA[1], LEICHOMBAM MENAN[2], and SRIMANTA RAY[1*]

[1]*Department of Chemical Engineering, National Institute of Technology, Agartala, Barjala, Jirania, Tripura (West), India*

[2]*Department of Civil Engineering, National Institute of Technology, Agartala, Barjala, Jirania, Tripura (West), India*

*Corresponding author. E-mail: rays.nita@gmail.com

ABSTRACT

The extensive production and consumption of pharmaceuticals has significantly increased the possibility of water contamination by pharmaceutical compounds. The chronic exposure and cumulative build up of these pharmaceutical compounds pose immense threat to humans and ecosystems. The conventional water treatment methods are largely ineffective in degrading these pharmaceuticals. The study reviews different alternative treatment employed for mitigation of pharmaceuticals in aqueous medium. Among different alternative treatment methods, electrochemical oxidation (ECO), which relies on generation of ions in aqueous medium, has been proven to be a promising and attractive technique for the effective oxidation of variety of organic contaminants. However, choice of electrode material largely contributes to the effectiveness of ECO process. Accordingly, a case study is presented on evaluation of electrodes fabricated from low-cost materials, namely, iron, aluminum, and carbon. Methylene blue (MB) was chosen as a model pharmaceutical. The effect of factors controlling the efficacy of ECO process, namely, pollutant concentration, applied current, and concentration of electrolyte, with electrode materials in decolorization rate of MB are thoroughly examined in the case study. A toxicity analysis of MB degradation on the red blood cell of domesticated chicken species (*Gallus gallus domesticus*) is also presented in the study. Mineralization study was conducted to confirm degradation of MB. Energy dispersive spectroscopy (EDS) was used as a tool to confirm the formation of oxidative species at the electrode responsible for degradation of pharmaceuticals. The case study demonstrates an alternative decolorization option using low-cost electrode materials which can significantly enhance the affordability and acceptability of the ECO process for degrading pharmaceutical pollutants and reports of using EDS as a tool to assess the degradation route.

20.1 INTRODUCTION

20.1.1 PHARMACEUTICALS

Health is considered as an important social and economic asset. In order to meet the growing need of the population and the demand for better health there is an increase in the demand of pharmaceuticals and personal care products. Pharmaceuticals with different physicochemical and biological properties and functionalities already have been manufactured and consumed globally. Pharmaceuticals are primarily classified based on the pharmacological properties which include antipyretics, analgesics, antibiotics, hormones, contraceptives, tranquilizers, etc.[64,86] These pharmaceuticals enter the aquatic environmental compartment through multiple pathways. Pharmaceuticals can enter water supplies through humans and animals excretion. The pharmaceutical compounds in human excretion may pass through sewage treatment plants (STPs) and then enter surface water; compounds in animal excretion may enter surface water by joining agricultural runoff, or enter groundwater by seeping through the soil.[98] Pharmaceuticals can also enter water bodies when patients, healthcare organizations, or pharmaceutical companies dispose of unused pharmaceuticals, and when pharmaceutical companies dispose of wastes generated during production. Pharmaceutical companies in the developed countries are subjected to the most advanced and comprehensive waste-treatment directives in the world and are extensively reviewed to ensure environmental safety.[12,80] In contrast, pharmaceutical companies in developing or underdeveloped countries discharge treated and untreated effluents on open land and into unlined streams.[41] Recent studies have indicated the presence of pharmaceutical and their metabolites in the effluents of STPs and subsequently in several water bodies.[14,20,97,86] The ecotoxicological consequences of various pharmaceuticals reported in the aquatic environment have also been reported.[76]

20.1.2 SIGNIFICANCE OF RESIDUAL PHARMACEUTICALS AND HEALTH EFFECTS

The risks associated with pharmaceutical contamination of the aquatic environment have become a major issue of concern for environmental scientists and engineers, as well as among the public. Pharmaceuticals are the chemicals that are designed to give a certain therapeutic (or biological) effect; therefore, certain environmental and public health risks can be anticipated from the exposure to the environmental pharmaceuticals. Besides, there are a few classes of pharmaceuticals that pose unambiguous impacts on the aquatic organisms, including microorganisms, phytoplankton, plants, crustaceans, fish, and insects, as well as on soil microorganisms and possibly humans.[42,53,54,84] These pharmaceutical classes include: cytostatic agents, immunosuppressive drugs, and some genotoxic antibiotics because of their evident cytotoxic, carcinogenic, mutagenic, and/or embryotoxic properties; human and veterinary antibiotics because of their pronounced microbial toxicity and the development of antibiotics resistance in environmental bacteria including human pathogens; natural and synthetic hormones because of their high efficiency, low-effect thresholds, and potential for endocrine disruption; halogenated compounds because of their resistance toward biodegradation and their mobility and persistence in the environment and the food web; heavy metal containing drugs and nontherapeutic medical agents because of the toxicity of the metals in certain oxidation states. In addition, the presence of other types of pharmaceuticals, such as analgesics and anticonvulsants, in water has potential public health issue. Although the concentrations found in water are often very low, but the long-term health effects are still largely unknown for the exposure to the trace pharmaceuticals and their metabolites, especially as a mixture of biologically active compounds.[50,82] Similarly, long-term exposure of aquatic organisms to trace pharmaceuticals in surface water may have some as yet known ecological impacts.

Acute toxicity of various pharmaceuticals including aspirin and ibuprofen were studied on aquatic organisms like fish and algae.[26] However, little is known about the chronic effects of pharmaceutical compounds on nontarget species, particularly invertebrates, where the relevant pathways have yet to be studied. Further complications occur when pharmaceuticals form a heterogeneous group consisting of compounds with diverse chemical properties and biological effects, with little known about their chronic effects.[32,78] Moreover, workers involved in the manufacturing of pharmaceutical products are exposed in the course of their work to the active pharmaceutical ingredient (API) in the products. Such APIs are designed to produce biological change in the human body, which is an unacceptable outcome in the pharmaceutical worker.[43] Due to the various health effects of the pharmaceuticals; the toxicology is one of the emerging research areas for the assessment of pharmaceutical toxicity in the environment.

20.2 TREATMENT OF PHARMACEUTICALS IN WASTEWATER

20.2.1 CONVENTIONAL TREATMENT

In general, the elimination of organic pollutants in aqueous solution needs one or more treatment techniques.[57] Conventional wastewater treatment methods such as coagulation–flocculation, sedimentation, filtration, etc., do not involve chemical transformations and generally transfer waste component from one phase to another, thus causing secondary loading of environment and waste disposal problem.[85] Biological methods are amenable only for biodegradable pollutants as in the case of domestic wastewater and industrial effluent in a limited ways. Several organic materials known to have ecotoxicity are difficult to treat by conventional treatment options; hence treatment methods are often specific to the pollutant being treated. Thus, wastewater containing pharmaceuticals cannot be fully treated by the conventional treatment methodologies. Therefore alternative techniques gain significance in treating the pharmaceutical effluents.

20.2.2 ALTERNATIVE TREATMENT

The incompetency of conventional treatments for complete removal of pharmaceutical compounds has placed the alternative techniques in the focus of research. The alternative techniques for the removal of pharmaceuticals include adsorption, ion exchange, and pressure filtration (Table 20.1). Another alternative treatment for the removal of pharmaceuticals is advanced oxidation process (AOP).

20.2.2.1 ADSORPTION

Adsorption is greatly dependent on the type of the adsorbent materials. The main characteristic of an adsorbent material is high porosity and large surface area. Activated charcoal is generally preferred as an adsorbent for removing organic contaminants. Granular activated carbons have been widely studied for the removal of pharmaceuticals. Various other adsorbents such as silica gel, activated alumina, and bauxite have also been studied. However, Westerhoff et al.[96] and Snyder et al.[81] reported that the adsorption process is not always effective for the complete removal of pharmaceuticals. Ionic strength, divalent cations, trivalent cations, and natural organic matter are noteworthy factors affecting the adsorption of pharmaceuticals. The adsorption of pharmaceutical wastewater was found to be dependent on water characteristics and the properties of pharmaceuticals (molecular charge, pKa, and hydrophobicity).[18]

20.2.2.2 ION EXCHANGE

Ion exchange mostly works on particle-free water and can fit any size treatment facility by scaling them accordingly. Ion exchange is mostly used for removal of hardness (by cation resin) or nitrate (by anion resin). In both cases, ions in the resin are restored using salt water. The use of ion exchange to remove pharmaceuticals is intricate.[27] Adams et al.[3] also reported that ion exchange is not a reliable method for the removal of pharmaceuticals.

20.2.2.3 PRESSURE FILTRATION

Pressure filtration employing various pressures and pore sizes of membranes which includes microfiltration (MF), ultrafiltration (UF), nanofiltration (NF), and reverse osmosis (RO) are used for removal of micropollutants from drinking water and wastewater restoration.[91] Many reports have been given on the rejection of pharmaceuticals on the NF/RO membranes.[51,100] The factors influencing the rejection of micropollutants in RO includes dipole moment, hydrophobicity, and molecular size of pharmaceutical compounds. The rejection of the pharmaceutical compounds is determined by solute–membrane and solute–solute interactions. However, Radjenović et al.[71] reported that negatively charged pharmaceutical compounds show very poor removal efficiency, whereas positively charged compounds were retained on the membrane with high efficiency. The membranes used in NF and RO for filtration are usually negatively charged that creates repulsion of negatively charged solutes and an attraction of positively charged ones which leads to high efficiency for positively charged pharmaceuticals to retain on the membrane layer and a lower removal efficiency for those that are negatively charged compounds.

TABLE 20.1 Treatment of Pharmaceutical Wastewater by Various Physical Treatment Methods.

Type of wastewater	Treatment method	Result	References
Pharmaceuticals, (antipyretic and analgesics)	Adsorption—mesoporous silica SBA-15	Removal efficiencies for the different pharmaceuticals range from 35.6 to 88.4%.	[17]
Pharmaceutical, (amoxicillin)	Adsorption—activated carbon/bentonite	% removal bentonite: 88.01%, activated carbon: 94.67%.	[69]
Antibiotics	Ion exchange/RO	Low exchange capacities 21–58%, the rejection rate for the antibiotics averaged 90.2%.	[3]
EDC and pharmaceuticals	MF/UF/RO/activated carbon	Activated carbon treatment shows the highest removal of greater than 90%.	[81]
EDC and pharmaceuticals	MF/RO	Removal efficiencies in the final recycled water were above 97%.	[5]
EDC and pharmaceuticals	UF/NF	NF retained EDC/pharmaceutical more than UF.	[100]
Pharmaceutical residues	NF/RO	Negatively charged pharmaceuticals: poor efficiency; positively charged: retained with high efficiency.	[71]

20.2.2.4 ADVANCED OXIDATION PROCESSES

Among other alternative treatments, AOPs are considered promising methods for the remediation of contaminated ground, surface, and wastewaters containing nonbiodegradable organic pollutants, like pharmaceuticals. AOPs involve the generation of highly reactive radicals (especially hydroxyl radicals) in sufficient quantity to effect water purification. Hydroxyl radicals are more powerful oxidants than the chemical agents used in traditional chemical processes. Hydroxyl radicals are also characterized by a nonselectivity of reactant for oxidation.[35] Accordingly different organic compounds are susceptible to be removed or degraded by means of hydroxyl radicals. Nevertheless, some of the simplest organic compounds, such as acetic, maleic, and oxalic acids, acetone, or simple chloride derivatives such as chloroform or tetrachloroethane, cannot be attacked by hydroxyl radicals.[13] Depending upon the nature of the organic species, two types of initial attacks are possible: the hydroxyl radical can abstract a hydrogen atom to form water, as with alkanes or alcohols, or it can add to the contaminant, as it is the case for olefins or aromatic compounds. The versatility of AOPs is also enhanced by the fact that they offer different ways of hydroxyl radical production, thus allowing a better compliance with the specific treatment requirements. However, a suitable application of AOPs to wastewater treatment often makes use of expensive reactants such as hydrogen peroxide and/or ozone, and therefore it is not an economic treatment option as the biological degradation.

20.3 ALTERNATIVE TREATMENT OF PHARMACEUTICALS—OXIDATIVE PROCESSES

The hydroxyl radicals in AOPs are very reactive and unstable and therefore, are continuously produced in situ by various hydroxyl radical production processes. The radical production involves either photochemical or nonphotochemical processes.

20.3.1 PHOTOCHEMICAL

The photochemical route of hydroxyl radical generation largely depends on the irradiation (light) source and intensity. The use of light increases the flexibility of the system, allowing the use of a variety of oxidants and operational conditions. Sunlight irradiation may also be used in some applications, but it must be taken into account that only 3–5% of ultraviolet (UV) light is present in the solar spectrum. Usually, UV light initiates the reaction of AOP due to high energy barrier of water splitting. Mercury lamps, amalgam lamps, or xenon arc lamps are often used as a source of UV.

20.3.1.1 DIRECT PHOTOLYSIS

Direct photolytic processes affect degradation of organic compounds in water and effluents without the addition of chemical reagents using the energy of photon in the UV regime through formation of hydroxyl radicals. Direct photolysis is important for compounds that react very slowly with HO· or do not react at all, namely, halogenated compounds. Several halogenated compounds have been reported to be degraded by direct photolysis using 254 nm irradiation.[46,55,79,90] Limitations of the process are: (1) low efficiency, (2) application only to compounds absorbing at 200–300 nm, and (3) only one target compound can be treated with reasonably good results. Accordingly, the photolysis is often combined with other conventional methods.

20.3.1.2 PHOTOLYSIS WITH SENSITIZER

In many cases, direct photolysis is reported to be favored by the presence of oxygen and substances which can act as photosensitizers.[30] In this sense, some dyes like phthalocyanine, methylene blue, etc., promote singlet oxygen formation.[97] The singlet oxygen is a powerful oxidizing agent and is reported to degrade organic matter.[90] However, one of the main problems of the sensitizer photolysis is the necessity of removing the dye from the water after the treatment. For this reason, attempts at immobilization to different supports have been reported recently, but this process leads to a decrease in the efficiency of singlet oxygen production compared with the sensitizer in a homogeneous water solution.[77]

20.3.1.3 PHOTOLYSIS WITH VACUUM ULTRAVIOLET

This process uses radiation of wavelengths lower than 190 nm, categorized as vacuum ultraviolet radiation (VUV). The excimer lamps are often used as source of VUV. The excitation leads in the majority of the cases to the homolytic water splitting and subsequent degradation of organic matter.[55,90] This process generates hydroxyl radicals in situ without the addition of external agents. Due to the high absorption cross section of water, the total incident radiation is absorbed within a very narrow layer around the lamp shaft. The quantum yield of reaction varies with the irradiation wavelength. Aqueous electrons (strong reductants) are also produced, but with a lower quantum yield, almost independent of the irradiation wavelength in the range 160–190 nm. The generated oxidants (HO·, HO$_2$·, and O$_2$·⁻) and reductants (H·, e⁻ aq., HO$_2$·, and O$_2$·⁻) make possible simultaneous reductions and oxidations in the chemical system. The process is highly efficient because VUV lamps generally have a high radiant power of illumination and water has a high cross section of absorption in the wavelength range. However, this technology requires the use of expensive quartz materials and has high energy requirement.[37–39,56]

20.3.1.4 HETEROGENEOUS PHOTOCATALYSIS

Heterogeneous photocatalytic process consists of utilizing a radiation to photoexcite a semiconductor catalyst generating hydroxyl radical or free holes. The process is heterogeneous because there are two active phases, solid and liquid. This process was widely experimented in the UV regime (wavelength shorter than 380 nm) as well as in the solar spectrum.[33] Many catalysts have been tested, although TiO$_2$ in the anatase form seems to possess the most interesting features, such as high stability, good performance, and low cost.[7,74] The disadvantages of the photocatalytic process include post-process separation of catalyst particle and fouling of the catalyst by the organic matter. Several phenolic compounds have been successfully degraded by photocatalytic process.[58,59,74]

20.3.2 NONPHOTOCHEMICAL

It is worthwhile to point out, that light-mediated AOPs, especially the homogeneous processes, are often not adequate for treating mixtures of substances of high absorbance, or containing high amounts of solids in suspension, because the quantum efficiency decreases through loss of light, dispersion, and/or by competitive light absorption. Accordingly, nonphotochemical processes were also widely explored for degradation of recalcitrant organic pollutants.

20.3.2.1 OZONATION PROCESS

Ozone is an effective oxidizing agent which reacts with most compounds containing multiple bonds, such as C=C, C=N, and N=N, but not with species containing single bonds (C–C, C–O, and O–H) at high rates.[36] At higher pH values, ozone reacts almost unselectively with all inorganic and organic compounds present in the solution.[83] Rising the pH of the aqueous solution increases the decomposition rate of the ozone that generates the superoxide anion ($O_2^{\cdot-}$) radical and hydroperoxyl radical ($HO_2^{\cdot}$). For example, the ozonide anion ($O_3^{\cdot-}$) is formed by the reaction between O_3 and $O_2^{\cdot-}$. The ozonide anion further decomposes to a hydroxyl ($HO^{\cdot}$) radical, such that, three ozone molecules will produce two $HO^{\cdot}$ radicals.[62] The rate constants of the hydroxyl radicals are typically 106–109 times higher than the corresponding reaction rate constants of molecular ozone. The oxidation of organic compounds may also occur due to the combination of reactions with molecular ozone and reactions with hydroxyl radicals.[62]

Several oxidative processes have been successfully applied for the removal of wide variety of recalcitrant pollutants including pharmaceutical and personal care products (PPCPs) as documented in different studies (Table 20.2).

TABLE 20.2 Degradation of Different Pharmaceutical Effluent in Oxidative Processes.

Type of pharmaceuticals	AOP used to treat the pharmaceutical process		References
	Type of AOP	Variation of AOP	
Amoxicillin	Ozonation process	O_3	[6]
	Photochemical process	Fe^{2+}, $FeO_x/H_2O_2/UV$, and solar	[66]
Carbamazepine	Ozonation process	O_3 and H_2O_2	[47]
	Photochemical process	UV/H_2O_2	[92]
	Combined process	O_3, O_3/UV, and H_2O_2/UV	[34]
Diclofenac	Ozonation process	O_3 and O_3/H_2O_2	[103]
	Photochemical process	UV/H_2O_2	[19]
		TiO_2/UV	[93]
	Combined process	$Fe^{2+}/H_2O_2/UV$	[73]
		$Fe^{2+}/H_2O_2/solar$	[68]
		Fe^{2+}/H_2O_2	[66]
		O_3, O_3/UV, and H_2O_2/UV	[34]
Ibuprofen	Ozonation process	O_3 and H_2O_2	[47, 103]
	Combined process	Fe^{2+}/H_2O_2	[66]
Paracetamol	Combined process	O_3 and H_2O_2/UV	[8]
		Fe^{3+} and $FeO_x/H_2O_2/UV/solar$	[89]
Naproxen	Ozonation process	O_3	[48]
	Photochemical process	UV/H_2O_2	[67]
	Combined process	Fe^{2+}/H_2O_2	[66]

TABLE 20.2 *(Continued)*

Type of pharmaceuticals	AOP used to treat the pharmaceutical process		References
	Type of AOP	Variation of AOP	
Sulfamethoxazole	Ozonation process	O_3	[87]
		O_3/H_2O_2	[47]
	Photochemical process	TiO_2/UV	[1]
	Combined process	$Fe^{2+}/H_2O_2/UV$	[40]
Iopromide	Ozonation process	O_3	[48]
	Photochemical Process	O_3/H_2O_2 and O_3/UV	[87]
Tetracycline	Photochemical process	TiO_2/UV	[4, 75]
	Combined process	Fe^{3+}, $FeO_x/H_2O_2/UV$, and solar	[11]
Diazepam	Ozonation process	O_3 and H_2O_2	[47]
	Combined process	O_3, O_3/H_2O_2, and H_2O_2/UV	[34]
Ranitidine	Photochemical process	TiO_2/UV	[4]
	Combined process	Fe^{2+}/H_2O_2	[1]

20.3.2.2 ELECTROCHEMICAL OXIDATION PROCESS

Electrochemical oxidation (ECO) is another variant of nonphotochemical AOP. ECO provides several advantages compared to other nonphotochemical processes. The inherent advantage is its environmental compatibility as it uses a clean reagent, the electron, and there is little or no need for addition of chemicals. ECO of pollutants can take place through two different oxidation mechanisms: (1) direct anodic oxidation, where the pollutants are destroyed at the anode surface and (2) indirect oxidation, where a mediator is electrochemically generated to carry out the oxidation. It has to be kept in mind that during electrooxidation of aqueous effluents, both oxidation mechanisms may coexist.

Direct oxidation of pollutants takes place in two steps: (1) diffusion of pollutants from the bulk solution to the anode surface and (2) oxidation of pollutants at the anode surface. Consequently, the efficiency of the electrochemical process will depend on the relationship between mass transfer of the substrate and electron transfer at the electrode surface. The rate of electron transfer is determined by the electrode activity and current density.

During indirect ECO, a strong oxidizing agent is electro-generated at the anode surface and then destroys the pollutants in the bulk solution. The most common electrochemical oxidant is probably chlorine which is formed by the oxidation of chloride at the anode, although the role of active chlorine in the oxidation of organic pollutants is not clear. The extensive use of active chlorine is due to the ubiquitous presence of chloride in wastewaters and to its quite effective action. Other common oxidants that can be electrochemically produced are hydrogen peroxide, peroxodisulfuric acid ($H_2S_2O_8$) and ozone. Metal catalytic mediators (Ag^{2+}, Co^{3+}, Fe^{3+}, etc.) are also employed for the generation of hydroxyl radicals, as occurs in the electro-Fenton scheme. However, the use of metal ions may result in an effluent with a higher toxicity than that of the initial effluent. Thus, this approach requires a separation step to recover the metallic species.[9]

The feasibility of ECO was first tested with a wide variety of synthetic wastewaters containing a diversity of target compounds. However, in recent years, research work focusing on the treatment of actual wastewaters has increased. Overall, the aim of these studies was mainly, to eliminate nonbiodegradable and/or toxic organic pollutants. Attempts have been made recently for removal of microcontaminants such as pharmaceuticals[72,88] (Table 20.3).

20.4 FACTORS AFFECTING ELECTROCHEMICAL OXIDATION PROCESS

20.4.1 EFFECT OF ELECTROLYTE CONCENTRATION

It is important to investigate the effect of electrolyte concentration since conductivity is an important parameter for ECO process, in addition wastewater usually contains considerable amount of salts. Many studies reported the use of sodium sulfate (Na_2SO_4) as electrolyte.[28,94] The common salt or sodium chloride (NaCl) can also be used as an electrolyte to increase the conductivity of the water. Other than the enhancement of conductivity, NaCl abate the effects of anions such as HCO_3^- and SO_4^{2-}. Accordingly the energy requirement can be reduced with the addition of NaCl with the increase in the conductivity.

20.4.2 EFFECT OF CURRENT DENSITY OR APPLIED CURRENT

Literature suggests that the current density or applied current is an important parameter in ECO process.[28] Research indicated that high applied current

TABLE 20.3 Degradation of Different Effluents in ECO Process.

Type of ECO	Industrial application	Efficiency	Pollutants	References
Electrochemical oxidation	Textile industry, food colorants, printing, and cosmetic mfg.	COD = 20.1% color removal = 15.8%	Azo dyes	[101]
	Dye industry, textile industry	COD = 80.3% color removal = 99.4%	Azo dyes	[72]
Anodic oxidation	Dye industry, textile industry	COD = 100% at 6–9 pH COD = 60% at 3 pH	2,4-dichlorophenol	[24]
	Refractory	COD = > 95%	1,4-dioxane	[23]
Electro-Fenton	Pharmaceutical, dye industry	Deg. = 75%	2,6-dimethylaniline	[88]
	Textile, pharmaceutical, cosmetic, and food industry	Color removal = 100%	Ponceau S azo dye	[29]
	Cosmetic, textile, and printing industry	Deg. = 80%	Methyl red azo dye	[101]
	Agricultural activities, pesticide	Deg. = 100%	Picloram	[65]
	Dye industry	Color removal = 98%	Acid yellow 36	[40]
Photoelectro-Fenton process	Pharmaceutical, dye industry	Deg. = 100%	2,6-dimethylaniline	[88]
	Pigment and dye industry	Color removal = 51.8%	C.I. basic yellow 28	[49]

results in faster deterioration of electrode material but also lead to side reactions including evolution of singlet oxygen.[52] The performance would be cost-effective at a low applied current but the process requires longer treatment time, while high applied current is much more efficient but expensive.

20.4.3 EFFECT OF POLLUTANT CONCENTRATION

The HO• radicals are produced in ECO process, the ratio of HO• radicals to pollutant concentration is known to affect the oxidation process. High concentration of pollutant may lead to shortage of reactive oxygen species. Moreover, an increasing of pollutant concentration would lead to more intermediate species formed, which were easily absorbed on the surface of anode and prevented the contact between pollutant and active sites.[28]

20.4.4 EFFECT OF ELECTRODES

For ECO, anode material plays a critical role on the organic pollutants degradation. It is well-known that electrode materials are conclusive for optimizing ECO process because of their impact on the effectiveness of mechanisms and reaction pathways.[22] Some electrode materials play a major role by performing powerful oxidation and formation of intermediate product while giving a final product of carbon dioxide, whereas some electrode are very slow in oxidation. Along with the material, the electrode surface also plays a major role in the interaction with the hydroxyl radicals during the oxidation.[31]

The different types of electrodes have been investigated for the ECO of organic pollutant, including platinum,[95] graphite,[16] boron-doped diamond (BDD), and various metal oxide electrodes such as IrO_2,[10] SnO_2, and PbO_2 (Table 20.4).[45,62,99] Out of which, PbO_2 and BDD, considered as "nonactive" electrodes, are the most attractive electrodes for their high oxygen evolution overpotential. BDD electrode has the advantages of superior electrochemical stability and a wide potential window for water discharge.[44] Nevertheless, the high cost and especially the difficulties to find an appropriate substrate for deposition of the diamond layer limit its large-scale application is restricted.[16]

TABLE 20.4 Different Pharmaceuticals Treated by Various ECO with Different Electrodes.

Type of pharmaceutical	ECO process with different electrodes	Result	References
Aspirin	Electrode: Pt or steel as cathode, plates of Pt or carbon fiber as anodes. Electrolyte: 0.1 N H_2SO_4 or 0.1 N NaOH	The biological availability of water was increased progressively by oxidation.	[95]
	Eectrode: Novel Ti/Sn–SbOx/Ni–PbO$_2$, Electrolyte: mol L^{-1} Na$_2$SO$_4$	Ni–PbO$_2$ electrode had the highest COD removal: 53.26%. TOC removal: 45.64%.	[99]
Amoxicillin	Electrode: BDD. Electrolyte: H_2SO_4. Current density: 100 mA cm^{-2}	COD removal: 92%.	[70]
Diclofenac	Electrode: Pt/stainless steel and BDD/stainless steel cells. Electrolyte: 0.05 M Na$_2$SO$_4$	Pt: good mineralization degree. BDD: the solution became alkaline, only attained partial mineralization.	[15]
	Electrode: BDD or Ti/Pt/PbO$_2$ as anodes and stainless steel foils as cathodes. Electrolyte: 0.035 M Na$_2$SO$_4$	COD removed 60–95% and TOC varying from 48 to 92% in 6 h. Higher values obtained with the BDD electrode.	[25]
	Electrode: TSF (magnetic TiO$_2$/SiO$_2$/Fe$_3$O$_4$ loaded), a counter electrode: Pt, and a reference electrode: a 15 W low pressure Hg lamp emitting at 253.7 nm	After 45 min PEC treatment, 95.3% of diclofenac was degraded.	[45]
Ketoprofen	Electrode: BDD/Pt with reference electrode Hg/HgCl, KCl at 25°C	Degraded ketoprofen to CO$_2$ and H$_2$O; poor mineralization at both BDD and Pt anodes in the presence of NaCl as electrolyte, while complete mineralization was achieved using Na$_2$SO$_4$ as electrolyte.	[99]

TABLE 20.4 (Continued)

Type of pharmaceutical	ECO process with different electrodes	Result	References
Paracetamol	Graphite bar as cathode and BDD/Pt as anode. Electrolyte: 0.05 M Na_2SO_4	Mineralization process accompanied with release of NH_4^+ and NO.	[16]
	Electrode: boron-doped diamond (BDD) and at Ti/SnO_2 anodes. Electrolyte: Na_2SO_4. Current: 100–800 mA	The approximate current efficiencies were 14% in the case of Ti/SnO_2 and 26% in the case of BDD.	[94]
Ibuprofen and naproxen	A counter-electrode: Pt, a working electrode: Bi_2MoO_6 particles deposited onto BDD surface, and a reference electrode: SCE; Electrolyte: 0.1 mg L^{-1} Na_2SO_4	Ibuprofen and naproxen can be rapidly degraded also efficiently mineralized.	[102]
	Two circular electrodes and stainless steel cathode, current density: 20–200 A m^{-2} at 20°C	Removal 55.1% ibuprofen at 2 h and naproxen: 94.9%.	[45]

20.5 A CASE STUDY

20.5.1 LOW-COST ELECTRODE

For the treatment of large wastewater plants, the use of high-cost electrodes may not give an economical result despite of high efficiency. Therefore there is a need for the search of lower cost electrodes which are inexpensive and can give higher efficiencies. Some reports have been given on the use of lower cost electrodes like stainless steel, carbon, and iron electrodes.[2,21] However, there are very few researches on the use of low-cost electrodes. In this case study, low-cost electrodes iron, aluminum, and carbon, have been employed. And they have been found to give very high efficiencies.

20.5.2 METHODOLOGY

Methylene Blue (MB) is selected as the experimental liquid due to its wide application in pharmaceutical industry as well as in chemical industry. The electrode materials iron (sheet with 0.5 mm thickness) and aluminum (sheet with 0.25-mm thickness) are procured from local market (commercial grade). A carbon prepared in house with low resistivity (of 2.5 ± 0.35 ohms/cm) is evaluated as electrode material. Electrodes are fabricated in-house (institute workshop).

The ECO experimental setup (Fig. 20.1) consists of a regulated DC power supply (A) connected with the electrodes (B) with an insulating copper wire (C). The electrodes (B) were then dipped into one electrolytic cell (D), filled by the experimental liquid (E). The experimental solution was continuously stirrer with the help of a magnetic stirring bar (F), and a magnetic stirrer (G). A regulated DC power supply (LQ6324, Aplab, Thane, India) capable of varying the current from 0–2 ampere (A) was used for applying potential across electrodes.

A 200 mL open cylindrical vessel with electrodes with effective cross-sectional area of 4000 mm^2 (0.004 m^2) is chosen for the experiment, such that the current density was kept at 250, 375, and 500 A/m^2 at 1, 1.5, and 2 A applied current. The separation distance between electrodes was maintained at 20 mm. The electrolyte or conductivity modifier, namely common salt (NaCl), was examined at molar concentration of 0.043, 0.086, and 0.129 respectively. The concentration of MB (pollutant) was varied from 100 to 500 mg/L. All experiments are conducted under ambient condition under stirring at rate 200 rpm using magnetic stirrer (5 MLH, Remi,

Mumbai, India). The experimental parameters were examined in one factor at a time approach.

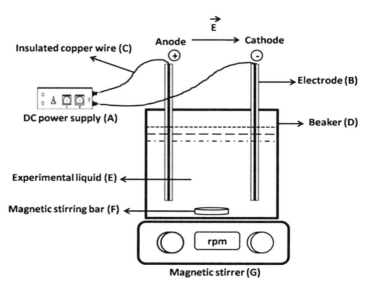

FIGURE 20.1 Schematic of the electrochemical oxidation cell.

The iron (Fe), aluminum (Al), and carbon (C) electrodes before and after ECO were characterized using a field emission scanning electron microscope (SEM) (Quanta 200, FEI Company, Hillsboro, OR), and energy dispersive spectroscopy (EDS) with GENESIS material characterization software (EDAX Inc., Mahwah, NJ) to analyze the stoichiometric composition of the electrodes. The decolorization of MB was monitored at 664 nm (λ_{max}) in UV/Vis spectrophotometer (Lamda 35, Perkin Elmer, India) with Winlab software (Perkin Elmer, India) and mineralization is studied using total organic carbon (TOC) analyzer (TOC-LPH, Shimadzu Asia Pacific, Singapore). The concentrations were computed through in-house calibration from respective instrumental measurements. The computed decolorization kinetics was used to compute the apparent rate constants for oxidation of MB are monitored. All chemicals used in the study are of research grade and obtained from Merck, India. The water used in the study is of ultra-pure grade (>18 Mohm resistivity) generated in-house (Option Q, Elga, India).

Blood samples for the toxicity study of MB were collected from local poultry shop from domesticated chicken species (*Gallus gallus domesticus*).

In total, 2% EDTA solution was used for the collection of blood. Red blood cells (RBCs) were separated from the plasma by centrifuging using 0.9% saline. The blood samples were viewed under the optical microscope (LB-212 Binocular Digital Microscope, Labomed Inc., Los Angeles, CA, USA) and images were taken at 100× magnification with ScopeImage (BioImager Inc., Maple, ON, Canada) software.

20.5.3 RESULTS AND DISCUSSION

20.5.3.1 KINETIC STUDY

The decolorization profile of MB for a pollutant concentration was monitored at λ_{max} (664 nm). A constant decrease of absorbance was observed with the progress in the ECO of MB. The decolorization data of MB well fitted the first order kinetics. The ECO of MB at different concentration levels followed the same rate order. The rate constant at the different factor level settings in the ECO process is tabulated in Table 20.5.

TABLE 20.5 Experimental Rate Constants of MB Decolorization at Different Factor Levels in ECO Process with Different Low-cost Electrodes.

Factor effect	Pollutant conc. (mg/L)	Applied current (A)	Electrolyte	Electrolyte conc. (M)	Electrode	Electrode gap (m)	Rate of decolorization 1st order	Regression coefficient 1st order
Pollutant conc.	500	2	NaCl	0.086	Fe–Fe	0.02	0.009	0.979
	200	2	NaCl	0.086	Fe–Fe	0.02	0.248	0.877
Applied current	100	2	NaCl	0.086	Fe–Fe	0.02	0.074	0.772
	100	1.5	NaCl	0.086	Fe–Fe	0.02	0.061	0.844
	100	1	NaCl	0.086	Fe–Fe	0.02	0.012	0.831
Electrolyte Conc.	100	2	NaCl	0.043	Fe–Fe	0.02	0.014	0.986
	100	2	NaCl	0.129	Fe–Fe	0.02	0.090	0.953
Electrode	100	2	NaCl	0.086	Fe–Fe	0.02	0.074	0.772
	100	2	NaCl	0.086	Al–Al	0.02	0.006	0.953
	100	2	NaCl	0.086	C–C	0.02	0.045	0.991
	200	2	NaCl	0.086	C–C	0.02	0.393	0.950

20.5.3.2 EFFECT OF APPLIED CURRENT

The applied current for ECO was varied across electrode (Table 20.5). The effect of current density for ED was studied for 100 mg/L MB with 0.086 M common salt. The applied current was varied from 1 to 2 A. Increase in applied current increases the availability of electron at the electrode surface. The electrons mediated the decolorization of MB. Accordingly, the decolorization of MB was observed to increase linearly with the increase of applied current.

20.5.3.3 EFFECT OF MB CONCENTRATION

The effect of MB concentration on ED was studied at 2 A applied current with 0.086 M NaCl for MB concentration varying from 100 to 500 mg/L. The decolorization of MB increases with increase in MB concentration from 100 to 200 mg/L, however on further increase in MB concentration decolorization decreases (Table 20.5). Increase in MB concentration initially increases the utilization of electrons but at higher concentrations available electrons at a specific current density limits the decolorization of the MB.

20.5.3.4 EFFECT OF ELECTROLYTE CONCENTRATION

The decolorization of 100 mg/L MB was studied at 2 A applied current density with NaCl salt as electrolyte at three different molar concentrations, varying from 0.043 to 0.129 M to investigate the effect of electrolyte concentration on decolorization and apparent rate. Increase in electrolyte concentration was observed to increase the decolorization rate of MB (Table 20.5). At increased concentration of electrolyte, the abundance of conducting ions favored better decolorization rate of MB.

20.5.3.5 MINERALIZATION STUDY

Mineralization study was conducted to confirm the degradation of MB associated with the decolorization. The mineralization study confirmed the degradation of MB during decolorization (Fig. 20.2). The extent of mineralization well correlated with decolorization level was observed at different parameter settings. Highest mineralization was recorded for 200 mg/L MB in 10 min time at 2 A applied current with carbon electrodes.

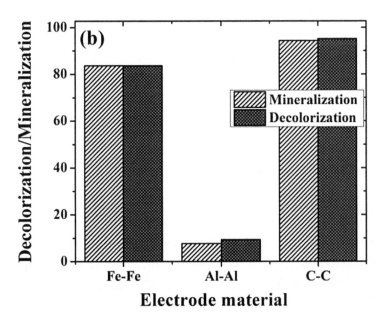

FIGURE 20.2 The effect of electrode material on mineralization of 100 mg/L MB at 2 A applied current.

20.5.3.6 COMPARISON OF ELECTRODES

The ECO studies were performed with electrodes constructed of different electrode materials at 2 A applied current with 100 mg/L MB using 0.086 M NaCl solution as electrolyte. Among the three electrode materials, namely, iron, aluminum, and carbon, that were evaluated, carbon recorded a maximum of 95% decolorization within 10 min time, followed by iron and later by aluminum (Fig. 20.2). The mass loss of carbon and aluminum electrodes (anode) was comparatively lower than iron and a maximum of 2–4% mass loss was recorded for iron electrode (Table 20.6). The counter electrode (cathode) suffers negligible mass loss for all three electrodes. Blank experiment (without current) was conducted to determine the absorption-based loss of MB. For all the three electrodes the absorption-based loss of MB was negligible, maximum being around 4% for carbon.

TABLE 20.6 The Change in Elemental Composition of Electrode for Different Electrode Materials on Electrochemical Degradation of MB.

Electrode materials	ECO parameter	Electrode	Elemental composition of electrode (based on EDS)	
		Mass loss after treatment	Before treatment	After treatment
Iron (Fe)	2 A, 60 min	2–4%	Fe: 78–79.5%	Fe: 59–61%
			O: 15–16%	O: 33–35%
Aluminum (Al)	2 A, 60 min	<1%	Al: 74–77%	Al: 66–69.5%
			O: 18–21%	O: 25.5–27%
Carbon (C)	2 A, 60 min	<1%	C: 85–87.2%	C: 70–72.6%
			O: 7–9%	O: 20–22.8%

20.5.3.7 CHARACTERIZATION OF ELECTRODES

The electrodes (Fe, Al, and C) before and after electrochemical degradation were characterized by EDS to determine the stoichiometric composition. The EDS spectrum (Fig. 20.3) showed that before electrochemical treatment the electrodes were primarily constituted of the respective element, namely, iron, aluminum, and carbon. However, stoichiometric fraction of oxygen at the electrode surface increased after treatment. The increment in weight percent of oxygen at electrode surface is 22–26% for iron and aluminum, and about 10–15% for carbon (Table 20.6). The presence of oxygen as constituent at electrode surface confirmed the formation of metal oxo-ions for iron and aluminum, and oxidative degradation mechanism of MB in the ED process.[52] A negligible decrease in weight percent of constituent element was recorded for aluminum. However, the weight proportion of iron and carbon was noticeably reduced after ECO process.

20.5.3.8 TOXICITY RESULT

For the toxicity study, RBC samples were treated with 20, 40, and 80 mg/L of MB respectively. RBC samples were also treated with aliquot of MB degradation collected after 0 min, 4 min, and 10 min of ECO. The RBC samples were thereafter viewed under the optical microscope to determine the changes in the morphology of the cells. The morphology of RBC samples treated with different MB concentration before and after electrochemical is presented in Figure 20.4.1. RBC treated with degraded samples collected at

Treatment of Pharmaceutical Wastewater: A Case Study

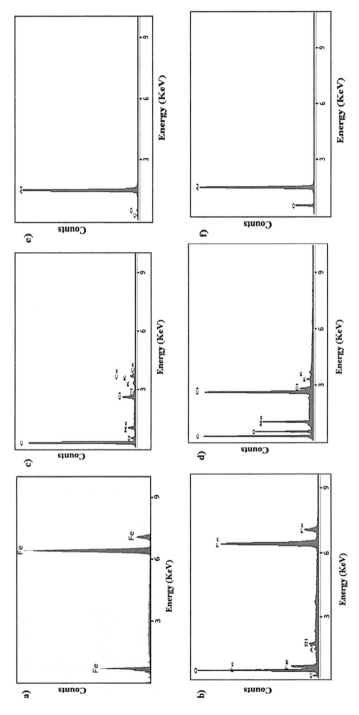

FIGURE 20.3 EDS profile of electrode before and after use in ECO experiment: (a) Fe—before use, (b) Fe—after use, (c) C—before use, (d) C—after use, (e) Al—before use, and (f) Al—after use.

4 min and 10 min during the degradation of MB are shown in Figures 20.4.2 and 20.4.3. The RBC treated with 20 mg/L of MB showed bluish coloration on the nucleus of RBC. As the concentration of the MB increases to 40 mg/L and 80 mg/L, the bluish color became darker. The coloration of the nucleus of RBC was observed to reduce when treated with MB degradation sample, collected after 4 min and 10 min of degradation in ECO process. However, the degradation products of MB resulted in aggregation and clumping of the RBCs. Higher initial concentration of MB resulted in more undesirable effect on RBC after degradation (Figs. 20.4.1 and 20.4.2).

FIGURE 20.4.1 RBC treated with different concentration of MB: (a) 20 mg/L, (b) 40 mg/L, and (c) 80 mg/L.

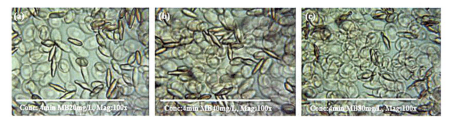

FIGURE 20.4.2 RBC treated with different concentration of MB degradation sample collected at 4 min: (a) 20 mg/L, (b) 40 mg/L, and (c) 80 mg/L.

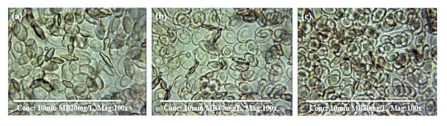

FIGURE 20.4.3 RBC treated with different concentration of MB degradation sample collected at 10 min: (a) 20 mg/L, (b) 40 mg/L, and (c) 80 mg/L.

20.6 CONCLUSIONS

The study reviewed the various options for treatment of an environmentally problematic pharmaceutical wastewater. AOPs were reviewed as an effective technique for degradation of variety of pharmaceuticals. ECO reportedly effective in treatment of pharmaceuticals was thoroughly reviewed for the effects of different operating variables, namely, applied current, electrolyte concentration, and pollutant concentration. A case study is presented on decolorization of a model pharmaceutical compound (MB) with low-cost electrodes. The decolorization of MB was correlated with mineralization to confirm the degradation at various levels of operating factors. Increased applied current linearly affected increased the decolorization of MB. The maximum decolorization of MB was recorded at electrolyte (NaCl) concentration of 0.086 M and pollutant (MB) concentration of 200 mg/L. Among three low-cost electrode materials evaluated in this study, carbon was observed to have highest decolorization and mineralization percent. Highest decolorization was recorded in the ECO process for 200 mg/L MB concentration at 2 A applied current with 0.086 M NaCl electrolyte in 10 min using carbon electrode. Thus, the present study demonstrated an alternative treatment option for recalcitrant and hazardous pollutants, such as, for pharmaceuticals contaminants, using low-cost electrode materials in ECO process. The application of low-cost materials as electrode can significantly enhance the affordability and acceptability of the ECO process as alternative treatment for various contaminants including pharmaceuticals.

ACKNOWLEDGMENT

The authors would also like to express their gratitude to the TEQIP-II, National Institute of Technology, Agartala, Tripura, India for the financial support and National Institute of Technology, Agartala, Tripura, India for infrastructural, and analytical facility and Manipur University, Manipur, India for characterization facility.

KEYWORDS

- pharmaceutical pollutants
- electrochemical oxidation
- low-cost electrodes
- methylene blue
- toxicity analysis

REFERENCES

1. Abellán, M. N.; Bayarri, B.; Giménez, J.; Costa, J. Photocatalytic Degradation of Sulfamethoxazole in Aqueous Suspension of TiO$_2$. *Appl. Catal. B Environ.* **2007,** *74*(3), 233–241.
2. Abhijit, D.; Lokesh, K. S.; Bejankiwar, R. S.; Gowda, T. P. Electrochemical Oxidation of Pharmaceutical Effluent Using Cast Iron Electrode. *J. Environmental Sci. Eng.* **2005,** *47*(1), 21–24.
3. Adams, C.; Wang, Y.; Loftin, K.; Meyer, M. Removal of Antibiotics from Surface and Distilled Water in Conventional Water Treatment Processes. *J. Environ. Eng.* **2002,** *128*(3), 253–260.
4. Addamo, M.; Augugliaro, V.; Di Paola, A.; Garcia-Lopez, E.; Loddo, V.; Marci, G.; Palmisano, L. Removal of Drugs in Aqueous Systems by Photoassisted Degradation. *J. Appl. Electrochem.* **2005,** *35*(7–8), 765–774.
5. Al-Rifai, J. H.; Khabbaz, H.; Schäfer, A. I. Removal of Pharmaceuticals and Endocrine Disrupting Compounds in a Water Recycling Process Using Reverse Osmosis Systems. *Sep. Purif. Technol.* **2011,** *77*(1), 60–67.
6. Andreozzi, R.; Canterino, M.; Marotta, R.; Paxeus, N. Antibiotic Removal from Wastewaters: The Ozonation of Amoxicillin. *J. Hazard. Mater.* **2005,** *122*(3), 243–250.
7. Andreozzi, R.; Caprio, V.; Insola, A.; Marotta, R. Advanced Oxidation Processes (AOP) for Water Purification and Recovery. *Catal. Today* **1999,** *53*(1), 51–59.
8. Andreozzi, R.; Caprio, V.; Marotta, R.; Vogna, D. *Water Res.* **2003,** *37*, 993.
9. Anglada, A.; Urtiaga, A.; Ortiz, I. Contributions of Electrochemical Oxidation to Wastewater Treatment: Fundamentals and Review of Applications. *J. Chem. Technol. Biotechnol.* **2009,** *84*(12), 1747–1755.
10. Bagastyo, A. Y.; Batstone, D. J.; Rabaey, K.; Radjenovic, J. Electrochemical Oxidation of Electrodialysed Reverse Osmosis Concentrate on Ti/Pt–IrO2, Ti/SnO2–Sb and Boron-doped Diamond Electrodes. *Water Res.* **2013,** *47*(1), 242–250.
11. Bautitz, I. R.; Nogueira, R. F. P. Degradation of Tetracycline by Photo-Fenton Process—Solar Irradiation and Matrix Effects. *J. Photochem. Photobiol. A Chem.* **2007,** *187*(1), 33–39.
12. Beachey, M. A. Pharmaceutical Hazard. *Pharm. Technol.* **2008.** http://www.pharmaceutical-technology.com/features/feature45434 (accessed Apr 29, 2014).

13. Bigda, R. J. Consider Fentons Chemistry for Wastewater Treatment. *Chem. Eng. Prog.* **1995,** *91*(12), 62–66.
14. Boyd, G. R.; Reemtsma, H.; Grimm, D. A.; Mitra, S. Pharmaceuticals and Personal Care Products (PPCPs) in Surface and Treated Waters of Louisiana, USA and Ontario, Canada. *Sci. Total Environ.* **2003,** *311*(1), 135–149.
15. Brillas, E.; Garcia-Segura, S.; Skoumal, M.; Arias, C. Electrochemical Incineration of Diclofenac in Neutral Aqueous Medium by Anodic Oxidation Using Pt and Boron-doped Diamond Anodes. *Chemosphere* **2010,** *79*(6), 605–612.
16. Brillas, E.; Sirés, I.; Arias, C.; Cabot, P. L.; Centellas, F.; Rodríguez, R. M.; Garrido, J. A. Mineralization of Paracetamol in Aqueous Medium by Anodic Oxidation with a Boron-doped Diamond Electrode. *Chemosphere* **2005,** *58*(4), 399–406.
17. Bui, T. X.; Choi, H. Adsorptive Removal of Selected Pharmaceuticals by Mesoporous Silica SBA-15. *J. Hazard. Mater.* **2009,** *168*(2), 602–608.
18. Bui, T. X.; Choi, H. Influence of Ionic Strength, Anions, Cations, and Natural Organic Matter on the Adsorption of Pharmaceuticals to Silica. *Chemosphere* **2010,** *80*(7), 681–686.
19. Calza, P.; Sakkas, V. A.; Medana, C.; Baiocchi, C.; Dimou, A.; Pelizzetti, E. Albanis, T. Photocatalytic Degradation Study of Diclofenac over Aqueous TiO_2 Suspensions. *Appl. Catal. B Environ.* **2006,** *67*(3), 197–205.
20. Carballa, M.; Omil, F.; Lema, J. M.; Llompart, M.; García-Jares, C.; Rodríguez, I.; Ternes, T. Behavior of Pharmaceuticals, Cosmetics and Hormones in a Sewage Treatment Plant. *Water Res.* **2004,** *38*(12), 2918–2926.
21. Chaturvedi, S. I. Mercury Removal Using Fe–Fe Electrodes by Electrocoagulation. *Int. J. Mod. Eng. Res.* **2013,** *3*, 101–108.
22. Chen, G. Electrochemical Technologies in Wastewater Treatment. *Sep. Purif. Technol.* **2004,** *38*(1), 11–41.
23. Choi, J. Y.; Lee, Y. J.; Shin, J.; Yang, J. W. Anodic Oxidation of 1, 4-dioxane on Boron-doped Diamond Electrodes for Wastewater Treatment. *J. Hazard. Mater.* **2010,** *179*(1–3), 762–768.
24. Chu, Y. Y.; Wang, W. J.; Wang, M. Anodic Oxidation Process for the Degradation of 2, 4-dichlorophenol in Aqueous Solution and the Enhancement of Biodegradability. *J. Hazard. Mater.* **2010,** *180*(1–3), 247–252.
25. Ciríaco, L.; Anjo, C.; Correia, J.; Pacheco, M. J.; Lopes, A. Electrochemical Degradation of Ibuprofen on Ti/Pt/PbO_2 and Si/BDD Electrodes. *Electrochim. Acta* **2009,** *54*(5), 1464–1472.
26. Cleuvers, M. Mixture Toxicity of the Anti-inflammatory Drugs Diclofenac, Ibuprofen, Naproxen, and Acetylsalicylic Acid. *Ecotoxicol. Environ. Saf.* **2004,** *59*(3), 309–315.
27. Clifford, D. A. Ion Exchange and Inorganic Adsorption. *Water Qual. Treat.* **1999,** *4*, 561–564.
28. Dai, Q.; Xia, Y.; Jiang, L.; Li, W.; Wang, J.; Chen, J. Enhanced Degradation of Aspirin by Electrochemical Oxidation with Modified PbO_2 Electrode and Hydrogen Peroxide. *Int. J. Electrochem. Sci.* **2012,** *7*, 12895–12906.
29. Desoky, H. S.; Ghoneim, M. M.; Zidan, N. M. Decolorization and Degradation of Ponceau S Azo-dye in Aqueous Solutions by the Electrochemical Advanced Fenton Oxidation. *Desalination* **2010,** *264*(1–2), 143–150.
30. Faust, D.; Funken, K. H.; Horneck, G.; Milow, B.; Ortner, J.; Sattlegger, M.; Schmitz, C. Immobilized Photosensitizers for Solar Photochemical Applications. *Sol. Energy* **1999,** *65*(1), 71–74.

31. Feng, L.; van Hullebusch, E. D.; Rodrigo, M. A.; Esposito, G.; Oturan, M. A. Removal of Residual Anti-inflammatory and Analgesic Pharmaceuticals from Aqueous Systems by Electrochemical Advanced Oxidation Processes. A Review. *Chem. Eng. J.* **2013**, *228*, 944–964.
32. Fent, K.; Weston, A. A.; Caminada, D. Ecotoxicology of Human Pharmaceuticals. *Aquat. Toxicol.* **2006**, *76*(2), 122–159.
33. Fernández-Alba, A. R.; Hernando, D.; Agüera, A.; Cáceres, J.; Malato, S. Toxicity Assays: A Way for Evaluating AOPs Efficiency. *Water Res.* **2002**, *36*(17), 4255–4262.
34. Gebhardt, W.; Schröder, H. F. Liquid Chromatography–(Tandem) Mass Spectrometry for the Follow-up of the Elimination of Persistent Pharmaceuticals During Wastewater Treatment Applying Biological Wastewater Treatment and Advanced Oxidation. *J. Chromatogr. A* **2007**, *1160*(1), 34–43.
35. Glaze, W. H.; Kang, J. W.; Chapin, D. H. The Chemistry of Water Treatment Processes Involving Ozone, Hydrogen Peroxide and Ultraviolet Radiation. *Ozone Sci. Eng.* **1987**, *9*(4), 335–352.
36. Gogate, P. R.; Pandit, A. B. A Review of Imperative Technologies for Wastewater Treatment I: Oxidation Technologies at Ambient Conditions. *Adv. Environ. Res.* **2004**, *8*(3), 501–551.
37. Gonzalez, M. C.; Braun, A. Vacuum UV Photolysis of Aqueous Solutions of Nitrate. Effect of Organic Matter II. Methanol. *J. Photochem. Photobiol. A: Chem,* **1996**, *95*(1), 67–72.
38. Gonzalez, M. C.; Braun, A. M. VUV Photolysis of Aqueous Solutions of Nitrate and Nitrite. *Res. Chem. Intermed.* **1995**, *21*(8–9), 837–859.
39. Gonzalez, M. C.; Braun, A. M.; Prevot, A. B.; Pelizzetti, E. Vacuum-ultraviolet (VUV) Photolysis of Water: Mineralization of Atrazine. *Chemosphere* **1994**, *28*(12), 2121–2127.
40. González, O.; Sans, C.; Esplugas, S. Sulfamethoxazole Abatement by Photo-Fenton: Toxicity, Inhibition and Biodegradability Assessment of Intermediates. *J. Hazard. Mater.* **2007**, *146*(3), 459–464.
41. Govil, P.; Reddy, G.; Krishna, A. Contamination of Soil due to Heavy Metals in the Patancheru Industrial Development Area, Andhra Pradesh, India. *Environ. Geol.* **2001**, *41*(3–4), 461–469.
42. Halling-Sørensen, B.; Nielsen, S. N.; Lanzky, P. F.; Ingerslev, F.; Lützhøft, H. H.; Jørgensen, S. E. Occurrence, Fate and Effects of Pharmaceutical Substances in the Environment-A Review. *Chemosphere* **1998**, *36*(2), 357–393.
43. Heron, R. J. L.; Pickering, F. C. Health Effects of Exposure to Active Pharmaceutical Ingredients (APIs). *Occup. Med.* **2003**, *53*(6), 357–362.
44. Hmani, E.; Elaoud, S. C.; Samet, Y.; Abdelhédi, R. Electrochemical Degradation of Waters Containing O-Toluidine on PbO_2 and BDD Anodes. *J. Hazard. Mater.* **2009**, *170*(2), 928–933.
45. Hu, X.; Yang, J.; Zhang, J. Magnetic Loading of $TiO_2/SiO_2/Fe_3O_4$ Nanoparticles on Electrode Surface for Photoelectrocatalytic Degradation of Diclofenac. *J. Hazard. Mater.* **2011**, *196*, 220–227.
46. Huang CP, Dong Ch, Tang Z. *Waste Manag* **1993**, *13*, 361.
47. Huber, M. M.; Canonica, S.; Park, G. Y.; Von Gunten, U. Oxidation of Pharmaceuticals During Ozonation and Advanced Oxidation Processes. *Environ. Sci. Technol.* **2003**, *37*(5), 1016–1024.

48. Huber, M. M.; GÖbel, A.; Joss, A.; Hermann, N.; LÖffler, D.; McArdell, C. S.; von Gunten, U. Oxidation of Pharmaceuticals During Ozonation of Municipal Wastewater Effluents: A Pilot Study. *Environ. Sci. Technol.* **2005,** *39*(11), 4290–4299.
49. Iranifam, M.; Zarei, M.; Khataee, A. R. Decolorization of CI Basic Yellow 28 Solution Using Supported ZnO Nanoparticles Coupled with Photoelectro-Fenton Process. *J. Electroanal. Chem.* **2011,** *659*(1), 107–112.
50. Jones, O. A.; Lester, J. N.; Voulvoulis, N. Pharmaceuticals: A Threat to Drinking Water? *Trends Biotechnol.* **2005,** *23*(4), 163–167.
51. Kimura, K.; Toshima, S.; Amy, G.; Watanabe, Y. Rejection of Neutral Endocrine Disrupting Compounds (EDCs) and Pharmaceutical Active Compounds (PhACs) by RO Membranes. *J. Membr. Sci.* **2004,** *245*(1), 71–78.
52. Kobya, M.; Demirbas, E.; Can, O. T.; Bayramoglu, M. Treatment of Levafix Orange Textile Dye Solution by Electrocoagulation. *J. Hazard. Mater.* **2006,** *132*(2), 183–188.
53. Kümmerer, K. Drugs in the Environment: Emission of Drugs, Diagnostic Aids and Disinfectants into Wastewater by Hospitals in Relation to Other Sources–A Review. *Chemosphere* **2001,** *45*(6), 957–969.
54. Kümmerer, K. Resistance in the Environment. *J. Antimicrob. Chemother.* **2004,** *54* (2), 311–320.
55. Legrini, O.; Oliveros, E.; Braun, A. M. Photochemical Processes for Water Treatment. *Chem. Rev.* **1993,** *93*(2), 671–698.
56. Litter, M. I. Introduction to Photochemical Advanced Oxidation Processes for Water Treatment. In *Environmental Photochemistry Part II*; Springer: Berlin Heidelberg, 2005; pp 325–366.
57. Mahamuni, N. N.; Pandit, A. B. Effect of Additives on Ultrasonic Degradation of Phenol. *Ultrason. Sonochem.* **2006,** *13*(2), 165–174.
58. Miao, X. S.; Bishay, F.; Chen, M.; Metcalfe, C. D. Occurrence of Antimicrobials in the Final Effluents of Wastewater Treatment Plants in Canada. *Environ. Sci. Technol.* **2004,** *38*(13), 3533–3541.
59. Minero, C.; Pelizzetti, E.; Piccinini, P.; Vincenti, M. Photocatalyzed Transformation of Nitrobenzene on TiO2 and ZnO. *Chemosphere* **1994,** *28*(6), 1229–1244.
60. Mowery, H. R.; Loganathan, B. G. Persistent Organic Compounds in Wastewater: Azithromycin and Urobilin Concentrations in Wastewater Treatment Plant Samples from Murray, Kentucky, USA. *Organohalogen Compd.* **2007,** *69*, 2961–2964.
61. Munter, R. Advanced Oxidation Processes–Current Status and Prospects. *Proc. Estonian Acad. Sci. Chem.* **2001,** 50(2), 59–80.
62. Murugananthan, M.; Latha, S. S.; Raju, G. B.; Yoshihara, S. Anodic Oxidation of Ketoprofen—An Anti-inflammatory Drug Using Boron-doped Diamond and Platinum Electrodes. *J. Hazard. Mater.* **2010,** *180*(1), 753–758.
63. Nikolaou, A.; Meric, S.; Fatta, D. Occurrence Patterns of Pharmaceuticals in Water and Wastewater Environments. *Anal. Bioanal. Chem.* **2007,** *387*(4), 1225–1234.
64. Nogueira, R. F. P.; Guimarães, J. R. Photodegradation of Dichloroacetic Acid and 2, 4-Dichlorophenol by Ferrioxalate/H_2O_2 System. *Water Res.* **2000,** *34*(3), 895–901.
65. Özcan, A.; Şahin, Y.; Koparal, A. S.; Oturan, M. A. Degradation of Picloram by the Electro-Fenton Process. *J. Hazard. Mater.* **2008,** *153*(1–2), 718–727.
66. Packer, J. L.; Werner, J. J.; Latch, D. E.; McNeill, K.; Arnold, W. A. Photochemical fate of Pharmaceuticals in the Environment: Naproxen, Diclofenac, Clofibric Acid, and Ibuprofen. *Aquat. Sci.* **2003,** *65*(4), 342–351.

67. Pereira, V. J.; Weinberg, H. S.; Linden, K. G.; Singer, P. C. UV Degradation Kinetics and Modeling of Pharmaceutical Compounds in Laboratory Grade and Surface Water via Direct and Indirect Photolysis at 254 nm. *Environ. Sci. Technol.* **2007,** *41*(5), 1682–1688.
68. Pérez-Estrada, L. A.; Malato, S.; Gernjak, W.; Agüera, A.; Thurman, E. M.; Ferrer, I.; Fernández-Alba, A. R. Photo-Fenton Degradation of Diclofenac: Identification of Main Intermediates and Degradation Pathway. *Environ. Sci. Technol.* **2005,** *39*(21), 8300–8306.
69. Putra, E. K.; Pranowo, R.; Sunarso, J.; Indraswati, N.; Ismadji, S. Performance of Activated Carbon and Bentonite for Adsorption of Amoxicillin from Wastewater: Mechanisms, Isotherms and Kinetics. *Water Res.* **2009,** *43*(9), 2419–2430.
70. Quand-Même, G. C.; Auguste, A. F. T.; Hélène, M.; Evelyne, L.; Ibrahima, S.; Lassine, O. Electrochemical Oxidation of Amoxicillin in Its Pharmaceutical Formulation at Boron-doped Diamond (BDD) Electrode. *J. Electrochem. Sci. Eng.* **2015,** *5*(2), 129–143.
71. Radjenović, J.; Petrović, M.; Ventura, F.; Barceló, D. Rejection of Pharmaceuticals in Nanofiltration and Reverse Osmosis Membrane Drinking Water Treatment. *Water Res.* **2008,** *42*(14), 3601–3610.
72. Raghu, S.; Lee, C. W.; Chellammal, S.; Palanichamy, S.; Basha, C. A. Evaluation of Electrochemical Oxidation Techniques for Degradation of Dye Effluents—A Comparative Approach. *J. Hazard. Mater.* **2009,** *171*(1–3), 748–754.
73. Ravina, M.; Campanella, L.; Kiwi, J. Accelerated Mineralization of the Drug Diclofenac via Fenton Reactions in a Concentric Photo-reactor. *Water Res.* **2002,** *36*(14), 3553–3560.
74. Ray, S.; Lalman, J. A.; Biswas, N. Using the Box-Benkhen Technique to Statistically Model Phenol Photocatalytic Degradation by Titanium Dioxide Nanoparticles. *Chem. Eng. J.* **2009,** *150*(1), 15–24.
75. Reyes, C.; Fernandez, J.; Freer, J.; Mondaca, M. A.; Zaror, C.; Malato, S.; Mansilla, H. D. Degradation and Inactivation of Tetracycline by TiO_2 Photocatalysis. *J. Photochem. Photobiol. A Chem.* **2006,** *184*(1), 141–146.
76. Richards, S. M.; Cole, S. E. A Toxicity and Hazard Assessment of Fourteen Pharmaceuticals to *Xenopus laevis* Larvae. *Ecotoxicology* **2006,** *15*(8), 647–656.
77. Schaap, A. P.; Thayer, A. L.; Blossey, E. C.; Neckers, D. C. Polymer-based Sensitizers for Photooxidations. II. *J. Am. Chem. Soc.* **1975,** *97*(13), 3741–3745.
78. Schindler, D. W.; Smol, J. P. Cumulative Effects of Climate Warming and Other Human Activities on Freshwaters of Arctic and Subarctic North America. *AMBIO J. Hum. Environ.* **2006,** *35*(4), 160–168.
79. Scott, J. P.; Ollis, D. F. Integration of Chemical and Biological Oxidation Processes for Water Treatment: Review and Recommendations. *Environ. Prog.* **1995,** *14*(2), 88–103.
80. Smith, C. A. Managing Pharmaceutical Waste. *J. Pharm. Soc. Wis.* **2002,** *5*, 17–22.
81. Snyder, S. A.; Adham, S.; Redding, A. M.; Cannon, F. S.; De Carolis, J.; Oppenheimer, J.; Yoon, Y. Role of Membranes and Activated Carbon in the Removal of Endocrine Disruptors and Pharmaceuticals. *Desalination* **2007,** *202*(1), 156–181.
82. Snyder, S. A.; Westerhoff, P.; Yoon, Y.; Sedlak, D. L. Pharmaceuticals, Personal Care Products, and Endocrine Disruptors in Water: Implications for the Water Industry. *Environ. Eng. Sci.* **2003,** *20*(5), 449–469.
83. Staehelin, J.; Hoigne, J. Decomposition of Ozone in Water: Rate of Initiation by Hydroxide Ions and Hydrogen Peroxide. *Environ. Sci. Technol.* **1982,** *16*(10), 676–681.

84. Sumpter, J. P. Xenoendocrine Disrupters—Environmental Impacts. *Toxicol. Lett.* **1998**, *102*, 337–342.
85. Sun, J. H.; Sun, S. P.; Sun, J. Y.; Sun, R. X.; Qiao, L. P.; Guo, H. Q.; Fan, M. H. Degradation of Azo Dye Acid Black 1 Using Low Concentration Iron of Fenton Process Facilitated by Ultrasonic Irradiation. *Ultrason. Sonochem.* **2007**, *14*(6), 761–766.
86. Ternes, T. A. Occurrence of Drugs in German Sewage Treatment Plants and Rivers. *Water Res.* **1998**, *32*(11), 3245–3260.
87. Ternes, T. A.; Stüber, J.; Herrmann, N.; McDowell, D.; Ried, A.; Kampmann, M.; Teiser, B. Ozonation: A Tool for Removal of Pharmaceuticals, Contrast Media and Musk Fragrances from Wastewater? *Water Res.* **2003**, *37*(8), 1976–1982.
88. Ting, W. P.; Lu, M. C.; Huang, Y. H. Kinetics of 2, 6-dimethylaniline Degradation by Electro-Fenton Process. *J. Hazard. Mater.* **2009**, *161*(2–3), 1484–1490.
89. Trovó, A. G.; Melo, S. A. S.; Nogueira, R. F. P. Photodegradation of the Pharmaceuticals Amoxicillin, Bezafibrate and Paracetamol by the Photo-Fenton Process—Application to Sewage Treatment Plant Effluent. *J. Photochem. Photobiol. A: Chem* **2008**, *198*(2–3), 215–220.
90. US/EPA. *Handbook of Advanced Photochemical Oxidation Processes*. 1998. EPA/625/R-98/004
91. Van der Bruggen, B.; Vandecasteele, C.; Van Gestel, T.; Doyen, W.; Leysen, R. A Review of Pressure-driven Membrane Processes in Wastewater Treatment and Drinking Water Production. *Environ. Prog.* **2003**, *22*(1), 46–56.
92. Vogna, D.; Marotta, R.; Andreozzi, R.; Napolitano, A.; d'Ischia, M. Kinetic and Chemical Assessment of the UV/H_2O_2 Treatment of Antiepileptic Drug Carbamazepine. *Chemosphere* **2004**, *54*(4), 497–505.
93. Vogna, D.; Marotta, R.; Napolitano, A.; Andreozzi, R.; d'Ischia, M. Advanced Oxidation of the Pharmaceutical Drug Diclofenac with UV/H_2O_2 and Ozone. *Water Res.* **2004**, *38*(2), 414–422.
94. Waterston, K.; Wang, J. W.; Bejan, D.; Bunce, N. J. Electrochemical Waste Water Treatment: Electrooxidation of Acetaminophen. *J. Appl. Electrochem.* **2006**, *36*(2), 227–232.
95. Weichgrebe, D.; Danilova, E.; Rosenwinkel, K. H.; Vedenjapin, A. A.; Baturova, M. Electrochemical Oxidation of Drug Residues in Water by the Example of Tetracycline, Gentamicine and Aspirin. *Water Sci. Technol.* **2004**, *49*(4), 201–206.
96. Westerhoff, P.; Yoon, Y.; Snyder, S.; Wert, E. Fate of Endocrine-disruptor, Pharmaceutical, and Personal Care Product Chemicals During Simulated Drinking Water Treatment Processes. *Environ. Sci. Technol.* **2005**, *39*(17), 6649–6663.
97. Wilkinson, F.; Helman, W. P.; Ross, A. B. Quantum Yields for the Photosensitized Formation of the Lowest Electronically Excited Singlet State of Molecular Oxygen in Solution. *J. Phys. Chem. Ref. Data* **1993**, *22*(1), 113–262.
98. World Health Organization. Pharmaceuticals in Drinking-water. 2011.
99. Xia, Y.; Dai, Q.; Chen, J. Electrochemical Degradation of Aspirin Using a Ni-doped PbO_2 Electrode. *J. Electroanal. Chem.* **2015**, *744*, 117–125.
100. Yoon, Y.; Westerhoff, P.; Snyder, S. A.; Wert, E. C. Nanofiltration and Ultrafiltration of Endocrine Disrupting Compounds, Pharmaceuticals and Personal Care Products. *J. Membr. Sci.* **2006**, *270*(1), 88–100.

101. Zhou, M.; He, J. Degradation of azo dye by Three Clean Advanced Oxidation Processes: Wet Oxidation, Electrochemical Oxidation and Wet Electrochemical Oxidation—A Comparative Study. *Electrochim. Acta.* **2007**, *53*(4), 1902–1910.
102. Zhao, X.; Qu, J.; Liu, H.; Qiang, Z.; Liu, R.; Hu, C. Photoelectrochemical Degradation of Anti-inflammatory Pharmaceuticals at Bi_2MoO_6–Boron-doped Diamond Hybrid Electrode Under Visible Light Irradiation. *Appl. Catal. B Environ.* **2009**, *91*(1), 539–545.
103. Zwiener, C.; Frimmel, F. H. Oxidative Treatment of Pharmaceuticals in Water. *Water Res.* **2000**, *34*(6), 1881–1885.

INDEX

A

Adipogenic induction
   ASCs, 363
      fluorescent dyes, 364–365
      phase-contrast micrographs, 364
      transmission electron microscopy, 365–366
Adipose tissue engineering (ATE), 357
Adipose-derived mesenchymal stem cells (ASCs), 357
   adipogenic induction, 363
      fluorescent dyes, 364–365
      phase-contrast micrographs, 364
      transmission electron microscopy, 365–366
   in vitro adipogenesis
      cell–material interaction, 362–363
      isolation and expansion, 358–359
      mesenchymal stem cells, multilineage potential, 359–360
      scaffolds, surface topology, 360–362
   in vivo adipogeneisis
      cell-collagen construct, H and E staining, 368
      H and E staining, 369–370
      histological analysis, 370–371
      histology, 367–368
      Masson's trichrome staining, 368–369
      paraffin processing, 368
      picrosirius, 368–369
      PMMA processing, 369–370
Adipose-derived stem cells (ASCs), 356
Alloy C-276
   electron beam welding (EBW), 168
   gas metal arc welding (GMAW), 168
   gas tungsten arc (GTA), 168
   gas tungsten arc welding (GTAW), 168
   materials and methods
      material and welding procedure, 169–170
   mechanical properties, 169
   mechanical testing, 176–180
   metallurgical characterization, 170
      mechanical testing, 171
   microstructure examination, 171–173
   pulsed current gas tungsten arc (PCGTA), 168
   pulsed current gas tungsten arc welding (PCGTAW), 168
   sem/eds analysis, 173–176
Antimicrobial characterization, techniques
   antimicrobial properties, 127
   colony forming unit counting technique, 126–127
   shaking flask method, 127
   zone of inhibition (ZOI) method, 125
Antimicrobial metals, 93
   antimicrobial surfaces, 95
   atomistic approach, 97–98
   benefits, 99–102
   copper, 94
   destroy undesirable viruses and bacteria, 95
   DNA and biological membranes, 96
   processing and manipulation, 95
   reactive oxygen species (ROS), 97
   redox active metals, 96
   thin film copper coatings, 99
Atomic force microscopy (AFM), 35
   advantage, 36
   different imaging modes
      contact mode, 41–42
      noncontact mode, 42–43
      tapping mode, 42–43
   instrument
      cantilevers, 38
      feedback and detector devices, 39
      motorized X–Y stage, 37
      piezoelectric scanner, 38–39
      piezoelectric scanner tube (PZT), 37
      sample, 37
   interaction forces, 39–40

measurement, modes
  constant force mode, 41
  constant height mode, 40–41
polymer composites
  applications of, 44–53
working principle, 37

B

Backscattered electron imaging (BEI), 6
Backscattered electrons (BSEs), 5
Barium strontium titanate (BaTiO3)
  applications, 86
    ferroelectric ceramic materials, 87
  dielectric
    coprecipitation method, 62–63
    crystal structure, 58
    domain theory, 70–71
    efficient dielectric supports, 68
    fundamental knowledge, 65–67
    hydrothermal synthesis method, 62
    permittivity constant, 67–68
    perovskite structure (ABO3), 58, 59–60
    phase transition, 59–60
    positive temperature coefficients of resistivity (PTCR), 58
    scanning electron microscope (SEM), 71–72
    sol–gel method, 63–64
    solid-state method, 61–62
    strontium-doped barium titanate, 59
    transmission electron microscope (TEM), 71–72
  dopants, effect of
    Ca doping, 81
    Cu doping, 80–81
    Fe doping, 78–79
    La doping, 80
    Mn doping, 79–80
    Ni doping, 79
    Sr doping, 78
  ferroelectric
    coprecipitation method, 62–63
    crystal structure, 58
    dielectric, fundamental knowledge, 65–67
    domain theory, 70–71
    efficient dielectric supports, 68
    hydrothermal synthesis method, 62
    permittivity constant, 67–68
    perovskite structure (ABO3), 58, 59–60
    phase transition, 59–60
    positive temperature coefficients of resistivity (PTCR), 58
    scanning electron microscope (SEM), 71–72
    sol–gel method, 63–64
    solid-state method, 61–62
    strontium-doped barium titanate, 59
    transmission electron microscope (TEM), 71–72
  microscopic studies, 84–86
    coprecipitation method, 62–63
    crystal structure, 58
    dielectric, fundamental knowledge, 65–67
    domain theory, 70–71
    efficient dielectric supports, 68
    hydrothermal synthesis method, 62
    permittivity constant, 67–68
    perovskite structure (ABO3), 58, 59–60
    phase transition, 59–60
    positive temperature coefficients of resistivity (PTCR), 58
    scanning electron microscope (SEM), 71–72
    sol–gel method, 63–64
    solid-state method, 61–62
    strontium-doped barium titanate, 59
    transmission electron microscope (TEM), 71–72
  preparation
    ceramic applications, recent advances in, 59
    coprecipitation method, 62–63
    crystal structure, 58
    dielectric, fundamental knowledge, 65–67
    domain theory, 70–71
    efficient dielectric supports, 68
    hydrothermal synthesis method, 62
    permittivity constant, 67–68
    perovskite structure (ABO3), 58, 59–60
    phase transition, 59–60
    positive temperature coefficients of resistivity (PTCR), 58

Index 495

scanning electron microscope (SEM), 71–72
sol–gel method, 63–64
solid-state method, 61–62
strontium-doped barium titanate, 59
transmission electron microscope (TEM), 71–72
study
  dielectric constant, 72–73
  dielectric property, 81–83
  ferroelectric property, 74–76, 83–84
  microscopic studies, 76–77
Biological and catalytic applications
  biomolecules-functionalized nanoparticles, 335–336
  fluorine-substituted heterocyclic nitrogen systems, 330–331
  functionalized nanomaterials
    aromatic heterocycles, 322
    blood brain barrier (BBB), 322
    boron nitride (BN), 323–324
    G protein-coupled receptors (GPCRs), 320
    hexagonal boron nitride (h-BN), 323
    ideal synthesis, 321
    ionic liquids (ILs), 322
  gold–sulfur nanoparticles
    bioactive thiolated molecules-conjugated, 333–334
    functional group in thiols (RSH), 332
    sulfhydryl (SH), 332
  metal-loaded boron nitride, 334–335
  nitrogen heterocycle system, synthesis and utility, 324–327
    multicomponent reactions (MCRs), 328–330
  organic nitrogen-containing heterocylic cations, 331–332
    inorganic fluorine anions (ILS), 332
Blends of dicyanate esters
  azomethine functionalized dicyanate esters, 277
  GPR blends, 277–280
  polyacrylonitrile, 277–280
  polyurethane, 277–280
  Schiff-base-functionalized dicyanate esters, 276–277

C

Cadmium (Cd), 380
  animal experimental studies, 382–386
  case study
    microscopic features of testis, effects, 386–389
    spermatozoa, effects, 386–389
  deleterious actions, 381
  histological examination, 381
  Leydig cell, 382
  sulfhydryl groups
    blockage of, 381
Case study
  CR(III)-picolinate on epididymis and thymus, 437–443
  microscopic features of testis, effects, 386–389
  spermatozoa
    effects, 386–389
Chromium, 428
  biological properties
    cellular and subcellular systems, 433
    dubious data, 433
    intracellular reductive, 432
    plethora, 432
    species, 431
  case study
    CR(III)-picolinate on epididymis and thymus, 437–443
  compositions and structural features, 429
  GTF, 429–430
  low-molecular weight chromium-binding substance (LMWCr), 430
  molecular level, 431
  nutritional supplements
    adversarial effect, 436
    American Diabetes Association, 436
    chemical formulations, 434
    essential nutrients, 434
    hypoglycemia, 435
    nutritional literacy review, 437
    veil-covered chromium, 435
Copper thin films for antimicrobial applications
  antimicrobial characterization, techniques
    antimicrobial properties, 127
    colony forming unit counting technique, 126–127

shaking flask method, 127
zone of inhibition (ZOI) method, 125
antimicrobial metals, 93
  antimicrobial surfaces, 95
  atomistic approach, 97–98
  benefits, 99–102
  copper, 94
  destroy undesirable viruses and bacteria, 95
  DNA and biological membranes, 96
  processing and manipulation, 95
  reactive oxygen species (ROS), 97
  redox active metals, 96
  thin film copper coatings, 99
characterization
  film thickness, measurement, 113–115
  mechanical properties of thin films, evaluation, 116–118
  residual stresses, 119–122
methods
  chemical vapor deposition (CVD), 107–108
  condensation, 102–103
  e-beam evaporation and deposition, 104–105
  electroless deposition, 108–109
  electroplating, 110–113
  sputter deposition, 105–106
  thermal evaporation, 102–103
microstructural characterization, 122–124
  techniques for, 124–125
preparation and characterization
  antimicrobial surface, 92
  growth of microorganisms, 93
  methicillin-resistant Staphylococcus aureus (MRSA), 92
Cyclic voltammetry (CV) studies, 141–143

D

Dielectric
  coprecipitation method, 62–63
  crystal structure, 58
  domain theory, 70–71
  efficient dielectric supports, 68
  fundamental knowledge, 65–67
  hydrothermal synthesis method, 62
  permittivity constant, 67–68
  perovskite structure (ABO3), 58–60
  phase transition, 59–60
  positive temperature coefficients of resistivity (PTCR), 58
  scanning electron microscope (SEM), 71–72
  sol–gel method, 63–64
  solid-state method, 61–62
  strontium-doped barium titanate, 59
  transmission electron microscope (TEM), 71–72
Diffracted backscattered electrons (EBSDs), 5–6
Dopants
  effect
    Ca doping, 81
    Cu doping, 80–81
    Fe doping, 78–79
    La doping, 80
    Mn doping, 79–80
    Ni doping, 79
    Sr doping, 78
Dyes
  acid–base properties, 343
  methods for removing, 345–346
  specific properties, 344–345
  technique used in, 343–344

E

Electron backscatter diffraction (EBSD), 227
  applications
    compositional gradients, 250–252
    determination of unknown phases, 252–253
    microtexture and twinning, 244–246
    multiphasic samples, study, 247–250
    orientation measurements, 238–244
    technique, development of, 238
    transmission mode, 253–254
  experimental details
    data analysis, 236–238
    measurement, 236
    specimen surface, 233–235
  materials characterization, role, 228–229
  materials studied, 232
  parameters, 232–233
  patterns
    analysis of, 230–232
    physics behind band formation, 230
  technique, 229–230

Index	497

Electron beam welding (EBW), 168
Electron gun, 7
   detectors for all signals
      Auger electron, 16
      backscattered electron imaging (BSEs), 14–15
      cathodoluminescence detection and imaging, 16
      electron beam induced current (EBIC) image, 15
      energy dispersive x-ray spectroscopy, 17
      image display and recording, 18
      secondary electron imaging, 13–14
      vacuum system, 18
      voltage-contrast imaging, 16
   electromagnetic lenses, 11–12
   FE electron gun, 8–9
   features, comparison of, 10–11
   sample stage, 12–13
   SE electron gun, 9–10
   thermionic emission electron gun, 8
Emerging contaminants (EC)
   anthropogenic activities, 398
   biological implications, 401
      Comet Assay, 402
      ctogenetic damage, 402
      electron microscopy, 405–412
      light, 405–412
      microscopy, 402
      microscopy in ecotoxicological sudies, role, 403–405
      nanoparticles (NP), 413–415
      organ-specific, 402
      pharmaceuticals, 412–413
   classes, 399
   ecotoxicology, 398
   nanoparticles (NP), 400–401
   pharmaceuticals
      wastewater treatment plants (WWTP), 399
   predicted environmental concentrations (PEC), 401
Energy dispersive X-ray (EDX), 140–141

F

Ferroelectric
   coprecipitation method, 62–63
   crystal structure, 58
   dielectric
      fundamental knowledge, 65–67
   domain theory, 70–71
   efficient dielectric supports, 68
   hydrothermal synthesis method, 62
   permittivity constant, 67–68
   perovskite structure (ABO3), 58, 59–60
   phase transition, 59–60
   positive temperature coefficients of resistivity (PTCR), 58
   scanning electron microscope (SEM), 71–72
   sol–gel method, 63–64
   solid-state method, 61–62
   strontium-doped barium titanate, 59
   transmission electron microscope (TEM), 71–72
Fluorescent in situ hybridization (FISH), 371–373
Functionalized nanomaterials
   aromatic heterocycles, 322
   blood brain barrier (BBB), 322
   boron nitride (BN), 323–324
   G protein-coupled receptors (GPCRs), 320
   hexagonal boron nitride (h-BN), 323
   ideal synthesis, 321
   ionic liquids (ILs), 322

G

Gas metal arc welding (GMAW), 168
Gas tungsten arc (GTA), 168
Gas tungsten arc welding (GTAW), 168
Gold-sulfur nanoparticles
   bioactive thiolated molecules-conjugated, 333–334
   functional group in thiols (RSH), 332
   sulfhydryl (SH), 332
Graphene-modified carbon microsurfaces
   ideal graphene
      reactivity of, 305–306
      working electrode material
         DNA analysis, 308–309
         dopamine (DA), 306–307
         glucose, 308
         heavy metal ions, analysis, 310
         paracetamol, 306–307
         protein analysis, 309–310

H

Hematoxylin–eosin (HE), 381
Histomorphometry, 373–374

I

Imaging modes
  AFM
    contact mode, 41–42
    noncontact mode, 42–43
    tapping mode, 42–43
In vitro adipogenesis
  ASCs
    cell–material interaction, 362–363
    isolation and expansion, 358–359
    mesenchymal stem cells, multilineage potential, 359–360
    scaffolds, surface topology, 360–362
In vivo adipogenesis
  ASCs
    cell-collagen construct, H and E staining, 368
    H and E staining, 369–370
    histological analysis, 370–371
    histology, 367–368
    Masson's trichrome staining, 368–369
    paraffin processing, 368
    picrosirius, 368–369
    PMMA processing, 369–370
Instrument
  AFM
    cantilevers, 38
    feedback and detector devices, 39
    motorized X–Y stage, 37
    piezoelectric scanner, 38–39
    piezoelectric scanner tube (PZT), 37
    sample, 37

L

Lead citrate, 381
Light microscopy
  wood, formation and decomposition
    cell wall biomass, 261
    cell wall structure, 259
    cells, biochemical changes, 262–266
    consecutive deposition, 260
    environmental factors, 258
    oak wood, ultrasound waves impact of, 267–271
    phloem and xylem, 262
    pine callus culture, study, 266–267
    radial diameter, 261
    temperature and precipitation, optimal values, 261
    tissue culture, 259
    xylem cells, 258

M

Microscopic studies, 84–86
  coprecipitation method, 62–63
  crystal structure, 58
  dielectric
    fundamental knowledge, 65–67
  domain theory, 70–71
  efficient dielectric supports, 68
  hydrothermal synthesis method, 62
  permittivity constant, 67–68
  perovskite structure (ABO3), 58, 59–60
  phase transition, 59–60
  positive temperature coefficients of resistivity (PTCR), 58
  scanning electron microscope (SEM), 71–72
  sol–gel method, 63–64
  solid-state method, 61–62
  strontium-doped barium titanate, 59
  transmission electron microscope (TEM), 71–72

N

Nano materials
  synthesis and characterization
    catalytic reduction of methylene blue, 347
    gold, preparation and characterization of, 347
    materials and methods, 347
    nanostructured biopalladium, preparation and characterization of, 347
    4-nitroaniline, catalytic reduction of, 348
    4-nitrophenol, catalytic reduction of, 348
Nickel hydroxide (Ni(OH)2) electrode
  battery electrodes, 134–135

cyclic voltammetry (CV) studies, 141–143
energy dispersive X-ray (EDX), 140–141
fabrication of electrodes
  polytetrafluoroethylene (PTFE) solution, 137
Fourier transform infrared analysis, 138–139
preparation
  β-nickel hydroxide electrodes with ZnO, 135–136
  zinc oxide, synthesis of, 135–136
scanning electron microscopy studies, 139–140
synthesis
  nickel sulfate (NiSO4), 135
  potassium hydroxide (KOH), 135
X-ray diffraction analysis, 137–138
Nickel sulfate (NiSO4), 135
Nutritional supplements
  adversarial effect, 436
  American Diabetes Association, 436
  chemical formulations, 434
  essential nutrients, 434
  hypoglycemia, 435
  nutritional literacy review, 437
  veil-covered chromium, 435

O

Organic cancer carcinogens from wastewater, removal
  dyes
    acid–base properties, 343
    methods for removing, 345–346
    specific properties, 344–345
    technique used in, 343–344
  entry into water bodies, 343
  nano materials, synthesis and characterization
    catalytic reduction of methylene blue, 347
    gold, preparation and characterization of, 347
    materials and methods, 347
    nanostructured biopalladium, preparation and characterization of, 347
    4-nitroaniline, catalytic reduction of, 348
    4-nitrophenol, catalytic reduction of, 348
  nanoparticle composition and size distribution
    methylene blue, catalytic degradation of, 350–351
    properties, 343
Orientation imaging microscopy (OIM), 227

P

Perovskite structure (ABO3), 58
Piezoelectric scanner tube (PZT), 37
Polyindole/poly(vinyl acetate) composites
  dielectric constant, 159
  experimental
    characterizations techniques, 150–151
    precursors, 150
    synthesis of composites, 150
  extinction coefficient, 157
  FTIR spectra, 152–153
  optical band gaps, 155
  optical conductivity (σopt), 158
  photoluminescence (PL) analysis, 153–154
  polymer composites, 148
    PIN, 149
    polyindole/poly(vinyl acetate) (PIN/PVAc), 149
  refractive index, 155
  solar energy, 148
  surface morphological study, 160–161
  UV–Vis spectrum, 154–155
  XRD patterns, 161–163
Polymer blends, microstructural analysis
  blends of dicyanate esters
    azomethine functionalized dicyanate esters, 277
    GPR blends, 277–280
    polyacrylonitrile, 277–280
    polyurethane, 277–280
    Schiff-base-functionalized dicyanate esters, 276–277
  epoxy/glass fiber composites
    bis(4-cyanato3,5-dimethylphenyl) naphthyl methane (CDPNM), 280–281
    epoxy-PU IPN, 280–281
    PBZ/JUTE, 283

PMR PI, 282–283
woven glass, 280–281
POSS nanocomposites
   cyanate ester/epoxy, 285
   eugenol-based PBZ/EPOXY/OAPS, 285–287
   optically active—PI/OAPS, 284
   PBZ-functionalized, 287–288
scanning electron microscope (SEM), 276
silica nanocomposites
   CNT nanocomposites, 296–298
   cyanate ester, 296–298
   GO nanocomposites, 299–300
   PBZ/silica nanocomposites, 290–293
   polytriazoleimide, 299–300
   unsaturated polyester (UPR), 289–290
   UPR/alumina nanocomposites, 293–294
   UPR/CACO3 nanocomposites, 295–296
   UPR/nano-ZnO nanocomposites, 294–295
transmission electron microscope (TEM), 276
Polymer composites, 148
PIN, 149
polyindole/poly(vinyl acetate) (PIN/PVAc), 149
Polytetrafluoroethylene (PTFE) solution, 137
Positive temperature coefficients of resistivity (PTCR), 58
Potassium hydroxide (KOH), 135
Preparation of BaTiO3
   ceramic applications
      recent advances in, 59
   coprecipitation method, 62–63
   crystal structure, 58
   dielectric
      fundamental knowledge, 65–67
   domain theory, 70–71
   efficient dielectric supports, 68
   hydrothermal synthesis method, 62
   permittivity constant, 67–68
   perovskite structure (ABO3), 58, 59–60
   phase transition, 59–60
   positive temperature coefficients of resistivity (PTCR), 58

scanning electron microscope (SEM), 71–72
sol–gel method, 63–64
solid-state method, 61–62
strontium-doped barium titanate, 59
transmission electron microscope (TEM), 71–72
Pulsed current gas tungsten arc (PCGTA), 168
Pulsed current gas tungsten arc welding (PCGTAW), 168

S

Scanning electron microscope (SEM), 3–4, 71–72
advantages, 23
applications, 23
backscattered electron imaging (BEI), 6
basic design
   electron gun, 7–17
disadvantage, 24
electrons and electromagnetic waves
   emission of, 6
fundamental principles
   backscattered electrons (BSEs), 5
   diffracted backscattered electrons (EBSDs), 5–6
   field emission (FE) gun, 5
   heat, 6
   Schottky emission (SE) gun, 5
   visible light (cathodoluminescence, CL), 6
   X-rays, 6
image quality, factors affecting
   accelerating voltage, influence, 19–20
   charge-up, influence, 21
   detector position, 21
   edge effect, influence, 20
   probe current, 20
   probe diameter, 20
   specimen direction, 21
   spot size, 21
   tilt angle, dependence, 20
   working distance, influence, 21
interaction volume, 19
optical microscope
   large aperture angle, 5

Index 501

polymer blends
  characterization of, 24–32
polymer nanocomposites
  characterization of, 24–32
  sample preparation, 22
    cutting, 23
    fracturing, 23
  secondary electron imaging (SEI), 6
  small aperture angle, 5
Scanning probe microscopy (SPM), 36
Scanning tunneling microscopy (STM)
  advantage, 36
Schottky emission (SE) gun, 5
Silica nanocomposites
  CNT nanocomposites, 296–298
  cyanate ester, 296–298
  GO nanocomposites, 299–300
  PBZ/silica nanocomposites, 290–293
  polytriazoleimide, 299–300
  unsaturated polyester (UPR), 289–290
  UPR/alumina nanocomposites, 293–294
  UPR/CACO3 nanocomposites, 295–296
  UPR/nano-ZnO nanocomposites, 294–295
Sol–gel method, 63–64
Split Hopkinson pressure bar (SHPB), 207
  alumina, 208
  commercial grade coarse, 222
  FESEM-based photomicrographic, 221
  glass fiber-reinforced polymer (GFRP), 218–220, 223
    sample, 211
  high strain rate experiments, 211–212
  layered, 210, 213–215
  monolithic alumina, 208, 213–215
  monolithic alumina specimens, 210
  PMMA rods, 211
  poly(methylmethacrylate), 216–218, 222
  postmortem examination, 221
  studies, 209
  XRD data analysis, 220
Study of BaTiO3
  dielectric
    constant, 72–73
    property, 81–83
  ferroelectric property, 74–76, 83–84
  microscopic studies, 76–77

Surface plasmon resonance (SPR), 190
  angular interrogation-based instrument, 197–198
  Kretschmann–Reather geometry, 194
  microscopy, 199
  molecular-specific interaction, 193
  otto geometry, 194
  sensing applications
    biosensing application, 202–203
    chemical sensing, 201–202
    gas sensing, 200
  wavelength interrogation-based instrument, 195–196
  wavevector, 192
Surface plasmon (SP), 191

T

Tissue engineering (TE), 355
Transmission electron microscope (TEM), 71–72, 276
Treatment of pharmaceutical wastewater
  alternative techniques
    adsorption, 465
    advanced oxidation processes (AOPs), 466–467
    ion exchange, 465
    pressure filtration, 465–466
  applied current
    effect, 480
  case study
    low-cost electrode, 477
    methodology, 477–479
  electrodes
    characterization, 482
    comparison of, 481
  electrolyte concentration
    effect, 480
  energy dispersive spectroscopy (EDS), 462
  factors affecting
    current density/applied current, 472–474
    electrodes, 474
    electrolyte concentration, 472
    pollutant concentration, 474
  kinetic study, 479

methylene blue (MB), 461
  concentration, effect, 480
  mineralization study, 480–481
nonphotochemical
  electrochemical oxidation (ECO), 471–472
  ozonation process, 469–470
pharmaceuticals
  active pharmaceutical ingredient (API), 464
  conventional wastewater treatment, 464
  health effects, significance, 463–464
  residual pharmaceuticals, significance, 463–464
  sewage treatment plants (STPs), 462–463
photochemical route, 467
  direct photolytic processes, 468
  heterogeneous photocatalytic, 469
  photolysis with sensitizer, 468
  photolysis with vacuum ultraviolet, 468–469
toxicity study, 482–484

W

Wood, formation and decomposition
  cell wall
    biomass, 261
    structure, 259
  cells, biochemical changes, 262–266
  consecutive deposition, 260
  environmental factors, 258
  oak wood, ultrasound waves impact of, 267–271
  phloem and xylem, 262
  pine callus culture, study, 266–267
  radial diameter, 261
  temperature and precipitation
    optimal values, 261
  tissue culture, 259
  xylem cells, 258
Working electrode material
  DNA analysis, 308–309
  dopamine (DA), 306–307
  glucose, 308
  heavy metal ions, analysis, 310
  paracetamol, 306–307
  protein analysis, 309–310